GLOBAL
PROJECT
MANAGEMENT
HANDBOOK

Other McGraw-Hill Books of Interest

Cleland, Gallager, & Whitehead • MILITARY PROJECT MANAGEMENT HANDBOOK

Barkley & Saylor • CUSTOMER DRIVEN PROJECT MANAGEMENT

Leavitt & Nunn • TOTAL QUALITY THROUGH PROJECT MANAGEMENT

Thamhain • MANAGING HIGH TECHNOLOGY PROJECTS

Turner • THE HANDBOOK OF PROJECT-BASED MANAGEMENT

Ritz • TOTAL CONSTRUCTION PROJECT MANAGEMENT

Ritz • TOTAL ENGINEERING PROJECT MANAGEMENT

GLOBAL PROJECT MANAGEMENT HANDBOOK

David I. Cleland   Editor

*Ernest E. Roth Professor and
Professor of Engineering Management
School of Engineering
University of Pittsburgh
Pittsburgh, Pennsylvania*

Roland Gareis   Editor

*Professor, University of Business Administration
and Technical University
Vienna, Austria*

McGraw-Hill, Inc.
New York   San Francisco   Washington, D.C.   Auckland   Bogotá
Caracas   Lisbon   London   Madrid   Mexico City   Milan
Montreal   New Delhi   San Juan   Singapore
Sydney   Tokyo   Toronto

Library of Congress Cataloging-in-Publication Data

Global project management handbook / edited by David I. Cleland,
  Roland Gareis.
        p.   cm.
    Includes bibliographical references and index.
    ISBN 0-07-011329-7
    1. Industrial project management. 2. International business
  enterprises—Management.   I. Cleland, David I. II. Gareis, R.
  (Roland)
  HD69.P75G56   1994
  658.4′04—dc20   93-30297
                                                          CIP

  2 3 4 5 6 7 8 9 0   DOC/DOC   9 9

ISBN 0-07-011329-7

*The sponsoring editor for this book was Larry Hager, the editing supervisor was
Peggy Lamb, and the production supervisor was Suzanne W. Babeuf. It was set in
Times Roman by McGraw-Hill's Professional Book Group composition unit.*

Printed and bound by R. R. Donnelley & Sons Company.

This book is printed on acid-free paper.

CONTENTS

CONTRIBUTORS

John R. Adams   *Western Carolina University, Cullowhee, North Carolina, USA* (CHAP. 14)

Salman T. Al-Sedairy   *King Saud University, Saudi Arabia* (CHAP. 5)

Henning Balck   *Institute of Project Methodology, Mannheim, Germany* (CHAP. 2)

Fred M. Bennett   *Australian Institute of Project Management, Curtin, Australia* (CHAP. 32)

Bopaya Bidanda   *University of Pittsburgh, Pittsburgh, Pennsylvania, USA* (CHAP. 21)

Wendy Briner   *The New Organisation, London, England* (CHAP. 15)

Anthony F. Buono   *Bentley College, Waltham, Massachusetts, USA* (CHAP. 17)

David I. Cleland   *University of Pittsburgh, Pittsburgh, Pennsylvania, USA* (CHAPS. 1, 20, 21)

Ladislav Chrudina   *Project Management Society, Prague, Czechoslovakia* (CHAP. 6)

Shriram R. Dharwadkar   *South Carolina State University, Orangeburg, South Carolina, USA* (CHAP. 21)

Adnan Enshassi   *Al-Fatah University, Tripoli, Libya* (CHAP. 24)

J. Evers   *Philips Project Management Training and Consultancy, Eindhoven, Holland* (CHAP. 12)

C. Michael Farr   *Air Force Institute of Technology, Dayton, Ohio, USA* (CHAP. 23)

Roland Gareis   *University of Business Administration and Technical University, Vienna, Austria* (CHAP. 4)

Colin Hastings   *The New Organisation, London, England* (CHAP. 15)

Per Willy Hetland   *Statoil, Stavanger, Norway* (CHAP. 30)

Michael Jelinek   *Drees + Sommer, Vienna, Austria* 28)

J. Kessel   *Philips Project Management Training and Consultancy, Eindhoven, Holland* (CHAP. 12)

Ilona Kickbusch   *World Health Organization Regional Office for Europe, Copenhagen, Denmark* (CHAP. 25)

Hans Knoepfel   *Rosenthaler + Partner AG, Zurich, Switzerland* (CHAP. 31)

Ervin László   *The Academy for Evolutionary Studies, Frankfurt, Germany* (CHAP. 3)

Mary Anne F. Nixon   *Western Carolina University, Cullowhee, North Carolina, USA* (CHAP. 9)

Aaron J. Nurick   *Bentley College, Waltham, Massachusetts, USA* (CHAP. 19)

Henry Padgham   *CH2M Hill, Denver, Colorado, USA* (CHAP. 29)

Rajeev M. Pandia   *Herdillia Chemicals Limited, Bombay, India* (CHAP. 18)

Jeffrey K. Pinto   *University of Maine, Orono, Maine, USA* (CHAP. 27)

Günter G. Rattay   *Primas Consulting, Manhartsbrunn, Austria* (CHAP. 6)

Peter Rutland   *Carrington Polytechnic, Auckland, New Zealand* (CHAP. 5)

Fritz Scheuch   *University of Economics and Business Administration, Vienna, Austria* (CHAP. 10)

Arnold Schuh   *University of Economics and Business Administration, Vienna, Austria* (CHAP. 18)

Georg Serentschy   *Austrian Aerospace Company ORS (Osterreichische Raumfahrt und Systemtechnik Gesellschaft) Vienna, Austria* (CHAP. 13)

Dennis P. Slevin   *University of Pittsburgh, Pittsburgh, Pennsylvania, USA* (CHAP. 27)

C. Joseph Smith   *The Ohio State University, Columbus, Ohio, USA* (CHAP. 8)

N. J. Smith   *UMIST, Manchester, United Kingdom* (CHAP. 26)

Peter Sutter   *Digital Equipment Corporation, Vosendorf-Sud, Austria* (CHAP. 16)

Hans J. Thamhain   *Bentley College, Waltham, Massachusetts, USA* (CHAP. 19)

Agis D. Tsouros   *World Health Organization Regional Office for Europe, Copenhagen, Denmark* (CHAP. 25)

John Tuman, Jr.   *Management Technologies Group, Inc., Morgantown, Pennsylvania, USA* (CHAP. 11)

Francis M. Webster, Jr.   *Project Management Institute, Cullowhee, North Carolina, USA* (CHAP. 22)

William Yeak   *Anderson Consulting, Columbus, Ohio, USA* (CHAP. 8)

Robert Youker   *Project Implementation, Bethesda, Maryland, USA* (CHAP. 7)

PREFACE

In the last thirty years there has been a tidal wave of interest in project management as a management philosophy to use in dealing with the many *ad hoc* activities found in contemporary organizations. Project management is clearly an idea whose time has come. A substantial body of theory exists in the field, reflecting the wide experience gained by practitioneers in many different industries and environments. Project management is recognized as a principal strategy and process to deal with the inevitable change facing organizations. The social, political, economic, technological, and competitive changes underway in the global marketplace require that any organization wishing to survive in the face of such change needs to understand how such change can be managed.

Business organizations in particular are facing awesome challenges in the intensely competitive global marketplace. Quality, productivity, costs, faster commercialization of products and services, cooperative research and development, and the dynamic changes being wrought by the "factory of the future" all can be dealt with through the use of project management philosophies and techniques. Add to these changes the continued erosion of "national products and services" to products and services that have to be designed, developed, produced, and marketed in global markets—the importance of a management philosophy to deal with such universal changes becomes apparent.

Project management has truly become "boundaryless"—cutting across disciplines, functions, organizations, and countries. The formation of "strategic alliances" to share project risk, resources, and rewards are becoming commonplace in the management of international businesses. Today, a truly domestic market does not exist; enterprise managers the world over must face the unforgiving global marketplace. Not only is the survival of enterprises at stake, the country's national and international competitiveness is at stake as well. In the past two decades the global economy has been transformed; vigorous new companies from countries in the Pacific Rim and elsewhere are challenging many of the traditional industries and the way of managing in these industries. The competitive pathway to be followed in the political, economic, and technological conversion of Eastern Europe and Russia to free market economies will be a pathway characterized by the use of project management strategies. The ability to develop and produce products and services faster, at lower costs, with higher quality, and meeting the criteria for both local and international markets have become key performance factors. To remain competitive in the global markets, companies must develop the ability to make incremental improvements in the technology embodied in the products and services to be offered in the markets, as well as in the organizational processes needed to conceptualize, create, design, develop, and produce value that provide total customer satisfaction.

Successful companies today are using project management processes to transfer technology from around the world and integrating those processes effectively into their products, services, and organizational processes. Global project management provides a solid foundation of management technology to create products and services that did not previously exist but are needed to remain competitive in the global marketplace.

In the fast evolving field of global project management, there was a critical need to pull together a practical reference to explain the new techniques of the field, provide understanding of the unique nature of global project management, and instill confidence in the user that truly contemporaneous global project management philosophies and strategies are available and can be used. This handbook provides that reference source for pragmatic how-to-do-it global project management information, tempered by that bit of theory needed to be consistent with the state-of-the-art of this important discipline. Anyone who wants to learn about global project management is faced with an abundance of published information. The best of this published information, integrated in this handbook and presented in a global perspective, provides a coherent and relevant prescription for the global owner. In a rapidly developing field such as project management, even experienced project "stakeholders" need a source that can help them understand some of the competitive changes in the world that are without precedent.

All project stakeholders—project managers, functional managers, general managers, project team members, support staff, and the many outside organizational units with which the global project manager must deal—will find this handbook useful. Students of project management may use the handbook as a self-study aid, for it has been organized to facilitate an easy and enjoyable learning process. Senior managers can, through a careful perusal of this publication, gain an appreciation, and respect, for what global project management can do to make their enterprises more competitive in the international markets.

This handbook is the result of the cooperative effort of many experts in the field, both in academic and in real world practice. The qualifications of these learned and experienced people are clear from the biographical sketch that is provided. The content of the handbook is broadly designed to be relevant to the general organizational contexts in which global project management is found.

Eight interdependent categories of project management are presented in the global context:

1. *Management of Change by Projects.* In order to cope with new challenges and chances in a dynamic business environment, companies more frequently and explicitly apply projects. Projects are instruments to manage change processes.

2. *National Development by Projects.* Major development programs in Eastern Europe and in developing countries are implemented by projects.

3. *Increasing Corporate Competitiveness on Global Markets by Project-Orientation.* The application of projects as temporary organizations for the performance of complex tasks contributes to the organizational flexibility of companies. In order to increase the competitive edge by successful project performance, companies require an explicit project management culture.

4. *Strategic Management and Strategic Project Management.* On the one hand projects are defined in order to implement company strategies; on the other hand side projects influence and contribute to the organizational learning and to the continuous adjustment of corporate strategies.

5. *Project Management to Concentrate Resources in Engineering and Manufacturing.* Simultaneous Engineering and Shared Manufacturing contribute to successful project performance.

6. *Cross-Cultural Project Management.* In projects, cooperation takes place across personal, corporate, and national cultures. Cultural management is a specific qualification required in projects.

7. *New International Challenges for Construction Project Implementation.* More complex perceptions of projects and new methods are required to successfully implement international construction projects.

8. *A Global Historical Project.* A fascinating review of one of the great explorers of all time. Captain James Cook's first voyage of exploration was a project in the truest sense of the word—a project that was at the cutting edge of science as well as imperial and commercial expansion.

Within the global marketplace the time between the creation of inception of a new technology in products and processes is decreasing. Until a few years ago, it took up to ten to twelve years or more for a scientific discovery—even a discovery leading to an incremental advancement of technology in products and processes—to wend its way from the point of origin to commercial use. Today much less time is required; global project management done within the context of concurrent engineering or simultaneous engineering, and through the organizational mechanism of product-process design teams is reducing that time dramatically. Global project management, done within the authority of strategic management, has become a common language for global enterprises to cooperate across organizational, cultural, and national boundaries in seeking mutual objectives and goals. It is to this purpose that the *Global Project Management Handbook* is dedicated.

ACKNOWLEDGMENTS

Many people contributed to this handbook. The authors who provided the chapters are an assemblage of experts in the business of global project management. Their contributions reflect a wide range of expertise and viewpoints in this growing and important field of project management. We thank, and are deeply indebted to, these chapter authors.

We thank our graduate students, who in many indirect yet meaningful ways added to the value created by this handbook. Our classroom discussions with these young scholars surfaced many ideas that became integrated into this publication. We also thank our many friends in the Project Management Institute and in INTERNET for the opportunity to discuss with them the strategy for the development of the handbook. Each of these friends contributed in some way to the substantive content of the book, as well as the intellectual processes needed to pull together this important "first of its kind" publication.

We are deeply indebted to Claire Zubritzky, who managed the overall development and administration of the handbook. As was to be expected, her professionalism, dedication, and optimism encouraged us during the long period from the book's concept through to its actual publication.

Special thanks to Dr. Harvey W. Wolfe, Chairman of the Industrial Engineering Department, and Dr. Charles A. Sorber, Dean of the School of Engineering of the University of Pittsburgh, who provided us with the needed resources and environment to pursue the creation and publication of this handbook.

Finally, we hope that the people who use this handbook will find it a useful and timely source for the development of the knowledge, skills, and attitudes needed to compete in the intensifying field of global project management.

David I. Cleland
Pittsburgh, Pennsylvania, USA

Roland Gareis
Vienna, Austria

P · A · R · T · 1

MANAGEMENT OF CHANGE BY PROJECTS

In order to cope with new challenges and opportunities in a dynamic business environment companies apply projects more frequently and explicitly. Projects are instruments to manage change processes.

In Chap. 1 David I. Cleland takes the reader on a journey to understand how project management is becoming "borderless" through a philosophy for managing across cultures. Projects are offered to provide a focal point to bring companies and nations together in the development of products and processes within the global business community.

In Chap. 2 Henning Balck emphasizes the global application of project management for new project types, such as organizational development projects, product development projects, and marketing projects. He is challenging traditional, "mechanistic" project management methods and asks for a systemic approach of project management for these new project types.

Ervin László refers in Chap. 3 to the social responsibility of the project manager versus the environment. The overall evolution of society has to be considered in the project context.

Roland Gareis introduces in Chap. 4 management by projects as the management strategy of project-oriented companies.

The explicit application of management by projects requires appropriate company structures and cultures. Networks of projects are presented as specific integrative structures of project-oriented companies.

CHAPTER 1
BORDERLESS PROJECT MANAGEMENT*

David I. Cleland

*University of Pittsburgh,
Pittsburgh, Pennsylvania*

> **Dr. David I. Cleland is the author/editor of 25 books and many articles in the fields of project management, engineering management, and strategic planning. He is a Fellow of the Project Management Institute and was appointed to the Ernest E. Roth professorship at the School of Engineering, University of Pittsburgh, in recognition of outstanding productivity as a senior member of the faculty. He is currently Professor of Engineering Management at the University of Pittsburgh.**

New and important applications of the discipline of project management are emerging in response to the changing boundaries of the world. Among the most impressive of these is the use of project management to deal with the multinational opportunities arising in a world altered by political and economic events and technological advances. With changes so encompassing and multilateral, the integrative approach of project management enables companies and nations to work together in an increasingly complex international business community. For project management this means a departure from its classical applications, namely, integrating and interrelating an organization's project activities across its departments or divisions through the use of matrix design.

Within an organization such divisions or departments may be thought of as borders. However, this chapter will present the more recent development of using project management to pull together the efforts of several organizations and nations to create new products and processes having competitive value for their partners. This use of project management will be described in terms of "borderless" projects. In the borderless project, technology, information, people, and management practices flow freely from one organizational or national entity to another.

A brief discussion of the various borders follows in order to demonstrate how encompassing the term "borderless" becomes in regard to project management.

---

*It should be noted that the theme for INTERNET '92 was *Project Management without Boundaries.* The title for this chapter and the resulting manuscript were developed before the INTERNET theme had been established. The similarity between the INTERNET theme and this chapter is obvious.

CLASSIFYING BORDERS

Internal Borders

There are traditional borders or walls that separate the various functions of an enterprise—engineering, marketing, manufacturing, procurement, finance, and so on. These are the traditional focus of project management for integrating project activities within an organization.

Project management continues to develop new forms that cross these internal borders, such as product and process design teams, production teams, task forces, and other similar organizational design strategies.

External Borders

Case I.   The existing boundaries between companies and other organizations involved in a cooperative project whose focus is to produce particular goods or services. When suppliers, A&E firms, regulatory bodies, government entities, subcontractors, and others decide to pool their resources to create something that did not exist previously, they must find ways to take existing borders into account.

Example.   Manufacturers seek the participation of their suppliers on new product and process design programs. For many of these manufacturer-supplier joint projects it is the beginning of a strategic relationship because manufacturers often give source contracts to their supplier partners. These contracts set strict quality, cost, and performance delivery goals for components that are delivered just in time to be put into the new product. At Deere & Company, the shop-floor workers now work with their counterparts at their suppliers to solve problems that might interfere with the schedule.

Case II.   Those existing boundaries that have to be hurdled when entering into a more basic partnership or strategic alliance with other companies, other institutions, or even with competitors.

Example.   In the 1980s competitors in many U.S. industries began pooling resources to do "precompetitive" research on the technology used for distinctive products. More than 250 R&D consortiums were created to pursue this cooperative research. U.S. companies have also formed joint projects to provide research links with competitors. For example, with the help of $120 million from the federal government, Detroit's Big Three auto manufacturers are working on a joint project for new battery technology for electric cars.

Case III.   Those situations in which national boundaries need to be crossed in building cooperative ventures between countries.

Example.   GE formed a joint venture project in 1988 with MABE, one of Mexico's largest appliance manufacturers. The following year the two companies opened a gas-range plant in San Luis Potosi. This year the plant will turn out 800,000 units.[1]

Projects do not cross only internal *or* external borders. In the large projects now working within the context of an international strategic alliance, the objectives are often realized by working across the internal borders of an organization as well as across the external borders of the other organizations involved and the international boundaries of the participating countries.

In the examples of borderless projects that follow, note the variety of borders involved and how many must be breached to produce products and processes having competitive value.

- Mazda's sportscar, the MX-5 Miata, was developed by a project team. The car was designed in California and financed in Tokyo and New York. Its prototype was created in England, and it was assembled in Michigan and Mexico, using advanced electronic components developed in New Jersey and fabricated in Japan.
- The Boeing Company is developing a new airliner, the 777. The plane will be designed in Washington State and Japan, assembled in Seattle with tail cones from Canada, special tail sections from China and Italy, and engines from Great Britain. Hundreds of interdisciplinary "design-and-build" teams improve Boeing's designs, reduce development costs, foresee and solve probable manufacturing problems, and get the aircraft to the market sooner.[2]

GLOBAL TRENDS

Several important global trends are already reshaping the competitive dimensions in the world and are also providing the opportunity for increased use of borderless project management:

- The dismantling of trade barriers within the European Community
- The U.S. and Canadian trade initiatives with Mexico
- The emergence of new markets brought about by the dismantling of the centrally planned economies of eastern Europe and the former Soviet Union
- The thrust of current GATT negotiations to bring down tariff and nontariff barriers to trade among all nations

These changes are occurring more rapidly than anyone could have imagined, and they will create enormous opportunities for project managers.[3]

Reflecting these global trends, cooperation with organizations across external boundaries is growing in importance. For example, in global competition where the uncertainties are so high, companies are finding it necessary to build alliances with other companies to share risk, resources, and rewards. In some cases, effective borderless project management brings about a cooperative effort between nations in order to accomplish common project purposes. The cooperative research now under way in the European Community 1992 is a prime example of external borders, case III, for the European nations have, through national cooperation, set out to develop new products and processes to enhance the competitive behavior of member companies and nations of the European Community.

As the borders of project management are extended, more complex forces come into play. These can either facilitate the role of the project team or make it more complex and challenging. Projects which extend across companies and countries will encounter differences in managerial values and philosophies, organizational and national cultures, and languages and usage. Common purposes of the project may become more difficult to attain because of the inevitable parochialism and provincialism encountered in such projects.

Yet the growing globalization of business makes borderless project management even more necessary. As nations work less independently and more cooperatively in ventures to attain common objectives and goals, the philosophy of borderless project management will increase in use, and, as a result, greater cooperation will be fostered among nations—more than has been known in the past. As nations, through the use of project-management techniques, cooperate more in business ventures, the likelihood of political isolation should be reduced accordingly. The probability of military confrontation should also be reduced, for as nations develop and share more common business and commercial interests, a bonding is likely to occur that will result in the probability of greater cultural and economic interindependence. As people from different companies and countries work together on project teams, there will be a technology transfer through personal and business interests, which will also draw everyone closer together.

CROSSING COUNTRY BORDERS

Project management is used to prepare companies to consider expanding into another country. Doing a credible job of research to determine the likelihood and rationale of locating in another country is an important first step in any alliance. Businesses are growing that help evaluate project opportunities in developing countries. For example, a nonprofit organization in Monterrey called Proexport has set up a project team that is analyzing all the companies on the Fortune 500 list in order to determine whether manufacturing or selling in Mexico makes sense for them.

A view of borderless project management is a broad, systems-oriented view of the world, wherever that view exists. This means an external focus, prescribed from the perspective of the global marketplace. A second important perspective of borderless project management is a dramatically increased employee empowerment and participation: training, decentralization of authority and responsibility, participation in process design by everyone, and the confidence of going with the judgment of individual employees at all levels in the enterprise. A third major force is the elimination of the boundaries themselves—the boundaries between organizational functions, between suppliers and customers, between companies in the industry and even competitors. The project work is to be organized and executed from the perspective of the products and processes needed to create project results, without any geographic limitations. It means that the project team, more than ever before, must identify the major work processes required to produce the project results and involve all of the organizations and people who contribute to those processes, irrespective of the organizational component or of the geographic location to which the people are assigned. The complex objective is to eliminate walls, boundaries, or borders in order to bring into focus the specific kinds of effort required to bring the project to a successful conclusion, one that will create value for the customer as well as to the project stakeholders.[4]

EXAMPLES OF BORDERLESS PROJECTS INVOLVING GLOBAL PARTNERS

There are many examples of such projects.

- Kohler Company, a Wisconsin-based manufacturer of plumbing equipment, recently completed a joint project with a Mexican partner in starting up a new $42 million toilet and sink manufacturing plant near Monterrey, Mexico.

- McDonnell Douglas, with an expected $2 billion investment from the Taiwan Aerospace Corporation, has initiated a project to develop the MD-12 airplane, an all-new long-range design intended to compete with Boeing's 747. Launching the plane will require about $9 billion, an initial investment of around $4.5 billion for design and development plus about the same amount to get the plane into full production. Taiwan Aerospace plans to build the wings for the MD-12.[5]

- In Europe two small steel companies developed a strategic alliance through the use of a project-management process to produce steel products for each other. Each agreed to manufacture products that the other previously produced inefficiently. As part of the alliance, each steelmaker was able to close some uneconomic mills. Besides costing less to produce, the products now have a higher and more uniform quality.[6]

- The new notebook computer produced by Sony shows how project management plays a key role in a strategic alliance between global partners. In 1989 Apple Computer enlisted Sony Corporation to design a new notebook-size Macintosh computer called the PowerBook 100. The president of Sony, Noric Ohga, gave the project top priority, with project manager Kihey Yamamoto having a free hand to hire engineers from any Sony division. The two companies faced the challenge of integrating cultural differences, a challenge made easier by their mutual excitement over building great new products that define new markets. When during the Persian Gulf War corporate travel restraints endangered the project's schedule, the project teams improvised teleconference meetings. These worked so well that Apple and Sony will use them in future project development efforts.[7]

- IBM has teamed up with Germany's Siemens, a competitor, to launch a joint production project of the next generation of memory chips. On another strategic project, IBM is working with Motorola, Inc., to develop a new chip-making technique for use by the turn of the century. IBM has also begun doing "technology audits" of the competitiveness of its suppliers versus their Japanese rivals. IBM was stunned by its findings. In the United States and Europe its handpicked vendors were becoming less competitive at an alarming rate.

JAPAN AND THE BORDERLESS PROJECT

Airbus Industrie officials are holding talks with Japan's leading aircraft company about helping to make the European consortium's next-generation jumbo jet. Whether or not the talks will lead to any firm agreement is moot, as Mitsubishi, Kawasaki, and other major Japanese companies have had long and close relationships with the Boeing Company of Seattle, Washington. As the costs of aircraft development projects skyrocket, Japanese capital and know-how will be crucial.[8]

Japanese investments in Europe, as of April 1991, totaled $54 billion. In addition, Japan established manufacturing footholds from Manchester to Milan. These investments have provided a great many opportunities for project management. Soon Europe will be the number 1 outlet for computers and other key products. With boundaries wiped from the map as 12 nations share in a vast single market, the Old World has become an even more fertile ground for borderless project management. The new Europe is a tempting opportunity for Japanese and U.S. companies as well as for existing companies on the continent, and Europe may well become the fulcrum of the world's economic power balance for the forthcoming century. It will be a booming base from which multinationals will consolidate their financial strength and the economies of scale to compete around the world. The opportunities for project management will be without precedent.[9]

In their use of borderless project management the Japanese use a logic of redundancy to manage product development. This is an overlapping process, where different divisions work together in a shared division of labor. Basically it is similar to the internal borderless approach of project management in the context of the matrix organizational approach. The Japanese use competitive product-development teams. A team is divided into competing groups that develop different approaches to the same project and then debate the advantages and disadvantages of their proposals. This encourages the team to look at the project from different perspectives. Eventually the project team develops a common understanding of the consideration of the project and launches a "best" approach selected from the competing alternatives.[10]

But working with the Japanese on a borderless project poses special challenges because the Japanese use a unique managerial style and a worker relationship based on a tradition that emphasizes collective values over individualism. Japanese managers invest for the long term and see worker participation in shop-floor matters. In contrast European and U.S. managers have generally shared similar cultural and political traditions that have stressed individualism, resulting in different management philosophies and style.[11]

AND SO GOES EUROPE

In the auto industry there are abundant projects under way in Europe to reconfigure the existing plants to the Japanese concept of "lean manufacturing." With only 7 years before the European Community lifts barriers against the full force of Japanese competition, car makers in Germany are rushing to make themselves more competitive. Volkswagen AG is applying Japanese production methods in all of its factories, including the one now under construction in the eastern German city of Zwickau. In Eisenach, once part of communist East Germany, Opel AG's new plant will be designed and operated by GM, Opel's troubled U.S. parent company, using the complete Japanese system of lean manufacturing. The Eisenach project, expected to cost DM 1 billion, is considered a crucial investment for Europe's auto industry.

The Eisenach project, striving to implement the total systems concept of lean manufacturing, involves planning for everything—the way the company will deal with its suppliers, manage inventories, encourage worker participation, manage total quality, and work at continually reducing costs. Six experts on Japanese production from all over the world have joined the project team. Design authority over the key sections of the plant has been given to those people who will later operate them. For instance, the person who designs the paint shop will later manage it when the factory starts operations. The strategic significance of the plant is keynoted in the comments of Tom LaSorda, the president of Opee Eisenach GmbH., when he states: "Our intention is to recreate the complete Japanese system on this site…the industry has no other choice if it wants to survive."[12]

The emerging integration of the economic community of Europe has already spawned a growth in borderless project management. In Europe, state and private corporations are investing massive sums in research projects in a wide range of technologies from telecommunications switches to wonder drugs. Europe's governments are funding cooperative research projects that pool the strengths of countries and the cooperating companies.

Since the mid-1980s European companies and research institutes have pooled their resources in 37 megaprojects, from genetic engineering to thermonuclear fusion. Approximately $20 billion is expected to be spent in Europe on cooperative research and development in the next several years, helping to create a single market out of the individual members of the European Community. For some of the smaller nations and companies, cooperative research and development is the only real route to technology development.

The vision is emerging of a Europe with shared political values, tied together economically and technologically, in part through cooperative research and development projects. The implications of one Europe as a competitive force in the global marketplace is awesome. A larger and more powerful Europe than anyone has realized is likely to emerge—a reality brought about to a significant degree by the effective use of borderless projects.

Gruppo GFT, based in Turin, Italy, is the world's largest manufacturer of designer clothing. Under the Gruppo GFT umbrella the company's employees in 45 small companies and 18 manufacturing plants make, distribute, and market approximately 60 designer and brand-name collections in 70 countries around the world. GFT is trying to create a "designer organization" able to adjust and adapt continuously to differences among markets and changes within markets. The GFT product project manager, the liaison between GFT and the fashion designer, translates the designer's vision into products, while maintaining sensitivity to the tradeoff between quality and cost.[13]

Italy's high-speed train project spanning the Italian peninsula in a large T running from Turin to Venice and from Milan down to Naples promises that the first trains will be operational in 5 to 7 years. The estimated cost of this project is $23.7 billion, more than twice the cost of the celebrated Eurotunnel project linking Britain and France. The high-speed train project has huge technical, economic, and financial hurdles. The Italian government has agreed to fund 40 percent of the project and to pay interest on the loans until the system becomes operational. The government has turned to banks to raise the outstanding capital through a combination of equity and debt on international financial markets. Thus far the deep-seated mistrust between the public and private sectors in Italy has not been evident on this project. Approval has been received from politicians in Italy and from bankers worldwide, who are scrambling to get in on the project.[14]

Project management both creates change and reacts to change in the competition arena.

THE CHANGE FACTOR

To be globally competitive requires big spending in research and development and on new equipment. It costs something in excess of $1 billion to develop a new generation of semiconductors and about $5 to 6 billion to develop a major new airliner. New products must be offered to demanding and unforgiving customers all over the world. In addition, the pace of change is accelerating, and product- and process development cycles are becoming shorter as new technology makes existing products obsolete— putting a premium on project management and flexible manufacturing. In the automobile industry the need to develop the Japanese techniques of lean manufacturing puts a premium on the ability to conceptualize, design, develop, manufacture, and service the innovative and competitive products that provide the customers with value. These changes in the global competition are fueling much of the need for project management: the management philosophy and process that bring this all about.

Europe's competitiveness in manufacturing will largely determine the quality of life for its people and its status as an economic power. Many remedial strategies will be required to further develop competitive strategies for manufacturing excellence by focusing on raising productivity and quality standards to those of the world's most competitive companies. To achieve these goals, European companies must develop teamwork, just-in-time inventory delivery, statistical quality control, and other integrated techniques of lean manufacturing. Project management is central to all of these strategies.

IN THE SPIRIT OF COOPERATION

Of course there has always been cooperation among companies and nations in the marketplace through licensing, joint ventures, trading companies, and similar united efforts. But most of these cooperative efforts were carried out after the products or services that formed the basis for cooperation had been developed. When this cooperation was carried out, organizations maintained their structural identities and therefore worked jointly as separate entities. Managerial and cultural interfaces were worked through the separate organizations, which were careful to maintain their separate entities. There was little "penetration of the veil" of the corporate or national body. Although there was a joint authority and responsibility, each partner maintained residual accountability for its contributed effort, and the relationship did not go beyond the boundaries of the companies into the working levels. Conflicts were viewed as between organizations or companies, not to be resolved at the working level but at the executive level, where the authority to resolve the conflict was lodged. Key decisions were reserved for the senior executives of the organizations working together. There was little if any empowerment of the working-level people. Cooperation was maintained only if the separate organizational entities were not threatened.

The traditional method of organizational and national cooperation as described in this chapter can be contrasted with the circumstances found in a borderless cooperative effort involving the concurrent development of products and processes.

THE DIFFERENCE

The formation of a joint borderless project team involves bringing people together who represent the different disciplines of the partners cooperating in the venture. A true intercompany and international project team is formed and organized. The interests of the project team, in accomplishing the objectives of the joint effort, tend to rise above the parochial interests of any of the members. Thus in the case of the development of cooperative research among the nations of the European Community, the objectives are to provide for the discovery and transfer of technology resulting in benefits for all of the member nations, not to serve the insular interests of just one or a few.

Borderless project management comes into play early in the development stages of the project, often during the preliminary design of the specifications for the project's products or services. The project team's objective cuts through each partner's organizational structure and draws people into the effort wherever located in the partners' organizational entities. There tends to be a loss of organizational identity in favor of a project identity. Managerial and cultural differences still exist, but the prudent project team recognizes these differences, plans for their resolution through team development or similar activities, and works on a continuous basis to reduce the damage that can be done to the project by cultural, linguistic, or traditional differences. When the sharing context of the project extends down to the working level, the project takes on a "superpurpose," going beyond the provincial objectives of the team partners. Adequate empowerment of the project team facilitates and eases the key decision processes on the project. For example, a product and process design team that works continuously in designing the product and processes also conducts an ongoing review of product and process design specifications. As a result, the number of formal design reviews is reduced, or even eliminated by the deliberations of the team as it works for the best design consistent with the project objectives. When the project team members work together daily on design considerations, there is limited need for the senior executives to participate in these decisions, unless there is conflict among the team members, or there has been a deterioration in the design that could lead to a potential compromise of the project objectives.

The borderless project team does whatever is necessary wherever necessary and however necessary, without any border constraints, in pulling the project together and managing it during its entire life cycle. The cooperation of many entities—functional, organizational, political, cultural, national—is needed to make the project successful. Borderless project management brings into being a "super" organizational unit that rises above the provincial interests of the partners and creates value for the partners as well as for the project's customers, suppliers, and other stakeholders.

In the management of any borderless project there will be key decisions that the partners will want to reserve for themselves individually. These key decisions are generally arrived at, however, after the project team has had "its day in court." In most successful borderless projects the key decisions usually come from specific recommendations of the project team. If the project team is surprised by the decisions of the partners or, conversely, if the project team surprises the partners on key decisions involving the project, damage will be done to the cultural ambience of the project team as well as to the partners themselves. Because such damage takes time to repair, and because the damage is never fully repaired, cooperation between the partners and the project team looms as an important factor for success.

WHY NOW?

For much of its history, project management has had to confront and hurdle a series of walls or borders—walls between organizational disciplines, walls dividing workers from managers and line from staff, walls between the project and its stakeholders, walls between the companies and between each company's customers and suppliers, and, finally, borders between countries. Now we see all these walls crumbling.

Global competition, strategic alliances, concurrent engineering, instantaneous communications, political upheavals, changing demographics, technological innovations, corporate restructuring and downsizing, worker empowerment, travel at the speed of sound, faxes, electronic mail—all these are eroding the boundaries of an earlier kind of project management. The once rigid walls or borders are yielding to the pressures of unprecedented social, political, economic, technological, and competitive change. And that change is everywhere, regardless of organizational function, company, culture, or country.

Borderless projects are used to further the capability of both organizations and countries. Among them are construction projects to upgrade or create new facilities and infrastructure, projects to provide for the development and acquisition of new product and process technology, projects to upgrade human and nonhuman resources, and projects to share resources with other enterprises. The result is that strategic alliances are being built with increasing frequency among suppliers, customers, and competitors to facilitate more effective competition. Indeed borderless projects are at the leading edge of new technologies and new economics in the world today. These projects play a vital role in dealing with, and even creating, change in the global marketplace.

But borderless projects and initiatives are not without their problems. Language and cultural barriers and differences in management philosophy are posing extraordinary challenges for the discipline of project management.

JOINT-PROJECT PROBLEMS

The GM venture in Korea where the company planned to marry engineering done at its German subsidiary to cheap Korean labor and U.S. marketing know-how to produce a low-priced subcompact car is nearing its end. Seven years ago GM and South Korea's Daewoo Group had agreed to initiate a joint project to build cars. The demise of the project is expected after years of acrimonious relations. The major forces that led to the failure included:

- A continuing clash of GM and Daewoo managers
- Evaporation of the hope for "cheap" labor of South Korea as an aroused labor force became expensive
- Unanticipated and underestimated obstacles caused by divergent business aspirations as well as by the different languages and cultures
- Disagreement over basic business matters, such as investments, to reverse the project's declining market share
- Claims of shabby treatment of the Daewoo CEO by U.S. partners
- GM's claims that Daewoo executives gave inadequate attention to product quality and to the labor disputes that repeatedly halted production at the Inchon plant
- Communication problems in translating German design changes so that the Korean suppliers and assemblers could follow directions

Even if strategies could have been developed to solve the problems plaguing this project, it would have been difficult to solve such problems since GM and Daewoo could not even agree on where to start. Whatever the outcome, like many divorced couples, both GM and Daewoo could wind up losers.[15]

Is borderless project management really different?

WHAT'S REALLY DIFFERENT?

What is different about borderless project management vis-à-vis the more "traditional" project management?

- The management of the borderless project cuts through organizational and national borders, searching out and applying the resources needed to make the project successful.
- Differences in cultures, mores, traditions, values, philosophies, and languages of

the project partners pose special challenges and, if not properly accounted for and managed, may contribute significantly to project failure.

• Attitudes of the project team members, the managers, and the professionals in each of the partner organizations take on added importance, particularly in recognizing the potential that macropolitical, social, cultural, legal, and economic forces can have on the project.

• The borderless project team tends to go beyond the partners' organization—a move that is necessary not only to fulfill the objectives of the project, but also to reach a project finish that contributes value to the partners, their customers, and the project stakeholders.

• The financial risks and implications of the project, both in the development effort and during the life of the products and processes that are created, can be immense, going far beyond the financial capabilities of any of the partners operating independently.

• Competition in the global marketplace is becoming, to a major degree, dependent on the ability to build borderless projects in order to develop new products and services by utilizing the support of organizational processes throughout the global community. In this way, organizations share the growing economic, political, social, technological, and competitive risks and, when successful, share in the rewards from cooperatively competing in a growing and unforgiving global marketplace.

Cleland, writing in the November 1991 *pmNETwork* magazine, challenged the readers to share their ideas of what lies beyond "project management."[16] This is a provocative and worthwhile challenge. One thing that is not beyond project management per se, but rather an extension of it, is the concept of borderless projects whose management transcends organizational and national boundaries. These projects provide the best practical means today for dealing with the changes impacting the political and economic entities of the world. To manage such a project is to cooperate in the management of resources to create something that does not currently exist. For many nations of the world today, borderless projects will play decisive roles in creating new economic, political, and social values, and they will be at the forefront of the changes needed by so many peoples and nations.

THE FUTURE

We have seen the use of project management in the formation of transnational strategic alliances in many different industries, including pharmaceuticals, automotive vehicles, robotic equipment, aircraft engines, and glass-refractory products. At present, in doing business in the former USSR and in eastern Europe, strategic alliances have been made for the development and production of military aircraft, military tanks, chemicals, and commercial aircraft. This has entailed the need for close cooperation of two or more parties working with suppliers of materials, components, equipment, and software. As competitive pressures increase globally through the emergence of new players that intensify the need for commercial products and services, earlier borderless projects are becoming even more intercompany and international in their scope.

The logic of the borderless project manager is to undertake activities anywhere in the world that will minimize the cost, accelerate the schedule, and maximize the technical performance of the project in the product and process development process.

As we move more into global competition, the management of projects to support global operations will take on more and more of a borderless characteristic. Project activities will be sited wherever it makes economic, marketing, and political sense in the international market. Project teams will be cosmopolitan in nature, and transactions to support the international project will come from different companies and countries. The more borderless the project, the more decentralized and diffused the authority and control will be. The project investment decisions will be made after an assessment of their global impact. Project development and investment decisions will be based on global competitiveness—not solely on a single company policy, but rather on what makes strategic global sense. No company or nation will be immune to the growing impact of borderless project management.

The growing importance of international networks in business matters means that companies will be forced to participate in more borderless projects, or face a stiff penalty in a marketplace that is becoming more insistent on new products developed to compete globally.

Industries and nations are being redefined through the power of information and computer technology. Closer relationships between suppliers and customers as sharing partners in the development of products and processes are blurring the distinctions between organizations and countries.

The borderless project manager will need to imagine possibilities outside of conventional categories, to envision actions that cross traditional borders, and to take advantage of interdependencies in the management of projects. More projects will be undertaken in which countries and companies will combine forces to create something of value for customers located throughout the global marketplace. What will matter more than ever is for the project team to think and work together, to pull together resources from many nontraditional sources, and to make workable new products and processes that make new connections.

In the emerging global economy it will continue to be difficult to say what is a "national" product. In earlier times, and not so long ago, products had distinct national identities, regardless of the number of borders that they crossed. Today products are developed, manufactured, marketed, and serviced using teams of people from many different international locations and are combined in all sorts of ways to serve customers' needs in different places.

CONCLUSION

Today the soaring costs and risks and the complexity of new technology are haunting the future of global companies that "try to go it alone." Even giants such as IBM, Boeing, and Siemens are moving more and more toward joint borderless projects to foster corporate cooperation by sharing resources and risks. The trend toward intercompany and intercountry cooperative projects has taken root. Exactly what form of industrial structure and competition will emerge from this trend, laws permitting, remains uncertain. The trend requires both ad hoc project management and more strategic or long-term management. It also requires increasing sophistication in the way projects are conceptualized and developed, and how the production, marketing, and aftersales service challenges are handled. Such cooperative projects using borderless teamwork, along with the entrepreneurship that has been a hallmark of competitiveness, will be important ingredients in a global industrial revival.

REFERENCES

1. N. J. Perry, "What's Powering Mexico's Success?" *Fortune,* pp. 109–115, Feb. 10, 1991.

2. R. Wartzman, "Boeing Company Is Girding for Dogfight over Market Share," *Wall Street J.,* Jan. 14, 1992.

3. Paraphrased from G. A. Peapples, "Competing in the Global Market," *Business Quart.,* pp. 80–84, Autumn 1990.

4. Paraphrased from W. R. C. Blundell, "Prescription for the '90s: The Boundaryless Company," *Business Quart.,* pp. 71–73, Autumn 1990.

5. H. Banks, "A Partner at Last?" *Forbes,* pp. 42–43, Dec. 23, 1991.

6. U. Gupt, "Tough Times Can Make Strategic Bedfellows—Alliances between Firms Promote Competitive Strength," *Wall Street J.,* Jan. 18, 1993.

7. B. R. Schlender, "Apple's Japanese Ally," *Fortune,* pp. 151–152, Nov. 4, 1991.

8. J. M. Schlesinger, "Airbus Industrie Said to Be Seeking Japanese Alliance," *Wall Street J.,* Nov. 19, 1991.

9. "The Battle for Europe," *Business Week,* pp. 44–50, June 3, 1991.

10. Paraphrased from I. Nonaka, "The Knowledge-Creating Company," *Harvard Business Rev.,* pp. 96–104, Nov.-Dec. 1991.

11. J. Hoerr et al., "Cultural Shock at Home: Working for a Foreign Boss," *Business Week,* pp. 80–84, Dec. 17, 1990.

12. T. Aeppel, "Opel Designs Car Plant on Japanese Lines," *Wall Street J.,* Jan. 21, 1992.

13. R. Howard, "The Designer Organization: Italy's GFT Goes Global," *Harvard Business Rev.,* pp. 28–44, Sept.-Oct. 1991.

14. S. G. Forden, "Italy's High-Speed Train Project Stays on Track with Public, Private Support," *Wall Street J.,* Oct. 14, 1991.

15. D. Darlin and J. B. White, "GM Venture in Korea Nears End, GM Betraying Firm's Fond Hopes," *Wall Street J.,* Jan. 16, 1992.

16. D. I. Cleland, "What Will Replace Project Management?," *pmNETwork,* p. 5, Nov. 1991.

CHAPTER 2

PROJECTS AS ELEMENTS OF A NEW INDUSTRIAL PATTERN: A DIVISION OF PROJECT MANAGEMENT

Henning Balck

Institute of Project Methodology,
Mannheim, Germany

Henning Balck studied architecture at the universities of Karlsruhe and Stuttgart, Germany. He has extensive working knowledge in domestic and international development projects with emphasis on building management and strategic project management in the fields of development, construction, and operation of "intelligent buildings." He has been project manager of several symposia on the application of chaos research in "management by projects." He is the author of several books and papers on systemic and evolutionary concepts in project management. Balck is the president of the Institute of Project Methodology in Mannheim, Germany.

PROJECT-MANAGEMENT CRISIS

During the second half of the past decade the demand for project management has increased dramatically—especially in large companies, where formerly only restrained interest had been demonstrated. After many years on the "back burner," project management appears in an unaccustomed limelight. A careful observer of the developing scene, however, can hardly stop making observations after a quantitative assessment of this success. Interested newcomers also represent new fields of application and, consequently, create changed requirements and expectations. Therefore it is not surprising if, after the initial enthusiasm, dissatisfaction, failure, and disappointment occur.

It has long been a well-known fact that project management is not a simple matter and that even in the classical areas of application, such as large-scale technical projects, failures are practically a daily occurrence. This has repeatedly been the subject of professional discussions.[1] However, during the past few years criticisms have been on the increase in different areas, namely, in research and development centers and in the field of software development. Projects as an original form of operation are typical for these areas. Typical as well is the "chaotic work style," which is still prevalent today in development projects. Only recently the picture started to change. Like a fire, the realization has spread that short lead times in the field of development and production are decisive factors for success in a competition where increasingly higher-quality products are being manufactured for increasingly smaller and short-lived market sectors. Enterprises which had been used to doing business in long-term economic cycles must reconsider their business strategies. In particular, companies with inflexible organizational structures and many levels of hierarchy will have to make adjustments, that is, they will have to become more flexible. This recent challenge will apply mainly to research, marketing, and product development. The whole crux of the reorientation required in enterprises and management structures can be expressed as follows: The right products must be introduced into the market at the right time. This is also the crucial point for an increased interest in the organizational form of project management. The introduction of project work breakdowns, network technique, project reports, and any additional, possibly computer-aided tools promises to solve several deficiencies in the traditional work organization:

- The time frame required for internal work processes must be reduced drastically.
- The administrative disruption within large organizational bodies during the process of temporary project organizations must be revitalized by "networking."
- The complex adjustment process by enterprises to changing or evolving markets must be organized. This can be achieved through the introduction of new technologies, a restructuring of the available organizational structures, the introduction of a new "management style," and other measures, which will be realized through projects.

If, based on the preceding, the introduction of project management becomes a strategic goal for the top management of a transforming enterprise, disappointments will occur if honest feedback is not obtained. The reason is simple: All methods and instruments combined under the label of "project management" were originally developed for and successfully employed at large technical projects. The present application to tasks in the structural changes of business and markets, however, is totally different.

Origins in Systems Engineering

A. D. Hall is considered the founder of this engineering-scientifically influenced method of work and organization. His book *A Methodology for Systems Engineering,* which was published in 1962,[2] is regarded as the first milestone in an increasingly differentiated path of development. The book centers on the problem of how complex technical forms or sociotechnical systems, such as entire factories, can be created in an orderly process.

The methods of systems engineering developed until now follow one classic guideline: the concept of work division. While this principle has been applied mainly to performance in traditional work processes, its application in systems engineering is being extended into phases of design and planning, or into developmental processes of innovative components. Herein lie both its strength and its weakness. Unmistakably, its strength lies in the fact that in deterministic processes, in particular, avoidable chaos can be reduced by creating project work breakdowns, network plans, or flowcharts, and then "executing" them. This effort, however, can get into trouble if those methods and instruments based on chaos reduction are to be applied to unavoidable or even creative chaos.

Origins in Operations Research

In the late 1950s the field of operations research (OR) developed at a rather rapid pace in parallel with systems engineering. The basic idea of OR is to optimize technical and social processes or structures by means of mathematical calculations. The fate of the world of OR is typical of a problem penetrating our whole industrial culture. At numerous universities and research centers, mathematicians and scientists have applied themselves to this field and have developed an incredible amount of calculation methods. In the meantime it has been established that the majority of these methods are ineffective as there is simply a difference between standards set for mathematical models and "realities," which rarely conform to the calculated world. Traditional OR procedures are based on exact definable initial conditions and constraints and will produce exact results. However, it is hardly ever possible to define these in indeterministic processes. Just the opposite: wherever the new and unexpected is prevalent, schedules, models, or concepts have to be "fuzzy" and uncertain. This fact is being recognized more and more by methodology researchers and basically puts to question the prevailing OR efforts.

There is, however, a further aspect that will render the rules of OR methods unacceptable. Complex processes of change and development are of irreversible, that is, historical character. Therefore every decision or business situation is—compared to a previous one—unique. This results in the fact that "planned" sets of schedules or structures are constantly being outrun by reality and are thus becoming obsolete. A typical example is the experience with the OR-produced "network plan." Although it cannot be denied that a skilled application of the network plan can result in valuable points of information, an increasing dissatisfaction is prevalent. The puzzling factor about the degree of distribution and the partial economic success of network-plan software is that it obviously comes not so much from the benefit gained from this instrument as from the unfailing expectations, that is, the belief in mechanistic mastery.

Bureaucratic Project Performance

In modern organizational sociology it has been established that the transition from early industrialization to the phase of mass production coincided with a gigantic bureaucratic process. This historical process peaked in the 1960s. However, this wave of bureaucracy produced major components of modern project management: handbooks, systemic and periodic reporting methods, documentation and information systems, and so on. It is undeniable that all of these instruments have justified values of their own and are, from case to case, valuable tools. There are, however, numerous examples where the application does not fit the task. The worst case is where project bureaucracy actually hinders creative dynamics, or causes a slowdown in performance.

Thus the old patterns and guidelines need to be revised. This applies even more so to the situation where projects of structural changes must be managed in markets and companies. However, the required reorientation is not a theoretical problem that could be outlined and solved within a sterile academic environment. The author is convinced that the presently beginning phase will create a new form of methods, and that this new form of methods will be created by the same group of people who have the requirements for and will apply these methods.

In this context this chapter is an appeal from one practitioner to others. However, with a possibly unusual emphasis: the appeal is for interested developers to form a network which, due to diversified synergies, will create a comprehensive and manifold methodology. This task cannot be performed by a single person, alone in isolated situations. This assessment is the result of the author's own positive experience with human networks stemming from four years of work on the subject of Reorientation in Project Management.

WITHDRAWAL FROM THE MECHANISTIC PARADIGM OF INDUSTRIALIZATION

The advancing technological revolution in the area of production has obvious parallels in the modification of the technical basis for office work. A common direction is the disintegration of permanently structured procedures and rigid hierarchies in favor of flexible and adaptable organizations. In their book, *The End of Work Division,* published in 1984,[3] Horst Kern and Michael Schuhmann highlighted the fact that since the early 1980s, this trend has become more and more pronounced. In an epilogue of the 1990 edition, however, the authors point out that this trend must still be viewed today as a confrontation between hostile role models.

Unmistakably, the more recent tendencies of high-tech industrialization are in contrast to the old images of automation. This includes the common image of "a factory devoid of people," or the analogous vision of "an office devoid of paper." Lately such images have been increasingly criticized, one of the reasons being the sometimes bitter realization that even large-scale efforts have not produced the desired results. Factory and office automation are both programmatic terms which have, since the beginning of their usage, been associated with a mechanistic understanding.

If the signs of the times are interpreted correctly, there are many indications that the basic direction of the industrialization process is changing.[4] Then it can be expected that these dynamics are not going to culminate in social megamachineries, as outlined in Mumford,[5] but will produce a technical culture, based on real-life conditions, which is closer to nature and life, and which will be more humane.

There is every reason to believe that the factory or office of the future will be created with the above thoughts in mind. Practitioners, in particular we as project managers, are well advised to rid ourselves of the constricting historical background of a mechanistic world image and rationalism. Without question the best method to help us correct our way of traditional thinking is "on-the-job training," that is, experiencing the real successes and failures in dealing with our everyday business endeavors.

NETWORKED PROJECT WORLDS PERMEATE COMPANIES AND MARKETS

Project Management Moderates Creative Work in Research and Development Projects

Investments in the field of research and development are increasing in all occupational areas at the same rate as the creation and application of innovations are becoming basic modes of contemporary economics. However, this statement is anything but a self-evident, casually noticeable fact. If the efficiency of research and development, and, thus, the possibility to utilize innovations effectively, will be the determining factor in developing companies, then the mobilized creativity will become the determining factor in general. This will result in a keen interest in any possibilities to influence creative processes. It is exactly at this point that the initial stage of project management will turn into an almost unique situation of golden opportunity: project management, seen as a benefactor, a beneficiary, and an accelerator of creative work, acting as a catalyst for discovery and invention.

However, in today's world of research and development this possibility is being blocked. First, a form of artistic mentality is dominating the areas of research and development. In the sense that "creativity needs chaos," management demands or needs will only be accepted, as peripheral conditions, if at all. In the sense that creativity should be unrestricted, any form of regulating and controlling function will generally be regarded as a disturbance by researchers and developers.

In the past years, and seemingly as a countermeasure against such management hostility, a trend has become apparent to execute strategically important research and development projects within a tight budget and time schedule, resulting in a number of problems for the executive as well as for the executing side. By maintaining the above attitude of pro or contra—or by even reinforcing it—this conflict cannot be resolved but can only be intensified. *Management of innovative processes can only truly succeed if the principles of innovative work also apply to management itself.* This implies that management must leave the field of dissected "managing" positions and must be open to deal with all aspects of developmental work.

Developing and Introducing New Products to the Market— Two Inseparable Project Subjects

Innovative products and newly developed processes are embedded more and more often in a comprehensive network of application requirements, risks, and sometimes unpredictable outcomes. It is significant that such innovations increasingly turn into the form of a system, that is, the innovations appear in a complex form or configuration of autonomous and related technical components and organizational interweaving. That leads to an important consequence: the purchase of new instruments, components, or equipment systems does not automatically guarantee a benefit. With the installation of a new refrigerator or coffee machine an immediate usefulness is achieved, whereas this is not the case with the installation of personal computers, communication networks, or—in particular—complex computer-integrated manufacturing systems. Such products are not merely "acquired," they are also "introduced."

Today the introduction of products and systems has become a complex problem for both the marketing and the investment departments. Therefore the benefits of innovations are dependent on the ability to integrate them into an already existing system. It is this process of systematic integration, of getting a system on its way, of overcoming obstacles and mishaps, of learning and adjusting that can only be successful in the form of a project.

The path of innovation—from blueprint to practical application—should thus be a two-step project: first to develop and then to introduce the innovation. Of course, this does not generally apply to every innovation, but it seems to be particularly true for strategically important technologies as, for example, computer technology, computer networking, factory automation, and office communication.

Organizational Development through Projects

The latter aspect, namely, the channeling or introduction of innovations into business settings and structures, is closely associated with a third project area: the development of the organization. In the same way management tasks are loosing their characteristic routines established during decades of economic development, and the change of company structures—even crisis management—is a daily occurrence in a manager's job, a change in understanding and executing leadership arises.

The new focus is on solving problems deriving from unpredictable and complex situations. Management dealings are becoming increasingly a matter of "sink-or-swim" situations in a turbulent business setting. The outcome can quite often only be predicted in a fairly diffuse manner. Rarely can the whole process be organized in a traditional manner, because we are dealing here with developmental processes in which a management of uncertainty and complexity is required. It can be characterized by such traits as the uniqueness of a given situation, the inability to predict status and project development, and a network determinism which is a constantly changing force during the course of actions.

Thus a structural change within a company or a department is, in retrospect, always a historical process, and all participating actors play roles in historically unique settings and constellations. This poses a central problem as well as a challenge for project management, as it could prove that the "chaos" described could be controlled and regulated within the framework of a flexible project organization. However, such controls and regulations must be understood in a new way. Projects considered from this point of view would have to be set up as an organizational form of change.

Projects as a Flexible Work Form

The previously detailed aspects with regard to technological and organizational changes could be considered as the inside view of what economists generally would call a structural change. As outlined, project actions could, to some extent, be considered a medium of such change. It is interesting to notice, however, that this change does not simply remove dysfunctional structures, nor does it only imply an adaptation to other environments and surroundings. Rather, it increasingly indicates that these processes of change, besides their transitional character toward new functionality, also contain a new and higher quality of industrial culture.

There have always been processes of change. One look at the rise and fall of human cultures and the evolution of all living and liveless forms is a confirmation of this fact. But a new aspect of the present transformation of industrial society is the highly increased tempo of change. Destabilization and stabilization alternate in a fast rhythm. Evolution in nature and in later periods of social history has been a slow process. Now we can "hear the grass grow." We are actors within an "evolutionary chaos."

This will have far-reaching consequences with regard to the basic principles in the fields of business and organization. Available are bureaucratic, military, and tayloristic organizational models which have a long line of traditions and are still prevailing today. If, as it appears, permanent change is a characteristic part of our future, then a future-minded form of organization can no longer be based on rigid structures. Therefore it is quite understandable that modern technology and current work organization are required to be flexible. In general, it can be stated that flexibility is seen as the basic concept for a revolutionary organizational culture and theory.[6]

Flexibility is the result of structural change and, at the same time, its precondition. Therefore there are two aspects to flexibility. It is the adaptation of economic and social transactions to a changed environment, and it acts as a propellant and catalyst of the dynamics of change itself. Thus projects are not only ways of change and development. Project management itself is a product of present change: the project as an open form of contract, or project management as a flexible method of production and service. There are numerous examples for the aforementioned statement in a variety of business branches. Interesting to note are tendencies in the area of mass production, where individualized products—close to the customer—are becoming the basis for production. Examples are order-specified computer chips or customized automobiles which are proudly called "unique" by the manufacturers.

PROJECT WORLDS IN THE LIFE CYCLE OF PRODUCTS

In general the life-cycle phases of products span from basic research to product recycling, and each phase contains interconnected autonomous project worlds. In each phase the polarities of technology and management have to be emphasized, the difference of which has been known for a long time and has been a widespread fact in our business world, manifested in the professional roles of engineers or scientists and business managers. This differentiation has now become a problem insofar as decisions of a business nature are more and more dependent on single aspects of a "managed" technical process. Product developers, researchers, and engineers have become the true initiators of economic success, because they have a direct influence on the creation and performance of products. This means that a management that concentrates exclusively on administrative tasks loses its purpose. The opposite thereof is linked and networked technological and management competence. But networking is not going to be easy, and considerable problems will have to be overcome.

The following four interrelations are future-oriented polarities which at present can be viewed only as classical controversies, but which, in the future, will become strategically more and more important for the success of businesses.

1. *The relation between formal organization and informal networks.* The trend is networking in flat structures. Formal regulations will become more flexible, informal relations will step out of the shadow and will increasingly become a perceived part of business systems.

2. *The relation between formal organization and projects.* The trend is that routine processes, which are typical for formal organizations, will decrease. Projectlike processes will increase. Projects will not be limited to departments and areas and will serve as integrating factors in structuring the division of work.

3. *The relation between single projects and project networks.* The trend is that single-project management will increasingly lose its isolation and will become part of a project network management. This will change the importance of single projects and will allow synergies to emerge between projects. However, it also increases the possibility of new types of conflict.

4. *The relation between informal networks and networked project organizations.* The trend is that informal networks, which have always existed, are a medium for "management by projects." They permit an easy start in organizing interrelated project organizations.

TOWARD A SYSTEMIC-EVOLUTIONARY METHODOLOGY OF PROJECT MANAGEMENT—POINTS TO PONDER FOR THE INTERESTED READER

Following is a collection of viewpoints and ideas which have come to play important roles in the author's professional experience during the past years. The selection represents several interconnected fields and controversies stemming from personal discussions and from symposia dealing with "Reorientation in Project Management."

Viewpoint 1

The most important mental resources of reorientation in project management are the general management theory and a corresponding reorientation in the business world (distinctive since the 1980s), and a revolution in the world of science (distinctive since the 1970s). In particular, the latter aspect seems to require further definition. Theories of self-organization and the related interpretations of evolutionary processes in the fields of physics, chemistry, and biology are understood by their main supporters as an overthrow of the traditional, two-centuries-old rationalistic world conception, with far-reaching implications for the understanding of rationality in the technical, economic, and political scopes of thought processing and performance. Well-known biologists such as Konrad Lorenz, Rupert Riedl, and Frederic Vester and physicists such as Hermann Hahen[7] or Ilja Prigogine[8] rattled the established scientific world with their publications and announced a new cultural epoch as a solution to the intellectual crisis. However, this fundamental change has to be evaluated very carefully and is not intended as a simple renewal of a future optimism. In publications by the above scientists, these chances of cultural renewal appear as a high-contrast image in front of the background of possible global catastrophes.

Since the early 1970s, management researchers at the University of St. Gallen for Economic and Social Science, namely, Hans Ulrich, Fredmund Malik, X. Gilbert, and J. B. Probst, have systematically influenced cybernetic ideas in a positive way and have, in particular since the early 1980s, picked up the message from revolutionary scientists. From originating syntheses, an extensive methodology with universal principles of managing and organizing has been created. Without doubt, many statements are still purely hypothetical. Thus both a changing general management theory as well as its basic scientific philosophy—a systems theory oriented toward self-organization and evolution—appear to be a source for new organizational forms and methods of logic and performance in projects.

However, supporters as well as opponents of this program should keep in mind that a one-sided transfer is not possible and not intended. In particular, the act of organizing and managing in projects or with projects is subject to each individual's type of performance. This requires independence in both the practical and the theoretical fields of project management. Conversely, a development in the professional field of project management will entail a reaction to the professionalism of top management.

Viewpoint 2

Of superior importance for a reorientation in project management is the systemic-evolutionary trend initiated by the St. Gallen researcher and management consultant Fredmund Malik.[9] His objective is to unearth a rationality which in its entity is part of a mechanistic understanding of labor. Malik's theory is based on the understanding that complex economic or even social systems have never been a product of such rationality nor can they ever be one. Organizations, like social systems, but mainly their interdependent networks in the form of "markets," have never been planned as such, nor can they, as a total, be part of any schedules. Malik believes that such systems are being generated by a process of self-organization, of which schedules and regulating actions are merely a part, that is, as an inherent influence and not as an external, unknown instance to the system.

Consequently, Malik divides management performance into two spheres:

- A performance sphere in which person-machine systems or person-person systems are formed rationally, that is, they are being created like products in a manufacturing process

- A performance sphere in which such forms disintegrate, enter a crisis or change, and become a creative basis for newly developing configurations

Malik's main interest is directed toward the latter sphere. Thus his reflections are centered around the problem of changeability of organizational structure and, in general, around the dynamics of revolutionary change.

Obviously projects, that is, irreversible and unique performance processes, are particularly suited to organize such processes. Therefore Malik's systemic-evolutionary management trend contains an almost natural affinity toward project management.

Of course, a comprehensive transformation of several aspects of the St. Gallen theory into the interests of project management has yet to be performed. In particular, there are two problem areas which are of main concern in a reorientation:

- What kind of consequences can be expected in top management from a new way of initiating and organizing projects?

- What kind of impact on project performance does a systemic-evolutionary management style have?

Viewpoint 3

In accordance with traditional concepts of systems engineering, the development of a machine or a technical system will be planned only as a step-by-step sequence of actions and will be executed in projects. The reality of everyday engineering work, however, does not correspond at all to the ideal of systems engineering. As a matter of fact, unpredictable and unexpected events or situations and innovative inspirations are typical for the development and introduction of new technical systems. Processes of self-organization and evaluation are interwoven into the daily business of engineers. However, the participants are hardly aware of the positive aspect of this fact. The understanding of technical professions is based on reliability and controllability. This makes professionals blind to accept "chaos" as a creative force.

Technical science is an applied science. Therefore it only makes sense that engineers take advantage of the scientific chaos research. In nature, ordered structures are the result of chaotic processes of evolution. Why would that be different in technical science?

Viewpoint 4

The traditional interest in project management is aimed at the problem of how to avoid unnecessary chaos. Even if this traditional expectation is being maintained, it poses an interesting question: how to reduce avoidable randomness through a higher system rationality? Here the aim would be to increase the reaction capability in a performing system. This means the capability to organize those situations in which unavoidable randomness can be properly handled during the course of the project while keeping the interest of the project in mind. From this point of view, traditional strategies of simplification, in particular a "reduction of complexity" commonly achieved by work division, are insufficient.

Viewpoint 5

Planned schedules, and even performance networks, are constantly becoming obsolete by a more or less chaotic reality. The appropriate answer is: "Raising the level of situational competence."

However, this means an increase in outlining effort as well as in communication. Consequently an increase in the value of the "human factor" and a certain decrease in the appreciation of technical tools are necessary. An increase in communication, for example, requires an increased readiness to discussions, group dynamics, and, thus, an increased requirement for social competence. Dealing with a raised level of complexity requires a raised level of participation in the organizational culture of project teams.

Viewpoint 6

Projects are action systems, that is, social systems. With regard to the problem of an urgently required theory of a method for both a strategic and an operative "management of change," the relation between project and project environment is coming into focus.

Looking into the project environment, project events appear to have an inside-outside tension. Important from the internal aspect is how project organizations can exist in turbulent environmental relations and, above all, can be successful. A resulting conclusion is project marketing, that is, a constant effort to "sell" project ideas, project objects, and project concepts within the appropriate environment.[10] The external aspect shows, for example, the relation between project management and both top management and hierarchy of power.

Viewpoint 7

Each project group can be understood as a configuration of characters. Therefore the prospect for success in group work does not only originate in the combined qualification profiles of professional competence. A group is also always a social system with a communicative and, thus, a sociological inherent dynamic force. This means that an important fact in project groups—as in families—is the matching or mismatching of people. It is, for example, unlikely that a team trying to solve difficult development problems will achieve remarkable results without one or several of its members being the "border-crossing type." Conversely, it seems misguided to fill a marketing position with a meticulous "stickler." Here it would be advisable to use a more relation-oriented person.

A closer look at interaction capabilities of people in project groups will lead to the question of how well these groups can deal with conflicts. It is a well-known fact that the communication in professional projects is dominated by business-related subjects and aspects. However, it becomes increasingly clear that this is no longer sufficient. The social competence of project managers as well as members becomes increasingly important.

The human relations network is a "soft structure." However, the social role of the engineer or business person which has evolved over time is, to a certain extent, insensitive to the reception of "weak signals." This means that there is a gap between work structures determined by professional borders or formal organizational regulations, and sociological interhuman relations retreating or even hiding behind behavioral connections.

But the realization is on the rise that mainly human capital and not amassed material resources should be regarded as the decisive potential for success in the economic competition. At a rather late point in time in the technical world, this turning toward humaneness in an often inhumane world of business and industry carries a high degree of embarrassment. The revolutionary change in the business world came unexpectedly and, as yet, has hardly been absorbed intellectually. For the time being this raises more questions than there are answers, and it causes more problems than there are solutions.

REFERENCES

1. H. Balck, "Projects as a Form of Change," in *Handbook of Management by Projects,* R. Gareis (ed.), Manz, Vienna, Austria, 1990.

2. A. D. Hall, *A Methodology for Systems Engineering,* Van Nostrand Comp., Inc., London, Toronto, Melbourne, 1962.

3. H. Kern and M. Schuhmann, *Das Ende der Arbeitsteilung?* (The End of Work Division?), C. H. Beck, Munich, Germany, 1990.

4. M. J. Piore and Ch. F. Sabel, *The Second Industrial Divide, Possibilities for Prosperity,* Basic Books Inc. Publishers, New York, 1984.

5. L. Mumford, *The Myth of the Machine,* vols. I and II, Harvest Books, Div. of Harcourt Brace Jovanovich, San Diego, Calif., 1964–1970.

6. T. Peters, *Thriving on Chaos,* Excel, a California Limited, London, 1988.

7. H. Hahen, *Synergetics. An Introduction. Nonequilibrium Phase Transitions in Physics, Chemistry, and Biology,* Springer, New York, 1978.

8. I. Prigogine and I. Slengers, *Order out of Chaos. Man's New Dialogue with Nature,* New Science Library, London, 1984.

9. F. Malik, "Evolutionary Management," *Cybernetics and Systems,* no. 13, pp. 153–174, Hemisphere Publishing, Bristol, Prenn.,1982.

10. R. Gareis, "The Management Strategy of the New Project Oriented Company," in *Handbook of Management by Projects,* R. Gareis (ed.), Manz, Vienna, Austria, 1990.

CHAPTER 3

THE EVOLUTIONARY PROJECT MANAGER

Ervin László

The Academy for Evolutionary Studies,
Frankfurt, Germany

Ervin László, Docteur ès-Lettres et Sciences Humaines of the Sorbonne and recipient of several honorary Ph.D. degrees, was born in Budapest, Hungary, in 1932. A former director of the U.N. Institute for Training and Research, he is a member of the Club of Rome, the World Academy of Arts and Science, the International Academy of Science, and the Académie Internationale de Philosophie des Sciences. László has served as professor of philosophy, systems science, and futures studies at various universities in the United States, Europe, and the Far East. He is the author or co-author of 51 books and over 500 papers and articles, Editor of *World Futures, The Journal of General Evolution,* Associate Editor of *Behavioral Science,* Founder-Director of the General Evolution Research Group, Administrator of Environnement Sans Frontière, Secretary-General of EUROCIRCON, the European Culture Impact Research Consortium, Principal Adviser to the Director-General of UNESCO, and Rector of The Academy for Evolutionary Studies in Frankfurt, Germany.

INTRODUCTION

As we enter the last decade of this century, we find ourselves at a crucial juncture in history. We are transiting into a new kind of society. What we are now living through is the transition from nationally based societies to an interconnected and increasingly seamless global socioeconomic system.

Standard answers are no longer to the point; classical assumptions concerning the nature of the contemporary world have collapsed. Not only have the rules of the game changed, the game itself is new. It is now mainly economic and environmental rather than political and military. This world is no longer an arena of struggle between capitalism and communism led by two superpowers; it is a more complex world, with more players.

The world over extensive, quantitative growth is giving way to intensive, qualitative change. This is a major shift. Extensive growth can occur by linear accretion—more of the same can be accumulated—but intensive development is nonlinear and transformative. It either evolves the system in which it occurs, or it destroys it.

We do not lack the intellectual and economic resources to tackle these problems. Scientific breakthroughs and technological ingenuity have given humankind the capacity to overcome all challenges; global networks and markets could bring the benefits to all. Motivation is now needed to face the problems and smooth the path of the present grand transition. The business sector has the greatest capacity for the constructive application of science and technology, and for assuring the beneficial functioning of worldwide networks and markets. Fate and fortune have combined to place unprecedented powers into the hands of contemporary enterprises, and, together with governments, these enterprises are now the fulcrum on which the success in the current transition will turn. Managers, especially project managers, have a unique role and great responsibility in this regard.

PROJECTS AND THE GLOBALIZATION PROCESS

The project manager is now operating in a business environment dominated by giant multinational and global actors. In the immediate postwar years, U.S. corporations in search of profit pioneered the shift from the national to the international level. The driver was the low cost of labor overseas. In the process they transferred considerable know-how and technology to the host countries. Competitive domestic industries arose in Europe, Asia, and Latin America. Before long, some of the overseas companies, especially in Japan, caught up with, and even surpassed, their U.S. counterparts in competitiveness, profitability, and market share.

Subsequently the worldwide flows of information became another driver of the globalization process. Because information can spread rapidly from company to company, it had become more difficult to build competitive advantage through the commercialization of inventions. In the early part of the century, research and development of new products led to profitable leadership positions that could be conserved first through patents and later through famous brand names. But when the knowledge base of innovation became globally and almost instantly accessible, imitation quickly led to the dissipation of profits in new projects. As a result, firms turned to new forms of growth and profitability. On the one hand they moved further downstream into consumer and service areas, and on the other they expanded horizontally through mergers and acquisitions. This created further growth in the size, scope, and geographic span of leading enterprises.

The creation of multinational companies constituted one stage in the globalization of business; the next stage was the emergence of nonnational, that is, genuinely global, corporations. By the late 1980s no country was holding a monopoly on technology, capital, talent, or innovation. In the continued search for profit and growth, internationally operating companies made breakthroughs in the laboratories of one country, placed shares with investors from others, and put the nationals of still others on a fast track to the top. Today the management of global players relies on cash-flow calculations rather than on national loyalties to decide where to shift production and capital. The multinational economies of scale are complemented by global economies of scope.

DECENTRALIZED ORGANIZATION STRUCTURES

In search of profits and competitive advantage, enterprises went international, then multinational, and then global. But in the course of this process, business has outgrown traditional modes of organization. If it is to remain efficient, and become a positive force in the contemporary world, it must shift from hierarchical to network modes of organization, and from centralized corporate planning to decentralized multiple-project strategies. Here is where the role of projects and project managers becomes important.

The globalization of business raises the specter of the global concentration of economic—even socioeconomic—power. The handful of global players that now dominate entire sectors could transform the world economy into a pyramidal structure. If the market leaders were to be hierarchically organized themselves, a rigid decision-making chain would emerge, concentrating from the shareholders and the board of directors to the top executives of the global companies, and diffusing from top management to middle-level managers, employees, subcontractors, service companies, clients, and ultimately to the great mass of consumers on all five continents. Aside from the ethics of subjecting the majority of the world population to decisions made within closed circles and in light of self-centered considerations of profitability and competition, there is also the question of efficiency. Centralized corporate strategies are likely to be inefficient in a complex and rapidly changing evolutionary environment. The computer simulation of evolutionary strategies by dynamic system theorists shows that success does not presuppose a hierarchical organization committed to carrying through a single master plan, but the availability of parallel processing capability in the system, that is, of active, simultaneous problem-solving activity by a collection of subsystems.

These insights argue loudly and clearly for replacing hierarchical structures defined as "tall" organizational charts by networklike structures described as "flat" charts. Such an "organizational revolution" is in fact occurring among the most advanced of the global companies. Like the motto of ASEA Brown Boveri claims, they are becoming "locals—worldwide." Despite the concentration of market power in their hands, the global companies are decentralizing their operations. Whereas the multinationals in the 1960s treated foreign operations as distant appendages for the production of goods designed and engineered in their home country with a hierarchical chain of command, the advanced global companies are network-like structures that do not command, but merely orchestrate, the efforts of their far-flung subsidiaries. The new globals are diversified ensembles of a wide variety of enterprises, many of which are of modest dimensions.

SOCIAL EVOLUTION AS PROJECT CONTEXT

Young people will tell you, "go with the force." There is no determinant "force" for project managers other than success in carrying out the projects entrusted to them, but success itself can depend on whether the project is "in tune" with larger trends and processes in the project's environment. Every project is at the same time an economic proposition, a technological challenge, and a social system. It impacts on a given social environment, and brings its benefits if objectives coincide with wider trends in that environment. The trends that hold sway today are not just local and temporal. There are deeper trends as well, trends defined and determined by the evolutionary dynamics of complex systems such as contemporary technological societies. Project managers would be well advised to familiarize themselves with these trends, for coordination with them can enhance the chances of success in the implementation of projects—and render the projects more responsive to the problems of contemporary societies.

Acquiring evolutionary competence in project management is not an unmanageable task. A new system of knowledge is now available concerning the basic trends that underlie the evolution of complex systems. The new sciences of evolution and their allies—cybernetics, information and communication theory, chaos theory, dynamic systems theory, and nonequilibrium thermodynamics—convey an understanding of the laws and dynamics of complex-system evolution, regardless of whether the systems are physical, biological, ecological, or human. The new understanding contradicts facile views that we would have reached the end of history, or that the future would be made by chance or mere puttering about. The course of history, it tells us, has a logic of its own; a logic that is not rigidly predetermined and yet not the plaything of chance. This logic governs the evolution of life on earth, and the evolution of matter in the observable universe. It also governs the evolution of human societies in the course of history. The evolutionary process is irreversible, chaotic, and nonlinear—but not unordered and haphazard.

RESPONSIBLE (PROJECT) MANAGEMENT

The evolutionary process is likely to continue in the future. A first extrapolation of the trend would suggest that societies will be more and more structured and complex, more effective and efficient in exploiting the energy flows of their milieu, and therefore more and more fluid.

Such a world, of course, cannot be structurally stable. Expectations that beyond the current crisis there would be an unaltering social, economic, and political order are mistaken. Societies will remain vulnerable, both in regard to the rapid and possibly radical changes in the values, worldviews, and expectations of their members, and to the equally rapid and radical changes that powerful new technologies are likely to create in their physical, social, and ecological environment. However, even if the future is likely to be as adventurous as the present, with sound foresight it could be made less hazardous. The present epoch of instability need not open the way to catastrophe; it could also yield to a new period of dynamic—rather than structural—stability. With sound evolutionary management, that period could be extended. Societies could be kept on a functionally stabilized dynamic plateau for a long time, if not, obviously, forever.

Management can be more than a personal play with chance. It can be a game played with the developmental probabilities associated with the evolution of the complex systems within which contemporary business and projects operate. Managers and project managers need to inform themselves and act responsibly. In the nineties responsible management no longer means merely efficiency, interpreted as doing a given job faster, nor does it mean simply effectiveness, in the sense of doing a better job. Responsible management also means more than securing strategic advantage—doing new jobs to increase revenues and competitiveness. Today responsible management means all this and more. It means accessing information on evolutionary processes and impacts, and using the information purposively, to coevolve the enterprise together with its wider socioeconomic environment.

Responsible managers are informed managers. Scientists are gaining a remarkably accomplished understanding of complex-system evolutionary dynamics, but they are powerless to influence the course of system evolution. Managers, especially project managers, operate at the leading edge of technological innovations and could have a crucial impact on the socioeconomic processes that decide the outcome of the great transition in which humanity now finds itself. They owe it to their companies, for the shared benefits of all.

NEW CHALLENGES FOR THE PROJECT MANAGER UNDER GLOBAL CONDITIONS

As reported in the September 18, 1991, edition of the *International Herald Tribune,* a major survey carried out by the Columbia University Graduate School of Business and Korn/Ferry International found that companies around the world share two major fears for the future: increased competition from abroad and the shortage of good managers who can meet the challenges of that competition. The fact is that the world is becoming more global and more competitive. Heightened competitiveness has created an unprecedented demand for good managers, who can ensure that their companies not only survive, but prosper as they move into the 21st century. Globalization and, in Europe, regional integration are the most far-reaching and profound change factors facing managers today. As Thomas P. Gerrity, dean of the Wharton School at the University of Pennsylvania, said, these factors are reshaping markets and competitive forces and demanding changes in the way managers think and work. In the future, and especially in Europe, as Bruno Dufour, general director of the ESC Lyon Graduate School of Business, pointed out, managers will have to move easily in many areas, including science, culture, and technological development. Consequently they must acquire information and skills beyond such standard specialties as finance, marketing, organization theory, and public relations.

These requirements apply to project managers more than to anybody else, since project managers are more directly exposed to shifting currents in corporate strategies as well as in the marketplace than managers operating within the comparatively stable hierarchies of major corporations.

The new levels of competition, together with worldwide economic restructuring, market shifts and uncertainties, environmental crises, and the accelerating rate of technological progress, oblige project managers to respond to the challenge of designing and implementing successful projects with concepts that are different from those that they used in the past. Conventional methods no longer fill the bill; even principles such as those distilled by Peters and Waterman in their "search for excellence" fail to provide the international and multicultural posture that has emerged as an imperative of successful project management. The economic, social, and political environment in which projects are now implemented has changed radically. In consequence attempting to manage impactful projects in reference to the introverted principles advocated by the mainstream management literature would be like concentrating all one's skills on flying an airplane and paying little attention to the airspace one is flying in. Today's pilots cannot be solely concerned with the functioning of their aircraft. They must also set a course in reference to climatic conditions, their current position, the projected destination, and, above all, the traffic around them. That traffic is intensifying; the airspace is becoming crowded and competitive. But there are no "air-traffic controllers" in the sphere of project management. The managers themselves must do their own navigating.

For effective navigation, information is the critical resource. Technological progress has endowed enterprises with enhanced powers and capabilities. However, they have also led to a rapid growth in corporate and social complexity. Project management is no longer an organized and orderly game where the players pursue preconceived plans to achieve predetermined ends, but an ongoing play with chance and probability in an environment where not only the chips and the players, but also the rules of the game, are subject to change. To assure reliable orientation in this unstable and probabilistic context, the expert and effective use of information becomes essential.

Project managers need to look outward to the world and not just inward to the problems of their own teams and clients. They must interact with the environment at multiple points at the same time, and continually update their strategies of implementation so as to exploit emerging opportunities and weed out obsolescence.

There are new criteria today of what constitutes good project management. In the sixties good management meant efficiency, and efficiency meant doing a given job faster. In the seventies good project management came to mean effectiveness: doing a better job. In the eighties the meaning of good management shifted to the ability to secure a strategic advantage for clients; but in the nineties good project management also calls for accessing, and acting on, relevant and up-to-date information on clients, markets, and even society at large. Good managers coevolve their project with its encircling social and business environment.

This means that project managers must become evolutionary managers. The contemporary systems sciences show that projects can be modeled as subsystems within larger socioeconomic systems. Their future is directly tied in with the fate of these broader systems. Good managers know that it is in their interest to sustain the broader system in the framework of which they operate. To evolve the system that sustains one's own project, even if it also sustains the projects of one's competitors, makes excellent sense. In a time of system transformation, altruism is pragmatic. Either the broader socioeconomic and corporate system survives, or everyone in the given industry sector risks becoming extinct.

A related requirement for project managers is to take into account a major benchmark of the late 20th century world environment: the convergence of states and economies on increasing levels of integration and organization. In Europe, national states converge on the continental level. The dissolution of the Soviet Union and Yugoslavia does not contradict this process. Convergence can only build on stable forms of political organization, and these states have been unstable because of an arbitrary integration of their diverse nationalities and republics. In order to reintegrate, they first had to disintegrate. The then unfolding process of convergence will be similar to that which already occurs in the rest of Europe, where sovereign nation states are joining together not only in an economic but also in a monetary union, and before the end of the century very likely also in a political union.

The importance of "going with convergence" is underlined by the recent rise of Yeltsin and fall of Gorbachev. Despite his undisputed accomplishments in introducing glasnost and launching perestroika, Gorbachev failed to understand the laws of evolutionary convergence. He attempted to reform an arbitrarily integrated system, instead of realizing that such a system must disintegrate before it could reintegrate. Yeltsin may not have realized this consciously, but this chosen strategy aligned him with the trend. Consequently perestroika, one of the largest and most daring projects of all times, has failed, while the commonwealth of independent republics, a suddenly emerging and hastily prepared project, became reality.

Going with convergence in the context of project management calls for forging forward-looking partnerships that transform one's erstwhile competitors into one's project's strategic partners. In the contemporary world even the best funded project cannot "go it alone." Strategic partnerships are required to take full advantage of new markets and new technologies, and the new set of regulations that applies to them. They are also required to avert having unscrupulous competitors take advantage of conscientious managers who are willing to undertake some necessary adjustments. Project managers must realize that cooperation often offers more advantages than competition.

Evolutionary project managers must also recognize the need for genuine partnership between men and women within the project team. Unlike the old corporate culture that viewed women as unskilled or semiskilled labor suitable only for routine tasks, evolutionary project managers bring women into all levels of decision making. Women have a unique ability to relate to people with flexibility and understanding, and thus to be effective and efficient in conducting business, especially when it involves crossing cultural or national boundaries.

CONCLUSION

The valuation of people, and of the environment in which they live, is the ultimate foundation of effective project management. This valuation must be embedded in the evolutionary context: evolutionary managers enter into a new social contract of responsible action with clients and with society at large.

Evolutionary project management is not utopian. Today's technologies of information and communication are altering perceptions of reality more than the printed word ever did. The rhythms of change in our unstable world are penetrating all minds. The new project managers now moving into executive positions question established values, attitudes, and practices, and search for new ways of thinking and acting. It is to be hoped that they will adopt evolutionary strategies and accept the responsibilities that go with them.

CHAPTER 4

MANAGEMENT BY PROJECTS: SPECIFIC STRATEGIES, STRUCTURES, AND CULTURES OF THE PROJECT-ORIENTED COMPANY

Roland Gareis

University of Economics and Business Administration,
Vienna, Austria

Roland Gareis holds an M.B.A. and a Ph.D. degree from the University of Economics and Business Administration, Vienna, Austria. He was a Fullbright scholar at the University of California, Los Angeles, in 1976, professor for construction management at the Georgia Institute of Technology, Atlanta, and a visiting professor at the Georgia State University, the ETH in Zürich, Switzerland, and the University of Quebec in Montreal. Since 1983 he has been the director of the postgraduate program "Project Management in the Export Industry" at the University of Business Administration and the Technical University in Vienna. He is president of Project Management Austria, project manager of the 10th INTERNET World Congress on Project Management, and manager of the research program "Crisis Management—Chance Management" at the University of Economics and Business Administration, Vienna. He is owner of Roland Gareis Consulting. He has published several books and papers on project management.

COMPANIES FOCUS INCREASINGLY ON PROJECTS

Projects Are Instruments to Cope with Complexity

Due to dynamic markets, new environmental and technological developments, and changes in paradigm, companies have to cope with new challenges and potentials (Fig. 4.1). In order to deal with an increasingly complex business environment, companies increase their complexity. Strategic business units are established, autonomous groups and quality circles are organized, and projects, as temporary organizations, are defined to perform unique and complex tasks.

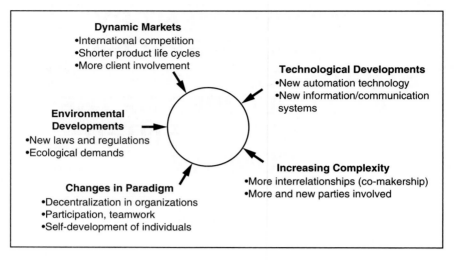

FIGURE 4.1   New challenges and potentials for companies.

The Project-Oriented Company Performs Different Types of Projects Simultaneously

In addition to the traditional contracting and research and development projects, new types of projects such as strategic planning and marketing and organizational development are undertaken. These new project types have specific characteristics:

- They have no external project owners and are therefore considered "internal" projects.

- They are socially complex, as their results often have an immediate impact on the strategies, structures, and cultures of the project-performing company.

- They are relatively small in size (100 to 500 work packages, duration of 1 year or less, cost of almost 1 million U.S. dollars). But they often employ heavy internal human resources.

- They are innovative and unique and not "repetitive," that is, standard technologies and procedures cannot be applied.

- In the early phases their objectives are often not determined precisely.*

In addition to the traditional project-oriented industries, such as construction, engineering, and electronic data processing, manufacturing, banking, tourism, and even administration turn to projects. From an inspection of Fig. 4.2 it becomes obvious that the "new" Project-Oriented Company performs small and large projects, internal and external projects, and unique and repetitive projects to cope with new challenges from a dynamic business environment.

The Project-Oriented Company has specific strategies, structures, and cultures to manage single projects as well as the network of projects performed simultaneously. It

---

*For the concepts of open, temporary, and concrete projects see Briner et al.[1]

Characteristics	From	To
Industry	Construction, engineering, EDP	Construction, engineering, EDP, manufacturing, banking, insurance, tourism, administration
Contents	Contracting, R&D, investments	Bidding, contracting, R&D, strategy planning, marketing, public relations, personnel development, organization development, investments
Size	Few large projects	Many small and large projects
Ownership	External projects	Internal and external projects
Experience	Repetitive projects	Repetitive and unique projects

FIGURE 4.2   New types of projects in new industries.

is characterized by having an explicit project-management culture: applying project-related incentives, presenting projects as temporary organization structures in the corporate organization charts, and documenting the project orientation in the corporate mission statement.

Management by Projects—the Central Management Strategy of the Project-Oriented Company

An increasing number of companies is resorting to projects. But projects cannot only be considered as tools to solve complex problems. Projects are a new strategic option for the organizational design of companies.

Management by projects is the central management strategy of the Project-Oriented Company.[2] By starting, performing, and closing down projects, Project-Oriented Companies try to achieve a dynamic balance, which is intended to ensure the continuous development and the survival of the company.

By explicitly applying management by projects as a management strategy the following organizational objectives are pursued:

- Organizational flexibility (projects as temporary organizations)
- Delegation and decentralization of management responsibility (lean organizations)
- Organizational integration (cooperation between different departments)
- Quality assurance (holistic project definitions)
- Goal orientation in the problem-solution process (projects as goal-determined tasks)
- Acceptance of project results (because of project team building and marketing)
- Continuous organizational learning and development through projects

Management by projects also allows pursuing personnel objectives. Different leadership approaches such as management by objectives, management by delegation, and management by motivation can be operationalized and integrated by projects. Project functions include motivation and personnel development. Membership in a project team is attractive, because projects allow one to perform new tasks, promote team-work, allow autonomy, require creativity, support feedback, and offer new opportunities after the project ends. Individual learning is promoted by projects because of the complexity of the tasks and the clearly defined project objectives.

Further, management by projects has a marketing dimension. Sometimes project-management know-how can determine the "unique selling position" of a company, and project-management services can be marketed to in-house and external clients.

By performing more and more projects as a way of doing business, companies implicitly apply management by projects. But only the explicit application of management by projects allows taking advantage of the benefits described.

The conscious application of management by projects as a management strategy will require certain adjustments in the company structures and culture, such as the following considerations:

- Adjustments in the organization and communication structures
- Development of new role perceptions of base organization and project personnel
- Development of new personnel planning methods (flexible, multiple job assignments)
- Issuance of new personnel qualifications (redundancy and variety)
- Provision of integrative measures (corporate vision and strategy, project-management culture)
- Acceptance of the autonomy, complexity, and dynamics of projects

THE PROJECT-ORIENTED COMPANY HAS A LEAN AND FLEXIBLE ORGANIZATION

Projects Are the Basic Structural Elements of the Project-Oriented Company

Traditionally the organizational structure of a company is designed to perform routine tasks in the most efficient way. The organization has to provide orientation to the company's personnel regarding the distribution of responsibilities, and it has to ensure stability and continuity in the relationships between the company and its supply and demand markets. In companies with little project orientation, projects are applied in addition to the traditional hierarchical line organization. By carrying out projects, these companies become flatter and more flexible (Fig. 4.3) because the spans of command are widening and the number of hierarchical levels is decreasing. More flexibility is created because of the potential to define and dissolve temporary project organizations.

In lean organizations one-dimensional supervisor-subordinate relationships are replaced by wider spans of communication. This requires a new communication quality. On the one hand, less time can be spent for operational directives, so a more strategic, visionary leadership approach is needed. On the other hand, new communication technologies can be applied.

Companies with high project orientation develop flexible networklike organizational structures. An example of a networklike structure is shown in Fig. 4.4. Many organizations can be perceived similarly if informal structures and projects are made visible and are formalized.

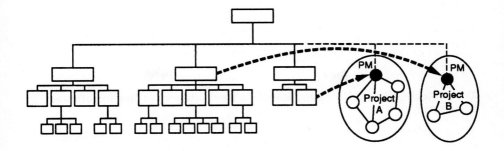

FIGURE 4.3 Flattening of organizational structure through projects.

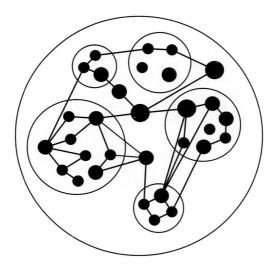

FIGURE 4.4 Networklike structure of an organization.

The extent of project orientation of a company's organization cannot be measured in absolute figures, but it can be looked at as a position on a continuum between a hierarchical line organization and a flexible networklike organization. The relationship between routine and project work determines the positioning of the company on that continuum. There is no optimum position between a line organization and a networklike organization. Each position has its functionality. But it can be observed that there is a trend toward flatter and more flexible structures.

The engineering company Fluor Daniel "uses project management as a way of conducting its business and, of late, as a way of improving its internal operations.... Fluor Daniel is able to conduct its business effectively in a decentralized, networked organizational atmosphere."[3]

The Project-Oriented Company Requires New Integrative Organizational Structures

The more projects a company performs simultaneously, the more differentiated becomes its organization and the higher becomes its management complexity. This complexity results from the complexity of the individual projects as well as from the dynamic relationships between them.

In order to support the successful performance of projects the organization requires integrative structures such as strategic centers, project steering committees, project resource pools, and centers of project management excellence. Many of these integrative structures are communication structures rather than organizational units.

There is a tendency of the functional departments of Project-Oriented Companies performing repetitive projects to develop into resource pools. The resource pool members are experts who are responsible for their project work. The resource pool manager is not responsible for the project work of the pool members. His or her responsibility consists in assigning pool personnel to the projects and ensuring that enough qualified personnel is available for the projects. Further he or she is responsible for the development and application of working standards and work ethics. "Traditional departments serve as guardians of standards, as centers of training and the assignment of specialists, they won't be where the work gets done. That will happen largely in task-focused teams."[4]

Projects performed simultaneously by a company can be perceived as a network of projects. Networks of projects are specific organizational structures of Project-Oriented Companies. By considering networks of projects as management objects the differentiated structures of the Project-Oriented Company are complemented by new integrative structures (Fig. 4.5).

Project Management Is a Basic Management Qualification in the Project-Oriented Company

In order to perform projects successfully, personnel management in the Project-Oriented Company has to meet specific requirements. To fulfill roles in the base organization and project roles, such as (internal) project owner, project manager, or project team member, managerial skills as well as functional skills are required. As most employees get involved in projects, project management becomes a basic management qualification and is no longer a specific expertise of a few project-management experts. Especially in small and medium-sized projects one person often takes on the roles of project manager as well as project team member (Fig. 4.6). Such multirole assignments offer synergies on the one hand, as integrative functions can be performed, and (interrole) conflicts on the other.

Because of varying project assignments, the planning of personnel resources needs to be done as zero-based budgeting. Quantitative and qualitative peaks in the work load often are managed by employing leasing personnel or by contracting work to external

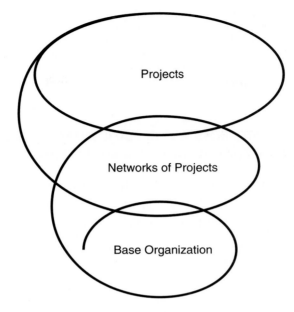

FIGURE 4.5 Structures of the Project-Oriented Company.

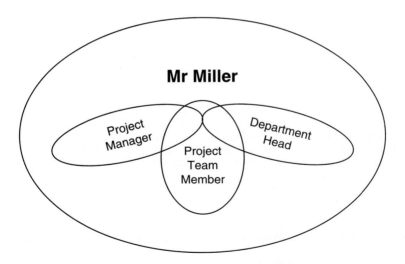

FIGURE 4.6 Multirole assignments in the Project-Oriented Company.

consultants or suppliers. Maybe in the future agreements with regard to personnel cooperation between companies within the same industry will become standard.

Job descriptions in the Project-Oriented Company need to be flexible as project assignments vary. Project work has to be an explicit element of general job descriptions. As the Project-Oriented Company has flat organization structures, possibilities for advancement within the hierarchy are limited. Project-related career plans (such as assignments to attractive projects) and incentive systems (such as project premiums, job rotations within project teams) are required. There are companies that require persons pursuing a general management position to have at least 3 years of project-management experience.

NETWORKS OF PROJECTS ARE SPECIFIC INTEGRATIVE STRUCTURES OF THE PROJECT-ORIENTED COMPANY[5]

Projects Performed Simultaneously Can Be Perceived as a Network of Projects

Networks of projects can be defined as a set of (relatively) autonomous projects being closely or loosely coupled. Networks of projects are not identical with programs. A program includes projects as well as a number of actions which are closely coupled by a common objective. The projects of a program might be perceived as a network of projects.

Networks of projects may consist of all projects of a company or of groups of different project types, such as contracting projects, acquisition projects, and research and development projects. The number of projects considered in a network of projects, their objectives and volumes, as well as their progress statuses vary. Start-ups and closedowns of projects lead to a dynamic network structure. On the other hand, networks of projects are stable since the types of projects considered, the types of relationships that exist, and the forms of communication applied in the network are relatively constant.

It is the objective of managing networks of projects to optimize the results of the overall company and not the results of individual projects. There might be conflicts of interest between the objectives of specific projects and the overall company objectives.

When managing networks of projects, in addition to internal objectives, external objectives are also pursued. If, for example, different projects are performed for the same client, the long-term client strategies and the basic client relationships need to be coordinated. If different projects use the same supplier, the general purchasing conditions may be optimized. Decisions regarding the start-up and closedown of projects are required, priorities have to be set among projects, and competitive and synergetic relationships between projects have to be identified and managed.

Managing Networks of Projects Requires a Formal Network Analysis

To decide whether or not a new project should be started, its benefits and costs have to be analyzed, and it has to be determined whether the company can cope with the additional project. A sound decision cannot be based on an isolated analysis of a single project but requires the consideration of the project as an additional element in the existing network of projects. The impact of closing down a project (such as transferring results and learning experiences of a project to other projects, new availability of resources) must similarly be related to the overall project portfolio.

In an analysis of a network of projects:

- A holistic view of the projects and their relationships is generated.
- Similarities and differences between the projects are determined.
- Competitive and synergetic relationships between the projects are recognized.

An analysis of a network of projects can be performed periodically—once a month or every other month, depending on the dynamics of the network. To better understand and communicate the complex structures of networks of projects, different types of analyses (contents, schedule, or resource analysis) and different forms of presentations (listings, portfolios, network graphs, bar charts, or tables) may be applied.

The bases for the analysis of a network of projects are the documentations of the individual projects. In order to be able to aggregate and compare the project data, common documentation standards are required. The contents analysis of a network of projects includes the grouping of projects and an analysis of the relations between them. In a listing of projects the projects of a network can be listed and differentiated by project types, such as internal or external, unique or repetitive, domestic or export, small, medium-sized, or large (Fig. 4.7).

Further, projects can be described by different relational criteria. Relevant criteria for the coordination of networks of projects are, for example, project volume, project progress status, priority of project, or attractivity of project.

Projects can be related to each other in a graph of the portfolio of projects. Portfolios can be developed for combinations of different criteria, such as profit and risk (Fig. 4.8).

For repetitive projects, such as research and development or contracting, progress milestones can be standardized. The progress status of each project can be determined and documented in progress charts. Then data from consecutive control dates can be related to each other (Fig. 4.9).

Interdependencies between projects can be presented in a graph of a network of projects (Fig. 4.10). Groups of projects can be formed and relationships between projects can be drawn and qualified. The qualification of competitive and synergetic relationships can be documented. (For example, if project A is successful, project D is

Project number	Project title	Internal or external	Unique or repetitive	Domestic or export	Small, medium, large	Contents
Org. 01	Office organization	Internal	Unique	Domestic	Large	Organization
Org. 02	EDP decentraliza-tion	Internal	Unique	Domestic	Large	Organization
⋮	⋮	⋮	⋮	⋮	⋮	⋮
R&D 01	Development of product A	Internal	Repetitive	Domestic	Medium	R&D
Mark. 01	Joint venture	Internal	Unique	Export	Medium	Marketing
Mark. 02	Acquisition of airport	Internal	Repetitive	Domestic	Small	Marketing
Contr. 01	Equipment factory	External	Repetitive	Export	Large	Contracting
⋮	⋮	⋮	⋮	⋮	⋮	⋮

FIGURE 4.7 Listing of projects.

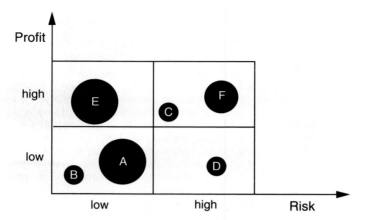

FIGURE 4.8 Graph of a portfolio of projects.

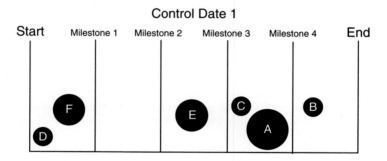

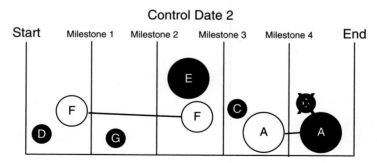

FIGURE 4.9 Progress charts of projects.

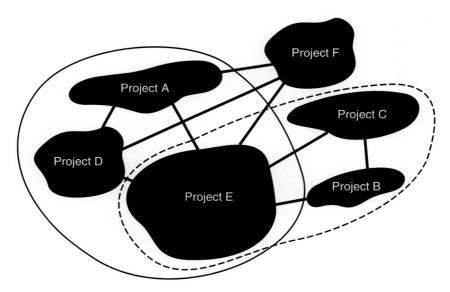

FIGURE 4.10   Graph of a network of projects.

jeopardized; certain results of project C are required to proceed with project B; project E and project F require the same scarce resource.)

It is the objective of the schedule analysis of projects to represent the durations and timing of the projects and to document important schedule dependencies between projects. The durations and the timing of the projects can be documented in a bar chart. Dates of milestones of projects which have to be achieved in order to proceed with other projects can be documented in a milestone list.

A resource analysis has to be performed when different projects require the same scarce resources. Traditional resource planning methods (multiproject management) are based on the critical path method (CPM) or precedence schedules. The assignment of the scarce resources is done at the activity level of the CPM schedule.

Often CPM-based resource planning is not applicable in projects because:

- The required level of detail for CPM planning and controlling is too high.
- A CPM schedule exists only for parts of a project.
- CPM scheduling is not applied at all in a project.

If no CPM plan exists, more global methods for resource planning based on rough estimates of the resource demands per period are required.

The explicit performance of the preceding analyses of a network of projects creates a new quality of information. Based on the analyses, the following measures can be applied to coordinate a network of projects:

- Redefining project objectives
- Changing project priorities
- Changing personnel assignments

- Leveling risks between projects
- Starting new projects
- Closing down projects
- Transferring know-how between projects

Social Networks (of Projects) Have Formal and Informal Communication Structures

In order to fulfill analysis and coordination functions, networks of projects require specific communication structures. To a certain extent the management of networks of projects can be considered as a top-management responsibility, and it can be institutionalized by establishing a steering committee of projects or a controller of projects. In steering committees of projects, managers of the base organization and key team members of the different projects should be represented.

Networks of projects can be perceived as social networks, which basically function because of existing informal communication structures. To promote these networking characteristics:

- Different roles in different projects can be assigned to the same person.
- Exchange-of-experience meetings between projects can be organized.
- Internal project presentations can be performed.

The application of general EDP-based project documentations, the promotion of self-organization processes in and between projects, the permission of horizontal communications, and the information of project team members about project strategies and company strategies are important communication tools in networks of projects.

With regard to EDP-based planning and controlling tools for the management of networks of projects, very little support is available. As mentioned, project-management software packages offering resource planning modules base this feature on CPM schedules, which often do not exist. Furthermore, in risk analysis, cluster analysis, and simulations a demand for EDP support still exists.

THE CULTURE OF THE PROJECT-ORIENTED COMPANY NEEDS TO AGREE WITH ITS STRATEGIES AND STRUCTURES

The Project-Oriented Company Is Characterized by the Existence of a Project-Management Culture

The culture of an organization can be defined by the set of values, norms, and patterns jointly developed and accepted by the members of the organization. The culture can be analyzed by observing the capabilities of the members of the organization as well as its artifacts, such as the corporate design and the handbooks.

In the continuously changing structures of Project-Oriented Companies an explicit cultural orientation is required to integrate projects into the company. By communicating the company's vision, mission, policies, and rules, management gives the orientation with regard to decision making and sets limits for self-organization (in projects).

An important "subculture" of Project-Oriented Companies is the project-management culture, that is, the different communication forms, roles, techniques, documentation standards, and leadership styles specifically applied in projects. The self-understanding of a Project-Oriented Company can be communicated internally and externally in the company's mission statement (Fig. 4.11).

---

§ 1: We are a project-oriented company.

We establish projects for the performance of complex tasks of small, medium and large volume. Our project management culture is developed continuously. We apply project management methods and procedures according to specific project requirements.

FIGURE 4.11   Central paragraph in the mission statement of a Project-Oriented Company.

There Is No "One and Only" Project-Management Approach

In the past Project-Oriented Companies have been performing large external projects successfully. When managers experienced in traditional project management now perform new project types, many realize that the management methods applied are no longer appropriate. For the successful performance of different projects differentiated project-management approaches are required.

Projects with an external project owner are usually more formalized than internal projects. Also external projects usually have a higher priority within the company than internal projects. So the access to company resources is handled differently. The larger a project, the more attention it will get in a company. The integration of the project manager into the company organization depends, among other factors, on the project scope. The higher risk of large projects usually requires more sophisticated controlling methods and a comprehensive project documentation.

Further there has to be an adequacy between the complexity of a project and the complexity of project management. Very complex projects might ask for very flexible (and therefore complex) organization structures but nonsophisticated scheduling techniques, which provide holistic information and allow easy adaption (Fig. 4.12). A high project complexity asks for explicit teamwork, which provides the required variety and creativity to solve complex problems. Teamwork is also specifically required for unique projects. Unique projects might be more attractive for team members than repetitive projects.

Furthermore, different project-management approaches might be applied in different project phases. Project phases can be differentiated by contents, such as engineering, procurement, or construction, or by processes, such as start-up, performance, or closedown. Different phases require different roles. (For example, the site manager plays a specific role in the construction phase.) Different planning methods may be appropriate in different phases (such as the bar chart for the engineering and procurement phase and the precedence diagram for the construction phase). On the other hand different leadership functions have to be performed in different process phases of a project (such as providing orientation in the start-up phase, giving feedback in the closedown phase).

It is strategic project-management function to decide on the appropriate management approach, which must be functional for the specific project phase. The decision has to be reflected upon and adapted when the project situation changes.

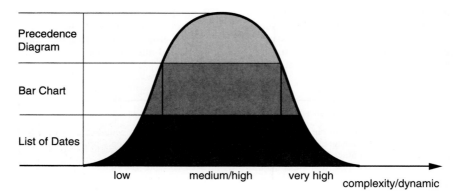

FIGURE 4.12 Application of scheduling methods according to project complexity.

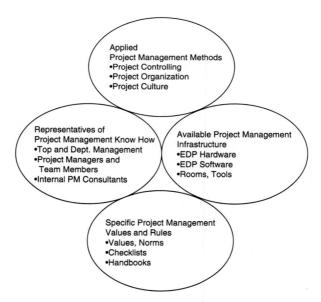

FIGURE 4.13 Elements of the project-management culture.

The Project-Management Culture Needs Continuous Development

The extent to which an explicit project-management culture exists in a company can be evaluated by observing cultural elements, such as the number of representatives of project-management know-how, the applied project-management methodology, the available project-management infrastructure, and the existence of specific project-management values and rules (Fig. 4.13).

The different cultural elements in project management are subjected to continuous evolution. Approaches to developing the project-management culture in a company are:

- Performance of project-management training
- Performance of project-management pilot projects ("training on the project")
- Involvement of external or internal project-management consultants to moderate project workshops and support project management
- Implementation of project-management software
- Standardization of the project-management methodology in project-management manuals

The demand and the potential for standardization in project management by developing standard work breakdown structures, defining standard milestones, and determining standard structures for project handbooks depend on the relationship between repetitive and unique projects. The greater the number of repetitive projects performed by a company, the more potential there exists for standardization. But there exists also the risk of losing the benefits of project work by standardizing too much and not allowing for appropriate autonomy and creativity in projects.

THE PROJECT-ORIENTED COMPANY HAS A GLOBAL PERCEPTION OF PROJECTS

Projects Can Be Perceived as Social Systems

Projects are tasks with specific characteristics. They are complex, relatively unique, risky, and important for the project-performing company. Project objectives are determined regarding scope of work, schedule, and budget, and are agreed on between the project owner and the project manager or the project team. Projects require a specific project organization. The performance of interrelated tasks involves different functional disciplines and asks for an explicit organizational design.

Projects can be perceived as social systems with distinct boundaries, with the ability to learn and to self-organize. A project can be considered as a (relatively) autonomous social system with a specific structure and culture. Elements of the project structure are, for example, project-specific roles and communication forms, project phases, and milestone dates. The set of values and the behavior patterns shared collectively by the project team, the project name and logo, and the form of the project documentation determine the project culture. The explicit development of project values is an important project-management instrument, as the values give orientation to project team members in all their actions and decisions (Fig. 4.14).

By the definition of project boundaries a project is clearly differentiated from its environment. On the other hand, a project can only be understood in its context. The project context comprises the decisions and actions made in the preproject phase, the consequences expected in the postproject phase, and the relationships between the project and its environment.

- Significant in contents and socially stimulating
- Tradition-minded and future-oriented
- Science and practice, mind and body
- Professional and innovative
- (Learning) experience and chance

FIGURE 4.14   Central values of an organization project of Congress.

A Holistic Definition of Project Boundaries Ensures Project Quality

A holistic definition of the project boundaries has to consider all tasks that are closely coupled as being part of the project. If tasks interrelating closely are not considered part of the project, the project performance will suffer. For example, a product development project should not just include the engineering, production planning, and pilot production tasks, but also the interrelated marketing, financing, and organization activities required to launch the new product successfully. Or an investment project might not just consider the engineering, procurement, construction, and commissioning tasks, but it might also define the interrelated personnel recruiting and development activities as well as the required organizational development activities as being part of the project in order to start up and operate the new plant successfully.

Along with the holistic definition of the project tasks goes the comprehensive definition of the project organization. All parties required to ensure the quality of the project work and the acceptance of the project results need to be involved. To be effective as a project team, the team members need to provide know-how capital, decision capital, and networking capital (Fig. 4.15).

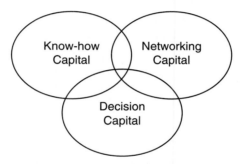

FIGURE 4.15   Composition of the project team.

The Mission of a Project Depends on Its Context

A systemic perception of a project differentiating the system from its environment promotes the explicit consideration of the project context. A comprehensive context analysis is the basis for strategic project decisions. In the analysis of the contents dimension of the project context the relationships between the project and the overall company strategies are considered. It has to be analyzed which company strategies initiated the project and how the project contributes to the achievement of the company strategies (Fig. 4.16). Furthermore, competitive and synergetic relationships between the project and other projects of the company have to be analyzed.

The analysis of the time dimension of the project context includes the analysis of events, decisions, and stakeholder relationships of the preproject phase and the analysis of expected consequences of the project in the postproject phase.

In a project environment analysis the social context of a project can be analyzed.*

---

*This is a further development of the stakeholder analysis.[6]

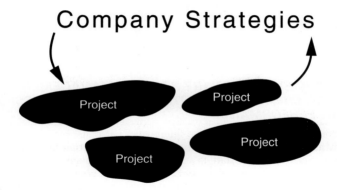

FIGURE 4.16 Relationships between projects and the company strategies.

The "relevant social environments" are defined, and the relationships between the project and each environment can be described by the definition of mutual expectations. Conflicts and potentials in the relationships can be identified, and strategies for the successful management of each individual relationship can be developed.

A relational project understanding, emphasizing the relations between a project and its relevant environments, permits reacting successfully to changes in a dynamic project environment (Fig. 4.17). Relevant project environments (such as users, suppliers,

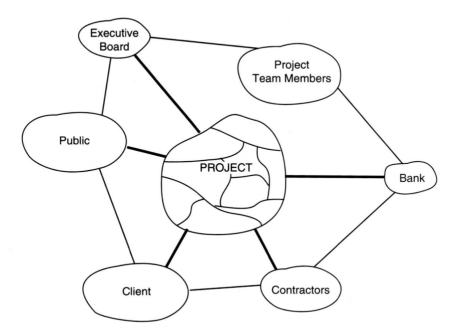

FIGURE 4.17 Interrelationships between a project and its relevant environments.

departments of the base organization) are supply and demand markets for a project. The project's relationships to these markets can be actively managed.

A SYSTEMIC-EVOLUTIONARY PROJECT-MANAGEMENT APPROACH CREATES NEW POTENTIALS FOR SUCCESSFUL PROJECT PERFORMANCE

Project Management Has a Strategic and an Operative Dimension

Important operative project-management functions are planning and controlling the scope of work, the project schedule, and the project resources and costs, managing project personnel, defining project roles and communication structures, and developing project-specific values, norms, and rules (Fig. 4.18).

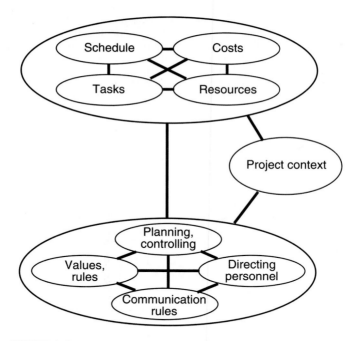

FIGURE 4.18   Functions and objects of consideration of project-management.

Project management has to relate to the project and to the project context, that is, to activities and decisions before the start of the project, to consequences of the project after the end of the project, and to relations to relevant social environments and to other projects. The explicit consideration of the project context is strategically oriented. Further, the selection of key project personnel and the selection of project-management methods adequate for the specific project under consideration are strategic project-management decisions.

Projects Require an Explicit Organizational Design

A project is created by its definition, it exists, and it gets dissolved. This temporary social system requires appropriate organizational designs during its life cycle. The traditional project organization models, the matrix project organization, the pure line project organization, and the influence project organization, were developed for companies structured in a line organization form, performing repetitive projects. In these models project-specific roles, especially the roles of project manager, project team member, and superior of a project team member, are defined, and varying formal authorities regarding the initiation of work, the disposition of resources, the control of work package quality, and the control of schedule and cost are assigned to these roles (Fig. 4.19). Depending on the extent of formal authority assigned to the project manager, these project organization models are differentiated.

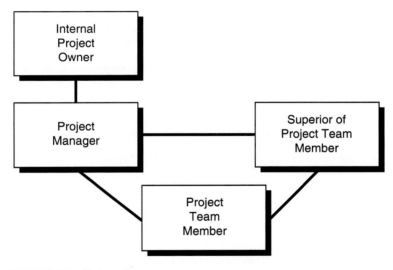

FIGURE 4.19 Project roles.

For companies performing different project types and applying resource pool concepts, new project organization models are required. Project-related roles need to be redefined. The role of superior of a project team member often no longer consists in performing work package control functions but, as resource pool manager, only includes the responsibility for the disposition of pool members. The responsibility for the quality of the performance of the work package lies with the project team member or with the overall project team.

For internal projects the explicit definition and performance of the role of internal project owner are required. Basic functions and responsibilities of the internal project owner are:

- Specification of a project's goals and strategies
- Assignment of the project manager and of key team members to a project
- Provision of an appropriate project infrastructure
- Information about the relationship between the project and the company's strategies
- Information about relationships between the project and other projects
- Determination of standards for project reporting and project documentation
- Conflict settlement, strategic decisions, and feedback to the project team
- Project marketing within the company and versus external environments

The role of the internal project owner can be performed either by an individual or by a group (such as a project steering committee).

When designing a project organization, a relational definition of the project roles is required. The expectations regarding the authority and responsibility of the project manager have to be in accordance with the expectations versus the other role players. For example, the decision authority of the project manager has to be related to the decision authorities of the internal project owner, the project team members, and the project team.

In designing project communication structures different project meetings having different objectives, different partners, and different frequencies have to be defined (such as project owner meetings and project team meetings). Further the timing of project workshops has to be decided (such as project start-up workshop, project status workshop, and project closedown workshop).

For the solution of complex problems teamwork is required. So in the project team-building process the objective must be to optimize the capabilities of the overall project team instead of maximizing the capabilities of individual team members. The adequate balance between redundancy and variety in the qualifications and cultures of the project personnel has to be found.

Considering the multirole assignments in projects, the multiple communication relationships between project team members, and the extent of teamwork, projects can be perceived as flat and networklike organization structures. Project organizations are dynamic. According to phase-specific demands the project structures have to be adapted.

Project Planning and Controlling Permits Constructing the Big Project Picture

The work breakdown structure and the time, cost, and resource plans are project models. By applying different planning methods, different points of view are taken with regard to a project. By combining these project models, a global project picture can be generated. Such a multimethod approach allows interrelating the contents of different project plans (for example, the action plan resulting from the project environment analysis with the work breakdown structure). By cross-checking the completeness of the project plans, project-management quality can be ensured.

In order to integrate data from different project plans, a common structural basis is required. The work breakdown structure allows one to relate time, cost, or resource

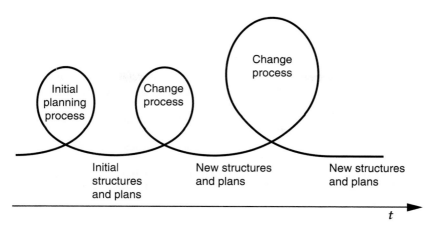

FIGURE 4.20   Cyclic project planning process.

data to different levels of work packages. So the demands for appropriate information from different members of the project organization can be met.

In order to reduce uncertainty, traditional project planning develops very detailed deterministic plans. This approach is based on the belief that better plans lead to better project performance, where "better plans" means more detailed plans. Such linear thinking assumes a stable project environment. In an environment of discontinuities, less detailed plans, but plans for alternative project scenarios and new planning methods, such as scenario techniques and discontinuity analyses, are required. By applying these methods, the complexity and the dynamics of projects can be met.

The development of project plans by the project team members, in cooperation with representatives of the project owner, suppliers, and authorities, results in a commonly shared project perception. Project plans then are no longer just a means to generate project data, but become a communication instrument and contribute to the development of the project culture.

Projects are subject to evolution. In order to relate to the continuous changes in projects, a cyclic project planning process is required (Fig. 4.20). Any resulting deviations from the original plans might be appreciated as a learning experience.

CONCLUSION

A company that frequently uses projects as temporary organizations can be perceived as a project-oriented company. In order to implement management by projects explicitly, appropriate corporate structures and cultures are required. Projects have to be complemented by new integrative structures, such as strategic centers, resource pools, steering committees, and networks of projects.

The project-management culture of a Project-Oriented Company is characterized by a global perception of projects, by a strategic and operative project-management approach, and by the promotion of project autonomy and evolution.

REFERENCES

1. W. Briner, M. Geddes, and C. Hastings, *Project Leadership,* Gower Publishing, Aldershot, 1990.

2. R. Gareis, *Handbook of Management by Projects,* Manz Verlag, Vienna, Austria, 1990.

3. J. R. Thatcher, "New Age Managers for Projects—Dealing with a 'World Turned Upside Down,'" *PmNETwork,* vol. 4, p. 9, May 1990.

4. F. Drucker, "The Coming of the New Organization," *Harvard Business Rev.,* p. 47, Jan./Feb. 1988.

5. R. Gareis, *Projektmanagement im Maschinen- und Anlagenbau,* Manz Verlag, Vienna, Austria, 1991.

6. D. Cleland, "Project Stakeholder Management," in *Project Management Handbook,* D. Cleland and W. King (eds.), Van Nostrand Reinhold, New York, 1988, p. 275 ff.

P · A · R · T · 2

NATIONAL DEVELOPMENT BY PROJECTS

Major development programs in Eastern Europe and in developing countries are implemented by projects.

In Chap. 5, Salman T. Al-Sedairy and Peter Rutland provide sharp insight into how project management can be used as a strategy to move forward in a developing country. Pointing out that, while the benefits of project management are substantial for a developing country, they note that it is necessary for government and industry to initiate the appropriate use of project management.

Ladislav Chrudina and Gunter Rattay introduce, in Chap. 6, the specific challenges and chances for development as integration instruments between Czech and Slovak Federal Republic (CSFR) companies and cooperation partners from Western Europe.

In Chap. 7, Robert Youker points out that a review of the results of project monitoring and evaluation of World Bank projects indicates that many of the key problems of implementation lie in the general environment of the project and are not under the direct control of the project manager. The author concludes that international project managers must set up a process to scan the environment, to identify potential problems, and to try to establish power relationships which can help them to manage the key actors and factors on which successful implementation of development projects will depend.

In Chap. 8, William R. Yeack and C. Joseph Smith catch the reader's attention quickly when they state, "Tangible and intangible costs directly related to failure of global projects are of such magnitude and complexity as to be inestimable." From this statement, the authors make the key point that project managers operating in a global context need to gain a working knowledge of world systems theories and the development background of the countries in which they operate or from which they derive project team members.

In Chap. 9, according to Mary Ann Nixon, China continues to emerge as a growing power in world markets. Chinese businesses have made overtures to industrially advanced countries to acquire business investments and technology improvement in China. As China continues to develop, there will be ample opportunity for project management strategies within that country, both for China and for the country offering to update Chinese businesses and technologies.

CHAPTER 5

PROJECT MANAGEMENT AS A WAY FORWARD IN A DEVELOPING COUNTRY

Salman T. Al-Sedairy
King Saud University

Doctor Al-Sedairy is currently Director of Research Center, College of Architecture, King Saud University, Saudi Arabia. He is author of several books and articles on architecture and project management.

Peter Rutland
Consultant

Doctor Rutland has taught in universities in England, Australia, Saudi Arabia, and New Zealand and operated his own management consultancy practice in England. He has worked in construction and architectural management fields for contractors and architects. His fields of specialization are project management, construction management, architectural management, and marketing professional services.

DEVELOPING COUNTRIES

Key Factors about Developing Countries

While developing countries undoubtedly present exciting opportunities and challenges for the implementation of the latest technological and managerial skills, these require special care and understanding in their application. Frequently, there exist different levels of economic development, economic structure, and national economic planning. Other factors, in many cases not easily recognized, which affect progress in developing countries may include protection of national sovereignty, conflict between traditionalism and modernization, memories of past (and present) abuses, and environmental conservation issues.

It is quite common in developing countries to have an inconsistent and extremely bureaucratic system of regulatory powers, with long chains of command involved in decision making. Power often resides in a few decision-makers who have limited access to facts, thus leading to considerable subjectivity in the decision-making process.[1]

Short-term perspectives, as a global phenomenon, impact more on a developing country than on a developed one. Accelerated change, uncertainty, and intensified world competitions require the developing country to utilize the optimum approach to constructing its infrastructure and building stock, with a view to continued maintenance and eventual replacement.

The economics of developing countries may range from poverty to affluence, and each presents its own problems. Rich countries, sourced from oil or agriculture, may devise their own self-funded development plans, aimed at producing self-sustained growth and, at least, a semiautonomous economy. The poorer nations can obtain loans, based on sound feasibility studies, for development projects to help move them up the development continuum. In both cases, however, the role of the government is preeminent when large-scale projects are needed to drive the development process, since only governments can command the financial and other resources needed to generate and support such projects.

Developing countries throughout the world share many common problems, amongst them[2]:

(i) Very high population growth, often far outstretching increases in productivity and financial growth.

(ii) The time required to create an "industrialized state" and the need to inject large amounts of private capital for it to be maintained when once begun. Industrialization is a common goal for most developing countries, but it carries with it many social and cultural problems that need addressing.

(iii) The need to develop human resources quickly so that the country can continue to develop everything else. This often leads to a conflict between education and training, since a government will undertake education programs and some training, but private industry must also carry a large proportion of the training needs on its shoulders.

(iv) Government administration and public administration, because of size and often rapid and uncoordinated growth, is not the best vehicle for managing the development of projects unless comprehensive training and updating is undertaken.

(v) An impatience to have nationals take over in managerial and technical roles. This often leads to premature transfers which in turn create problems for project development.

(vi) "Good management" techniques may be at odds with tradition and sociocultural background, which play a very important part in the lives of managers and determine much of their behavior. Conflict often results.

The activities involved with the production of projects in a developing country must address the issues raised above, and more. This chapter looks at some of them in more detail, and indicates appropriate approaches to minimize their detrimental influence.

Projects

The very nature of a developing country involves creation of very large, often complex projects which are undertaken in single or staged phases. While it is difficult to categorize projects in developing countries, they do exhibit a number of distinctive features.

(*a*) Since often much has to be completed in as short a time as possible, the front-end time on projects may be restricted.

(*b*) They are often complex and large, and require cooperation and coordination between many government agencies and national and international bodies.

(*c*) The technological and scientific aspects may be emphasized, often at the expense of management.

(*d*) Experience gained, particularly in the project management field, is not readily transferred to subsequent projects for their benefit.

(*e*) Errors in estimates of time and cost may be large, since there may be no standards or precedents to follow.

(*f*) The need for parallel transfer of skills to local employees is considered essential in most projects.

(*g*) Product and service suppliers are often from different countries, and thus use differing standards, fittings, etc.

(*h*) Project organizations are often formed for specific projects, and then disbanded on completion.

(*i*) Main control centers may be outside the developing country, thus causing delays in communication and decision making.

In general, however, projects in developing countries rank as very important elements in their growth, and the construction sector may contribute anywhere from 7 to 35 percent of the gross national product (GNP) in developing countries.[3] Thus project management development must rank high among the needs of a developing country.

Each of the features *a* to *i* above are critically affected by an overriding factor in developing a country's project management, and that is the influence of the culture and traditions of the country. In a developed country these have reached a reasonably sophisticated level of definition, comprehension, and acceptability. But this is not often the case in a developing country, and the following section deals with this critical factor more fully.

Social and Cultural Factors

In the early stages of the development of project management it is fair to say that tools and techniques predominated (e.g., network analysis) and were quickly followed by the information systems movement, which led to computer-dominated project management systems. This was particularly evident in the 1960s, and continued into the mid-1970s before the large gap between the system and the people—and the consequent problems that it generated—became evident.[4]

The fundamental problem lies in the clash between the culture of the overseas nations that are designing/constructing/maintaining projects and the culture of the developing country itself. For example, the German approach tends to be systematic and highly structured,[5] but this may clash with the more laid-back, loosely structured approach to problems in the Middle East.

Thus it is essential to recognize those aspects of national culture which may affect the project management process and then take them into account when establishing a project management approach. Three key issues about intercultural relations are given below:

(i) There is a need to exercise self-control and to deliberate steps when managing in different cultures; this will overcome a subconscious belief that everyone shares one's value systems.

(ii) Strong views of cultural stereotypes are held by many managers, often based on little or no real exposure to these cultures; e.g., many believe that British are cold and snobbish.

(iii) The development of ethnocentrism, the belief that one's own culture is the best, is a central issue in cross-cultural management activities.

When considering culture as an element in project management, it is critical to assess how a culture determines an individual's approach to at least the following components:

(i) *Emotions:* Are they easily changed? Are they kept hidden? Are they important?

(ii) *Time:* Is it considered equivalent to money? Is it a strict controller of actions? Is it highly valued?

(iii) *Power:* Is personal power important? Where is the power base?

(iv) *Social behavior:* Are rituals important? Is frankness preferred? Is formality rated highly?

(v) *Conflict resolution:* Is direct confrontation acceptable? Does compromise figure highly in the culture? Is aggression or gentleness the norm?

(vi) *Personal achievement:* Are individuals clearly personally goal-oriented?

(vii) *Decision making:* Is centralization or dispersion common? Are committees prevalent? Are decisions taken objectively or subjectively?

(viii) *Group dynamics:* Are groups important in management? Are groups self-determined or predetermined? How are they usually structured?

(ix) *Risk:* Is risk an acceptable part of management? How is risk assessed? Are risk and uncertainty clearly separated?

(x) *Bureaucracy:* Does it prevail? Is it clear or confused? Can it be circumvented easily?

(xi) *Trust:* Are relationships built on trust? Is trust openly exercised? How much trust can be exercised?

These are just some of the components; a more complete cultural awareness matrix can be found in the literature on negotiating skills.[6]

Project managers and top management agree that there is nothing more important to a host country than the foreign manager's respect for its social and cultural values, particularly in developing countries. This chapter now proceeds to look at two key participants in project management, namely the client and the contractor, with particular reference to developing countries.

THE CLIENT

Over recent years, types of clients and their expectations have changed considerably, and in a developing country it is essential for the nature of the client to be understood.

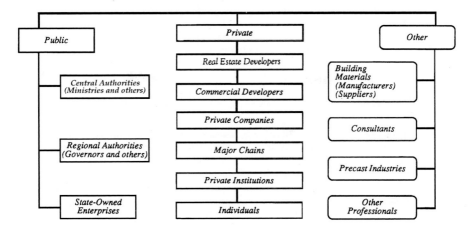

FIGURE 5.1   Client types by category.

Sharp[7] classifies clients as (*a*) those who spend their own money and occupy the building, (*b*) private companies, (*c*) public liability companies or large corporate organizations, (*d*) government bodies. This useful categorization is further extended by Head,[8] who employs the main headings of *public, private,* and *others.* When these approaches are applied to a developing country, the breakdown could approach that shown in Fig. 5.1.

There is no question that, in a developing country, the major client is public, with the government exerting a great influence on the socioeconomic environment through its origination and control of projects. Thus the public client (government) plays a key role in industrialization because it commands the financial and other resources needed to handle the large-scale projects. As development proceeds, certain areas of investment may be reserved by government for the private client to encourage the private sector.[9]

No matter what client type is operating, the project management activity must maintain great flexibility, since the client often does not have a well-defined organization structure and policy is not clearly defined or transmitted. In the case of government clients, this may appear in individual departments, or in the interaction between various departments in a ministry. The instability of the political and economic situations in a developing country is a serious problem, since change of key personnel without notice can drastically affect plan formation and implementation.

The central authorities, who are responsible for long-term strategic and development planning, must also develop appropriate supporting policies to match the physical development. These include educational, fiscal, and trade policies. If these are not adequately prepared, projects may be curtailed when it is seen that they are not going to be used effectively.

Nearly all client types in developing countries exhibit the major problems of deficiency of management skills in staff and weakness of administrative systems. This has been supported by many studies carried out by the United Nations, as can be seen from the following excerpt from a report: "On the whole, experience has shown that much progress in plan implementation could be achieved through the improvements in public administration such as: (*a*) The setting up of well-staffed agencies or units for the preparation of feasibility studies, on the one hand; and for the planning and follow-up of project implementation, on the other hand; (*b*) The stream-lining of administrative procedures to accelerate payments, facilitate the issue of import licenses and speed up decisions regarding expatriate labor, and the making of advance preparations, including the training of project personnel."[10]

THE CONTRACTOR

The construction industry occupies an important place in any country's economy. It provides an appreciable share of the gross domestic product, generates a high proportion of gross fixed capital formation, and is relatively labor-intensive. Such an important industry should be well understood and defined, and yet it continues to nurse a fundamental problem, historic in origin, which stems from systems and procedures related to the distinct division between responsibility for planning and design and the responsibility for construction. The inefficiency of various administrative and legal procedures involved in taking a design from its inception to final construction has been studied and reported on,[11] yet little seems to have been achieved by way of improvement. The main split continues to be between design and production, a split unique to the construction industry.

In a developing country, this split may be exaggerated when large geographical distances, fundamentally different cultures, and huge technological skills gaps exist, as often they do. Large contractors, both foreign and local, will have similar financial and organizational structures, and are used to the contractual framework, although the local firm may rely heavily on subcontract labor. The small contractor in developing countries is accorded a very low status in comparison to the large counterpart. The small contractor may be seen as "an unpatriotic, dishonest businessman who, given half a chance, would either use shoddy materials, leave out some parts of structure, make unjustified claims or abscond with advances or loans paid to him, or influence consultants to certify unjustified payments to him."[12]

As a developing country moves along its development continuum, the role of the large international or large local contractor diminishes, and the importance of the small local contractor grows. There is often no medium-sized local contracting group for some time in a developing country, and thus much work, normally done by that group, is undertaken by the small contractor. However, the concept of competitive bidding is widely practiced in developing countries, leaving the small contractor little room to maneuver in tendering. The design is established and materials and equipment prices are almost set for all tenders, so the small contractor is left with only a few areas on which to make a profit. These are (i) reduce overhead, (ii) improve labor productivity, (iii) improve site management, (iv) clever purchasing, and (v) use bidding and risk assessment theory to improve chances. Of these, only iii, iv, and v are really viable in a developing country, and yet these are the areas where the small local contractor is very often least qualified.

The historically developed system of design separated from production, even in a developed country, causes many serious problems and inefficiencies. In a developing country these are further aggravated by cross-cultural issues, some of which have been dealt with in the section "Developing Countries."

It can be seen that the peculiarities of developing countries impinge greatly on the production of building/civil engineering projects, particularly in the client and contractor fields.

The concept of project management will now be addressed, with particular reference to developing countries. It will then be considered as a way to improve the development processes of developing countries.

PROJECT MANAGEMENT

Project management is widely applied across all industries, in small and large organizations and for small or big projects. Project management is particularly useful wherever considerable coordination is required between several parties, and thus is well-suited to the construction industry. However, it must be accepted that project management is not automatically the answer for all situations, as it brings with it some interorganizational difficulties which may prove hard to overcome in traditional organizations. Stuckenbruck[13] recommends that, before a firm commits itself to the project management concept, 15 questions be asked, and, if there are several yeses, it should be used. The list below is a modified approach to Stuckenbruck's list for a firm, and applies to a developing country.

Questions to Ask When Considering the Use of Project Management in a Developing Country

1. Are the projects very large?
2. Are the projects technically complex?
3. Are the projects true systems in that they have many separate parts or subsystems which must be integrated to complete the operational whole?
4. Is each project a part of a larger system that must be closely integrated?
5. Does the client body strongly feel the necessity of having a single point of information and responsibility for a total job?
6. Are strong budgetary and fiscal controls required?
7. Are tight schedules and budgetary constraints foreseen?
8. Are quick responses to changing conditions necessary?
9. Do projects cross many disciplinary and departmental boundaries?
10. Are there many complex projects being conducted concurrently?
11. Is there likely to be a conflict between parties concerning the project?
12. Is the client committed to firm completion dates?
13. Is it likely that changing conditions may seriously affect projects before completion?
14. Are there major purchases and procurements that must be made outside the country?
15. Are there major portions of the system which must be subcontracted outside the country?

Project management uses a systems framework, and is the application of a systems approach to a complex task or project whose objectives are explicitly stated in terms of time, costs, and performance parameters. Each project may be one of several sub-

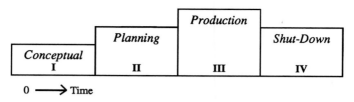

FIGURE 5.2   Stages in the project life cycle.

systems and thus the subsystems must be managed in an integrated way for the effective accomplishment of the defined objectives.

A project will progress through a series of stages which have been defined by many authors. For a developing country, these stages may be described as shown in Fig. 5.2.

The conceptual stage involves goal setting, definition of scope of work, classification of objectives, and establishment of a formal organization structure to oversee projects. In the developing country, this is where the vital aspects of long-term development plans are considered and established. At the planning stage the established organization from the conceptual stage can identify specific subprojects and generate work breakdown structures to achieve overall development plan objectives. This is where, in a developing country, the establishment of an effective command/control system becomes very important. The production stage involves the management of all aspects of the subproject contracts once let and the necessary associated reporting and updating of information as work proceeds. The shutdown stage, often called *close out,* is the critical handover period when the operational aspects of subprojects are classified and the vital aspect of ongoing management is considered. This stage also involves the consideration of how the resources employed on finishing projects can be redeployed effectively.

It is important for a developing country to realize that Stage III will consume a major portion of the country's development budget, sometimes in the order of 60 to 70 percent, and that Stage II could well consume 20 to 25 percent. The lowest *financial* consumption is usually Stage I, *but* this does not reflect upon the level of care needed at this stage, since carryover effects are very large in a developing country's project management systems. The well-known curve produced by Azud[14] (Fig. 5.3) is extremely relevant in developing country project management, and thus conceptual decisions on cost issues are worthy of *considerable* deliberation.

	Conceptual Stage	Planning Stage	Production Stage	Shut-down Stage
MAX.
Degree of
influence
over cost
control by
Client.
MIN.	Project Approval	Level of cost control by Client.		Cost Reports

FIGURE 5.3   Indication of the declining influence on cost control by client in project life cycle. (*Adapted from G. Azud.*[14])

Project management has three simple objectives. These are

(i) To ensure a project is finished on time

(ii) To ensure a project is completed within its cost budgets

(iii) To ensure a project meets the functional and technical performance standards as defined in the initial goals (conceptual stage), and thus satisfies the end users

Inherent in these simple statements are a myriad of complex details, problems, and issues needing attention, yet in a developing country it is *always important to keep these three issues at the focus of project management activity,* and not to be waylaid by excursions into new techniques and elaborate equipment that may well degrade the level of object achievement.

Project management, as a systems concept applied at a country's level, can offer more than just the normal benefits of its use. In any situation where dynamic and rapid change is occurring, then project management will be beneficial, particularly if technological change and specialization are involved. In a developing country, the prime benefits of project management should be in:

(i) Cost savings, both initial and ongoing

(ii) Compatibility of provisions from widely different sources

(iii) Establishment of a flexible organization system to accommodate future changes

(iv) The creation of a vehicle for an effective management information system to be generated

(v) The improvement of educational levels in general, and in particular in technological areas[15]

(vi) A better organization structure in the country, which will bear it in good stead as its development continues

In fact, several benefits have been identified in this area by R. A. Johnson et al.,[16] including less duplication of functions, more cooperation between various units, better follow-up, and more familiarity with the use of computerized information systems.

The contrast between the initial somewhat relaxed, nonaccountable culture in a developing country, in which some problems are ignored and others are recognized as permanent, and the very detailed planning, checking, and controlling processes of project management creates a very beneficial surge to a developing country in the areas of accountability, risk analysis, and improved productivity. The overall discipline imposed by project management hits at the very issues which, if left unaddressed, may prevent the success of initially good development programs.

Despite the many benefits to be gained from the application of project management in developing countries, there are many problems facing such an application. Some of these are listed below:

(*a*) At a conference in Saudi Arabia in 1989,[17] the following items were identified as key contributors to problems facing project management:

- Lack of integrated teamwork, epitomized by parties to the project acting as independent and conflicting bodies
- Lack of cooperation between legislative, executive, and monitoring sections of the government
- Lack of communication between agencies
- Unclear definitions of authority, responsibility, and accountability

- Lack of detailed and clear procedures, which leads to overemphasis on, and excessive time allocation to, routine matters, thus reducing time spent on important management functions
- Lack of recognition of project management as a profession
- Ill-defined contractual and arbitration systems

(b) Dennis A. Rondinelli[18] in his research identified the following problem areas as those that occur most often on projects in developing countries:

- Ineffective project planning and preparation
- Faulty appraisal and selection processes
- Defective project design
- Problems in start-up and activation
- Inadequate project execution, operation, and suspension
- Inadequate or ineffectual coordination of project activities
- Deficiencies in diffusion and evaluation of project results and follow-up action

(c) H. K. Eldin[19] includes the following in his review of difficulties faced in attempting to apply modern project management techniques in developing countries:

- Variety and frequency of changes
- Lack of supporting policy measures
- Inadequacies in coordination and follow-up action
- Deficiencies in public administration
- Shortages and failures of contractors
- Lack of sufficient and accurate information
- Lack of cooperation
- Lack of trained personnel

It can be seen that many problems are common to all of the three above lists, and many are fundamental issues for a developing country. The very last item of Eldin's list (c), "lack of trained personnel," and, specifically, the lack of good project managers, is probably one of the major problems facing any developing country. It is the project managers, as much as project management tools and techniques, who are at the focal point of implementation of effective project management in developing countries. The role of the project manager will be looked at in the following section, where the way forward for a developing country is considered.

THE WAY FORWARD

Total Project Management System

In order to provide an understanding of how project management can open a way forward for a developing country, it is important to present an appropriate total system. This can then be adjusted and manipulated according to the stages of development and the various factors discussed in the opening section of this chapter. Figure 5.4 shows one such appropriate system which has worked well for the overall construction activities in a development-plan-oriented developing country. Three examples of how the subprojects were managed, using variations on the total system shown in Fig. 5.4, follow.

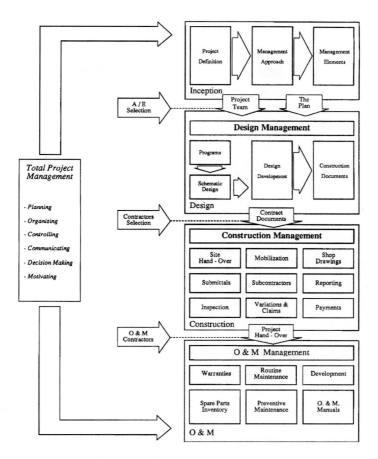

FIGURE 5.4   Total management system for construction (*adapted from Salman Al-Sedairy, "Large Scale Construction Projects," Batsford, London, 1985*).

Examples

When the decision is taken to use project management as a system for development in a developing country, the ramifications spread throughout all levels. Such a decision was taken in Saudi Arabia around 1975 when it was near the peak of its tremendous growth and development, and some 30 percent of its gross domestic product (GDP) was given over to construction development activity.

Example 1.   The need for Saudi Arabia to develop an internal industrial base was recognized, and two new cities were designated and designed as the cornerstone for this purpose. They were Jubail on the Arabian Gulf and Yanbu on the Red Sea, both

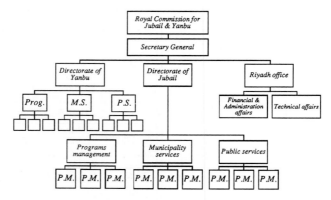

FIGURE 5.5 Royal Commission organization structure.

linked by a corridor of raw product pipelines. Jubail will serve a population of 250,000, and Yanbu, 130,000. Both are designed to support petrochemical, mineral-based, and light industries. In order to expedite these, a Royal Commission was established, and a project management system was instigated whereby each city was designated a *project,* with the integrative command center based in the country's capital city, Riyadh. Figure 5.5 shows the schematic organization structure; permeation of project management throughout the structure is evident.

Example 2. This facilities program was a major overall project program broken down into a number of subprojects. This was handled through an externally imposed project management system via a matrix structure. Figure 5.6 shows the organization and the process.

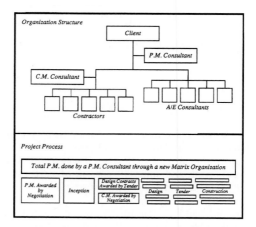

FIGURE 5.6 A $2 billion facilities program in Saudi Arabia.

Example 3.   This is a specific building project, or one of the projects that are usually contained within the developmental project management system. The project management system in this case was in-house client-operated, and wholly contained within the client body. Figure 5.7 shows the organization and process used.

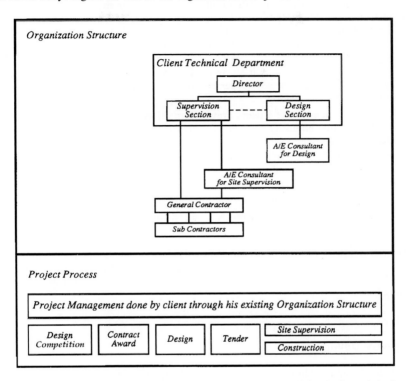

FIGURE 5.7   Program for a $300 million government office building in the capital of Saudi Arabia.

The Project Manager

Virtually all the research, authorities, and reports indicate that the role of the project manager is critical to effective project management. This role is central to the planning and production stages of all projects, and how the project manager deals with these stages has a direct bearing on the success or failure of the project.

The project manager role is clearly seen in Fig. 5.4 in relation to the whole project management system. From this can be recognized the key managerial skills required of this role.

These are mainly integrative in nature, for the work activities, the human resources, and the finances. In addition, integration should be used to resolve conflict; this is a vital aspect of the project manager role. This role stands at many boundaries between different systems in the total project management process, and thus is constantly required to maintain communication channels across these boundaries.

The project manager's roles and responsibilities are well defined by Adams and Campbell in their booklet for the Project Management Institute.[20]

As Example 3 in this chapter shows, the project manager often works in a matrix organization structure, particularly in developing countries. In this context, the project manager will be required to undertake the following activities:

(i)   To distribute the work schedule to other departments in the matrix and check on performance. This is essentially an identification and allocation activity to ensure the whole is correctly divided into parts. This is often very difficult in a developing country because of the separation of the departments, both physically and mentally.

(ii)   To maintain client contact at all times. In a developing country, this may place great stress on the project manager because of the factors discussed in the opening section of this chapter.

(iii)   To lead the members of a team, drawn from different departments, without having a clear line authority over them. This means the project manager must use the appropriate leadership style for the developing country's level of organizational development.

(iv)   To negotiate doable and mentally acceptable work schedules and budgets with the various departments.

(v)   To be constantly aware of status and profitability levels on all projects.

(vi)   To be on the lookout for new projects and to convert inquiries into serious commissions.

There is no doubt that the project manager plays a key role in the developing country if it is to effectively use a total project management system approach, hence the need for training of indigenous managers as discussed below in "Project Management Training."

Contractor Development

Foreign large- and medium-sized contractors always play a major part in the early development stages of a country, but the time eventually comes when the indigenous contractors will have to take over much of the construction work, and obviously the ongoing maintenance programs. This calls for a conscious training and education program by the country for its local contractors, particularly the large number of small ones that arise as development proceeds. However, training alone is insufficient unless it is paralleled by a government interest and support for the contracting industry which extends beyond the level of mere financial support.

Areas where this support can be given are

(a) Risk Allocation.   In developed nations, contractors usually carry all (or nearly all) the risk associated with projects, and have developed sophisticated ways of risk assessment together with adequate resources to offset losses with gains elsewhere. However, for the embryonic indigenous contracting industry of a developing country, such a one-sided risk allocation can be fatal for the contractors, and harmful to the developing country's growth. It is possible to restructure the "traditional" risk allocation model, and the World Bank[21] indeed suggested the following changes as a possible starting point:

(i)   Damage to the works resulting from natural causes should be repaired at the client's expense.

(ii) The client should take responsibility for providing data on subground conditions and then compensate the contractor if the data prove incorrect.

(iii) The contractor should not be required to be responsible for checking the drawings for errors.

(iv) Penalties for late completion should be relatively small.

It would be interesting to see what effect such changes would have on the performance of local small contractors, at both the estimating and production stages.

(b) Modification of Contract-Letting Procedures.   The nature of contract documentation and estimating procedures expected by clients militates against the local small contractor's chance to compete successfully. The level of sophistication required to price designed projects is often absent from local contractors' systems, leaving them disadvantaged against foreign, well-experienced competitors.

The government can support the local contracting industry by:

(i)   Using tender documents that relate to the local contractors' method of pricing, rather than to an international, design-oriented set of documents.

(ii) Considering alternative design approaches from the contractor which are based upon a greater knowledge of practical construction methods.

(iii) Relaxing of requirements for guarantees and bonds in certain areas, or provision for such bonds etc. through the government.

(iv) Limiting the use of imported materials with their associated financial exchange rate problems.

(c) Provision of Contractor Registration Procedures.   This enables the indigenous contractor to identify just what is needed to be able to tender for projects of defined scope, and then to strengthen those areas in the organization that are weak when compared with registration requirements. A contractor classification (or grading system) is used in many developing countries, but unfortunately the *standards* required to qualify for approved lists are not often known to the *contractors*, who respond with a "shot in the dark" approach to application for approval. Such a registration must be continually monitored and upgraded to ensure that new applicants are not discouraged and that existing approved organizations have not slipped below the required standards.

(d) Efficient Compensation Programs.   Poor cash flow is the major reason that indigenous contractors either cannot obtain work or fail to complete that which they gain. An unsteady cash flow may well result from the contractor's own poor organization, but quite often in developing countries it derives from erratic, cumbersome payment procedures by the client.

(e) The Level of Project Generation.   Peaks and troughs in the supply of projects present a severe strain on the local small contractor, who can ill afford to resource-up for a project with little or no guarantee that another will be forthcoming in the near future. Governments can consider their development plans with a view to providing more smaller projects within one large one, thus providing a chance for the local contractor to participate in what otherwise would be beyond its capabilities. A

developing country often lacks any form of federation of individual organizations, and so whole industries have no cohesive vehicle for representation to the government. The small local contractor, therefore, has to deal with the mass of government bureaucracy alone, with no buffer or support. One way forward is to create a "unified buffer" between the local contractor and the government, which is often also the client.

Thus, the local contractor can be represented, assisted, and supported by an organization which draws on a corporate strength from all the small contractors. This has been done by some African countries, to good effect. The government or client generally provides:

Projects

Documentation

Payment

and expects:

Performance on time/cost

Risks to be accepted

No changes to original project

The local small contractor has a need for:

Even workload

Financing

Contract procedure changes

Efficient payment systems

Modified risk levels

The *buffer organization* will attempt to resolve conflicts and differences that arise from the above, for the benefit of the local contracting industry.

For the local contracting industry to undertake a major portion of new and maintenance projects, it must find competent project managers and construction managers. If all the areas above were acted on by the government, then the local contractors must have the managerial skills to assimilate these changes and capitalize on them. This aspect of training will be covered in more detail in "Project Management Training," below.

For the total project management system depicted in Fig. 5.4 to be effective in a developing country, the issues concerning the relationship between the indigenous contractors and the government/clients need serious rehearsing.

Project Management Tools

During the conceptual, planning, production, and shutdown stages of projects, there are various tools of project management that are particularly valuable in developing countries. These are fully explained in detail in the vast literature on project management.[22-27] Some of these are discussed below in relation to their importance for a developing country.

(a) Feasibility Studies.   Within the context of national planning programs, there are a number of key feasibility issues to be addressed. Among them are

- The effect of funding availability on the feasibility of development plans
- The relationship of capital planning to identified feasible projects, and hence the need for the country to commit required resources to feasible projects
- Full preparatory studies are needed, often on a sectorial basis, in order that financial and technical feasibility studies can be undertaken
- The projects' outputs must be defined in order to determine the level of feasibility

Very often in developing countries, many aspects of feasibility studies are overlooked, bypassed because of time pressures, or simply not considered necessary because the culture of the country does not include such "forward assessment" thinking processes. This apart, there is no doubt that the *full* range of feasibility study tools should be used by developing countries from the overall country developmental stages to the individual project stages, and great benefits will accrue.

(b) Objective Definition (Project Specification).   At the development plan stage, the objectives are broad and general, but closely linked to capital budgetary and feasibility studies on a countrywide scale, bearing in mind the issues in *a* above.

At the project level, the specification should identify factors relating to the project site; contractual matters; commercial issues, including time scale and financial matters and organizational responsibility determinations; design and technical standards; the total organization system; target-oriented programs; and quality assurance systems.

At both levels, this is a critical project management tool for a developing country to utilize as it sets a discipline upon the client (government) to consider in advance many issues which could cause bottlenecks as the project proceeds.

(c) Cost Estimates.   Despite the uncertainty of some long-term estimates, and the lack of faith shown in them by many officers in government, it is essential for cost estimates to be prepared. It is better for a developing country to adopt an existing standardized approach to cost estimating first, and then adapt it to its particular culture, rather than to attempt to develop a new one from scratch. This way, feedback data are at least recorded in a systematic manner and can be used as a datum for future estimates.

(d) Time-Scale Estimates.   Project duration and milestone establishment should be based on feedback data from previous projects rather than on detailed planning at this stage. It is here, in a developing country, that the cybernetic project management approach has real benefit in generating realistic duration estimates.

(e) Cash Flow Analysis.   This is a critical project management activity in developing countries, particularly when there is a need to balance the funding among the competing projects, often of equal importance. It is essential to understand the cultural factors affecting cash flow analysis in the developing country, particularly the issues of interest and inflation, which subsequently affect all discounted cash flow (DCF) calculations.

(f) Work Breakdown.   The key to successful work breakdown in developing countries is to establish *manageable* subprojects, or tasks, within the existing and proposed organizational systems. This is where the concept of total project management systems is necessary. The example of the new industrial city project given in the section titled "Examples" shows clearly how such a work breakdown can operate.

(g) Work Scheduling.   This is the area where custom-built or off-the-shelf computer programs for project management and control are valuable. From the early time scale bar charts, the detailed logical network diagrams are developed. The use of precedence diagrams is very beneficial, but still meets resistance in some developing countries. The use of probabilistic networks is necessary only on projects with a high degree of uncertainty, and they are not really trusted by the project teams. The key to the use of the vast array of computerized scheduling and updating packages[28] is in the format of the *output*. Conversion from networks to bar chart form is recommended, and the "exception principle" concept should be widely employed. The manager in a developing country who is responsible for making decisions based on the original schedule and subsequent updates does not have the time (and sometimes the expertise) to interpret poorly formatted, overcopious computer printouts.

(h) Budgetary Establishment.   The work breakdown and work schedule stages provide much more detailed information on which to prepare overall budgets. It is essential in developing countries to reconcile the new budget estimates with the original cost estimates given at the conceptual stage. This enables corrective practical steps to be taken to limit expenditure if the original overall financial objectives are exceeded, before the projects have been taken too far.

(i) Project Organization Systems.   It must be remembered that in developing countries there are rarely soundly established organization structures and recognized administrative standard procedures, so these must be established from the outset of project management. Even if there is an established organization structure, there may be a need to adjust it to fit the proposed development plans and subsequent projects. Flexibility is the key here, and a constant vigil should be maintained in order to reduce restrictive bureaucratic processes from limiting the project development.

(j) Objectives-Review Procedures.   The concept of project management focuses on the achieving of objectives, thus one key activity is the constant review of progress against the initial objectives. Any threat to these objectives as the project develops must be acted on immediately, particularly if changes in the organizational system generate variations to the project.

(k) Quality Assurance.   The concept of quality assurance is of increasing importance to developing countries and must be incorporated into the project management systems that they establish. The literature base for this subject is expanding,[29] and, even though it stems from mechanical engineering in established countries, it has much to offer developing countries.

(l) Postproject Evaluation.   This is highly critical in developing countries as it provides the much-needed basis for the planning and design of future projects. Final costs, technical specifications, work scope, as-complete documentation, and maintenance programs must be carefully recorded *in a format* that allows them to be used at the conceptual and planning stages in the future. A developing country has a great opportunity to create an integrated data feedback system in a standardized format, and this must rank highly in the project management process.

The above project management tools and activities incorporate a plethora of financial, technical, and management techniques, mostly computerized, which can provide the project management team with the necessary data on which to make the necessary decisions for project development. But they are ineffective if not built into an integrated project management system.

Project Management Training

Probably the major problem facing the effective application of project management in developing countries is the need to train the national workforce and include the subject in the country's educational programs.

Whatever training programs are developed, they must:

(i) Enable the trainee to relate the theoretical concepts to the local realities. The superimposition of theories, systems, and technologies, all of which are supported by a high-tech, sophisticated society, only serves to confuse trainees and not help them.

(ii) Encourage creativity and innovation, and a critical synthesis approach to problem solving. The focus should be on problem solving, not analysis, within the existing structure, while continually striving to improve it.

(iii) Highlight the integrative nature of project management and the necessity for highly developed social and managerial skills alongside technical and mechanical ones.

The first step toward effective project management training is to have the subject accepted as part of university and technical college curriculums. Once accepted it may be developed into a separate option, or expanded to form a postgraduate program which builds on an undergraduate core of general management and science.

The training of project managers can continue from the graduate stage through practical project implementation by using a structured experience program.

For the part of the indigenous population that is not able to pass through the university system, a modular series of project management packages should be developed and run both in-house for client, consultants, and contractors and in the country's tertiary educational institutions. If such institutions do not exist, then foreign consultants can be hired to design and run such programs, but always ensuring that they conform to the three points made at the beginning of this section.

The major needs for training in project management are

(a) To train project managers, who will be the focal point of project development and implementation, and have highly developed managerial skills.

(b) To train computer-literate managers and technical staff who will provide the support for the integrated project management system.

(c) To train development administrators who can plan and coordinate the country's project cycles as an integrated process.

(d) To train specialists in areas specific to the management of international projects. Such areas as economics, economic assessment, feasibility studies, environmental impact studies, finance, financial appraisal, planning, scheduling, control systems, human resource management, and technology transfer will be included in such specific training programs.

It is the responsibility of the government in the developing country to *actively* support the project management system as its way forward for better development.

Industry and commerce can give support by the creation of a project management association (linked to an established professional project management organization), which can take on some of the issues discussed under benefits, problems, and training in this chapter.

In conclusion, the benefits outlined above under "project management" are substantial for a developing country. To receive them it is necessary for the government and industry and commerce to instigate an integrated project management system—one that does not merely copy what exists elsewhere, but that reflects the needs of the country and will provide a way forward for it to grow and achieve its long-term objectives.

REFERENCES

1. W. A. Dymsza, "Trends in Multi-national Business and Global Environments; Perspective," *Journal of International Studies,* pp. 31–33, Winter 1984.

2. H. K. Eldin and Ivars Avots, "Guidelines for Successful Management of Projects in the Middle East: The Client Point of View," *Proceedings of Project Management Institute Symposium,* Los Angeles, pp. 11-m.1–11-m.5, 1978.

3. S. Al-Sedairy and P. Rutland, "Project Management as a Determinant of, and Influence on, Organizational Modes of a Developing Country," *Proceedings of 10th INTERNET World Congress on Project Management,* Vienna, June 1990.

4. E. Gabriel, *Dimensions of Project Management,* A. Reschke and H. Schelle (eds.), pp. 71–72, Springer-Verlag, New York, 1990.

5. E. Gabriel, *Dimensions of Product Management,* A. Reschke and H. Schelle (eds.), p. 73, Springer-Verlag, New York, 1990.

6. D. W. Hendon and A. Hendon, *How to Negotiate Worldwide,* Gower Technical Press, Aldershot, England, pp. 76–84, 1989.

7. D. Sharp, *The Business of Architectural Practice,* William Collins & Sons, London, pp. 12–13, 1986.

8. J. D. Head, *Managing, Marketing and Budgeting for the A/E Office,* Van Nostrand, New York, p. 96, 1988.

9. S. Al-Sedairy and P. Rutland, "Performance and Process Mis-Matches in Project Management Departments," paper submitted to *Construction Management and Economics,* February 1991.

10. H. K. Eldin, "Human Element Constraints in Applying Modern Management Techniques in Developing Countries," *Proceedings of Project Management Institute Seminar Symposium,* Montreal, p. 280, 1976.

11. H. Banwell et al., *The Placing and Management of Contracts for Building and Civil Engineering Works,* HMSO, London, 1964.

12. G. Ofori, "The Construction Industry in Ghana," mimeographed *World Employment Programme Working Paper,* restricted, ILO, Geneva, 1981.

13. L. C. Stuckenbruck, *The Implementation of Project Management,* Addison-Wesley, New York, pp. 17–18, 1981.

14. G. Azud, "Owner Can Control Costs," AACE, 1969.

15. K. G. Kohler, "The Transfer of Management Know-How to Developing Countries," *Management International Review,* vol. II, no. 6, June 1971.

16. R. A. Johnson, F. E. Kast, and J. E. Rosenzweig, *The Theory and Management of Systems,* McGraw-Hill, New York, 1967.

17. *Proceedings of the Conference on the Procedures for the Production of Public Buildings,* King Saud University, Saudi Arabia, 1989.

18. D. A. Rondinelli, "Why Development Projects Fail: Problems of Project Management in Developing Countries," *A Decade of Project Management,* J. R. Adams and N. S. Kirchof (eds.), Project Management Institute, Drexel Hill, Pennsylvania, pp. 295–300, 1980.

19. H. K. Eldin, "Human Element Constraints in Applying Modern Management Techniques in Developing Countries," *Proceedings of Project Management Institute Seminar Symposium,* Montreal, pp. 279–281, 1976.

20. J. R. Adams and B. W. Campbell, *Roles and Responsibilities of the Project Manager,* Project Management Institute, Drexel Hill, Pennsylvania, 1988.

21. B. Balkenhol, "Small Contractors: Untapped Potential or Economic Impediment?," mimeographed *World Employment Programme Research Working Paper,* restricted, ILO, Geneva, 1979.

22. D. Lock (ed.), *Project Management Handbook,* Gower Technical Press, Aldershot, England, 1987.

23. E. A. Stallworthy and O. P. Kharbanda, *Total Project Management,* Gower Technical Press, Aldershot, England, 1983.

24. J. Stinson and E. Gardner, Jr., *Management,* McGraw-Hill, New York, 1989.

25. Internet, "Project Management Tools and Visions," *Proceedings of 7th Internet World Congress,* Copenhagen, 1982.

26. S. Shaheen, *Practical Project Management,* Wiley, New York, 1987.

27. F. L. Harrison, *Advanced Project Management,* Gower Technical Press, Aldershot, England, 1985.

28. *RIBA Computer Software Selector,* RIBA, London, 1990.

29. G. Atkinson, *A Guide through Construction Quality Standards,* Van Nostrand Reinhold, Wokingham, England, 1987.

CHAPTER 6

MANAGEMENT OF INTERNATIONAL PROJECTS IN THE CSFR*

Ladislav Chrudina

Project Management Society,
Prague, Czechoslovakia

Günter Rattay

PRIMAS Consulting
Vienna, Austria

Ladislav Chrudina holds a Commercial Engineering (Ing.) degree and a Ph.D. degree from the Prague School of Economics. He was project manager at the Military Design Institute in Prague and an expert for development and construction project management at the state administration. He was a lecturer on project management at the Prague School of Economics from 1986 to 1989 and at the Technical University in Prague from 1988 to 1990. He is a founder and president of the Project Management Society in Czechoslovakia and the author of several books and articles on project management. Since 1990 he has been a deputy to the minister for economy of the Czech Republic.

Günter G. Rattay graduated from the University of Economy and Business Administration in Vienna, Austria. He was the manager of projects in the fields of research, product development, preparation and conducting of events, and development of marketing plans, among others. He is a member of the board of the Projektmanagement Austria-Institut, the Austrian National Association of INTERNET, and a member of the Executive Committee of INTERNET. He also is the director of the education and research program "Project Management for International Projects in Central Europe." He is a consultant in the field of project management and with project-oriented companies.

Editors' Note: The reader is aware of the political, military, and economic changes under way in the Czech and Slovak Federal Republic (CSFR). This chapter was written before these changes were initiated. We have chosen to leave this chapter as it is, in the hope that stability will return to this part of the world.

THE ECONOMIC AND POLITICAL SITUATION IN THE CZECH AND SLOVAK FEDERAL REPUBLIC (CSFR)

History

The CSFR was founded in 1918 as the successor of the disintegrated Austrian-Hungarian empire in the territory of the Kingdom of Hungary, inhabited mostly by a Slovak population. With the exception of the past 40 years, the history and the culture of the nations inhabiting the territory of the CSFR have been shaped in the context of western European history. Nevertheless, the historical development of the two nations—the Czechs and the Slovaks—manifests certain characteristics.

The cultural and the political integrity of each of the two republics—the Czech Republic and the Slovak Republic—should therefore always be taken into consideration.

Between the two world wars the CSFR reached a high level of political, economic, and cultural development and could rank with the most advanced countries of Europe. After the end of World War II "the iron curtain" assigned the CSFR to the group of eastern European countries. The rapid "socialization" of the country, as well as the fact that its advanced industrial potential was fully subjected to the needs of the eastern bloc, deformed to a great extent both the system of values and the economic structures of the country. The institutions of market economy, including private ownership, were completely destroyed. After the "velvet revolution" of November 1989 the CSFR has been reestablishing the traditional values of European democracy and a market economy, and restructuring its economy.

Political Reform

Immediately after the November revolution the foundations were laid and basic conditions created for the rise of a pluralistic democracy. The abolition of the monopolistic primacy of the Communist Party, as well as the fairly liberal legislation concerning the foundation and the status of political parties, started a real boom period for new political parties and movements after 40 years of very strict limitations.

The political situation in the CSFR is, however, at least for the time being, far from stable. The political structure of the country in the sense of a classic parliamentary democracy is still only being shaped. The same applies to the differentiation of the Czechoslovakian society. A restoration of the old Communist regime is, however, not assumed by even the most catastrophic scenarios.

Economic Reform

After a year of discussions and preparations, new legislation was passed by the Czechoslovak Federal Assembly effective January 1, 1991, marking the beginning of the transition of the country to market economy. The system of central planning was completely abolished, prices were liberalized, and legal foundations laid for private enterprise, including liberalization of the inflow of foreign capital. The so-called internal convertibility of the Czechoslovak crown has been achieved and privatization started.

It is the aim of the economic reform to establish in the CSFR, within 3 to 5 years, a viable system of market economy with its institutions and legislation as compatible as possible with the system of the European Community.

It is of course quite clear that there is a difference between the creation of the organizational and legislative conditions and an actually functioning market economy. The latter does not only need the institutional and legislative framework, but also time for entrepreneurs to be born and trained, to change customer behavior, and to raise market relations.

It is certainly rather difficult to estimate the length of time that this process will require, but a realistic prognosis should reckon with at least 10 years. That process can, and should, be accelerated by internationalization of the Czechoslovak market, through foreign capital, and through the transfer of know-how and entrepreneurial and managerial culture.

The vast project to train Czechoslovak managers, stated under the auspices of various governments and communities as well as private foundations, aims at speeding up that process. Direct participation of the Czech and Slovak businesspeople and managers in joint ventures will undoubtedly prove to be the most effective type of transformation.

The Legislative Framework

The economic reform must of course be based upon new legislation. And this is in a way the most complex part of the process of transformation of the Czechoslovak economy, because the new legislation required is far from being drafted by the reform-minded economists and lawyers only, but by all three parliaments and by political consensus. This very often pushes the original intentions of the reform in another direction—the direction often less advantageous to the reform.

The contemporary legislative framework has two principal characteristics—legal security and the legislation that is being drafted.

Legal security, invariably guaranteed by the constitution of the country, includes:

- Inalienability of private property, and the right to freely dispose of that property
- Liberal acts concerning pricing
- Protection of foreign investments

The legislation being drafted should be conceived to be fully compatible with the legislation of the European Community. In view of the quick rate of the legal reforms, it does not seem to be of use to list all the laws that are being prepared.

The first stage of the reform is to be terminated in 1992 by new tax acts to be passed effective January 1, 1993, introducing to the CSFR the system of taxes currently used in Europe (total-cost-related base plus value-added tax). The rate of taxation is rather difficult to estimate, but the aim is for it to be also at the "European level." The legal reform will eventually result in a uniform commercial code.

Psychological Aspects of the Reform

The following joke used to be told in the CSFR not so long ago: What is the difference between an expert in a market economy and an expert in a planned economy? If you present a problem to experts in a market economy, they will immediately start trying to find the ways of solving it, because they will conceive of it as their great chance. If you present the same problem to experts in a socialist economy, they will immediately start collecting a lot of proofs to the effect that the problem cannot be solved, because they will conceive of it as a risk and trouble.

It would be very dangerous for us to accept generalization. In the excommunist countries there are certainly a number of people who are active and enterprising, and who do not rely on someone to direct their lives. Nevertheless, foreign partners will often meet with facts that do not fully coincide with their ideas and practice.

In the first place let us mention the lack of self-reliance and the aversion to discharging responsibility for decisions, found with the managers of state enterprises and officials at the ministries. The decisions are adopted only very slowly. It is recommended that the western partners try from the very start of negotiations to get together all decisive partners (there will invariably be more of them than in the West) and create an atmosphere of joint conception of the project.

In the second place it is the ignorance of the Czechoslovak partners concerning the procedures of commercial negotiations, including inadequate information on the legal systems of the countries of the market economy. Until very recently all the negotiations with foreign partners were the privilege of a few officials at the ministry of foreign trade. At present various consulting agencies are being established in the CSFR. Their presence at the negotiations of a project is thus to be expected. The level of their services can, however, vary very considerably. As far as the larger projects are concerned, a consultant paid by the government can also be engaged.

The third problem is posed by the more or less nonexistent "entrepreneurial ethic," which is only just being formed. The foreign investor may meet with a partner who does not represent the interests of the enterprise and promotes only his or her own individual interests. In exceptional cases the foreign investor may also meet with attempts by Czechoslovak individuals to seize managerial posts in the joint ventures being established. Despite the temptation to use the services of these people, it will certainly be better to reject them. They are very likely to manifest again their lack of earnestness in the future. And on top of that, the information that the foreign partner will get from such people will always be oriented only toward the benefit of the informant, not toward the benefit of the enterprise.

The longing of some individuals to become rich overnight is very strong after a fast of 40 years and the exchange rates favor the western currencies. There are therefore much better possibilities of corrupting people in the CSFR than elsewhere in western Europe.

Fourth, if the foreign investor enters into negotiations with the representatives of a state enterprise, political pressure can very often not be excluded. That risk is lower as far as the public stock companies are concerned, and the situation is the most stable in private firms. A discreet inquiry about the stability of the partner in Czechoslovakia can save the foreign investor a lot of trouble. The situation is, however, getting more stabilized at present.

The Manager a Person without Prestige?

A peculiarity that is quite understandable for a Czechoslovak national, but fairly surprising for a foreigner, can be the social prestige of project managers and professionals. Whereas the employees in the Civil Service—cabinet ministers, directors, heads of departments—rank very high, the survey carried by the Sociological Institute of the Czechoslovak Academy of Science shows that managers are ranked even below occupations such as teachers, nurses, and miners.

The low prestige of the managers is due to their complex role in the past. At the time when running the Czechoslovak national economy by directives was the preferred way, they did not work as "moderators of working teams" but rather as "stewards," the latter being a most ungrateful role.

All this also belongs to the "cultural heritage" of the past period. The low prestige of the managers is also reflected in their relationship with the administration of the enterprise: they are not considered important partners, only employees doing their jobs.

THE STATE OF THE ART OF PROJECT MANAGEMENT IN THE CSFR

During four decades of the centrally planned economy in the CSFR the understanding of "management" and "the manager" and their roles has developed quite differently from that in a market economy. As a result of these very different types of economies, extremely different attitudes, behavior patterns, and personalities have been created. Even new terminology has appeared, which could be a cause for misunderstanding because the meaning of a word can vary from one economy to the other. This situation persists, and probably will persist for some time. This creates some conflicts, which could only be bridged if the differences were recognized by all parties of international projects.

Of course, project management is not a totally new topic in Czechoslovak economic life. Even during the period of the centrally planned "nonmarket" economy, projects were developed and managed. The framework of central planning was favorable (in some cases) for the compulsory utilization of some typical project management tools, such as network analysis and multiprojecting. It seemed that the centrally planned economy was a good environment for project management, and some people believed it.

Generally speaking, this is the starting point of project management in the CSFR. The state of the art of project management, observed from different points of view, is defined in the following.

Project Definition. This phase is greatly neglected and almost unknown. In the centrally planned economy this phase was not considered as a part of project work, but it was included in the system of the state plan preparation. It is generally well known that this system was very rigid and bureaucratic, indoctrinated by ideology and politics. Project managers were solely involved in the phase of project implementation. This is why their knowledge is based only on three areas: time, cost, and performance management.

Confused Terminology. There is a very vague understanding of such basic words as "project," "project preparation," and "project management." Not long ago the term "project" was used only for a certain set of design documentation for construction projects. For that reason the majority of Czechoslovak experts understand project management as the management of building erection.

Teamwork. Explicitly stated, there is not only inferior teamwork, but even poor knowledge of team creation techniques, particularly among those experts who used to work in a system of central preparation and execution.

Decision Ability. Four decades of conducted economy resulted in a very unpleasant situation. There is a very low ability of managers to reach decisions and take responsibility. At present this state is accelerated by an unstable and incomplete legal framework.

Risk Management.   There is a direct connection between the decision-making ability and the state of the art of risk management. The centrally planned economy did not recognize risk as a requisite feature of decision making. The methods aimed at defining, limiting, and controlling the risk were not used. The level of knowledge and implementation of this management key area is very low.

Project Start-Up.   Projects have been launched "abruptly," without proper preparation. Distrust of organization and planning grows out of the inferior ability to manage risks, ignorance of teamwork, and the poor division of responsibility. At present this feeling is heightened by instability of the environment and the prevailing distrust of any possibility to project workers' activities. But even improvisation requires proper preparation.

Time Management.   There is a good acquaintance with time management techniques, particularly those based on network analysis. It was always the first priority of the state plan to reach the planned goal as soon as possible. The great campaigns for rapid realization of the state plan are still well remembered. Nevertheless, experience with time management gained especially with the network planning of construction projects differs from that in a market economy. Tasks have very often been led to their ends without doubt about their purpose and without feedback.

Cost Management.   Due to the transformation from central planning to a market economy, experience and knowledge of cost management is losing value. Everything undergoes substantial changes, even calculation and budgeting. The tough practice of cost management will vary from the current routine, and all new professional skills will be needed.

Performance Management.   Performance management (quality management) has a good background in the area of construction projects. In this area project management is based on excellent experience, as many projects have been run under very difficult circumstances. Difficulties have been as follows: neglected supply network, rigid contract law, irregular deliveries, unstable quality, and the like.

Communication Ability.   Generally speaking, the art of communication is underestimated. Reports and statements are usually verbal, and graphic presentation is nonexistent. This fault goes back to the times when it was advantageous to say nothing with a lot of words. Good communication was prevented by persistent distrust between participants. This problem is linked to inferior teamwork ability.

PC Skills.   Today conditions are changing very rapidly. Of course, the average level is lower than the European standard.

OPPORTUNITIES AND RISKS OF INTERNATIONAL PROJECTS IN THE CSFR

Types of Projects of Cooperation

According to the new liberal laws, foreign investment capital can participate in all business areas in the CSFR. Only the production of weapons and drugs is prohibited.

Although there can be many alternative ways to invest in the CSFR, a few examples follow.

Direct Investment.   These activities can be conducted without Czech or Slovak partners. Despite this fact, significant attention and care are required to create a culture compatible with the Czech and Slovak environment. Czech or Slovak experts should be engaged. As they are employees and not partners, the team cannot be considered a "joint team."

Joint Venture—Branch Company.   This form involves a joint project with no less than one foreign and one Czechoslovak partner. Both partners remain independently owned legal entities and create a new branch—a jointly owned (daughter) legal entity. A culture compatible with the Czech and Slovak environment, but a common understanding of the project must be established. The Czechoslovak governmental policy does not support and encourage joint ventures when the Czechoslovak partner is state-owned. The private sector, on the other hand, is strongly favored.

Joint Venture—Mixed Company.   This type of joint venture is established when a share of property of an existing Czechoslovak enterprise is bought by a foreign investor. It means that the ownership of Czechoslovak enterprise turns over. This possibility exists only when the Czechoslovak enterprise has been denationalized and reconstructed as a joint corporation owned by the state. These projects are aimed at creating common marketing and common production strategies, but the strategy of the total privatization need not be of interest to both partners. Czechoslovak management staff in particular can use such a project to avoid total privatization of the enterprise. For that reason, the government monitors these projects very carefully.

Opportunities and Problems in the CSFR

Official publications stress the different advantages of investing in the CSFR, such as well trained but very cheap labor forces, low debt, relatively good infrastructure, and so on.

What are the Czechoslovak problems and where are the gaps that should be filled?

Environmental Protection.   It is generally known that the environment in the CSFR has been greatly neglected and may be in catastrophic condition. Projects aimed at stabilizing the current state as well as projects aimed at developing and implementing environmentally safe technologies might be very successful.

Small and Medium-Sized Businesses.   In the CSFR small and medium-sized businesses do not exist at all. The development of these types of enterprises based on private ownership will be of great significance and will be supported by the government.

Industry.   Currently industries have excellent technicians but poor management, low productivity, and nonexistent marketing strategies. Thus technological modernization and restructuring are of great significance.

Tourism.   The CSFR offers well-accessible great natural beauty, as well as historical, architectural, and urban attractions of great cultural value. On the other hand, undeveloped, poorly managed accommodations and other services hamper the tourism industry.

Banking.   Undeveloped structure, insufficient capacity, and neglected services are characteristic. Opening financial markets creates a deep gap between demand and capacity of the existing banking system.

Agriculture.   Up to now agriculture has been based on cooperative mass production. Unemployed strips of land along the Austrian and German borders can be used to create business based on a combination of tourism and natural food production (family farms with vacation rentals).

These examples illustrate the great opportunities for joint ventures in the CSFR. The Act on Large Privatization, passed by the Federal Assembly on February 26, 1991, specifies the procedure of the denationalization and privatization of large state-owned enterprises. Great challenges and opportunities for foreign investors, as well as for domestic entrepreneurs, have now been created.

Changes forthcoming under the Act of Small Privatization offer many opportunities for investment, too. Of course, the first round of auctions for small and medium-sized businesses, which took place in January 1991, involved only Czechoslovak citizens. In the course of the auctions local bodies will offer nearly 70,000 restaurants, retail shops, services, and local industries in exchange for cash. Those business units which did not sell for at least 50 percent of their assessed value will be carried over to a second round, which will include foreign investors.

The initial period of joint ventures is the period of the small pilot projects. A large proportion of the ventures involved one-person enterprises, however, and commitments of less than 5000 crowns. Only 37 involved investments greater than 10 million crowns. After March 1991 more serious projects of up to 10 million crowns were offered.

Risks of International Projects in the CSFR

When looking for investment opportunities in the CSFR, you also have to take into account the risks. The following are often mentioned by foreign investors.

Risks Concerning Property and Legal Aspects.   Due to the political and economic changes within the CSFR, structures also change often. As a result the contact person who represents the Czechoslovak partner often changes, and almost signed contracts lose their validity. The situation is much more stable if the partners are private firms and denationalized joint stock companies. Investors should check the current status of their Czechoslovak partners (state-owned enterprise, cooperative, joint stock company, private business) and the timing of the privatization project.

Risks Concerning Financial Aspects.   Prices of real estate and rents for offices are growing. Some partners from the CSFR are not able to pay and foreign official institutions (such as the Austrian Kontrollbank) will no longer insure such risks. Investors should collect all necessary information on the financial situation of their CSFR partners. A domestic auditor has to be engaged, because foreigners cannot understand the statistical data from the centrally planned period.

Risks Concerning Different Cultures.   Czech and Slovak managers and officials will behave as they were used under the old system. To transform this behavior into a more market-oriented one is a long-term objective. Behavior that is led by efficiency of performance, cost, and time is not as obvious as it should be. Most Czech and

Slovak partners are not interested in adventures but want to enter into long-term business relationships.

PROJECT MANAGEMENT AS SUCCESS FACTOR FOR INTERNATIONAL COOPERATIONS

Project Management to Integrate the Different Cultures of International Companies and CSFR Companies

As a consequence of the different political and economic systems in the past, the existing culture of CSFR organizations is very specific and has little similarities with foreign companies. Although the cultures of western companies and their CSFR partners are different, both have the same objects of consideration, which is their common project (such as a joint venture, a privatization project, or a reorganization).

Both parties are able to integrate their different cultures by developing a common understanding on the object of consideration and on the management of this object. The first step of integrating different cultures is to establish a project team, which consists of Czechoslovak as well as western team members. Periodic team meetings and the establishment of technically oriented subteams, which are also mixed, may further deepen the cultural exchange.

Project management may therefore be the approach to integrate different cultures. If both partners (foreign investor and CSFR manager) develop a common language and understanding of project management, the success will be more likely. A project-management training course where the whole team further develops its skills may help to bridge the cultural gap.

Adequate Use of Project-Management Methods

The following project-management methods seem to be adequate for international projects in the CSFR.

Work Breakdown Structure.   A work breakdown structure according to a phase concept may help to recognize project crises at an early stage and to manage them better (repetition of phases, modification of objectives, project interruption, stop).

The work breakdown structure in Fig. 6.1 is an example of the planning and establishing of an international joint-venture company.

Project Phase Orientation.   Clearly defined project phases will reduce the complexity of the overall project by providing the advantage of cyclic procedures (Fig. 6.2). Figure 6.3 lists the tangible outputs of each phase.

In summary, the advantages of such a phase-oriented project structure are as follows:

- Creation of well-defined portions of the overall project to decrease complexity.
- Reduction of project risks by defining stop-or-go decisions between the different phases.
- Possibility of repeating phases if necessary.

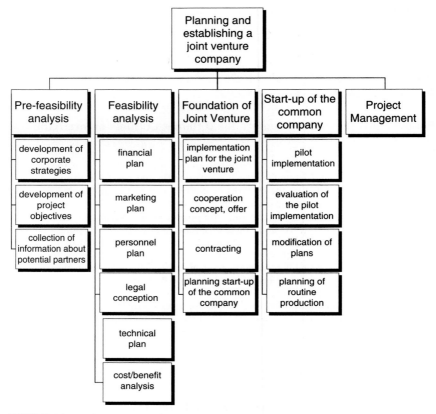

FIGURE 6.1 Work breakdown structure of joint-venture project.

- Closedown of a project phase offers the possibility of deciding anew on how to proceed during the next phase.
- Increase in flexibility by adjusting the strategies and basic rules to the specific demands of each phase.
- Project organization may be adapted specifically to the needs of each project phase. For example, in a prefeasibility phase strategically oriented team members are of importance, but during the foundation and start-up phase the team has to perform detailed operational work. At least the project manager has to remain the same during the entire project.

Of great importance seem to be the phase transitions, where one phase is closed down and the next phase started:

The cyclic procedure shown in Fig. 6.4 provides periodic comparisons between the project strategies and the actual situation at the end of each project phase. In that way a strict goal orientation of the entire process is guaranteed.

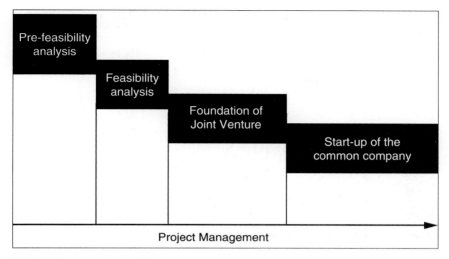

FIGURE 6.2   Main phases of joint-venture project.

Phase:	Tangible Outcomes:
• Pre-feasibility analysis	• Accepted written company strategy • Accepted written project objectives • List of potential partners • Written evaluation of potential partners
• Feasiblility analysis	• Cost / Benefit analysis • Detailed plans (financial plan; ...)
• Foundation of Joint Venture	• Contract of cooperation • Implementation plan
• Start-up of the common company	• Evaluated objectives • Responsibilities assigned to the local management staff • Trained personnel

FIGURE 6.3   Tangible outcomes for each project phase.

Project-Management Training and Pilot Projects

Project-Management Training and Pilot Projects are effective means to establish common views and a common terminology between Czechoslovak and western partners. Well-trained partners, especially in the field of project management, increase the chance of successful cooperation.

Because of an urgent need of Austrian companies as well as CSFR associations for good cooperation and a similar understanding of project management, the Projektmanagement Austria-Institut is establishing a specific training program. The training program is aimed at raising the professionalism of eastern European managers involved in any kind of joint-venture projects with western European companies.

This can be done mainly by matching the different management languages and

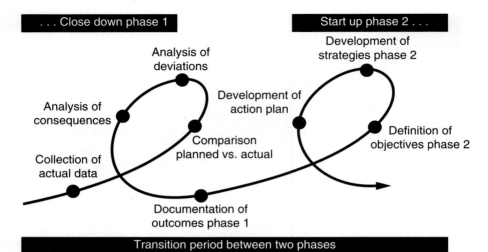

FIGURE 6.4 Transition period between two phases.

establishing a common (project) management terminology. The program includes topics such as market orientation, methods and tools of project management, teamwork, and implementation of actual ongoing projects of the participants, such as

- Privatization projects in the CSFR
- Joint ventures between Austrian, CSFR, and Hungarian companies
- Marketing concepts for a CSFR region

A second means for improving professional cooperation is a pilot project. Implementation of investments, a concept for organizational development, a market survey, and the like could be the aim for the pilot project. Thus both partners develop their common language and views not only through participating in training courses, but also through "training on the project."

To make such pilot projects successful, a few aspects should be taken seriously:

- The pilot project should be encouraging for both parties, but not hazardous.
- The project team should be assisted by an external consultant or moderator, who inspires learning processes by periodic discussions.

SUMMARY

Because of the specific economic and legal situation in the CSFR, entering these new markets by establishing joint ventures or by founding company subsidiaries constitutes complex tasks and a real challenge for western companies.

Professional management of such projects will definitely increase the likelihood of success for both sides, the western investors and their CSFR partners. An important success factor for eastern European projects is a common understanding of basic values, of project-management terminology, and of tools by both business partners.

Participation in common project-management training or in a common pilot project may help to establish a good team.

CHAPTER 7

MANAGING THE INTERNATIONAL PROJECT ENVIRONMENT*

Robert Youker

Bethesda, Maryland

Mr. Youker is an independent consultant and trainer in project implementation operating from Bethesda, Maryland. He was an Adjunct Professor of Project Management in the Engineering Management School of George Washington University and also taught at Harvard University, the University of Wisconsin, the Asian Development Bank, A. D. Little, and the University of Bradford, England, among others. Previous experience also included president of Planalog Management Systems and analyst with Xerox Corporation and Checchi & Company.

INTRODUCTION

What are the most important activities a project manager can undertake to ensure successful project implementation on international projects? The literature of project management places great emphasis on planning and management tools for the project manager to use to control time, cost, resources, and quality of performance. However, a review of the results of project monitoring and evaluation on World Bank projects indicates that many of the key problems of implementation lie in the general environment of the project and not under the direct control of the project manager.[1] Of all the alternative ways on which a project manager can spend time, which activities are most important for successful implementation?

The answer to this question is not easy. Specific actions will naturally be related to the specifics of each project situation. But are there any general conclusions we can draw? Can we learn from the successes and failures of other project managers? This chapter is based on World Bank experience, but the conclusions are also applicable to other international and domestic projects.

PROBLEMS ON INTERNATIONAL PROJECTS

The World Bank expends extensive resources on both the ongoing monitoring of project progress and the subsequent evaluation of project results. For ongoing projects, the

---

*This chapter was originally published as "Managing the International Project Environment," *International Journal of Project Management*, Vol. 10, No. 4, November 1992. Used by permisson.

bank staff prepares regular supervision reports. Twice a year the bank conducts a review of the status of implementation of its entire portfolio of projects. Also periodically, the bank and individual countries review the progress of implementation for all projects in that country. At the end of the implementation period for each project, a completion report is prepared by the country and the bank. The Operations Evaluation Department reviews these reports and in some cases conducts an evaluation in the field of the impact of a project. Thus the bank has extensive evaluation information, both ongoing and after the fact, on the results of implementation.

What do these evaluations tell us about the critical factors for the success of project implementation and the actions required by project managers? The major overriding conclusion from analyzing all of these various evaluations is that many of the most important problems of implementation lie in the general environment of the project and are beyond the direct control of the project manager. Project managers do need to define objectives, prepare plans, schedule resources, and control progress, but these internal management mechanisms alone are not sufficient for successful implementation. Successful project managers must also try to manage or influence the key actors and factors in the project environment. The project manager must look outside the project and look ahead to anticipate problems and to develop contingency plans.

SPECIFIC PROBLEMS

The World Bank supervision and evaluation processes have identified the following types of problems:

- Shortage of local counterpart funds. (The government treasury does not have the money it promised to finance local expenditures such as purchase of land.)
- Inability to hire and retain qualified human resources, especially managerial and technical personnel. (Government personnel policies and procedures do not mesh with the needs of a temporary project.)
- Ineffective transfer of technology and difficulty of building and sustaining institutional capacity.
- Difficulty in changing the policy environment, such as pricing.
- Inadequate accounting, financial management systems, and auditing.
- Shortage of supplies and materials due to overall economic problems.

The nature of the problems will vary by country and type of project, but the fact remains that all of these problems are in the general environment and not under the direct control of the project manager.

The solution of these problems requires innovative and proactive behavior on the part of project managers. Project managers will, in effect, have to become diplomats and to work out alliances with various sources of power to gain influence and to assure effective project implementation.

Thus we see that the project manager's job is not confined to controlling events within the project organization. Often as important to the eventual outcome—frequently even more so—are the project's linkages with the external environment. This is particularly so when, as is inevitably the case in a developing economy and society, that environment is changing rapidly.

THE SYSTEMS APPROACH

The development since World War II of a systems perspective and a strategic management context has generated increasing interest in viewing the organization as a subsystem that operates within a larger system (or environment) and interacts with other subsystems. This analysis is useful in project management, where the project is viewed as surrounded by an environment both within and outside its parent organization.

The basic elements in the systems approach are the system itself, its boundary with the environment, and the inputs and outputs that link it to its environment[2] (Figs. 7.1 to 7.3). Different terms are used by different authors, but here we shall settle on the term "project environment." (Other terms are "domain,"[3] "set,"[4] and "context."[5])

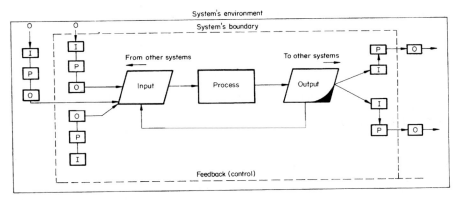

FIGURE 7.1   System parameters, boundary, and environment.[2]

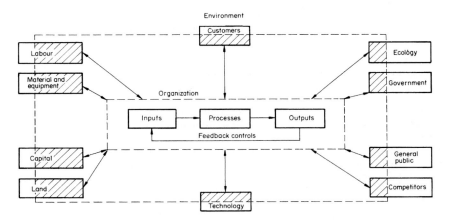

FIGURE 7.2   Organization, its resources, and its environment.[2] Crosshatched boxes indicate degree of control or resource. Solid-line boxes indicate degree of independence or environment. Dashed lines are boundaries demarking system from its environment.

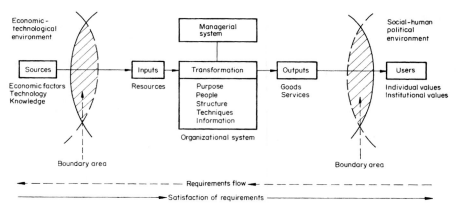

FIGURE 7.3 Organization as open input-output system.

What is this project environment? A dictionary definition is a useful place to begin. The environment, according to the *Random House Dictionary,* is "the aggregate of surrounding things, conditions or influences." Thus the environment includes virtually everything outside the project: "its technology (i.e., the knowledge base it must draw upon), the nature of its products, customers and competitors, its geographical setting, the economic, political and even meteorological climate in which it must operate, and so on."[6] These factors—and especially changes in them—can affect the planning, organizing, staffing, directing, and so on, which constitute the project manager's chief functions.

The key conclusion to be drawn from the systems perspective and the concept of the project environment is that the project has relationships with other subsystems or other organizations in the environment. These create dependency relationships on both sides. To be ultimately successful, therefore, the project manager must look outside the project. He or she must study and try to manage or adapt to the relevant external forces on which the project's success depends.

HOW TO ANALYZE THE PROJECT ENVIRONMENT

Galbraith[7,8] introduces the concept of uncertainty as the key factor in analyzing the project environment. Uncertainty becomes a problem for the project manager because of the dependency relationship between the project and the uncontrolled elements in its environment. The greater the degree of dependence and the greater the degree of uncertainty, the greater the problem for the project manager. So the basic purpose in analyzing the project environment is to define potential problems, to assess the probability of their occurrence, and to try to solve them.

To do this, the project manager needs to:

- Scan the project environment
- Identify the relevant actors and factors
- Define the degree of dependency in the relationship

- Estimate the nature of the uncertainty and the probability of something going wrong
- Analyze the degree of power the project has to control the key actors and factors
- Identify potential problems (high dependency, high risk, and low power)
- Develop contingency plans to deal with potential problems by analyzing stakeholders' purposes and planning linkages to increase power and influence

CATEGORIZING THE RELEVANT PROJECT ENVIRONMENT

The relevant environment will clearly be different for each project. The specific relevant environment for a project is determined by three factors peculiar to the project[8]:

1. The product (service, market, and user) context of the project
2. The technology involved
3. The physical location

For example, the construction of a concrete building in Albany, New York, will most likely make the cement factories in nearby Hudson, New York, key elements in the project environment because they will probably become suppliers of inputs to the project.

Since the environment is defined as everything surrounding the project, scanning the environment is not easy. The models in Fig. 7.4 provide several alternative ways to scan the environment.

1. Elements that are suppliers of inputs, consumers of outputs, competitors, and regulators[9]
2. Elements that are physical (e.g., climate), infrastructural (e.g., power supply), technological (e.g., plant genetics), commercial, financial, or economic (e.g., banks), psychological or sociocultural (e.g., attitudes toward credit risk), or political or legal (e.g., local government)
3. Elements that are hierarchical and sometimes geographical, such as government at various levels (national, regional, or local)
4. Elements that are actors (e.g., individuals, groups, institutions) or factors (e.g., attitudes, trends, laws)

ENVIRONMENTAL SCANNING

Clearly not all the elements in the project's environment will be crucial to its success. Identifying those which are relevant and important—by systematically scanning the project environment—should be an important part of the project manager's job. This environmental scanning (sometimes also known as environmental analysis or environmental mapping) leads the manager to identify the important elements in his or her environment.

This latter—the identification of the key actors in the project environment—is sometimes known as "stakeholder mapping," that is, mapping out which people or groups have a stake in the project's success or failure.[10]

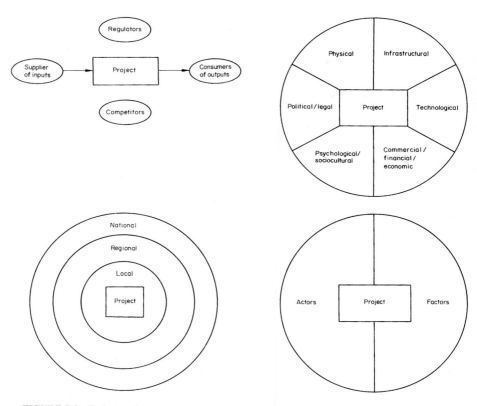

FIGURE 7.4   Project environment.

Whatever terms or analytical models are used, the process of scanning the environment to identify and then monitor those elements that can affect the project is extremely important for the manager. Schemes such as those presented in Fig. 7.4 can help the project manager to be systematic about it, but the relevant elements must be identified separately for each project and rated for their degree of importance to project success.

Scanning refers to the exposure to and perception of information. The more information the manager has available, the greater his or her chances of identifying the key elements in the project environment. The means of scanning can vary from "an undirected, fortuitous, and subconscious observation to a purposeful, predetermined, and highly structural inspection."[11]

THE POWER FIELD

Another important analysis for the project manager is to think in terms of various levels of power over external actors and factors. By power we mean the ability to get

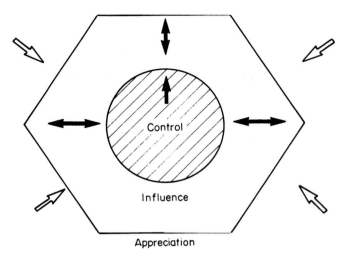

FIGURE 7.5   Power field.

someone to do something you want done. Although power is a continuum, Smith and coworkers, who developed this model, find that dividing power into three levels, as follows, is useful: control, influence, and appreciation[12,13] (Fig. 7.5). Control is the ability to give orders and expect that they will be carried out. Influence is less power than control. While your own actions will affect the achievement of the objective, full achievement will also require the actions of others over whom the project manager does not have direct control. Appreciation means no power or influence, but just knowledge or awareness of the potential impact of the actor or factor. For example, a project manager may control a clerk, have influence with an engineer in the Ministry of Works, but only be able to appreciate his or her dependence on the staff of the Ministry of Finance.

THE PROCESS OF ENVIRONMENTAL ANALYSIS

The first step after scanning is to list all of the actors and factors that will, or could, have an influence on the success of project implementation. This could include actors such as a cement factory or a railroad or factors such as rainfall or the foreign exchange rate. The second step is to lay out a map of the actors and factors in three concentric circles by the relative degree of power (control, influence, or appreciation) over these key elements (Fig. 7.6). The third step is to evaluate the degree of dependence of the project on the various actors and factors. The fourth step is to estimate the likelihood or risk of something going wrong. These four steps will identify serious potential problems, where the project is highly dependent on an outside factor which has a high probability of going wrong, over which the project manager has no control or influence. For example, a tree-crop agriculture project, such as macadamia nuts, might be highly dependent on the supply of a special insecticide which is not manufactured within the country. If there is a high risk of bugs and a strong possibility that the product cannot be acquired locally, then the project manager has a potential problem.

Once the manager has scanned the environment and identified the potential problems, the work is not over. The project manager must:

- Continue to scan for new elements that may affect the project outcome
- Monitor those elements identified as key ones to detect changes in them that will affect the project
- Evaluate the project's linkages with the key elements of the environment and manage these linkages

THE TURBULENT ENVIRONMENT

Emery and Trist[5] have coined the term "environmental turbulence" to relate the increasing pace of change in the world to the task of managing organizations, relationships, and dependencies. Indeed with current social, political, economic, financial, and technological upheavals the relevant environment is almost certain to change over the life of a 2- or 3-year project. Examples which come immediately to mind are the war in the Middle East, interest rates, microcomputers, and oil prices, all of which have surely influenced many projects.

In environmental analysis the project manager must also look at trends over time and must try to spot and analyze dynamic factors which do not yet or are only beginning to affect the project. Interestingly enough, this job is not as hard as it seems. The basic forces often are readily identifiable for those who want to see them. The curve of lower costs of microcircuits has been a steady trend for several years. The Ayatollah Khomeini had been preaching his philosophy for many years in both Iraq and Paris, yet most decision makers ignored the evidence.

MANAGING KEY ACTORS AND FACTORS

The means of managing key environmental factors involves both structural (organizational) and process strategies. The basic strategies involve:

- Collecting information on what is happening
- Identifying problems that cannot be controlled
- Developing influence and power in an attempt to "manage" key factors in the environment

Structural linkages include formal organizations, coordinating committees, and liaison managers. Process changes include plans, reports, and team building.

It is probably easiest to demonstrate the importance of environmental scanning with a specific example. Consider an agricultural project financed by the World Bank in a Latin American country. Among other things, the project seeks to:

- Support on-farm development of about 1000 ranches and small livestock farms through the provision of credit

Managing the international project environment

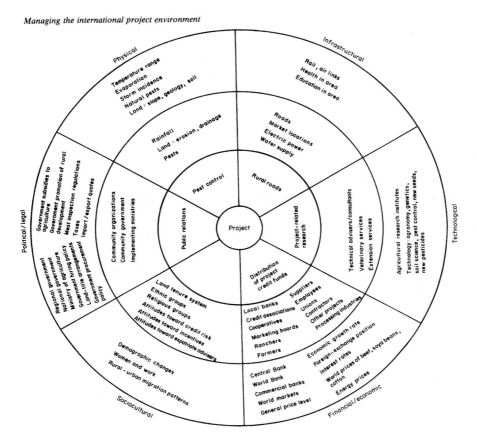

FIGURE 7.6 South American livestock and land-settlement project—environmental scan. Key actors and factors are indicated by boxes and circles, respectively.

- Consolidate the development of 16 land-settlement schemes in one region of the country by providing credit to 3000 small mixed farms (mainly cotton and soybeans) and constructing rural roads

Figure 7.6 demonstrates many of the elements in the project's environment. Undoubtedly there are others. The key actors are indicated by boxes and key factors by circles.

PROBLEM-SOLVING CONTINGENCY PLANNING

Once having identified a potential problem, what does the project manager do? The project manager should develop contingency plans for solving the potential problems.

This brings us to the fifth and last step in the environmental analysis. How can the project manager increase his or her relative degree of power over those key actors and factors on which the project is highly dependent? The project manager needs to establish linkages to increase the degree of control or influence. Can we sign a long-term contract? Establish a buffer stock? Open up an alternative supply? Obtain permission to import? What can the project manager do to increase control or influence?

The key point here is to try to understand the objectives and purposes of the key actors or stakeholders on whom the project's success is dependent. Stakeholders act to achieve their objectives. The project manager will increase his or her influence if he or she can understand the stakeholders' basic purposes and can relate them to desired outcomes.

SOURCES OF POWER FOR PROJECT MANAGERS

What sources of power does the project manager have available? In the diplomatic process of forging alliances, how can the project manager entice cooperation? The sources of interpersonal power have been defined by French and Raven[14] in five categories as reward power, coercive power, referent power or charisma, expert power, and legitimate power. Reward, punishment, and the power of formal authority of one's position are easy to understand. Expert power is the ability to influence through our specialized knowledge. We might add information as a sixth form of power.[15] Referent power or charisma is the influence that comes from persons wanting to identify with or to follow another person.

Effective project managers know their sources of power and use them to achieve their aims. Research on NASA project managers in the Moon Program by Gemmill and Wilemon[16] of Syracuse University showed that effective project managers used a wide variety of forms of power, while ineffective managers tended to rely on formal authority or the threat of punishment. In a new book on power and influence, John P. Kotter of the Harvard Business School says[17]:

> Exceptional managers understand that true organizational power is based much more on inspirational leadership than on executive rank and status. These managers have achieved their stature by establishing the power bases that are essential to the exercise of leadership.

What does this mean in practice for managers on international projects? When the project manager needs to gain influence with an important actor in the environment, he or she may have to use a conscious approach to using the available forms of power. Developing a personal relationship through social interaction is one obvious form of developing influence.

CONCLUSIONS

Successful project managers will do a good job of planning, scheduling, and controlling the implementation process. But they must also set up a process to scan the environment, to identify potential problems, and to try to establish power relationships which can help them to "manage" the key actors and factors on which successful implementation will depend.

This analysis of the key elements of the environment is not going to solve all of the problems, some of which are truly intractable. But at a minimum, it will give early warning of those problems.

The job of a project manager is not easy. But there is one great advantage. A project has a clear goal, with defined objectives of time, cost, and performance. The desire among the whole project team to accomplish that goal is one of the key sources of power for the project manager. One role of diplomacy is to widen the circle of those who feel they are members of the project team. One useful technique is the Project Launch Workshop, which has proven to be a very effective method for team building on new projects to increase identification and commitment.[18]

The workshop consists of one- or two-day sessions to go over the objectives and plans for the project with the key stakeholders. In some cases this could be a short session with high-level political figures, or it could be a longer session with the working team, or both. The purpose is to interest the key stakeholders in the objectives of the project in order to gain their support. Such support is essential in avoiding or solving the key problems in the project environment.

REFERENCES

1. World Bank, "Twelfth Annual Review of Project Performance Results 1986," Washington, D.C., 1987.

2. P. P. Schoderbek, A. G. Kefalas, and C. G. Schoderbek, *Management Systems—Conceptual Considerations,* Business Publications, Inc., 1975.

3. J. D. Thompson, *Organizations in Action,* McGraw-Hill, New York, 1967.

4. W. Evan, "The Organization Set: Toward a Theory of Interorganizational Relations," in *Approaches to Organizational Design,* I. D. Thompson (ed.)., University of Pittsburgh Press, Pittsburgh.

5. F. E. Emery and E. L. Trist, "The Causal Texture of Organizational Environments," *Human Relations,* p. 21, Feb. 1965.

6. H. Mintzberg, *The Structure of Organizations,* Prentice-Hall, Englewood Cliffs, N.J., 1979.

7. J. Galbraith, *Designing Complex Organizations,* Addison-Wesley, Reading, Mass., 1973.

8. J. Galbraith, *Organizational Design,* Addison-Wesley, Reading, Mass., 1977.

9. W. R. Dill, "Environment as an Influence on Managerial Autonomy," *Administrative Science Quart.,* pp. 409–443, Mar. 1958.

10. D. I. Cleland, "Project Stakeholder Management," *Project Management J.,* p. 36, Sept. 1986.

11. F. J. Aguilar, *Scanning the Business Environment,* Macmillan, New York, 1967.

12. W. E. Smith and R. Youker (eds.), "Organizing for Development," in *A Power Framework for Project Management,* PM Network, Project Management Institute, Washington, D.C., Aug. 1991.

13. W. E. Smith, B. A. Toolen, and F. J. Lethem, *The Design of Organizations for Rural Development Projects—A Progress Report,* Staff Working Paper 375, World Bank, 1980.

14. J. R. P. French and B. Raven, "The Bases of Social Power," in *Studies in Social Power,* D. Cartwright (ed.), Institute of Social Research, University of Michigan, Ann Arbor, 1959.

15. R. Youker, *Power and Politics in Project Management,* PM Network, Project Management Institute, May 1991.

16. G. R. Gemmill and D. L. Wilemon, "The Power Spectrum in Project Management," *Sloan Management Rev.,* pp. 15–25, Fall 1970.

17. J. P. Kotter, *Power and Influence,* Free Press, New York, 1985.

18. R. Satin, "Project Launch Workshops," World Bank, 1982.

CHAPTER 8
CAPTURING GLOBAL PROJECT INVESTMENT AND DEVELOPMENT POTENTIALS

William R. Yeack
Andersen Consulting

C. Joseph Smith
The Ohio State University

William R. Yeack is a Senior Manager with Andersen Consulting. He is associated with the San Francisco and Tokyo offices and specializes in large, complex global management information system projects. His expertise includes both the financial and the utilities industries and managing teams comprised of as many as 12 nationalities operating concurrently on three continents.

C. Joseph Smith is a Fellow of the Department of Anthropology, Ohio State University, and Senior Researcher with the Center for Creative Leadership, Cross-Cultural Leadership Program Development, Greensboro, N.C. Expertise includes using anthropological field methodology to develop global project-management applications.

INVESTING IN GLOBALIZATION

Tangible and intangible costs directly related to the failure of global projects are of such magnitude and complexity as to be inestimable. It is being reported that premature return rates for expatriates range from 16 to 40 percent, with estimated replacement costs ranging from $55,000 to $200,000 per manager. In the case of American expatriates in Japan, one estimate suggests failure rates as high as 80 percent.[1] These figures are based on long-term expatriate assignments. If shorter-term, project-based assignments could be included, the figures would be likely to reach another order of magnitude.

The sample population used to derive these figures is taken primarily from American-based multinationals.[*] Therefore the figures are relatively specific to American management experience in global activities. There might well be a high degree of national variability in the success rates of individuals on global assignments

---

*Citations used in this article to derive these figures were taken from the literature.[2–6]

if global comparisons were made. Or there may be little variability, indicating that the factors common to operating in a global context currently pose a significant degree of difficulty to all humans.

Since the figures measure failure on the basis of early return, it can be assumed that there are assignment performance failures that are not included because the individuals are not returned prematurely. At times a failure cannot be detected, much less prevented, in the field because the organization has made no provision for obtaining the information. This can cause a subsequent assignee to "fail" due to inherited circumstances. Organizational failure to insert on-site support and monitoring mechanisms into the global assignment process creates a serious gap, leading to individual and organizational failures in critical global activities.

As a consequence of the disturbing findings of such reports, American management has undertaken a complex of activities in the educational and organizational sectors to improve international effectiveness. In general, the international community is aware of an acute overall need to develop multiculturalism of its respective national populations. As the authors of a comprehensive study of Japanese society put it: "In the process of writing, a concern with how to develop multicultural perspectives has emerged as important."[7] Developing a multicultural perspective is certainly the key to global project-management success. In the long term, if talk of educational reform is implemented,[8] upcoming generations will theoretically arrive in the workplace with a multicultural perspective. In the short term, organizational training and development must compensate for deficiencies in the existing population.*

Since there is no way to estimate the real costs of failure in terms of individual and organizational damage or lost opportunity costs, it can only be assumed that when seeing the figures currently being reported, one must presume an exponential difference between a reported figure and the real cost of failure to achieve greater effectiveness in global contexts.

Perhaps there is another serious matter of "perspective" involved in global business issues. Reports focus on "failure costs" as losses. However, if global projects are approached in terms of investment rather than direct cost, and couched in terms of individual and organizational development opportunity, the issues take on a new light in terms of constructive potentials for return on investment.

It should also be noted that these issues are rarely considered from the individual perspective, so it is unreported how much might be lost (invested) at a personal level in the course of real-estate transactions, tax consequences, and various other personal expenses that may accrue from an international assignment. Although organizations offer compensation packages that attempt to minimize any loss at the individual level, numerous individuals have reported personal costs that were unforeseen by the individual or the organization. (At the other extreme, organizations can be too generous in compensating individuals on overseas assignment. If individuals on assignment display a standard of living too far above that of their local peers, the disparity can be an adverse factor in establishing effective working relationships with the host population. In some cases, assignees enjoy a standard of living higher than their normal situation at home, which may be offered as an inducement to the family members, or may be predicated on the fact that some countries require expatriates to employ a certain number of domestics.)

At the individual level, experience also runs the gamut from a high degree of personal accomplishment, satisfaction, and self-development to the extreme end of the

---

*Porter and McKibbin[8] provide the first major study of American management education since the Carnegie studies 25 years earlier. Major emphasis is given to recommendations for internationalizing curricula.

spectrum, including endangerment of life during local and international political upheavals, and psychological or physical distress of the assignee or family members. In a discussion of the intense feelings inherent in international assignments, it was stated[9]:

> Further, the incidents they discuss are more often negative than positive. People seem more able to bring forth remembrances of disappointments, anger, fear, and frustrations than remembrances of happiness and contentment. Still, people remain enthusiastic about their sojourns and frequently recommend that more people should live in other cultures.

If individuals are better prepared and receive more on-site support, a significant shift toward the positive end of the spectrum is likely to occur, which will result in better performance overall.

Global assignments are treated as distinct events rather than as an open-ended fact of organizational or professional life. Often, if the assignment does not require actually moving to a foreign country, there is no real preparation process in place to support individual performance. When project assignments do require moving to a foreign country, preparation is often an onslaught of new information rather than refinement of an incremental development process. Very few university or organizational training programs are currently international to a degree congruent with the realities of international activity. Most often the first experience is a trial by fire.

At present, preparation processes tend to be grafted onto existing training and development frameworks. In some cases international training is detached entirely from the organization and takes the form of hiring consultants or sending individuals to external international training programs. There has been a sudden proliferation of "experts" in intercultural training with considerable unevenness in quality. For example, one executive wrote[10]:

> Americans who want to do business in Japan realize that there are different modes of behavior and want to learn them. They therefore attend a myriad of seminars on understanding Japan and pay external consultants to teach them "proper behavior." Most are not getting much for their money. I recently attended a seminar given by an internal Japanese consultant of a major American company who offered the following advice: "Use business cards on every occasion." "Prepare to spend time and money on entertainment." "Be involved with after-hours social contact, dinner and drink." "Be aware of Japanese sensitivities."
> When asked for specific ways to do the above and to explain why each is important, the presenter had no answers.

While university and organizational training programs are endeavoring to embed internationalism in their training and development processes, it will take considerable time to thoroughly implement globalization.

There is also a real possibility that current management frameworks are unsuited to global activities to a degree that an altogether new paradigm must be established. One of the best accounts to bring into question the very essence of cultural factors in management, most particularly American management (which has been exported extensively), occurred in Edgar Schein's discussion of organizational culture's interrelationship with leadership[11]:

> I remember vividly a confrontation between a German manager and his American col-

leagues. The Americans were kidding the German about his formal heel clicking, head nodding to superiors, bowing, and shaking hands. They wondered how he ever got any work done what with all the ritual. One day he pointed out angrily to his "informal, open" American friends that when he went to his boss's office every morning he did click his heels, shake hands, and bow; and then, in a completely open manner, he told his boss the *truth* about what was going on in the company pertaining to that day's business. The implication was not lost on his audience that their surface openness often masked their concealing of critical information.

International projects often call fundamental assumptions into question and can thus highlight basic flaws in values, processes, systems, and structures that may be negative factors in domestic as well as international project performance. In this example the assumptions surrounding American informality and openness are challenged and communication and truth are highlighted. The mindless positivism that dominates American management is engendered in the incessant repetition of a maxim, "don't bring me a problem, bring me a solution," which stifles communication of issues within American management. It implies that management is not involved in solving problems and must only be alerted to situations if the individual has devised the solution. (Otherwise there is a tendency to "kill the messenger" who simply articulates a problem rather than its solution.)

Another fundamental value that is highlighted in global context, which profoundly affects organizational processes, systems, and structures, is competition. On a global project, team members who have been enculturated within a competitive paradigm can pose difficulties for themselves and the project if they cannot adapt to the cooperative requirements of a global project. The integrity, flexibility, and tolerance required to function effectively in a global context are not fostered within an excessively competitive paradigm.

The current population of project-management professionals operating in an international context must make creative use of the resources that exist in the here and now. If adequate resources do not exist within the organization, consideration must be given to purchasing them from outside resources. (Even though it may be difficult to determine the quality of such resources, something will be better than nothing.) Resources determined as requirements to address the multicultural issues of global projects should not be viewed as part of project expense, but rather are individual and organizational development investments. Initial planning stages of international projects must clearly define and articulate to top management *all* the resource requirements of the project that are critical to success. Given the status of most individuals and organizations currently involved in global activities, top management should think that there is something wrong if a project has determined no such requirements.

These special needs include preparation training requirements of project management and team members. Global projects entail two major role components—the guest and the host. Most current preparation programs are focused on the guest role. Yet the population representing the host role is often actually larger and just as critical to the success of intercultural relations. It may be up to the project manager to ensure that the host culture team members are prepared to receive their guests.

In one case, host team members of a multicultural project representing 12 different nationalities were interviewed concerning the potential for improvements in global project management. Individuals repeatedly pleaded for training in how to work with their "guests." In this particular case the individuals specifically stated that, of all the cultures present, their greatest difficulty was in relations with Americans on the team. Individuals from the other cultures also articulated particular difficulties in dealing with Americans. This would suggest that the Americans (with individual exceptions, which the team members themselves pointed out) were perhaps least prepared as a group to operate effectively in this multicultural context. While the Americans were singled out to some extent, it should be pointed out that the Japanese members of the team also experienced a high degree of difficulty based on the frequency of early returns of team members.

This highlighted the degree of difficulty that might be added to the assignment if the host country is a developing country and the guest team members are from industrialized nations. Even though the country involved is a vacation spot for the Japanese, there was difficulty in adapting to the disparity in hygienic standards during the course of a long-term stay. Physical illness and psychological distress (most often, failure to recover from culture shock) requiring early return home thus caused repeated, unexpected personnel disruptions during the course of the project. In the opposite circumstance, with guest project members from developing countries being hosted in industrialized nations, there may be other types of difficulties, such as overcoming reactions rooted in difficulties of transitioning back into native context, or perceiving treatment in the host country as condescending. People often remark that they feel they are treated "as if they were stupid" or "not human."

Project managers operating in a global context need to gain a working knowledge of world systems theories and the development background of the countries in which they operate or from which they derive project team members. In the case of many developing countries, the industrialized countries frequently represent previous colonial or occupational powers, and the project manager may encounter deep-seated (often unconscious) hostilities on the part of guests or hosts. Disparities between countries can have both obvious and subtle effects on the project. Understanding the kinds of "everyday forms of resistance" that can become second nature can be extremely useful in project management.[12]

An anecdote from an individual from a developing nation who was on assignment in Japan was insightful from the standpoint of the subtlety of such differences as well as the microchanges that occur through such global events. While making our way to the subway in Tokyo, the individual explained that, in his country, littering was normal behavior. Shortly after arriving for work in Japan, the individual became conscious of his personal behavior and realized the extent to which it deviated from local behavior. He went on to say that he not only curbed the behavior in the host country, but had also stopped the behavior when at home. It is not difficult to imagine this individual passing the behavioral change on to close associates at home. Over time and with increasing prevalence of global activities, such microchanges achieve a cumulative effect.

The current population of project-management professionals with international experience is the most valuable resource (asset) an organization has toward globalization efforts. The investment return on that experience should be captured within the organizational and individual development process. Unfortunately the potential return on this investment is diminished or lost by ineffective repatriation processes that too often lead to the experienced individual's departure from the organization. Even if repatriation support exists within an organization, it is often not directly linked to organizational development that will lead to better selection and preparation for subsequent global managers.

An investment focus also highlights the fact that global projects provide significant generalized development opportunities because their logistical issues and the intensity of psychological, social, and cultural processes provide worst-case scenarios. In reality, project managers do not have to cross any geographic borders to find ambiguity and tough tradeoffs associated with real contradictions and the need for acquiring skills in negotiation and gaining credibility with "strangers."

All project managers must manage self, manage risk, manage to acquire and disseminate information, manage to communicate with diverse populations, manage dynamic human systems, manage disparity, manage tangible and intangible resources, manage formal and informal systems, manage politics, manage across organizational and functional boundaries, and manage ultimately to get something done. In an international context all these elements are pushed to an extreme. It would seem highly pragmatic to assume that generalized lessons can be derived from these types of projects.

In one case, a British architect who practiced in Africa for 10 years stated: "Managing to get anything done at all was quite a feat, much less getting anything done to code." Or, as one project manager completing an assignment in the Amazon put it when asked what other lessons he had derived from the experience: "Other things? I don't know, so many I can't describe them, but overwhelmingly a sense that if I could survive this, nothing would ever hurt that way again."[13]

Even though the strategies devised in such global contexts often involve generalizable principles for negotiating, communicating, and problem solving in nonglobal situations, most organizational processes currently do not capture the lessons these experiences provide.

It is vital that global project managers communicate their needs and findings to top management, and inform human resource processors within their organizations. Project managers often wait for human resources to come forth with programs rather than serving as initiators or catalysts for programs that meet organizational and individual needs. Human resource functions often become isolated from organizational realities because human resource professionals do not venture directly into the field of organizational activity. (It is rather interesting that projects can reach team sizes of over 100 and have no full- or part-time on-site human resource support, yet an organization that reaches 100 employees will generally have a human resource specialist.)

One way to bridge the disassociation of project management and human resources is to use global projects as development projects for internationalization programs. We have found that it requires minimal resources to overlay a development project on an organizational project. By having a human resource–related professional (or team) conduct a development project using actual projects, they gain first-hand on-site experience of the realities of such events. This enables them to develop more directed and content-specific programs for the organization. Project team members can make significant contributions to the development project with minimal expenditure of time and effort, and will generally derive immediate benefit from participation. When designed properly, this type of development project addresses the multicultural issues of particular concern to the project manager during the course of the project. We have found, in fact, that this is a notably more effective means of internalizing training and development than can be achieved through conventional training and development delivery mechanisms that are classroom-based.

In the case of global projects, project managers must currently accept responsibility for ascertaining the level and degree of support that will be required for multicultural issues during the entire life cycle of the project. In some cases a concurrent development project may be the most effective solution. However, in all cases realistic planning and estimating for global projects incorporates cross-cultural factors that will affect communication, training, time lines, and resource requirements, including the possibility of a formalized full- or part-time team member role of cultural broker.

Cultural brokers have been employed in critical negotiation processes with considerable success, and this approach can be applied to project management. The role of cultural broker is to surface and resolve the cross-cultural issues that can lead to breakdowns in the overall process of achieving goals. Because the role of cultural brokers is overtly defined by the parties involved, the individual assigned the role can function as a communication clearinghouse and take proactive measures to mitigate potential problems. Usually the individual selected for this role has specific expertise or skills such as bicultural and bilingual training or experience. It is becoming increasingly common for more than two cultures to be involved in critical organizational activities, which can make it more difficult to find the necessary cultural and language skills embodied in one individual. (The academic discipline currently providing the most focused training for individuals who can broker a multicultural context would be anthropology.) Too often the project manager undertakes to play the role of cultural broker (along with all his or her other roles) and is unqualified or cannot devote adequate attention to the task.

While many projects may not be able to justify a concurrent project or the use of a dedicated specialist on a full-time basis, project managers can often make provisions to insert mechanisms to address cross-cultural issues at critical points in the project life cycle, and to make more effective use of project team members who have skills and experience that contribute to the intercultural aspects of the project.

When using project team members as "cultural brokers," the project manager should be aware of a potentially serious pitfall in this approach. Individuals who have functioned in this capacity report potential difficulties in relationships with other team members who perceive them as having special access to management and project information due to their bicultural and bilingual skills. These individuals are often included in meetings and are privy to higher levels of information than their peers in rank. Also, if the role is not formalized, these individuals can have difficulty in gaining the credibility and authority that are necessary to effectiveness.

Early stages of the project team development process are always critical, and it can be difficult to recover from getting "off to a bad start." This is intensified in global projects. Guests are often undergoing culture shock, which can be of varying degree and duration, and hosts are forming their opinions while their guests are coping with adjustment. It is during this stage that informal communication processes on the project can be compromised to the extent that they never adequately form. Global project teams can stabilize in such a way that dialogue is never achieved and team members can operate for long periods of time on the basis of gross national stereotypes rather than achieving refinements leading to effective working relationships among individual team members.

It is very natural for individuals from the same culture to spend time together, especially if they are in the guest role. Compatriots provide comfort from the strain of communicating in a second language and adjusting to a foreign environment, or adjustment to hosting foreigners within the environment. However, this can lead to cliques that fractionate the project along national boundaries in a dysfunctional manner. Project managers should be alert to their own tendencies toward this type of behavior since the example they set will often be followed by the project team. In the early stages of the project, investing time and resources in mechanisms such as special activities that encourage individuals to foster relationships across national boundaries can be important to preventing unhealthy levels of clique formation, which can develop into countercultures that undermine project effectiveness.

HOW DOES A PROJECT MANAGER PREPARE FOR A GLOBAL PROJECT?

At the crux of this issue is the fact that no general framework is established for managers who must operate globally that will allow them to adapt readily to the specifics of a given project assignment.

There are currently 168 nation states. More importantly, there are an estimated 3000 to 5000 distinct ethnic groups (depending on the criteria used to define ethnicity). It is not possible to address all the possible permutations of ethnic and national diversity that can comprise a specific project. An attempt will be made to highlight some general points that can be adapted by project managers to their specific experiences.

One of the first tasks of a global project manager is to understand the project context in terms of the obvious national differences, but also in terms of any specific ethnic issues that may be important factors. For example, in one case a number of years ago an individual assigned to establish a new plant in Africa mistakenly hired individuals from warring tribes. A series of murders on the shop floor ensued. Needless to say, the individual had to acquire a better knowledge of the local environment and modify hiring practices accordingly.

The specific mix of the project team will be a significant factor. The proportional representation of each country, which country is host, the distribution of decision-making powers, the control over resources, and the history of global project experience at the national and individual levels will be unique to each project. A global project may have as little as one member from a different country. If that single member is a guest, holds ultimate decision-making power, and controls project resources, it is quite a different circumstance than a project with team members who may have been assigned early in their careers because they have specific expertise or need a first overseas experience.

Real and perceived disparities between countries can affect project relations in overt and subtle ways. Project managers from dominant countries must be particularly sensitive to issues related to "saving face" when working as guests. Relations with peers and superiors in the host country can be highly sensitive and delicate matters of negotiation and diplomacy.

One example of how seemingly insignificant issues can be matters of symbolic import occurred when a project manager from an industrialized country arrived for work in a third-world country. The guest manager had considerable power and control over resources, many times that of his host country's structural peers (and superiors). Within a few days of arrival the manager noted that there was never any toilet paper in the office bathrooms. He asked his peers to explain this circumstance and received a reply to the effect that the country's dire financial straits were such that if toilet paper were placed in the stalls, it would immediately be stolen. The individual informed his peers that he intended to provide toilet paper on the floors on which his team members worked and would be responsible for the expense. He proceeded to call a meeting with the janitor, instructed him to install tiny locks on the dispensers, asked for the janitor's home telephone number, and informed him that he would personally call him at home if he ever found the dispensers empty. The individual had unwittingly embarrassed the individuals of the host country and had caused his peers (and superiors), who could not match the working conditions for their subordinates, to lose face by making toilet paper an issue. It might have been wiser to keep a roll of toilet paper in his desk drawer and to instruct other guest team members to do likewise rather than highlighting the disparity in economic conditions of the two countries involved in the incident. Little things often acquire disproportionate symbolic import in such contexts.

As project managers are asked to operate in contexts further and further removed from past experiences and personal comfort zones, organizations must provide increasing levels of support for development and operating conditions. However, if the organization has not yet caught up with the realities of international operations, it is up to the project manager to compensate on behalf of the project team.

At the inception of a project, top management has made decisions (presumably on a long-term as well as on a short-term basis), entered into agreements (which they have reason to believe are beneficial to the enterprises or governments involved), committed resources (hopefully not "bet the farm" without good reasons), made preliminary plans, fixed some goals with their counterparts, superiors, and subordinates (hopefully the right individuals are involved in a cooperative and nonhostile atmosphere), and selected you as project manager for a global project (hopefully after very careful consideration of your character, your background, your strengths and weaknesses as they know them, your potential for development, and your family or personal circumstances as they relate to the demands of the global project).

During the process of assuming the position of global project manager, the individual involved must assess the situation intensively. An investigative journalism framework is always useful to any assessment process, particularly at this stage: who, what, where, when, how, and why?

If the project manager has not been involved in the early stages of the project development process, it is important to link to the project history by means of intensively interviewing all the key individuals. Ascertaining who was involved, what transpired, what is to be done, where the project fits in the larger context, where geographically (spatially) it is to take place, when it must be done, how top management expects to get the project completed in terms of the resources required (and available), why the organization needs to do this project in the first place, and why they picked you to do it, is critical to initiating effective project preparation.

Once the project manager has performed this investigation, he or she should have a clear idea of the vision, mission, and politics of the project from the top-management perspective.

During the assumption of the project-management role stage, the project manager and top management should explore the potential for beneficial by-products from the project—what else can be derived from the resource expenditures? It is common to overlook the true potential of many projects, particularly when possible benefits relate to longer-term, less tangible, and more opportunistic outcomes than the present project goals will affix. For example, as suggested earlier, at this stage it may be decided to achieve organizational development through a concurrent human resource project. At the very least, the project manager should assume that his or her own self-development will be an outcome of the project experience.

Global projects are peak performance situations requiring all the skill, experience, persistence, stamina, and resourcefulness the individual can muster. Ruthless self-examination and assessment of the situation are ongoing components of the global project-management process.

Since global projects often entail extensive travel, relatively long assignments overseas, or hosting foreign team members, the project manager's fitness for this aspect of the project is a serious issue. Physical fitness, personal circumstances including family, character, and personality, as well as managerial skills and experience can be tested and strained to the limits during a global project.

Assuming the project manager has ascertained critical information during the preliminary investigation stage concerning the national and ethnic cultures to be involved, he or she must now consider how to prepare him- or herself, the team, and the overall environment, and determine the impact the cross-cultural factors will have on the planning, implementation, and conclusion stages of the project. It is not possible to assign fixed values to cross-cultural factors, nor will it ever be likely that more than rough rules of thumb can be derived. This aspect of the project is highly situationally specific. One thing, however, is certain: if the project is global, cross-cultural factors will be operating on the dynamics of the project, and some allowance and contingency must be made for that fact. The fact that the project is global must be formally incorporated into project risk management processes.

We find that this risk factor is best managed by inserting cross-cultural management measures at the planning stage. Along with the other gateways that are usually established for the project process, assessment activities and action plans for addressing the cross-cultural factors are as important as technical checkpoints. Formalizing the cross-cultural aspect of the project demonstrates management's concern for this issue, sends appropriate signals to team members and top management, and provides motivation for project team members to address and successfully resolve the cross-cultural issues that will be encountered during the project life cycle. Formalizing a project team member role such as cultural broker, conducting a recognized organizational development project, overlaying a cultural risk management methodology on any formalized methodology in place for managing projects, conducting periodic training and development, or, at the very least, adding cross-cultural performance as a personnel assessment factor, and including cross-cultural requirements in personnel goal-setting processes will serve to highlight the global nature of the project. Project managers should clearly communicate performance expectations and responsibilities for cross-cultural factors to all members of a global project team.

If this is the first global project for the project manager, he or she must carefully consider the management framework he or she has employed in the past. Is it adequate to the new assignment? What adjustments and additions to management skills must be made to meet the demands of a global project? As mentioned earlier, the fundamental paradigm under which the project manager is accustomed to performing must be questioned because that paradigm is inherently culture-bound at a number of levels—national, ethnic, industry, corporate, functional, technical, educational, and familial. The project manager, as well as each individual team member, is the product of a complex composite of enculturation processes throughout the course of a lifetime. Enculturation provides individual members of the culture with a set of working assumptions that makes the day-to-day business of life more predictable and efficient. It is particularly not an easy task to make the most unconscious of those assumptions conscious.

A working set of dimensions is used by anthropologists to make the investigation of fundamental cultural assumptions manageable. The primary listing of dimensions includes:

1. The nature of time (past, present, near future, or future orientations; discrete increments or fluid; short, medium, or long time units)
2. The nature of space (private or public)
3. The nature of human nature (basically good, mixed, or basically bad)
4. The nature of human activity (being, controlling, or doing orientations)
5. The nature of human relationships (groupism or individualism; authoritarian/paternalistic or collegial/participative)
6. The nature of reality and truth which forms the basis for decisions (moralistic or pragmatic)
7. The nature of human relationship to the environment (reactive where environment is dominant, symbiotic, or proactive where environment is dominated)

The possibilities of each dimension can be seen as a continuum running from one extreme to the other. Each culture has a distinctive pattern, and the consequence of orientations will be reflected in the behavior of individuals and will have an effect on global projects.

Time, for example, is truly relative. Many examples of cross-cultural difficulties relate to this dimension. Project planning must incorporate the potential differences in how cultures relate to the very notion of "deadline" and must often establish a realistic compromise for scheduling purposes. Decisions made in terms of a long time frame often take longer than short-term decisions. The very distinction between short-term and long-term can be notably different. Subtle differences such as notions of punctuality for work or social events can create difficulties if individuals do not adopt a more flexible attitude in cross-cultural situations. In some cases the logical direction of shift on the continuum can be obvious. In relation to time, members of a culture with a short-term, fixed-increment orientation will likely experience nothing but frustration if they do not make a shift in the direction of longer-term and more fluid notions of time when guests in such a culture or if they are managing a team dominated by this orientation.*

The project manager should reflect on these dimensions throughout the life cycle of the project. This framework is a useful tool, and it can be a powerful explanatory device for cross-cultural situations. However, it is not a panacea and should be used with judgment. Positions of cultural members regarding these orientations can readily shift based on specific situations, and this tool can exacerbate the already difficult issues of stereotyping that are inherent to global projects.

Flexibility is required on the part of everyone involved, and it is also important not to allow "culture" to become an excuse for the status quo. The project manager must be realistic about the degree to which he or she can change management style or force others to change. It is often more important for the manager to be congruent in his or her behavior than performing some miraculous chameleon transformation. Congruence provides team members with a measure of predictability, which is desirable. Successful global project management makes wide allowances for differences and seeks to capitalize on the diversity of a global team rather than homogenizing it into a bland amalgam or force team members to parody a dominant cultural style.

Successful global project managers demonstrate a high level of integrity. Individuals in any culture can spot a lack of integrity, and all the language skills and knowledge of local customs in the world will not compensate for such a lack.

It is up to the project manager as a key individual, who is responsible for organizational resources, for the outcome of the project, and for the individuals involved in producing the outcome, to perform a ruthless self-examination. Global projects highlight strengths and weaknesses at both the organizational and the individual levels. The intensity of the psychological, social, and cultural processes encountered in global projects cannot be adequately described in oral or written form. Individuals must ultimately find through experience the heights and depths of their personal capabilities. In terms of personal growth and development, a global project is a unique opportunity.

Conventional wisdom suggests that experience is indeed the best teacher. Although learning from experience is a constant human activity, individuals are highly variable in their ability to adapt quickly and resourcefully to experiential lessons, and to gain experience through external and self-educational processes. Many individuals would like to think they do not make the same mistake twice. In a global project it can be difficult to even know whether you made a mistake at all if you make serious mistakes in setting up formal and informal communications within the team.

In the case of global project management there is the issue of language acquisition. It is a well-known fact that, for most, the capacity to acquire new languages is a func-

---

*It is not possible within the scope of this chapter to give full treatment to the dimensions cited. Comprehensive discussions of dimensions can be found in the literature.[11,14–19]

tion of previous exposure to other languages, innate ability, and age. Can you learn a language through tapes, intensive courses, immersion, private tutoring, or at all? If it is to be a multicultural team, does it even make sense to divert personal resource[9] to learning a language beyond a courtesy level?

Determining the degree of language skills that will be required in the project is an essential task for the project manager, and can become complex when the project is multicultural. It was notable that during the Gulf War, which was certainly the epitome of a successful global project outcome, General Schwarzkopf made sure that the American liaison he assigned to each country in the coalition spoke that language.

In another case, project management had tutors come to the office on a regular basis to teach project members both English and Japanese. Project team members were given the opportunity to enhance their skills in either of the languages, which would provide them with a permanent skill that would be useful to the individual and the organization beyond the project at hand. In the same case, the project manager (an American working in Japan) learned that there were important cultural rules, involving when it was even appropriate for him to use the Japanese he had learned. Although he had expended an intense effort to acquire rapidly a working knowledge of the language, it was made very clear that he gave offense if he attempted to use it in certain situations. He had to make use of a translator for the sake of appearances. So much for the general assumption that others will always be impressed and delighted when someone speaks their language.

As mentioned earlier, team members who perform ad hoc roles of translator for project management can experience difficulties with peers and superiors who are lacking language skills. Project management should be alert to this potential for disruption. The team member involved is not likely to directly communicate this difficulty to project management. If project management is not aware of its role in eliciting this delicate information, formal and informal communication can eventually be compromised through the relational difficulties experienced by the translators. When language skills on a project team are highly variable, team members can feel they are not gaining full access to information, miscommunications can occur easily, and individuals who function as translators can be perceived as having special access incongruent with their rank level. This phenomenon was reported by individuals who were from group-oriented cultures where it is highly undesirable to stand out or not stay within one's place. The difficulties these individuals were experiencing in being set apart from their group were extremely serious to them.

This also generally highlighted the difficulties inherent in adapting reward, recognition, and promotion procedures to multicultural teams. When project management is from an individualistic culture, it is easy to unwittingly create difficulties for team members from group-oriented cultures. This was characterized as the "crab problem," meaning that in a basket of crabs, if one crab attempts to climb out of the basket, the other crabs will catch hold of him and pull him back. Project managers must keep in mind that when the project is completed, individuals may have to live with the consequences long afterward. Individuals from group-oriented cultures with different approaches to career progression who may be singled out for special responsibilities and promotions (especially due primarily to language rather than technical or functional skills) within the team may be put in an extremely awkward, stressful position that may damage rather than enhance their careers within their own cultural context.

Project management must carefully consider the effect language will have in the project. Special resources may be required for oral and written translations, meetings may take longer, decisions may take longer, written communication may require translations which can cause delays, and miscommunications may cause errors which cause delays. The cumulative effect of a complicated linguistic context for the project should be considered in project estimating.

Formal communication always receives attention from project management. Determining who is involved in official distribution channels of written communication, establishing meeting schedules, and determining meeting participants will be affected by the cultural factors of the project. In global projects it is likely that increased meeting frequency is desirable, particularly at the early stages of the project. The project manager must also take into consideration the cross-cultural etiquettes of who participates in meetings. Meetings are an important resource in a global project for fostering cross-cultural communications, preventing miscommunications, and encouraging a healthy informal system to develop among team members. Written communication can be seriously affected by variability in language skills. Long documents may require summarization, or written communication may require more reinforcement through oral mechanisms. Often individuals who can communicate orally in a foreign language will have a struggle to read in that language. As a consequence, written communications may not be read adequately, or may not be read at all. More emphasis on visual communications is required in global projects. When presenting information in written form, incorporate graphics, and when presenting information orally, do not hesitate to stop and draw a picture.

An often overlooked tool for cross-cultural communication effectiveness is the use of metaphors. Metaphors will not only help clarify communications, but can also help ascertain divergence in cultural values or perspectives on a situation. Metaphors could well be the most powerful aspect of human language. Given the role of metaphors in such processes as transferring knowledge, cultivation of intimacy, generating meaningfulness, understanding, knowledge and truth, equilibration to new experiences, categorization, and problem solving, the link between metaphor and power does indeed seem to exist. The global project manager would do well to consciously and conscientiously tap into that power by cultivating his or her own metaphorical communication skills and paying close attention to the metaphors of those on the team.

In one international organization all training and development programs were produced and conducted in the United States. One criticism from international participants particularly focused on the overuse of typically American metaphors throughout the course of the training programs. Not only is the prevalence of sports and military metaphors in American management a barrier to effective communication, but in many cultures these metaphors are fundamentally offensive. Among the terms mentioned specifically in this case were sports metaphors such as on deck, at bat, off base, root for, in the batter's circle, and slam dunk. Many countries do not engage in the sports from which these metaphors are derived. In one document analyzed, European and Asian personnel resorted to food-related metaphors: decisions were cooked, concepts were swallowed, people were ripe for assignments. In another case, a merger between an American and an Asian company was being investigated for determining the degree of success perceived by both parties. To the Americans it was a merger; to the Asians it was a marriage. Needless to say, a marriage in which one of the partners does not realize there is a marriage is likely to be a troubled one. This metaphorical discrepancy between the two companies aptly characterized the difficulties they were experiencing in this situation.

Being conscious of metaphorical habits that may not be effective in a cross-cultural context, being alert to the metaphors used by others and adapting to their metaphorical framework, and being conscientious about devising appropriate metaphors for the rhetorical aspect of project management communication is invaluable to improving cross-cultural communications. (Project managers are often involved in delivering rhetoric intended to inspire the team, and they make use of metaphors for these purposes more often than they realize.)

Humor is another area of human communication that has not received much attention in global management literature. A sense of humor can often be important to establishing relationships and negotiating delicate situations. Guests and hosts on a global project should not take themselves too seriously and realize that they will make silly mistakes. It is important to realize that notions about humor vary across cultures, and sensitivity to differences may indicate the need to curtail joking that might give offense. In one Asian country, for example, teasing is prevalent. Virtually everyone has a nickname, and there is often a humorous anecdote behind the acquisition of the nickname. Teasing is often used to soften a "scolding," or to exhort someone to perform properly. This teasing might be taken too seriously by a member of another culture and create difficulties, or, once appropriate levels of relationships have been established, it might be an important tool for project management in dealing with members of this culture.

Humor can be a consideration when selecting project team members. If the project manager has observed that an individual has a particular penchant for bigoted or exceedingly vulgar humor, there could be serious questions concerning how the individual will perform in a cross-cultural context, and it may be necessary to point this out as part of the individual's preparation. (Individuals often use humor as a permission to express deep-seated prejudices.)

Humor can also function as an early warning system to project management. Oral humor and graffiti that might begin to appear in the environment can indicate that stress and tension levels are escalating beyond acceptable levels, or that segments of the team are experiencing dysfunctional relations. Humor (and "play") can also be used by project management to reduce tensions and defuse aggressions.

Informal communication can be the greatest challenge to a global project manager. By definition informal systems are not "managed." However, they can be guided (most often through the process of setting an example), and this can be a particularly critical area to global project success. Project management can attempt to provide situations that are conducive to forming the relationships important to developing an informal system. Special events centered around the lunch hour or after work can be helpful in the early stages of team building, and can be useful in celebrating project milestones and maintaining informal communication and morale throughout the course of the project. It is important to consider cultural differences in the amount of afterwork activity considered appropriate. In some cases the host country may not normally engage in a significant degree of afterwork socializing. This may make it more appropriate to schedule events during the lunch hour. However, project team members who are guests may need the support of social activities initiated by project management, even though they may be from cultures in which afterwork socializing is not the norm. These events provide good opportunities to highlight the cultures present on the project in a positive manner. Team members from the various cultures can be asked to plan an event that highlights their country—its customs, food, clothing, entertainment, and so forth.

Finally, the project manager must take seriously the fact that he or she is the leader and role model in forming effective formal and informal communication within the project context. Team members will take their cues from project-management behaviors. The leading and role modeling aspects of managing any project are always important. Whether from the perspective of personality or of technical virtuosity, this aspect distinguishes the project manager from other team members. When the project is global, it is more likely that weaknesses in these areas will contribute to project failure, and strengths to success.

CONCLUSION

We have highlighted some pragmatic issues related to undertaking the management of global projects that we find are not often mentioned in literature on the topic. Until training and educational processes establish a multicultural framework that enables future generations to operate more naturally in a global context, global project management is primarily a matter of individual self-education and learning through direct experience.

Our commentary is based on the field development approach recommended in the text, which we find is the optimal solution at this stage of global project-management activity. If specialized resources are not available, the project manager can establish an ancillary activity for the team, requiring them to meet at specific stages in the project life cycle, report the issues and solutions, and perhaps develop a final report at the conclusion of the project that maintains a focus and a dialogue on the multicultural aspects of the project.

Understanding cultural issues requires individuals to stretch themselves intellectually and behaviorally. Models that oversimplify cross-cultural issues or descriptions of a given culture are not adequate to most real situations. Figure 8.1 provides an example of the kind of model more appropriate to the task. Mouer and Sugimoto derived this multidimensional model to depict significant aspects of Japanese society.* However, many of its elements can be adapted to other national contexts; the overriding point is that it is multidimensional.

National and ethnic culture is a complex, multidimensional phenomenon with an extensive historical basis. The dual function of culture is to provide internal integration and external adaptation for groups of people. Culture both constrains and facilitates human behavior at macro as well as micro levels of activity.

Our findings show that there are fundamental qualities that transcend cultural boundaries. Successful project managers possess a high level of personal and professional integrity. We find that the qualities inherent to success in global contexts also transcend time. They are aptly summed up in a book on deportment written over a century ago[20]:

> Politeness is benevolence in small things. A true gentleman [or lady] must regard the rights and feelings of others, even in matters the most trivial. He respects the individuality of others, just as he wishes others to respect his own. In society he is quiet, easy, unobtrusive, putting on no airs, nor hinting by word or manner that he deems himself better, or wiser, or richer than any one about him. He never boasts of his achievements, or fishes for compliments by affecting to underrate what he has done. He is distinguished, above all things, by his deep insight and sympathy, his quick perception of, and prompt attention to, those small and apparently insignificant things that may cause pleasure or pain to others. In giving his opinions he does not dogmatize; he listens patiently and respectfully to other men, and, if compelled to dissent from their opinions, acknowledges his fallibility and asserts his own views in such a manner as to command the respect of all who hear him. Frankness and cordiality mark all his intercourse with his fellows, and, however high his station, the humblest man feels instantly at ease in his presence.

While a global project manager most certainly has considerable homework to do in

---

*Mouer and Sugimoto[7] provide an exceptionally comprehensive discussion of Japanese society, a good example of the level of knowledge that should be acquired by global project managers when preparing to deal with another culture.

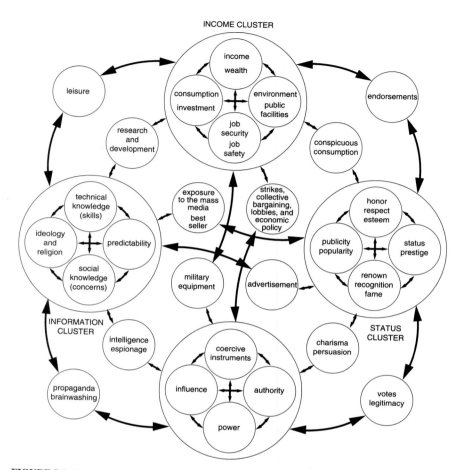

FIGURE 8.1　Interactive model for societal rewards.

acquiring information pertinent to the context in which he or she must operate, perhaps, at the personal level, world-class project management may, after all, be simply a matter of class. In a real sense, the guest-host aspect of global projects calls for the most basic precepts of hospitality that are common to all cultures. All cultures have standards of "good manners," largely for the purpose of facilitating successful negotiation of familiar as well as unfamiliar situations and positive interactions with strangers as well as friends and loved ones. While the level and intensity of global activities occur on an unprecedented scale, global activities are nothing new to humans, and all cultures have devised a set of ideals for human behaviors that transcend cultural boundaries.

REFERENCES

1. H. B. Gregersen and J. S. Black, "A Multifaceted Approach to Expatriate Retention in International Assignments," *Group and Organization Studies,* vol. 15, pp. 461–485, Dec. 1990.

2. T. E. Adams and N. Kobayashi, *The World of Japanese Business,* Kodansha International, Tokyo, Japan, 1969.

3. J. S. Black, "Workrole Transitions: A Study of American Expatriate Managers in Japan," *J. International Business Studies,* vol. 19, pp. 274–291, 1988.

4. L. Copeland and L. Griggs, *Going International,* Random House, New York, 1985.

5. K. F. Misa and J. M. Fabricatore, "Return on Investment of Overseas Personnel," *Financial Executive,* pp. 42–46, Apr. 1979.

6. R. Tung, "Selection and Training of Personnel for Overseas Assignments," *Columbia J. World Business,* vol. 16, pp. 68–78, 1985.

7. R. Mouer and Y. Sugimoto, *Images of Japanese Society,* Kegan Paul International, London, 1986.

8. L. W. Porter and L. E. McKibbin, *Management Education and Development: Drift or Thrust into the 21st Century?,* McGraw-Hill, New York, 1988.

9. R. W. Brislin, K. Cusher, C. Cherrie, and M. Yong, *Intercultural Interactions: A Practice Guide,* Sage Publ., Beverly Hills, Calif., 1986, p. 248.

10. R. H. Reeves-Ellington, "Using Business Skills and Cultural Skills for Competitive Advantage in Japan," in preparation.

11. E. H. Schein, *Organizational Culture and Leadership,* Jossey-Bass, San Francisco, Calif., 1985, p. 47.

12. J. C. Scott, "Everyday Forms of Resistance," in *Everyday Forms of Peasant Resistance,* F. D. Colburn (ed.), M. E. Sharpe, New York, 1989, pp. 3–33.

13. M. W. McCall, Jr., M. M. Lombardo, and A. M. Morrison, *The Lessons of Experience: How Successful Executives Develop on the Job,* Lexington Books, Lexington, Mass., 1988, p. 15.

14. N. J. Adler and M. Jelinek, "Organization Culture: Culture Bound?," *Human Resources Management,* 1986.

15. G. Hofstede, *Culture's Consequences: International Differences in Work-Related Values,* Sage Publ., Beverly Hills, Calif., 1980.

16. G. Hofstede, "The Cultural Relativity of the Quality of Life Concept," *Academy of Management Rev.,* vol. 9, no. 3, pp. 389–398, 1984.

17. G. Hofstede, B. Neuijen, D. Daval Ohayv, and G. Sanders, "Measuring Organizational Cultures: A Qualitative and Quantitative Study across Twenty Cases," *Administrative Science Quart.,* vol. 35, pp. 286–316, 1990.

18. H. W. Lane and J. J. Distefano, *International Management Behavior: From Policy to Practice,* Nelson, Canada, 1990.

19. E. H. Schein, "Innovative Cultures and Organizations," in *Management in the 1990s,* Sloan School of Management, Massachusetts Institute of Technology, vol. 90s, p. 88-064, Nov. 1988.

20. J. H. Young, *Our Deportment,* F. B. Dickerson & Co., Publishers, Detroit, Mich., 1879, p. 23.

CHAPTER 9

A RISK ANALYSIS OF CURRENT PROJECTS IN THE PEOPLE'S REPUBLIC OF CHINA

Mary Anne F. Nixon

Western Carolina University,
Cullowhee, North Carolina

Mary Anne F. Nixon received the J.D. degree from the North Carolina Central School of Law Evening Program in 1984. She is an assistant professor of business law at Western Carolina University, where she teaches in the undergraduate Business Administration and Law program and the graduate Master of Project Management program. Her professional development activities have been directed toward analyzing current legal issues and their impact on the business world. These research interests are divided into two distinct categories: the legal aspects of project management in the national and international arenas and the conduct of business with the People's Republic of China. Dr. Nixon also specializes in teaching and conducting workshops in project management and applied contracting principles.

INTRODUCTION

In the last 14 years the Chinese have made overtures to technologically and industrially advanced countries in an effort to initiate business investment in China. Project managers are or should be extremely interested in these activities, since the development of any joint venture or the establishment of any business in China is a project in and of itself, and may lead to additional follow-on activities involving project management expertise.

The Chinese themselves have come to respect the principles embodied in project management. In a survey of 436 Chinese middle managers conducted as recently as 1988, 32.4 percent reported they could "get to the top" most quickly through expertise in general management and administration. This is a major change in Chinese attitude, with only 27.3 percent perceiving politics as the way to advance. At the time of the survey, the government had pledged to further support the reform by promoting workers, by placing equal emphasis on job competence and Party ideology, and by improving management training and techniques.[1]

This new Chinese attitude toward foreign business activities and investments in China has resulted in many successful activities and a rapid growth of the Chinese national economy, particularly in the private sector. In the spring of 1989 the Tiananmen Square incident occurred. Despite Chinese assurances that business would continue as usual while the government dealt with its internal problems, western countries, including the United States, levied heavy economic sanctions. The perceived level of risk in investing in China appeared too high for many U.S. businesses, whose investment strategy became very cautious.

Despite major differences in values and culture, a process of adaptation appears to be developing for both the Chinese and the countries they are approaching for technology and capital. In November 1990, Foreign Minister Qian Qichen visited President Bush, ending the U.S. post-Tiananmen policy banning high-level meetings with China. More recently Mr. Li Peng, the Premier of the State Council of the People's Republic of China, met with President Bush at the United Nations. These and other meetings can be interpreted to signify the easing of tensions between the two countries and the reestablishment of the process of accommodation. Those countries and companies wishing to initiate or continue business transactions with China at this time will have to assess current risks and potential future gains carefully if they hope to benefit from joint business ventures in China.

China's eighth Five-Year Plan (FYP, 1991–1995), announced in December 1990, was not as specific as the seventh Five-Year Plan, which envisioned a 40 percent increase in the total volume of imports and exports.[2] However, the current goals should require continued purchases from the United States, especially in energy, telecommunications, electronics, and transportation, even in light of the latest reforms. American investment in China should also generate new opportunities for U.S. sales of machinery, raw materials, technology, and services. With accelerating opportunities in the Chinese investment scene, many U.S. companies are continuing to investigate the formation of various types of business relationships with the Chinese.

RISK MANAGEMENT CATEGORIES

Risk management is a systematic process in which risk factors are identified, evaluated, and planned for. It is a formal approach to sorting through numerous and varied uncertainties in order to identify critical issues and provide contingencies for dealing with the difficulties that may arise. A risk analysis of a project conducted as a joint venture in China must be conducted with an understanding of the potential benefits from the viewpoint of both parties involved, since the potential benefits provide the only practical framework for determining whether the undertaking is worth the risks involved for both parties.

In classifying risks according to cause rather than effect, various risk categories can be identified according to the perceived ability to manage those risks effectively. Risks can be classified as (1) external and unpredictable, (2) external and predictable but uncertain, (3) internal and nontechnical, (4) technical, and (5) legal sources of project risk. Four of the five functions are discussed in this chapter as they relate to conducting projects in China. Technical functions specific to a project's technology and design are beyond the scope of this chapter, but must be considered in any risk analysis process designed to evaluate the potential for success of any specific project undertaking.

External Unpredictable Occurrences

What exactly do recent developments mean to businesses interested in continuing or commencing projects in China? The current political and economic situation can be classified as an external unpredictable variable, in which unanticipated government intervention took place as a result of an unexpected social side effect of changes in the economy. These events must be interpreted in light of the history of rapid economic growth that has taken place during the last decade.

- In late 1978 the Chinese Communist Party (CCP) took steps to reform the economic structure of the People's Republic of China. These economic advances, referred to by the Chinese as "market socialism" or by foreign observers as "open-door policy," were gradual and continuing under the leadership of Deng Xiaoping, the 80-year-old elder statesman and head of the new Party Advisory Commission. These goals and reforms were accompanied by a decentralization of the Ministry of Foreign Economic Relations and Trade (MOFERT), which is the powerful central economic decision-making organization that exercises power over many investment decisions in the provincial governments.

- On October 11, 1987, the State Council promulgated its new "provisions for the encouragement of foreign investment." These 22 articles, which were effective on the date of issuance, reduce the costs involved in investing in China by lowering different kinds of fees, use preferential measures already accorded to the "special economic zones" (SEZs) and 14 coastal cities to guide investment toward export-oriented or hi-tech enterprises and toward inland Chinese cities, and give foreign corporations more operating autonomy.[3]

- In the plenary session held in late 1987, Deng stepped down from his seat of power, Li Peng became Premier, and Zhao Ziyang was elected General Secretary of the Chinese Communist Party Central Committee. The western world waited to see if the assurances given for the continued implementation of Deng's liberal policies supporting foreign business investment would materialize. That policy continued, resulting in an overheated economy with excessive market demands, leading to high rates of inflation. Ordinary citizens benefited from the reforms and the open-door policy, but they also faced major problems of rising prices and inflation, corruption in some parts of the Party and government, and (from the Chinese perspective) inequity in the distribution of goods and income.

- China attempted to slow its economic growth to 10 percent or less and cut its 1989 investment in fixed assets to 20 percent of the actual investment in 1988. Certain non-production-oriented projects, such as the construction of hotels and theaters, were halted. Banks were ordered to stop providing loans to these projects, and departments under state control ceased to provide necessary raw materials, electricity, and construction licenses. However, those projects using imported and foreign funds received favorable protection, especially those which had reached world advanced technological levels or those in energy, transportation, telecommunications, and raw materials.[4] Yet when the workers and students experiencing the improved materialistic and entrepreneurial advantages of the economic reforms publicly demanded more reform, more freedom, and more infrastructural support, the "hardliners" Deng Xiaoping (now chairman of the Central Military Commission) and Premier Li Peng moved quickly to terminate such ideas: Tiananmen Square, 1989. Zhao Ziyang was removed from office.

- The 1991–1995 Five-Year Plan is to retain current economic reforms while looking for methods of retaining economic activity under centralized government control.

The underlying problems persist: widespread unemployment, growing income disparities, continued support of state-owned corporations,* and dependence on government subsidies in the form of cheap rent, food, and transportation. Much of China's economy operates outside the plan, but its chief function is to allocate credit and approve major construction projects. These projects are widely expected to be located in central China in an effort to allocate evenly the prosperity experienced by the development of the coastal cities and special economic zones. Foreign investment is still welcomed with continued government selectivity and the stated goal, "...based in the truth of Marxism...prices, finance, taxation, banking, planning, investment and labor and wages...should center around the goal of establishing a new economic system."[6]

- Deng Xiaoping ordered the Communist Party to uphold his market reform policies with the statement that "anyone who attempts to change the basic line...will be overthrown because the common people will not permit him to do so." He offered a six-point instruction Central Committee Document no. 2, which urges "more daring" economic experiments, seeking faster growth, and urging the party not to overemphasize political stability. Deng has emphasized the fact that the increasing influx of foreign investment will not lead to the downfall of Communism in China. He was critical of attempts to label policies socialist or capitalist. "The correct approach is to judge whether something is helpful to developing the productive forces...strengthening the nation and improving living standards."[7]

An atmosphere of great anticipation exists in China today as the yet unannounced date of the 14th Party Congress draws near. In this Congress, China's policies in relating to global changes which have taken place in Europe and in Russia will be decided. The battle between the economic reformists led by Deng and the conservatives led by Chen Yun continues. The decisions to be made are so critical to China's future, Deng has emerged from his position of controlling from behind the scenes to a highly public attempt to use his immense political strength in support of continuing his fundamental policy of economic reform into the next century. According to standard procedures and barring any emergencies, the 15th Party Congress will not be held for 4 years. This essentially will be the last Congress the "old guard" from both sides will be able to attend. Therefore each side is attempting to put its people in the major positions of leadership. The World Bank and the Overseas Private Investment Commission (OPIC) as well as all businesses contemplating investment in China will be closely scrutinizing the decisions and personalities emerging from the 14th Party Congress.

Essentially these current events can be interpreted to mean that project managers and companies anticipating projects in China must expect that the demands for reform, freedom, and infrastructural support that climaxed in Tiananmen Square are likely to continue and have their effects on projects in China. The level of risk of such interruptions is likely to depend on the type of project to be undertaken, but risks exist for all projects to some degree. It is essential for companies to consider such political movements in evaluating the level of the risk in terms of the type of project undertaken, that is, those projects seen to contribute to production and export industries have a relatively low risk of being interrupted by such activities, but those dealing with internal services such as hotels and theaters are likely to be at high risk.

---

*"Some 60 percent of state industries already are in the red, the government is going broke, and China's grizzled leaders are too unpopular to bail the government out through their usual practice of force feeding state bonds to workers."[5]

External Predictable but Uncertain Occurrences

The external predictable but uncertain risk management category includes operational functions such as the market risk of changes in availability and cost of raw materials, maintenance, the demand for goods or services, competition, consumer demand or rejection, fitness of the product for the purpose intended, and the willingness of buyers to honor purchase agreements. This category also includes environmental and safety issues, currency changes, inflation, taxation, and social impact issues. The recent student and worker demonstrations discussed as part of the "external unpredictable" category could be placed in this category as well, depending on just how predictable the demonstrations are seen to be.

According to a recent article in *Fortune,* the fastest growing economy on earth, Guangdong Province's Pearl River Delta, has averaged a real annual growth of approximately 15 percent. Unlike with the economy in the rest of China, Tiananmen Square caused only a momentary slowdown in this area, which produces nearly 5 percent of the total industrial output and 10 percent of China's exports. This area is being used as an example of China's impressive economic potential. Per capita income today is roughly double that of China as a whole, and city leaders expect it to increase more than threefold by the end of the decade, from $800 to $2800 annually.[8]

In China, as in most countries, taxation is a broadly accepted fact of life. The Chinese taxation system, however, is very complex and difficult to understand. Potential foreign investors often express great frustration and confusion in trying to determine the costs of establishing a business in China. Questions include what taxes apply, who levies them, when and where collection is made, and what exemptions or reductions apply. The answers to these questions can be explained by identifying the types of taxes, the tax organization, the tax preferences, and the procedures of tax administration.

Generally speaking, tax treatment for foreigners and foreign investment is more favorable than that applicable to domestic enterprises. Despite the tax preference offered, some foreign investors still hesitate to initiate projects in China. According to some American businesses, the reasons for the declining interest in China as a competitive investment market are more complex and extensive than the simple project taxation classifications. Problems include soaring costs, seemingly arbitrary tax and tariff levies, inadequate legal protection, and price gouging. In addition tight foreign exchange controls, limited access to the Chinese market, bureaucratic foot-dragging, lack of qualified local personnel, and lack of infrastructural supports for such essentials as water and electricity have been discussed as reasons for this loss of confidence in the Chinese market.[9]

An additional complication is the negotiation phase of the joint venture. Although many companies will say that the major problems with the Chinese do not come into play during the negotiation phase of the venture, there are an equal number of experienced executives who will disagree. Negotiations have been known to be prolonged for 3 to 6 years.* In China, business is based on personal relationships. "Who you know matters more than anything else. The Chinese never say 'no' to a proposal, but you can end up being disappointed. You have to go behind the scenes to get the real

---

*Gareth C. C. Chang, 43, president of a subsidiary called McDonnell Douglas China, was born in China, educated in the United States, and an experienced manager of risky projects. He spent more than 6 years negotiating the contract for his subsidiary. Ironically he, too, had inside connections with the current leadership of the country. His father was the aide of the Nationalist Chinese leader who was engaged in negotiations with the Communist Chinese. His counterpart for the Communists was Deng Xiaoping.[10]

answer...."* Chinese and many foreign parties also have differing views on what the agreement means once it is actually reached. Americans, for example, believe that the give and take process of negotiation culminates when both parties have maximized their positions. The Chinese, however, believe that the agreement simply binds the parties to a common endeavor in which each side will make continuing demands on the other.[11] The Chinese have a phrase which characterizes the misconceptions of many American companies which are oblivious to the realities of doing business in China: "*tong chuang yi meng*—same bed, different dreams."[9]

The dreams of foreign investors of the future will be influenced by the current political reform. Li Peng and others have recreated a centralized economy by trimming private enterprise efforts and using laws and administrative measures to subdue, but not stop, freewheeling provinces.[12] It would appear that the external predictable risks of dealing with joint ventures in China are relatively high. It is predictable that the conditions of social change, labor issues, political uncertainties, foreign currency exchange, and other issues may change during the project, with potentially significant impacts on the project's schedule, cost, and performance. Knowing that these difficulties are likely to occur, the project manager may or may not be able to develop appropriate contingency plans for dealing with such issues. Evaluating the risk involved must therefore depend on the specific nature of the project and the necessary level of interaction with the local Chinese economy.

Internal Nontechnical Occurrences

The three main concerns within this category of risk analysis are schedule, cost, and cash-flow interruption. The traditional scheduling problems include delays due to regulatory approvals, labor shortages and stoppages, material shortages and late deliveries, unforeseen site conditions, start-up difficulties, unrealistic scheduling, and lack of access. In particular, lack of adequate transportation appears to be a major problem for many projects in China. Inadequate transportation affects the ability of workers to commute to work sites, causes delays in the delivery of resources and materials needed to conduct project efforts, and contributes to the difficulty in coordinating activities across different locations.

Efforts to improve these problems are under way in many areas. Lockheed has recently set up a joint venture at Guangzhou airport to maintain the China Southern Airlines. GEC Alsthom of France and AEG Westinghouse, a German-U.S. joint venture, are currently bidding on a $350 million contract to provide cars, signals, and related equipment for a new subway system.[8] It is clear that the Chinese recognize these issues and are taking steps to correct them. A 20,000–25,000-km highway network is being planned to link the entire nation, with expressways and high-speed first- and second-grade motorways connecting all major cities, industrial centers,

---

*Quote from Desmond Wond, chief operating officer of the New York–based World Trading and Shipping Ltd. (WTS), in an interview by Daniel Southerland, *Washington Post* Foreign Service, "U.S.–China Joint-Venture First Under U.S. Law." The newly created Good Earth Development Corp. is the first U.S.–Chinese joint venture established in the United States under Delaware law. One of the Chinese partners will be the Kanghua Industrial Co. Ltd., whose chairman is Deng Pufang, son of China's Senior Leader, Deng Xiaoping. The point Pufang is making concerning the negotiation of contracts is that he has an insider's advantage by virtue of his family connections to ease all aspects of dealing with his country.

transport hubs, and port cities. This network will not be completed, however, for some 50 years.[13]

Costs are particularly difficult to predict in the Chinese environment, especially in the area of labor and business operating expenses. The Chinese bureaucracy has created a labor problem for foreign companies. The Foreign Enterprises Service Corporation (FESCO) has in the past screened workers politically before assigning them to foreign companies, and set their wages higher than in most Asian economies. High and unpredictable levels of compensation plus the other operational costs of doing business in China can easily result in a non–competitively priced final product. As one commentator observed, "it costs Nike more to make shoes in China than in Maine, and if Chinese-made Peugeot trucks were exported, they would cost more than those made in France!"[9]

Companies were horrified to discover that the People's Republic was among the world's most expensive business centers. Maintaining a manager in Beijing cost approximately $250,000 a year, which was 40 percent more than the same position would cost in London. Apartments were scarce and rented for $6000 or more a month, payable a year in advance. A light meal without drinks was $9. So little office space was available that companies rented hotel space for business use. However, after the troops opened fire on the demonstrators in Tiananmen Square, tourists disappeared from Beijing. The top hotels, such as the China World Hotel, envisioned as an elegant refuge for refined tourists, wealthy traders, and globe-trotting tycoons, is renting rooms for about $50 a night.[14]

This is not to say that the Chinese alone are solely responsible for the difficulties plaguing these joint-venture projects. Foreign investors also contribute to the poor business environment by spending too little time in strategic thinking, too little money, or both. Under the mistaken belief that all they had to do was provide technology and token management instead of strategic planning and firm, long-term commitments, many companies created their own problems. The much publicized failure of the American Motors Jeep joint venture, for example, stemmed as much from inadequate financial planning and unrealistic capitalization as from the problems encountered in China itself. Many joint ventures can be compared to arranged marriages, with the central bureaucracy providing the multinational corporation groom with a venture partner bride. Mismatches are not surprising when the capabilities and potential of both parties were either unknown or misjudged elements.[11]

While there are significant internal nontechnical issues to deal with, however, conditions do seem to be improving. Perhaps we can look forward to a cooperative partnership in which the Chinese do all that is possible to reduce predictable internal nontechnical uncertainties. This would certainly be the ideal situation since the major advantage of predicting a difficulty lies in taking corrective action before the difficulty becomes insurmountable.

Legal Occurrences

China's highest government agency dealing with legal affairs, the State Council's Legal Bureau, has quickened its pace in enacting new rules and regulations and has actively supervised the enforcement of existing ones. Most of these regulations focus on economic activities and seek to facilitate China's cooperation with foreign investors. Provisions concerning foreign investment take the form of laws. Almost all the tax laws concerning foreign investment and foreigners take the form of laws passed by the People's Congress. The domestic tax laws are considered temporary regulations passed by the State Council or the Ministry of Finance. They do not take the formal form of a law. In order to promulgate a law, a draft of prospective legislation is issued from the Legal Bureau at the state level to the Legal Bureau at the provincial level, requesting commentary and soliciting opinions and input from interested parties. Generally this information is sought from the relevant government agencies, for example, the provincial tax bureau, for tax legislation.

Foreign businesses can voice their opinions through these agencies. These opinions are then directed back to the Legal Bureau, who in turn forwards them to the Financial and Economic Committee of the People's Congress. The Legal Bureau is active in supervising and inspecting this process to ensure that the existing regulations are being implemented in the factories. In addition to these reforms, the bureau also arbitrates any dispute between a provincial government and a central ministry over a conflict in provincial and central regulations on the same issue.[15]

The People's Congress enacted a patent law in 1985* and a copyright law in 1987. The Joint Venture Law, established in 1979, granted legal entity to joint-venture enterprises, enabling them to sue or be sued. Joint ventures are subject not only to the Joint Venture Law, but to all other Chinese legislation as well. Thus in the absence of specific provisions in a joint venture's contract of association or in the Chinese Joint Venture Law, Chinese general law will apply to the organization and operation of the venture.[17]

The enterprise income tax was reformed in 1991† when the two existing systems of taxing enterprise income were combined into one. The Income Tax Law of the People's Republic of China for Enterprises with Foreign Investment and Foreign Enterprises, commonly referred to as the foreign taxation system reform, was an effort by the National People's Congress to improve the investment climate in China for international companies. As a result of this reform, all forms of foreign investment, including the joint venture, the cooperative enterprise, and the wholly foreign-owned enterprise, enjoy the same preferential tax treatment.

Another positive change is the fact that the Supreme People's Court is addressing the problem of the system's failure to carry out the court's decisions in economic cases. According to its figures, 20 percent of all judgments in economic cases were not executed. While debtor inability to pay was the primary reason the payments ordered were not made, intervention by administrative leaders also existed. Administrators had, in some instances, ordered banks not to cooperate with the court

---

*The Chinese patent law went into effect April 1, 1985. The office receives about 5000 applications from foreigners a year, representing 50 countries. Japan ranks first, followed by the United States and Germany. Germany has signed the only existing agreement to date on patent cooperation projects and has contributed financially to the construction of the patent office itself. China has sent 150 specialists to Germany to receive training in the field, in addition to that received from German specialists sent to China to lecture and give technical assistance. Foreign patents granted cover machine building, electronics, chemicals, metallurgy, and instruments.[16]

†Between 1980–1981 and 1991 two separate income tax laws applied to foreign investment in China: the joint venture income tax law of 1980 and the foreign enterprise income tax law of 1981. The problems with this system stemmed from the fact that different forms of foreign investment were taxed differently. Joint ventures were taxed at a flat rate of 30 percent, whereas cooperatives and wholly foreign-owned enterprises were taxed at progressive rates, ranging from 20 to 40 percent. In addition there were great discrepancies in tax preferences: a joint venture scheduled to operate for more than 10 years regardless of industry type was exempted from the income tax for the first 2 profit-making years. Thereafter it was taxed at one-half the statutory rate for another 3 years. On the other hand, only those wholly foreign-owned and cooperative enterprises engaged in agriculture, forestry, and animal husbandry were exempted from income tax for the first profit-making year. They were then taxed at one-half the statutory rate for another 2 years. The latter form of business was at a disadvantage, and complaints were expressed concerning the different tax treatments.

when ordered to transfer money from the account of the losing party to that of the winner. Although the court lacks adequate personnel to see that its decisions are fully implemented, it is significant that the court is now making a concerted effort to enforce its decisions and to prosecute those persons who refuse to comply under existing criminal statutes.[18]

Other complaints of multinational corporate investors are being addressed individually by the appropriate authorities. The CCP Central Discipline Inspection Commission has acknowledged that the problems of kickbacks, bribes, or "gifts" have become rampant in some areas and departments, and the commission has expressed its strong dissatisfaction. It has decided that "only by cracking down severely on bribe-takers...is it possible to ensure the healthy development of China's economic contacts with foreign countries and the smooth progress of reforms and the implementation of the open policy." Under this decision, any Party member who has taken a bribe in violation of the law will be expelled from the Party regardless of rank. Further, Party members who have taken bribes in any form must confess their offenses or they will be given Party disciplinary actions.[19]

The much publicized problem of nepotism still exists. As in the United States, no laws exist against this very prevalent practice. It was not until 1985 that workers had the ability to request a change of employment from the position in which they had been placed. Today factories wishing to recruit workers file an advertisement for jobs listing the skills required. If interested and qualified, workers can go to the applicable unit or the factory. Today, in a new spirit of competition and freedom, the demand is that family members still being placed in high-paying and prestigious job positions by their high-ranking political official relatives be removed, and that qualified persons, regardless of their political ties, be given the chance to compete for the positions based on ability.

Attempting to create and make the national laws work while they are being developed and while the free-enterprise system is being expanded is a challenging task at best, but making the national laws work can set an example for the provincial and local laws, which must also be consistent with free-enterprise activity. It is the local and provincial laws, however, that are likely to become a major issue in the foreseeable future. With the major provinces and cities able to institute their own approaches to supporting free enterprise, it is clear that there will be a wide variety of legal implications to deal with in any large project conducted in China. The shortage of legal talent qualified in Chinese law and available to business investors is another major complicating factor. While changes in the Chinese national laws are widely documented and publicized in the world press, provincial and local laws tend to be major unknowns which must be researched carefully. The risk is therefore likely to vary widely, depending on the particular locations, the technical nature of the project, and the availability of legal assistance qualified not only in the national but also in the provincial and local laws affecting the specific project. Even where such talent is available, the cost of obtaining it must be carefully analyzed.

CONCLUSION

China's new open-door policy aimed at developing joint-venture projects with foreign investors has clearly created some difficulties for both the Chinese and the business people involved in the projects. Nevertheless the Chinese are actively attempting to manage their activities and control the difficulties faced by investors. How extensively these reforms will change the investment climate in China for multinational corporations in the months and years to come remains to be seen. The mayor of Tianjin, Li Ruihuan, has urged foreign business people to take advantage of the opportunities being offered by the new reforms. In doing so, however, the mayor has also reminded us that China should not be viewed only as a market. China, too, has its interests. "While supplying loans, technology or equipment, you are expected to help us conscientiously and earnestly raise the competitiveness of our products and gain a footing in the international market, or make our products genuine import substitutes, or buy our other products, or give orders for product processing....In a word, we would like to have two-way trade, cooperation to mutual benefit."[20]

In response to the concern of many potential investors in the Chinese market that the "open door" will ultimately slam shut in our faces, the following assurances were made: "In the struggle against bourgeois liberalism, there was a saying about 'retreat' in the reforms and the open policy...the process of reforms and opening to the outside world may be fast or slow at different times, but in general, it undoubtedly should be further expanded. Retreat is absolutely out of the question."[21] Deng Xiaoping stated: "The way we have gone is the right way. A policy cannot be changed as long as it is supported by the people."[22] Tiananmen Square, the reaction to a period of "slow" implementation of economic reforms, strongly demonstrates that there is pressure to maintain the pace of new venture project development in China. Indicators since that time have been for continued and increased encouragement to foreign investors to participate in the economic development of China.

Potential investors may wish to take the following issues into consideration:

- The political developments surrounding the 14th Party Congress in 1992
- New and continued social and economic policies: policies encouraging foreign investment on the provincial and state levels
- Human resource distribution and the general education level of workers as they relocate from the northwest to the east coast and southwestern China
- Transportation and infrastructural support available
- Enforcement of newly implemented laws protecting intellectual property and highly developed technology
- Mutual benefit of the project to all countries involved: the specific nature of the project and level of interaction with the Chinese economy

In recent telephone interviews with three executives directly involved in current joint-venture projects or negotiations to develop such projects in China, a consistently positive attitude was noted toward the prospects of continuing the Chinese business relationships in the future. Even though these corporations were experiencing some of the same problems discussed in this chapter, the executives attributed the difficulties either to inexperience or to trying to develop projects in a period of transition and reform. The future of foreign investment in China remains optimistic overall. Business projects in the People's Republic still have a vast potential and are likely to become even more rewarding for both the investors and the Chinese as their respective commercial and political interests are recognized, and as the various organizations involved have the opportunity to identify and mitigate the risks involved in working with one another. Identifying and clarifying these risks, however, is a first essential step in the evaluation of a potential joint project if the foreign business is to survive, much less flourish, in China.

REFERENCES

1. "Chinese Managers Face Major Changes," *Research Beat,* University of Michigan School of Business Administration, p. 5, Fall 1988.
2. *Asia 1986 Yearbook,* p. 127.
3. Zhang Yuejiao, Department of Treaties and Law, MOFERT, "Better Terms for Foreigners," *Intertrade,* p. 57, Nov. 1986.
4. "Rectifying the Economic Order Is Our Top Priority—Zhao," *China Daily,* p. 4, Oct. 29, 1988.
5. "A Yen for Yuan Makes Mr. Yang a Man of 'Millions'," *Wall Street J.,* p. 1, col. 4, Dec. 27, 1990.
6. "China's Leaders Approve a Plan for Next 5 Years," *Wall Street J.,* p. 4, col. 3, Dec. 31, 1990.
7. A. S. Tyson, "China's Deng Rallies for Reform," *Christian Science Monitor,* p. 2, Mar. 10, 1992.
8. F. S. Worthy, "Where Capitalism Thrives in China," *Fortune,* pp. 71–75, Mar. 9, 1992.
9. J. P. Sterba, "Great Wall: Firms Doing Business in China Are Stymied by Costs and Hassles," *Wall Street J.,* p. 12, col. 3, June 1986.
10. J. Sullum, *Fortune,* p. 47, Aug. 18, 1986.
11. Steven R. Hendrys, "The China Trade: Making the Deal Work," *Harvard Business Review,* p. 78, July-August, 1986.
12. "Hu's Death, May's Big Events, and the Future of Reform," *Business China,* vol. 15, no. 8, p. 57, Apr. 24, 1989.
13. Gao, Jin'an, "Highway Network to Link Nation," *China Daily,* p. 3, Mar. 8, 1989.
14. "Yugoslav 'Tourist' Flood into China, Pack Their Bags after They Get There," *Wall Street J.,* p. A6, col. 1, Dec. 27, 1990.
15. "Legal Bureau Cracks the Whip," *China Daily,* p. 3, col. 1, July 13, 1987.
16. "Applications for Patents up by 54%," *China Daily,* p. 1, col. 7, July 15, 1987.
17. Chinese Joint Venture Law, 1979, article 2.
18. "Implement Law Court Decisions," *China Daily,* p. 4, col. 1, July 16, 1987.
19. "Party to Expel Members Who Take Bribes," *China Daily,* p. 1, col. 6, July 4, 1987.
20. "Tianjin to Raise $1Bn for Growth," *China Daily,* p. 1, col. 1, June 9, 1987.
21. "Effective Publicity Vital to Promote Reforms—Zhao," *China Daily,* p. 1, col. 1, July 10, 1987.
22. Xie, Ming-Gan, "China's Open-Door Policy and Sino-American Trade," *Business Horizons,* p. 11, July-Aug. 1987.

INCREASING CORPORATE COMPETITIVENESS ON GLOBAL MARKETS BY PROJECT ORIENTATION

The use of projects as temporary organizations for the performance of complex tasks contributes to the organizational flexibility of companies. In order to increase the competitive edge by successful project performances, companies require an explicit project-management culture.

Fritz Scheuch and Arnold Schuh analyze the specific demands of international marketing projects in Chap. 10. The cooperation in projects between the base organization of global companies and their local distributors is analyzed.

John Tuman recognizes in Chap. 11 that the key to survival in the global marketplace is speed, flexibility, innovation, entrepreneurship, and a strategy that enables the organization to dominate its market. He suggests that organizations dealing in the global marketplace must band together in projects and in networks to meet the needs of a changing world. Through projects and networks the values, beliefs, and ideals of divergent groups of nationalities build an organizational culture that transcends national interest and traditions.

J. Kessel and J. Evers describe in Chap. 12 the approach of implementing project management in the decentralized Philips concern in The Netherlands. They stress that the project-management subculture has to be related explicitly to the different company cultures within the organization.

In Chap. 13 Georg Serentschy analyzes the specific challenges for personnel management in small project-oriented companies which cooperate in large international space programs.

Regarding the explicit development of a project-management culture within the industry, John Adams considers project management as a profession. He notes in Chap. 14 that all of the characteristics of a profession are in place and that major global companies are beginning to recognize project management for what it is, proactively developing their people to work in the project-management environment. Adams concludes by noting that the project-management profession must work toward increased recognition of the need for professionalism.

CHAPTER 10

INTERNATIONALIZATION AND MARKETING PROJECTS

Fritz Scheuch and Arnold Schuh

University of Economics and Business Administration,
Vienna, Austria

Fritz Scheuch holds an M.B.A. and a Ph.D. degree from the University of Economics and Business Administration in Vienna, Austria. Since 1979 he has been a full professor at the university's Department of Marketing. He is also academic director of the postgraduate program "Project Management in the Export Industry," offered jointly by the University of Economics and the Technical University of Vienna. At present he is rector of the Vienna University of Economics and Business Administration. He is the author of several books and papers on industrial marketing management, service, and export marketing.

Arnold Schuh holds an M.B.A. and a Ph.D. degree from the University of Economics and Business Administration in Vienna, Austria. Since 1986 he has been assistant professor at the university's Department of Marketing. He was a visiting professor and lecturer at the College of Business and Economics, University of Kentucky, United States, during the fall term of 1990. The focus of his teaching and research is on international and strategic marketing management and marketing channel management.

INTRODUCTION

In the last decade the competitive environment of the world markets has changed fundamentally. Major developments have been the rise of the Pacific Rim economies, the decision of the European Commission to create a single European market by the end of 1992, the opening of eastern Europe, and the emergence of global markets. All these developments have greatly affected international business decisions, such as the direction of geographic extension or the location of production facilities, as well as sourcing and marketing decisions.

Among these developments the emergence of global markets had the strongest impact on international business (Levitt[1]). In industries such as consumer electronics, pharmaceuticals, and soft drinks the leading companies started to follow a global strategy. These companies have discovered that the same basic need can be met with a global approach (Keegan,[2] p. 33). Similar demographic and economic trends have been leading to a convergence of consumption patterns in industrialized countries. Technology-driven products overcome cultural barriers and shape consumer behavior (as, for example, computers). In many cases the global markets are created more or less by marketing effort, such as the soft drink market. The emergence of distributors operating on a regional or global scale and of regional communication media (cable, satellite, television) forced international marketers to integrate their activities across different countries.

Considering these changes in the environment of the world markets, the terms "international marketing" and "internationalization" took on a new meaning. In the past, international marketing was primarily associated with either exporting from a home base or, in the case of multinational corporations, with the establishment of local operations and with localized product and marketing concepts developed by these subsidiaries.

The evolution of a new company type, the global company, is a very recent phenomenon. Global companies seek to serve basically identical markets which appear in many countries of the world. Global marketing deals with the interconnection of the many country strategies and the subordination of these country strategies to a global framework.[2] This framework ensures that marketing activities are focused on global market opportunities and that they take advantage of this fact by leveraging products, systems, and experience.

The concept of internationalization as a process in which firms gradually increase their international involvement, expressed as steps in the "establishment chain" and as the number of foreign markets covered, has to be adapted to the new realities of international business too (Johanson and Wiedersheim-Paul[3]). The importance of company ownership as a means to control operations in foreign markets is declining. Successful examples in the textile, casual wear, and consumer electronics industries show that by licensing, by contract manufacturing, or by original equipment manufacturing agreements control can be exercised to a similar extent.

New forms of cooperation such as strategic alliances are on the rise in international business. In an alliance, two entire firms pool their resources, usually complementary ones, in a collaboration, and both partners expect to profit from the other's experience (Jeannet and Hennessey,[4] p. 291). Cooperating with foreign partners is a way to ensure distribution access and allows the quick buildup of a worldwide presence that is necessary to market a new product with a short product life. The shortening of product life cycles in high-technology industries requires simultaneous product introductions in world markets to recover investments. The traditional "waterfall approach" has become obsolete as markets in these industries no longer develop sequentially and is replaced by the "shower approach" (Keegan,[2] p. 31).

Considering these changes in international business project management has also become a new focus. While in the past the export of capital goods (such as power stations or railway construction) or the setup of foreign operations (such as sales subsidiaries, local production operations, and joint ventures) were the dominating project management tasks in international marketing, now the internal coordination of international marketing activities and the cooperation with partners are typical and predominant. The traditional organizational structures are not adequate for the management of the increased complexity in international marketing. The traditional pyramid organization with its one-dimensional geographic, product, or functional organization structure has reached its limits in international business (Roux-Kiener[5]). As a response to their experiences in managing international businesses, international companies have developed hybrid structures, matrix structures, and task forces, coordinating groups or project teams. In particular, the project team approach provides the kind of coordination that is necessary to facilitate cross-national marketing activities (Bartlett and Ghoshal[6]). In addition, an increasing number of marketing tasks is no longer performed by company personnel. Partners assume certain tasks in distribution, product development, or promotion on the basis of cooperation agreements. Control by directives is ruled out in such a network organization. Project management is a key element in managing these complex relationships.

This chapter attempts to show the characteristics of international marketing projects, to define organizational types of international marketing projects, and to give examples of how project management is used in international marketing practice.

MANAGEMENT OF INTERNATIONAL MARKETING PROJECTS

Project Management

Project management as a management function can be analyzed from different perspectives:

* Following the traditional approach, project management is the grouping of tasks for a temporally limited job. These tasks include time planning, cost accounting, monitoring of project steps, project documentation, and so on.

* Considering the organizational aspects of project management, project tasks can be analyzed with regard to their integration into the organizational structure and with regard to the distribution of tasks. Project management can be introduced as an ad hoc committee, a task force, a project team, or a business unit in matrix organizations. The form of organizational integration affects the division of competence and responsibility between project managers and functional managers, the reporting lines of project members, and the way conflicts are resolved in the organization.

* A recent development is the systemic-evolutionary view of project management. Projects are seen as social systems that consist of members of the base organization and of external project members, such as representatives of stakeholder groups. Projects are part of a complex organizational and environmental system. The relevant project environment influences the objectives and progress of the project. Projects are shaped by evolutionary processes. The subject of the project and the implementation are variable. All project elements have to reckon with deviations; they have to be flexible and adaptable.

Adopting the systemic-evolutionary view, projects are characterized as follows (Gareis,[7] p. 41):

* "A project is not just defined as a unique, complex task and/or a specific organization form but is perceived as a social system, with dynamic boundaries, with the ability to learn and to self-organize.
* A project can be clearly differentiated from other systems by its specific culture. The set of values and behavior patterns collectively shared by the project team is an essential project management instrument. It gives orientation to all parties involved. Culture develops evolutionary by learning and communicating.
* The strategy and the mission of a project are defined by its context. They are defined by the decisions and actions in the pre-project phase, and by the relationships between the project and its environments.
* A relational project understanding, emphasizing the relations between a project and its relevant environments, is the basis to successfully react to changes in a dynamic project environment.
* Relevant project environments (e.g., users, suppliers, departments of the base organiza-

tion) are supply and demand markets of the project. The project's relationships to these markets can be actively managed."

Marketing Projects

The number of marketing tasks that assume project character is on the rise in companies operating only in the domestic market as well as in international companies. Classic marketing projects are product development and product introduction, the establishment of branch offices or sales subsidiaries, the setup of joint ventures, the development and control of advertising campaigns, and so on. Marketing projects differ from the rather process-oriented project tasks which are typical of the capital goods business (power stations, railway systems, and so on) in the higher degree of determination of future marketing mix decisions and in the impact on customer markets and relevant environments. Following this line of thought, marketing projects are characterized by a strategic component and by a particular communicative constellation, that is, marketing projects are confronted by a complex structure of different interest groups. These groups encompass official authorities, mass media, marketing intermediaries, suppliers, competitors, and customers (Patzak,[8] p. 310; Cleland,[9] p. 96).

The relationships to these interest groups cannot be seen as a simple two-way communication between communicator and addressed public because frequently communicative interdependencies between the different parties exist. The isolated view and handling of such relationships, as for example, in a product introduction project between product manager, sales force, and retailers, does not fit the complexity of the communication system. Interest groups which may criticize the environmental compatibility of a new product and which, at the same time, show strong public relations activities, must not be neglected. Project managers have to take into account the various interdependencies between the different groups to achieve their project objectives. Marketing projects therefore have to be analyzed from a systemic view of project organization and management.

Marketing projects are of strategic character because either the decision to form a project team is the result of corporate strategic planning or the scope of the expected consequences gives the project a strategic character. For illustration take the example of the introduction of product innovations on a worldwide scale, the modification of corporate design elements, or the development of a new corporate identity after a merger.

Due to the long-term impact of marketing decisions, the high complexity has to find an adequate entry into project planning. Quite often complex marketing tasks are performed without organizing them as projects. In this case product managers or customer group managers assume implicitly the function of a project manager. In order to improve efficiency, project tasks should be clearly defined and assigned to project members. Figure 10.1 shows a linear responsibility chart for the functional relationships of a project within a matrix organization. Authority, responsibility, and accountability have to be clarified to avoid conflicts.

International Marketing Projects

Cross-national marketing activities, such as multicountry product development, market testing, product introductions, setup and modification of distribution systems, or the development of international advertising campaigns, require coordination between

Activity	General Manager	Manager of Projects	Project Manager	Functional Manager
Establish department policies & objectives	1	3	3	3
Integration of projects	2	1	3	3
Project direction	4	2	1	3
Project charter	6	2	1	5
Project planning	4	2	1	3
Project—functional conflict resolution	1	3	3	3
Functional planning	2	4	3	1
Functional direction	2	4	5	1
Project budget	4	6	1	3
Project WBS	4	6	1	3
Project control	4	2	1	3
Functional control	2	4	3	1
Overhead management	2	4	3	1
Strategic programs	6	3	4	1

Code

1: Actual responsibility
2: General supervision
3: Must be consulted
4: May be consulted
5: Must be notified
6. Approval authority

FIGURE 10.1   Linear responsibility chart of project-management relationships. (*From Cleland,[9] p. 155.*)

the organizational units of the base organization and the local distributors or partners of an international company. These planning, coordination, and adaptation tasks across several country markets and operating units can only be established in the form of a project-management organization. The complexity of such tasks, the necessary involvement of all concerned persons and organizational units, the dependence on global, regional, and local forces in the decison-making process, as well as the time aspect of the objectives favor a project approach.

Complexity and flexibility are the features of these project tasks. They result from:

- The organizational constellation of the international marketing system, which typically consists of various modes of operation such as sole agents, sales subsidiaries, joint ventures, licensing or franchising organizations, and local manufacturing units
- The requirement to adapt marketing programs, although economies of scale will be gained by standardization
- The necessity to coordinate programs with independent partners, as it happens in strategic alliances or in joint ventures (Dülfer,[10] p. 24)
- Differences in personal attitudes and beliefs, culture, language, income levels of project members, or simply distance (Archibald,[11] pp. 153–158).

Organizational Models for International Marketing Projects

International marketing projects can have different organizational forms.

Process-Oriented Model.   In this approach the project members are still integrated in the base organization. However, every marketing project has to follow a determined reporting and coordination procedure in which all involved functional departments or product divisions at international headquarters and the subsidiaries take part. Very often only certain procedures are standardized, such as market research processes or price calculation.

Team Model (Task Force).   In this case the implementation of an international marketing project is transferred to a project team. Team members are borrowed from the base organization on a full- or part-time basis. It is important to clarify the process of project execution: where and when are periodic meetings held, who becomes chairperson of the project, for instance, in the form of a lead-country manager, and how is the distribution of competencies between country organizations and central functions or between global and regional headquarters? The standardization of marketing programs (products, brands, advertising campaigns, and so on) as project goals carries an enormous potential for conflict in the international organization. Tasks such as information gathering; preparation for final decisions, coordination, and mutual advising; decision making; and implementation are again complex subprojects.

International companies are increasingly faced with the situation that several marketing projects have to be carried out simultaneously. The staffing, the coordination of these projects, and the overall control create a demand for a new position within international organizations—the multiproject management.

PROJECT MANAGEMENT IN THE INTERNATIONAL MARKETING PRACTICE

Coordinating International Advertising

Global Standardization versus Localization.   The question of whether the standardized or the localized approach is the more appropriate form of campaign planning in international advertising is an ongoing controversy among international marketers and advertisers. In its extreme form, standardized international advertising is understood as the practice of advertising the same product in the same way (appeal, message, copy, media) in all markets of the world. Since the realization of a totally standardized global advertising concept is nearly impossible, the discussion centers around the appropriateness of variations within advertising contents from country to country.

While the decentralized and customized approach was the norm in international campaign development in the past, new developments such as the convergence of consumer behavior in the industrialized countries, intensified global competition, and a changing media scene offer greater opportunities for the cross-national standardization of brand and advertising concepts.

The emergence of the "world consumer," namely, consumer groups in different countries that show the same purchasing behavior and attitudes toward products, allows marketers to form homogeneous cross-national target groups and to create uniform marketing programs for these segments. Examples for global marketing strategies can be found in the industrial goods, consumer electronics, prestige goods, soft drink, and fashion sectors (Sorenson and Wiechmann[12]; Ohmae,[13] p. 147).

Global competition is another force that determines international advertising. The appearance of the same set of competitors in all major markets supports the development of global strategies. In international brand positioning not only the customer wants and interests have to be considered, but also the positioning of the competitors' products. The persistent pressure to reduce costs is a further consequence of global competition. Media costs and the production costs of advertising materials and creative works are regarded as prime targets of cost savings in the packaged-goods industry (Onkvisit and Shaw,[14] p. 51; Görke[15]).

In addition, the changing media scene, especially in western Europe, is forcing international marketers to coordinate their communications strategies. Confronted with an increasing media overspill in Europe, national advertising campaigns should not be developed without taking into account the impacts on other affected markets (Holzmüller[16]). The media overspill is very obvious in the case of cable and satellite television, but terrestrial broadcasting stations (television and radio) and print media also play important roles. More and more consumer product companies are beginning, for instance, to coordinate their product introductions and relaunches among the German speaking countries. By harmonizing the advertising campaigns irritations of customers who otherwise might be confronted with different, maybe contradictory, brand concepts can be avoided.

The proponents of the standardized approach refer to the following advantages[1,2,15]:

- Economies of scale through reduced advertising costs
- Higher efficiencies of advertising campaigns due to the concentration of budgets
- Creation of a uniform brand image throughout Europe or the world

However, these arguments should not nourish the impression that localization in international advertising has become obsolete. The barriers to standardization are still numerous. The main forces that favor local adaptations of international campaigns are differences in language, culture, stage of product life cycle, media availability, usage of media, and advertising regulations. Given such striking differences in the advertising situation, using a customized advertising campaign will very likely result in a higher communication effectiveness. Advertising blunders can only be avoided when communications strategies are oriented toward the receiver, who is the most important link in the communication process.

Frequently cited failures of badly handled standardized campaigns such as commonplace slogans, motivational problems with the local staff (not-invented-here syndrome), or unappealing brand positionings could be circumvented by a better participation of local advertising personnel during the campaign development (Tostmann,[17] p. 57; Green et al.[18]). As Peebles et al.[19] (p. 96) put it, the fault lies in the implementation not in the validity of the standardized concept. The solution to this dilemma is a more cooperative project-management-oriented style in advertising campaign management.

The Programmed Management Approach. The programmed management approach is seen as a useful managerial framework for the coordination of international advertising campaigns.[19] It represents a six-step program that permits headquarters to standardize certain elements of the national campaigns while at the same time it promotes individual market inputs (Fig. 10.2). The approach maintains strong central control but allows flexibility in execution. Interaction between home office and local subsidiaries takes place at every single stage of the campaign development. The end result is not necessarily a single standardized campaign for all countries involved, but mostly a highly harmonized concept that meets strategic marketing and communications objectives.

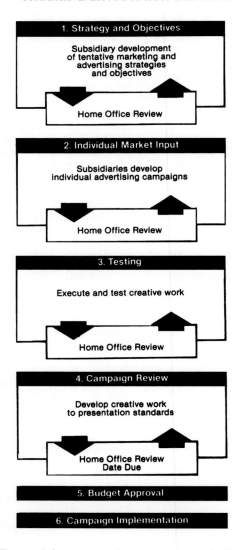

FIGURE 10.2   Framework for programmed management approach. (*From Peebles et al.,*[19] *p. 30. Reprinted with permission from* Journal of Marketing.)

Goodyear International, Inc.'s 1976 campaign is cited as an example for the successful implementation of the programmed management approach (Peebles et al.,[19] pp. 31–34). Goodyear used the approach to direct and coordinate the development and implementation of its 1976 advertising campaign in 11 western European markets. The entire campaign planning was organized as a project. The procedure and planning dates were laid down in a management flowchart which covered the period of April to September 1976 (Fig. 10.3). The responsibilities of the home office and of the

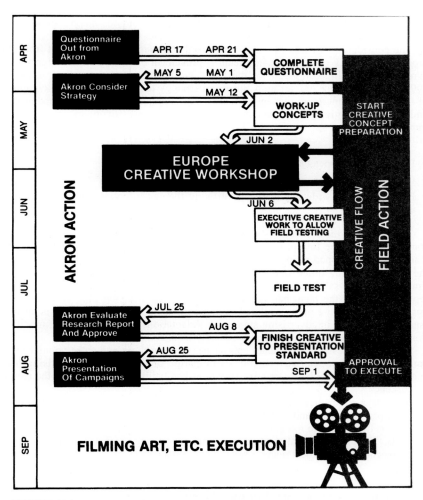

FIGURE 10.3 Management flowchart, 1976 advertising campaign development at Goodyear International, Inc. (*From Peebles et al.,[19] p. 32. Reprinted with permission from* Journal of Marketing.)

European subsidiaries, which comprise local input regarding strategy and objectives as well as home office evaluations and approvals, are also indicated in the chart. The chart was distributed to each European subsidiary and to each country's network advertising agency.

Similar approaches can be found in other companies too. Mercedes Benz AG, the German luxury car manufacturer, departed from the traditional practice of national campaign development with subsequent bilateral coordination between headquarters and local subsidiaries in favor of a multilateral team approach. Now the guiding principle in advertising campaign development at Mercedes Benz AG is to base the concept on as many commonalities as possible and to allow only as much individualization as necessary.

The considerable efforts in coordination and communication that result from this procedure are only manageable in a team approach. In order to devise a highly standardized advertising campaign for the European introduction of the S-class cars in 1991 Mercedes Benz AG set up a European strategy team. The team was composed of three to five functional managers (marketing, communications) from headquarters and one communications manager from each affected European subsidiary or importer as well as agency people. The European strategy team carried the responsibility for:

- Developing European advertising approaches
- Deciding on a European corporate identity
- Finding Europe-wide applicable advertising themes in cooperation with the lead agency

The result of these meetings was a briefing for the lead agency, which was the starting point for the European advertising campaign. On the basis of the briefing the lead agency developed a European campaign that synthesized the national aspects of the Mercedes Benz brand image into a common denominator, a European identity. Finally the proposal was presented to the strategy team for approval. Each country representative of the team had the same saying in the decision process. It was also possible to insist on a more localized campaign if facts were supporting this position (such as continuity in advertising approach). Adaptations varied from slight modifications of headlines or texts to nearly completely localized campaigns.

These examples show that team approaches become more and more popular in the context of pan-European advertising. Participation of the national advertising managers tends to produce better creative strategies and facilitates execution in the national markets.

Nowadays companies that are operating Europe-wide do not really have the choice between pan-European and locally developed campaigns. The market and competitive conditions force them to integrate their advertising strategies on regional or even global levels. Therefore the quality of international advertising campaigns is rather a matter of execution and project management than of conception.

Coordinating International Product Policy

In the area of product and brand management international firms face the challenge to coordinate their product and brand strategies in national markets as well. European companies with a multinational orientation and structure are hit by the changing competitive conditions in the European Community above all. Following a country-by-country approach in an emerging common-market area might impair the viability of those multinational companies. The removal of trade barriers, the homogenization of consumer wants and distribution structures, as well as the existence of globally or pan-European oriented competitors force those companies to exploit synergies and economies of scale to stay competitive.

The standardization of marketing programs is a typical measure that is taken to increase the competitiveness. Objects of standardization in the product area are the product itself (features, formula, design), product positioning, and packaging. The product management will start the process by checking the product mix for candidates which lend themselves to standardization. In a second step product elements (formula, brand name, packaging) will be analyzed according to their international transferability and acceptance (local regulations, trademark and patent laws, consumer wants). In a final assessment the costs and benefits of a standardized approach will be compared.

Considerig the situation—coordination across national markets and, as a consequence, across local management—a rather centralized planning mode would fit for this task. But at the same time it is very obvious that this decision process necessitates the participation and cooperation of local staff. Without the involvement of the national product managers in the planning process, important local inputs would perhaps be missing and troubles in the implementation of the international product strategy could arise.

Coordinators, coordination committees, or project teams are proven instruments for conflict resolution. Two organizational concepts which are differing by their degree of institutionalization are presented: international brand teams and the lead-country concept.

International Brand Teams. In multinationally oriented companies the coordination across the foreign subsidiaries should be achieved without harming the existing culture and structure of the firm. In this case international brand teams can be established. The international brand team normally comprises all local brand managers involved as well as technical staff, marketing managers, and communications managers from headquarters. Depending on the assigned role, these teams can have the function of either discussion circles or committees with decision-making authority. In the former case—in many companies the first step to an international participation—information is exchanged between team members. The importance of discussion circles lies in the socialization and the symbolic effect rather than in achieving big cost savings or perfectly standardized international marketing programs. In the latter case these groups assumed competencies and responsibilities for the formation of international marketing strategies. If these brand teams are headed not by a manager from headquarters but by the general manager and his or her brand group, the term lead-country concept is used. (Other common terms are key-market subsidiary, global-market mandate, or competence center.)

Lead-Country Concept. The lead-country concept represents a form of decentralized coordination (Fig. 10.4). In order to avoid the overload of top management with coordination tasks, the lead-country group assumes the coordination task for a brand or product group on a worldwide basis or within a certain region, such as the European Community. This group is supported by the brand managers from the other subsidiaries and by functional managers from headquarters.

A famous user, if not inventor, of the lead-country concept is Procter & Gamble (P&G), the American laundry and personal-care company. P&G formed so-called Eurobrand teams for two important brands in its European business, the disposable diaper Pampers and the liquid detergent Vizir. In the Vizir case the German subsidiary

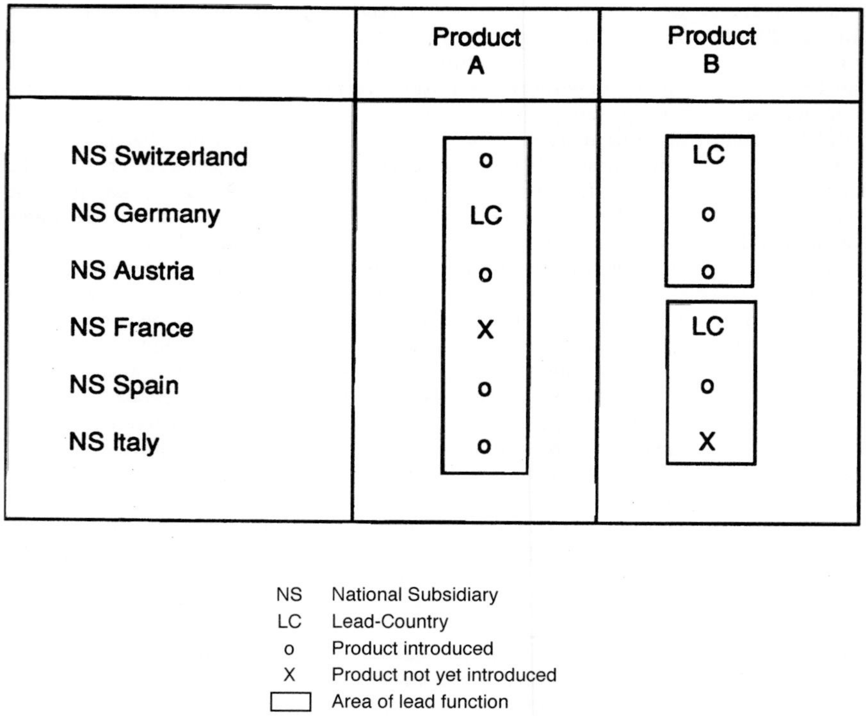

	Product A	Product B
NS Switzerland	o	LC
NS Germany	LC	o
NS Austria	o	o
NS France	X	LC
NS Spain	o	o
NS Italy	o	X

NS   National Subsidiary
LC   Lead-Country
o   Product introduced
X   Product not yet introduced
☐   Area of lead function

FIGURE 10.4   Lead-country concept.

assumed the role of lead country. The Eurobrand team planned and decided on the following issues (Bartlett and Ghoshal,[20] p. 89):

- Product and market testing
- Package design
- Advertising theme
- Marketing strategy

This organizational concept enabled P&G to launch Vizir in six new markets within a year. This was the first time P&G had ever rolled out a new product in that many markets in so brief a span. It was also the first time the company got agreement in several subsidiaries on a single product formulation, a standardized advertising theme and packaging line, and a sole production source.[20]

The selection of the lead country can be guided by the following criteria:

- Success, expertise, and experience in managing the brand nationally (marketing competence of the lead subsidiary)
- Strategic importance of country market (competitive situation, global lead market for product line)

- Country-of-origin effect (such as cosmetic products from France)
- Location of production unit

The advantages of the lead-country concept are obvious. The decentralized coordination of a European brand strategy avoids overstrain of the management in the regional head office. The concept implies, psychologically, an upgrading of the local subsidiaries. It represents a power shift within the organization that might help to overcome the traditional vertical headquarters-subsidiary relationship. The national subsidiaries are more than a mere executive arm and, hence, can gain a specific profile within the group. As long as the participation of all involved national brand managers in the decision-making process is ensured, motivation will be high.

In comparison to the international brand teams, however, the lead-country concept has an authoritarian character. It might turn out that centralization is only transferred to the subsidiary level. That is especially true when the lead country forces the sister companies to accept its successful national brand strategy and, as a consequence, is not willing to adopt its "tried and proven" strategy to different conditions (Kashani,[21] p. 95). This negative effect can perhaps be reduced if the lead-country concept is applied to more than only one or two brands, as it frequently happens now. If different subsidiaries assume the lead role for different brands, the built-in interdependence would show very clearly the need for reciprocal cooperation. In addition, information and planning systems have to be designed in such a way that the lead subsidiaries have access to all necessary data in order to perform their tasks.

Conducting and Coordinating International Market Research

International market research has become a crucial element in the formation of international marketing strategies. International product managers and advertising managers who are responsible for the development of brand and marketing strategies across a number of countries need data of their target markets for decision making. Decisions range from strategic issues—international brand positioning, selection of target markets, and target groups—to more tactical issues such as the transfer of local advertising campaigns to other markets or packaging decisions.

There are two groups of decisions in international marketing: intracountry decisions that concern the search for the optimum marketing program in a specific country, and intercountry decisions that focus on country selection and on the integration of country-level operations on a regional (e.g., Europe, Latin America) or global level.

Marketing research for intracountry decisions can be conducted on a country-by-country basis, and hence the research design is very similar to domestic marketing research. The aim of research is to collect specific country market data (economic situation, competitors, market segments, consumer behavior, distribution channels) in order to find the most appropriate strategy for a national market.

The second type of decisions, the intercountry decisions, cannot be handled on a country-by-country basis. Decisions such as country selection, sequence of entry, allocation of funds among operating units, or development of regional or global marketing programs require a research design that allows comparisons between markets. The search for similarities among customers in different countries is a topic in global marketing. Members of cross-national target groups have more in common with each other than with many of their fellow citizens. The existence of such cross-national target groups with respect to certain product classes (business travelers—airlines, hotels, rental cars; teenagers—soft drinks, consumer electronics, casual wear) is a prerequisite for the standardization of marketing instruments. In order to support a rationale, decision-making studies in this field have to establish comparability and equivalence of research in different countries.

Apart from finding the adequate research design, international market researchers have to cope with the following problems (Douglas and Craig,[22] p. 16):

- Complexity of research design due to operation in a multicountry, multicultural, and multilinguistic environment
- Lack of secondary data
- High cost of collecting primary data
- Difficulties in establishing the comparability and equivalence of data and research conducted in different contexts
- Intrafunctional character of many international marketing decisions
- Problems associated with coordinating research and data collection in different countries

Especially the last issue is of interest in this context. It is clear that multicountry market research requires a project organization. Different actors with different tasks are involved. If the client is a multinational company, the international head office, the regional headquarters, and national subsidiaries may be involved. Local subsidiaries may be involved in funding as well as in research agency selection and control. It is possible that an advertising agency or a consulting firm may commission the study on the client's behalf (Barnard,[23] p. 47).

On the research supplier side the client can choose between multinational research companies, research chains, and local research suppliers. It is advisable to use a lead agency whose task is the coordination of all activities among the units involved.

Barnard[23] (pp. 54–61) describes a typical sequence of tasks in multicountry research studies (Fig. 10.5). In this example a coordinating research agency is used which employs local agencies for data collection. The project starts with a request or briefing of the client. Except for the simplest studies, the coordinating agency will contact the local agencies and will ask for a feedback. In the best case the local agency will not only give a cost and timing estimate but also comments on the proposed method and on their own resources and planned work commitments. Experience proves that a checklist should be used to avoid ambiguity. This checklist should comprise all relevant job details, and all identifiable tasks should be clearly allocated between the coordinating agency and the local agencies.

In the consolidated proposal or bid the coordinating agency states clearly the tasks to be undertaken and explains the cost structure of the bid. After discussion, subsequent modifications, and approval of the modified bid, the coordinating agency develops the plan for execution. As the coordinating agency takes full responsibility for the quality control of local fieldwork, the local agencies have to follow a set of agreed procedures. Such requirements are, for example, minimum standards for each completed interview, minimum number of interviews, or monitoring of fieldwork progress at prearranged dates. Similar coordination tasks arise in data processing (such as form of data transmission, formatting). At the final reporting the coordinating agency has to adapt the study and the presentation to the preferred styles and conventions of the client company (for example, roles of reports and presentations; balance of verbatims, tables, and commentary; local reports or international summary or both).

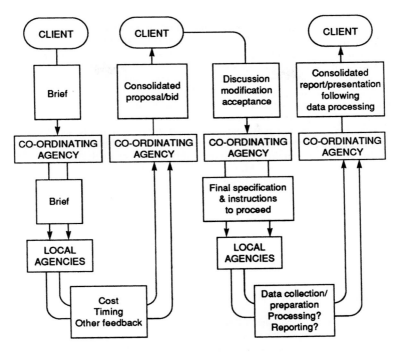

FIGURE 10.5   Sequence of tasks in multicountry study. (*From Barnard,*[23] *p. 55.*)

SUMMARY AND CONCLUSIONS

Changing competitive conditions on the world market create new challenges for the management of international businesses. Primarily the shortening of product life cycles coupled with dramatically rising costs of investment, the globalization of competition, the increasing preference for cooperation or alliances as the mode of market entry and market penetration, as well as new emerging media and distribution structures require an adequate response in international marketing management.

As a consequence, management tools have to be developed that allow a reduced response time to market developments, facilitate communication and contact between different organizational units, and ensure cooperation and input from all persons involved. The complexity of this management task is enhanced by the international dimension. International markets differ in their volume, products offered, price levels, competitive situation, channel and media structure, legal restrictions, as well as in their customer preferences. International marketers permanently have to analyze and decide if an extension of marketing concepts from one country to another is possible or if a localized approach is preferable. For these decisions they need current information, inputs from local markets, advice of product, functional, and country experts, and the cooperation of all persons involved.

The traditional pyramid organization is not able to cope effectively with this situation, which is typical for international companies. Project management is one of the tools that facilitate cross-national marketing activities. Multicountry research studies, Europe-wide product launches, the development of an international corporate identity strategy, or the coordination of a global sponsoring strategy demand the project approach. While in the past the project approach was primarily used in the planning and execution of capital goods exports or the setup of local subsidiaries, now the internal coordination and integration of international marketing activities is the main purpose.

From the international marketer's point of view, future research in project management has to focus on project control and multiproject management. The control of various project teams with different goals, tasks, competencies, and team structures will become more important. With an increasing number of projects the emergence of interdependencies between projects is more likely and the evaluation of project performance will also become a topic. Moreover, a project coordination function permits the pooling of valuable know-how, experiences, and resources.

REFERENCES

1. T. Levitt, "The Globalization of Markets," *Harvard Business Rev.,* pp. 92–102, May-June 1983.

2. W. Keegan, *Global Marketing Management,* 4th ed., Prentice-Hall, Englewood Cliffs, N.J., 1989.

3. J. Johanson and F. Wiedersheim-Paul, "The Internationalization of the Firm—4 Swedish Cases," *Management Studies,* pp. 305–322, Oct. 1975.

4. J. P. Jeannet and H. Hennessey, *International Marketing Management,* Houghton Mifflin, Boston, Mass., 1988.

5. A. Roux-Kiener, "Organizations: A Step to Transform the Traditional Pyramid," in *Handbook of Management by Projects,* R. Gareis (ed.), Manz Wirtschaft, Vienna, Austria, 1990, pp. 111–118.

6. C. Bartlett and S. Ghoshal, *Managing across Borders,* Harvard Business School Press, Boston, Mass., 1989.

7. R. Gareis, "Management by Projects: The Management Strategy of the 'New' Project-Oriented Company," in *Handbook of Management by Projects,* R. Gareis (ed.), Manz Wirtschaft, Vienna, Austria, 1990, pp. 35–47.

8. G. Patzak, "A Morphological Model of Project Management," in *Handbook of Management by Projects,* R. Gareis (ed.), Manz Wirtschaft, Vienna, Austria, 1990, pp. 306–316.

9. D. Cleland, *Project Management—Strategic Design and Implementation,* TAB Books, Blue Ridge Summit, Pa., 1990.

10. E. Dülfer, "Projekte und Projektmanagement im internationalen Kontext," in *Projektmanagement International,* E. Dülfer (ed.), C. E. Poeschel Verlag, Stuttgart, Germany, 1982, pp. 1–30.

11. R. Archibald, "Overcoming Cultural Barriers in Corporations to Effective Management by Projects," in *Handbook of Management by Projects,* R. Gareis (ed.), Manz Wirtschaft, Vienna, Austria, 1990, pp. 153–160.

12. R. Sorenson and U. Wiechmann, "How Multinationals View Marketing Standardization," *Harvard Business Rev.,* pp. 38–167, May-June 1975.

13. K. Ohmae, *Macht der Triade—die neue Form weltweiten Wettbewerbs,* Gabler, Wiesbaden, Germany, 1985.

14. S. Onkvisit and J. Shaw, "Standardized International Advertising: A Review and Critical Evaluation of the Theoretical and Empirical Evidence," *Columbia J. World Business,* pp. 43–55, Fall 1987.

15. W. Görke, "Globaler Glamour," *Absatzwirtschaft,* no. 6, pp. 50–51, 1985.

16. H. Holzmüller, *Internationaler Werbe-Overflow. Entscheidungsdeterminante im Marketing von Konsumgütern,* Verlag Harri Deutsch, Frankfurt am Main, Germany, 1982.

17. T. Tostmann, "Globalisierung der Werbung: Faktum oder Fiktion," *Harvard Manager,* no. 2, pp. 54–60, 1985.

18. R. Green, W. Cunningham, and I. Cunningham, "The Effectiveness of Standardized Global Advertising," *J. Advertising,* no. 4, pp. 25–30, 1975.

19. D. Peebles, J. Ryans, and I. Vernon, "Coordinating International Advertising," *J. Marketing,* vol. 42, no.1, pp. 28–34, 1978.

20. C. Bartlett and S. Ghoshal, "Tap Your Subsidiaries for Global Reach," *Harvard Business Rev.,* pp. 87–94, Nov.-Dec. 1986.

21. K. Kashani, "Beware the Pitfalls of Global Marketing," *Harvard Business Rev.,* pp. 91–98, Sept.-Oct. 1989.

22. S. Douglas and S. Craig, *International Marketing Research,* Prentice-Hall, Englewood Cliffs, N.J., 1983.

23. P. Barnard, "Conducting and Co-ordinating Multicountry Quantitative Studies across Europe," *J. Market Research Soc.,* vol. 24, no.1, pp. 46–64, 1982.

CHAPTER 11
CULTURAL STRATEGIES FOR GLOBAL PROJECT MANAGEMENT

John Tuman, Jr., P.E.

Management Technologies Group, Inc.,
Morgantown, Pennsylvania

John Tuman, Jr., P.E., is Senior Vice-President with Management Technologies Group, Inc., a U.S. consulting firm specializing in project management, organizational development, and information technology. Mr.Tuman is responsible for providing consulting and training services to corporations in the United States and overseas.

INTRODUCTION

The globalization of business is a reality of our time. Companies that want to thrive must devise strategies and programs for volatile, complex markets. One strategy is for companies in different parts of the world to form associations of mutual interest, that is, integrate elements of their unique capabilities and their resources to create a network organization. Network organizations are created to gain global leadership in specific markets.

This type of global undertaking demands a new management approach. Diverse political, economic, legal, and national factors require companies to implement new organizational concepts and sophisticated management systems, and to develop a new breed of manager. In addition, companies must constantly innovate to meet the needs of a changing global market. This requires companies to develop and manage organizational cultures which are attuned to the new environment. In the global marketplace, organizational culture is a valuable resource. If this resource is used wisely, it can provide competitive advantage.

This chapter focuses on two issues critical to success in a global environment: (1) the design and implementation of network organizations and (2) the establishment of a process for systematic innovation and entrepreneurship in a network organization. The catalyst that makes these concepts a reality is project management.

OVERVIEW OF GLOBAL CHANGE: AN AMERICAN PERSPECTIVE

Since the end of World War II the world economic scene has undergone profound changes (Fig. 11.1). The United States emerged from World War II as the most pow-

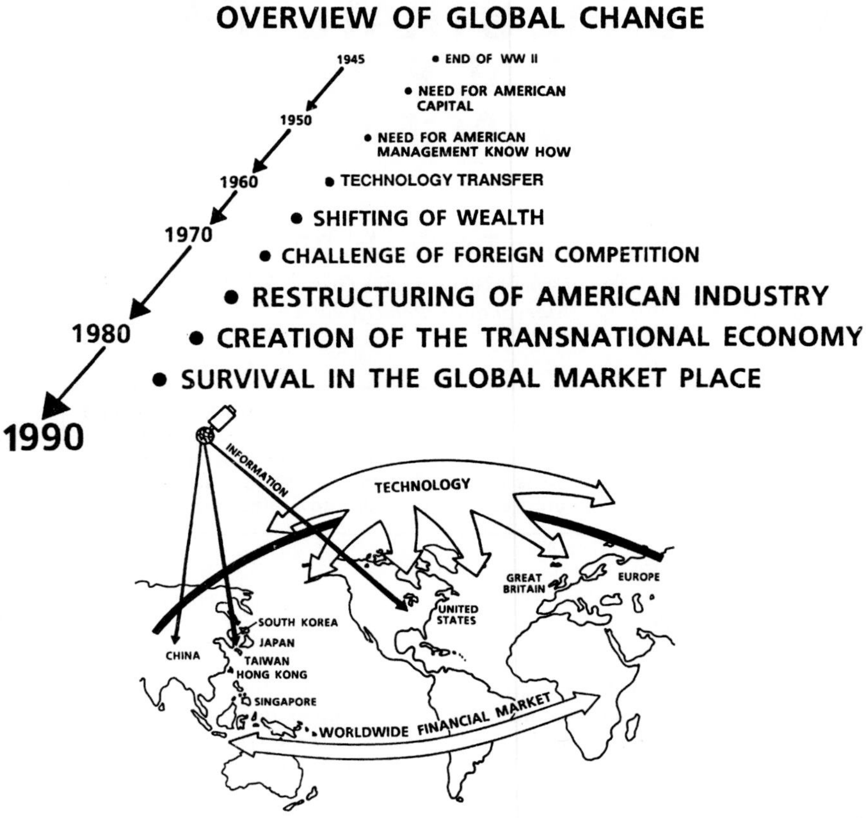

OVERVIEW OF GLOBAL CHANGE

FIGURE 11.1 Overview of global change, 1945 to 1990.

erful industrial nation in the world. Prosperity and economic growth immediately following the war years was unprecedented. By 1950 the U.S. gross national product was about 45 percent of the global product.[1] These were truly wonderful times; it was great to be an American. We secured freedom for the world and now we were showing the world how to build a better life. After the U.S. dollar, the most sought after item was American management know-how. We were the technologists, innovators, and entrepreneurial leaders of the world. And we shared freely of what we had.

During the 1950s and 1960s America provided the financial aid and the technical and management know-how to rebuild the economies of many nations. Americans also carried most of the burden of the cost of the cold war. By the 1970s the free industrialized nations of the world had rebuilt their industries, and the United States began to feel the impact of foreign competition. During the 1970s and the 1980s, American industry had to undergo massive restructuring. Millions of people were dislocated as smoke stack industries began to disappear—America was evolving into a service-based economy. By the 1990s American business was adjusting to the requirements of a global marketplace.

THE GLOBAL MARKETPLACE

Today the U.S. economic share of the global product is about 22 to 24 percent.[1] As a result of our generosity, and the hard work of the striving nations, Americans have witnessed the growth of new economic forces. Nations such as Japan, Germany, and Italy have regained and even exceeded their prewar economic capabilities. Significant wealth has shifted to the oil-producing nations of the Middle East. And third-world nations such as Singapore, Hong Kong, Taiwan, and South Korea have emerged to become economic and industrial giants. Even the East Bloc nations are moving to participate in the global marketplace. We can define the global marketplace as the environment where no supplier is too remote and no customer too foreign.[2]

Many factors have contributed to the development of the global marketplace. Information technology makes instantaneous financial transactions possible, and new ideas can move quickly to different lands. Also, process technology can be transplanted quickly to take advantage of the cheapest labor and the friendliest markets. The mobility of capital, technology, and information inspires a kaleidoscope of change in the global environment. This changing environment presents unparalleled problems as well as opportunities for managers everywhere.

The Harsh Realities of Global Competition

The global marketplace has created a new type of competition. Drucker[3] refers to this new rivalry as "adversarial trade." According to Drucker, competitive trade focuses on creating customers, but adversarial trade aims at dominating an industry or a market.

A good example of adversarial trade is Japan's approach to the global marketplace. The Japanese began rebuilding their industries after World War II by concentrating on excelling in a few areas. They paid special attention to improving quality and productivity, and they took advantage of their low labor cost. After a time, the Japanese had virtually eliminated any serious competition in certain product lines—such as consumer electronics, cameras, and optics—around the globe. Using similar techniques, other newly industrialized nations began to establish strong footholds in a number of markets. As capital and technology spread rapidly around the world, these newly industrialized nations developed capabilities and strength to eliminate most competition in their target markets. They also made it virtually impossible for newcomers to enter their markets. The message is clear: in order to survive and prosper, companies must dominate their markets.

Today the only market for most companies is a global market. As one chief executive of a medium-size U.S. low-tech company put it, "any business I can reach by telephone is a potential customer as well as a potential competitor." Thus in order to survive, it is imperative to maintain a market presence in any area of the world that offers opportunity for global leadership. A good example of the new global business strategy being put into practice is General Electric's business restructuring. GE has moved out of markets where it cannot be a leader worldwide. It has sold off a number of domestic operations, even very profitable ones, and bought overseas businesses that provide opportunity to become a leader in the global marketplace.[3]

The strategy initiated by GE is being duplicated in a variety of ways (Fig. 11.2) by many medium-size and small firms. By forming alliances, the smaller companies can establish a market presence anywhere on the globe. By carefully amalgamating their

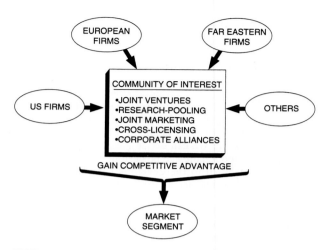

FIGURE 11.2   Approaches to global marketplace.

technical, productive, and management capabilities, smaller companies can establish a premier market position just as well as large multinational corporations.[4]

The need to form an alliance or a community of interest has little to do with the size of the company. It has more to do with the need to maintain a leadership position in a particular business. According to Drucker, "to maintain a leadership position in any one developed country, a business—whether large or small—increasingly has to attain and hold leadership positions in all developed markets world-wide." The need to maintain leadership in all the developed markets worldwide is a direct result of global competition, the dissemination of process technology to all developed nations, and advances in information and communication technology.[4] Information speed, accessibility, and costs are such that no developed nation enjoys a substantial competitive advantage for long.

It is clear that, as the technological and productive capabilities of competitor nations begin to converge, market leadership will go to those who innovate, manage well, and function effectively as integral parts of a multinational project-like organization. To succeed, alliances have to create a new type of organization, an organization that can transcend national interest, traditional methods of operation, and ingrained corporate norms, values, and beliefs. These organizations will need a unique culture, one that is specifically suited to doing business in a global environment.

WHAT IS ORGANIZATIONAL CULTURE?

Organizational culture can be defined as the collective set of values, beliefs, assumptions, and symbols that consciously and unconsciously guides the way a firm conducts its business.[5] In the most simplistic sense, culture is collective mental programming.[6] Organizational culture is rooted in the history of the organization; it is learned, and it is passed from older to newer members (Fig. 11.3). The strongest elements of an organization's culture have stood the test of time.

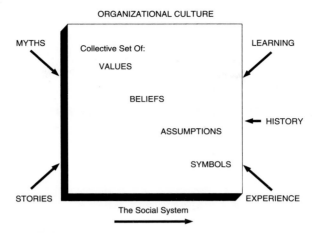

FIGURE 11.3   Organizational culture.

Why Culture Is Important

A firm can apply new technologies, processes, and procedures to optimize, simplify, and automate work tasks; it can also speed up and improve information flow and accessibility. In addition, the firm can simplify its structure, reduce levels of management, and speed up decision making. All of these changes should help the firm improve its effectiveness and efficiency; frequently, however, they do not.[7]

The difference between firms that manage themselves well and firms that do not depends on the perceptions, values, beliefs, and motivations of the people involved— in other words, the culture of the organization.

An organization's culture is implicitly linked to the strategy, structure, systems, processes, and procedures that integrate the organization's value chain, infrastructure, and management control and reporting systems (Fig. 11.4). If the union between the firm's culture and systems is complementary, then the organization functions effectively and efficiently with minimal formal control. Thus the right kind of culture can lead to high performance.[8] The "right" culture can be propagated in any organization. Furthermore, culture can be managed like a project, and this type of project can lead to competitive advantage.[9]

In the global marketplace the right kind of culture is one that sustains a dynamic flexible organization, stimulates innovation and entrepreneurship, and uses the synergy of the team to accomplish its mission. Creating this type of culture is a job that project management is well qualified to carry out.

The Culture of Project Management

For project management, interdependence, initiative, team work, freedom, responsibility, and self-motivation are highly valued characteristics. Only project management ranges across the organization to plan, direct, and control resources to get the job done with little or no formal authority. People who work in project management enjoy the

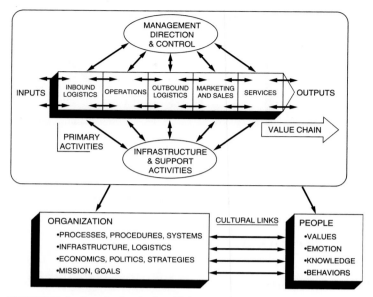

FIGURE 11.4   Organizational cultural links.

challenge and freedom as well as the turbulence of the environment. This environment produces an organizational culture that fosters innovation, entrepreneurship, vitality, and the drive necessary for competitive advantage. Typically, individuals who are comfortable in this type of environment have the physiological and psychological characteristics that are needed to thrive in an unstable, unpredictable, high-risk setting. These are the individuals who can provide the values, beliefs, and energy for a new type of global organization. Structure and systems provide the form and process for competitive advantage, but people and culture provide the intellect and the will to succeed. In the global environment project management will have to be the architect and manager of unique organizational structures and cultures.

THE NEW GLOBAL PROJECT MANAGER

The mission of global project management is to integrate the capabilities of diverse companies to achieve leadership in a specific area of the global market. Simply stated, the job is to build a new type of international organizational alliance and facilitate its operation. This new organization will be formed from independent companies—which are typically managed through a process of formal rules, procedures, and well-defined levels of authority—and given direction through a mode which stresses flexibility and cooperation based on shared capabilities, goals, and visions. This concept is a sharp departure from the rigid monolith of the past and provides the foundation for the kind of vibrant organization specifically tuned to the challenges of the global marketplace.

The goal of the global project manager is to bring together companies from different parts of the world to do something that they could not do on their own—obtain leadership in a global market.[10]

LIMITATIONS OF THE TRADITIONAL ORGANIZATION

The traditional corporate organizational structure, that is, the vertical or hierarchical structure, has served us well in the past. Its design objective was efficiency through subdivision of work and specialization. However, the traditional organization was not designed for innovation or change. These tend to promote instability, and instability fosters inefficiency in the traditional organization. More important, the traditional organization tends to grow in size when it has to deal with new and complex problems. Hence organizational cost goes up while responsiveness goes down (Fig. 11.5).

Large hierarchical structures are becoming outdated in an information environment. Speed, synthesis, and flexibility are the critical factors—not order, repetition, and stability.[11] In order to address the need of the corporate community operating in the global marketplace, we need organizations which can operate in an unpredictable, rapidly changing, competitive, and highly information-intense environment. Most of all, we need organizations which are specifically designed to provide competitive advantage.

CREATING NETWORK ORGANIZATIONS FOR THE GLOBAL MARKETPLACE

A network organization can be created by selectively integrating the capabilities of several companies—which can be located in different countries in various parts of the globe—to obtain leadership in a specific area of the global marketplace (Fig. 11.6). The participating companies remain independent. However, elements of their organization are committed to supporting, either partially or fully, the mission of the new network organization. For example, certain of the participating organizations may consign their research and engineering capabilities to the venture; others may assign their marketing and sales capabilities; still others may contribute their production and distribution strengths. The daily operations of these individual entities would remain under the jurisdiction of the parent company. However, the axis for management control and coordination would be the responsibility of the global project-management organization. In turn, the global project-management organization would be accountable to the owner's group.

The global project manager's job is akin to managing a matrix organization. However, the difference lies in the complexity and breadth of responsibility. Whereas traditional project management is concerned with accomplishing cost, schedule, and technical objectives—operating primarily within the framework of a given organizational culture—global project management deals with the creation and management of a totally new organization—with many different cultures—for the express objective of dominating a global market.

Creating a network organization by integrating components of several existing organizations involves the following process. (1) The most capable organizational elements among the participating companies are identified and nominated to participate in the new venture. (2) An infrastructure is built, implemented, and maintained to sus-

THE TRADITIONAL ORGANIZATION

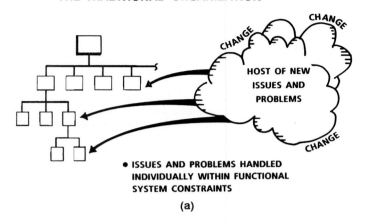

● ISSUES AND PROBLEMS HANDLED
INDIVIDUALLY WITHIN FUNCTIONAL
SYSTEM CONSTRAINTS

(a)

THE TRADITIONAL ORGANIZATION
EXPANDED TO MEET THE CHALLENGES

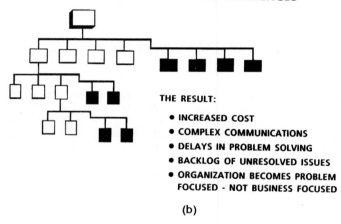

THE RESULT:

● INCREASED COST
● COMPLEX COMMUNICATIONS
● DELAYS IN PROBLEM SOLVING
● BACKLOG OF UNRESOLVED ISSUES
● ORGANIZATION BECOMES PROBLEM
FOCUSED - NOT BUSINESS FOCUSED

(b)

FIGURE 11.5   (*a*) Traditional organization. (*b*) Traditional organization
expanded to meet challenges.

tain the new organizations. (3) A team-building process is instituted to create the organizational culture and commitment that are needed to sustain the new concern.

The global project-management organization establishes a power base for its
authority. In most cases this power base will be derived from the authority granted by
the owner's group, the use of interpersonal influences by the global project team, and
the power that this team derives from the information and control system that they
assemble and operate.

The global project manager's most important job is to organize all the participating
elements of the network organization to achieve competitive advantage.

THE NETWORK ORGANIZATION

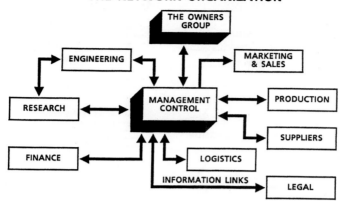

EACH BLOCK REPRESENTS THE CONTRIBUTION OF THE PARTICIPATING ORGANIZATION.
ALL THE INDIVIDUAL ELEMENTS OF THE NETWORK ORGANIZATION ARE TIED TOGETHER
BY INFORMATION LINKS THROUGH THE INFORMATION TECHNOLOGY SYSTEMS OF THE
MANAGEMENT CONTROL FUNCTION. A PROJECT MANAGEMENT ORGANIZATION MANAGES
THE TOTAL NETWORK ORGANIZATION. THE OWNERS GROUP ESTABLISHES THE OBJECTIVES,
FORMULATES POLICY AND SHARES THE RESULTS ACHIEVED.

FIGURE 11.6   Network organization.

Framework for Competitive Advantage

An organization gains competitive advantage when it finds new ways to outperform
its rivals. Any business that can perform better, faster, and cheaper than its opponent
will grow profitably.

Some measure of competitive advantage has been achieved by companies who
have gotten rid of underperforming business units, streamlined their operations,
reduced management levels, and given more attention to their customers. However,
the problem is far more complex. Real competitive advantage requires speed. Speed is
the dominant factor in achieving competitive advantage: speed in collecting information, speed in formulating plans and making decisions, and speed in taking action.
Fast-response companies manage successfully what Hout and Blaxill[12] refer to as the
OODA loop—the cycle of observation, orientation, decision, and action—so coined
by the Air Force to explain the difference between successful and unsuccessful combat pilots. Fast-response companies know what is going on at all times, and they
respond quickly (Fig. 11.7). These companies create a unique organizational architecture that enables them to outrun the competition in all facets of the business environment.

BUILDING FLEXIBLE, HIGHLY RESPONSIVE NETWORK ORGANIZATIONS

The basic building blocks for a flexible, highly responsive network organization are
structure, systems, and people (Fig. 11.8). Organizational structure should be gov-

FAST RESPONSE COMPANIES

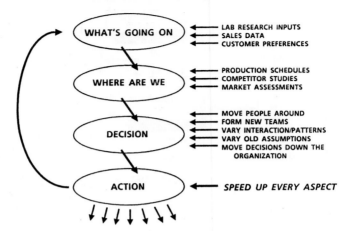

FIGURE 11.7   Fast-response companies.

ORGANIZATIONAL RESPONSIVENESS DEPENDS ON:

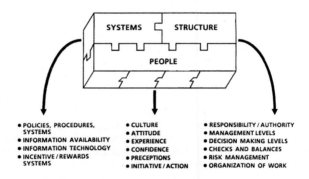

FIGURE 11.8   Organizational responsiveness.

erned by the strategy and vision established by top management. The structure of the organization should provide for the best arrangement of people and work to carry out the strategy. In turn, systems—especially information systems—integrate the organization so that it can perform its work as efficiently as possible. And finally, people, the human dimension of the organizational architecture, provide the innovation, entrepreneurship, knowledge, skill, and determination to accomplish management's strategy and vision. Each of these building blocks needs to be shaped, integrated, and directed to provide competitive advantage.

Enhancing Organizational Structure

An organization's structure can be enhanced in three areas: (1) the value-added chain, (2) the infrastructure, and (3) management direction and control. Enhancements made in one or all of these areas can lead to competitive advantage. However, real competitive advantage is gained when improvements are made on a system basis, that is, each component improvement is linked to a corresponding improvement in another organizational component. In the past, most organizational improvements revolved around reductions in the labor content of the product or service. Today improvements must be directed at elements of the total organization.

Improving the Value Chain.   To give an organization competitive advantage, it is important to enhance its total value chain. The value chain as described by Porter and Millar[13] is a series of interdependent activities that bring a product or a service to the customer. A company's value activities include primary activities—physical creation of the product, marketing, distribution, and services to the buyer—and support activities—legal, personnel, R&D, and infrastructure, which are needed to sustain the company's primary activities. A company's value chain is a system of interdependent activities, connected by linkages which must be coordinated and optimized (Fig. 11.9). Opportunities for gaining competitive advantage are created when a company finds ways to improve and transform activities in the value chain and the linkages between them. Some of the most dramatic gains in competitive advantage have been achieved by companies who have used advanced information technology to enhance not only the value chain and its linkage, but also the very nature of their product or service.

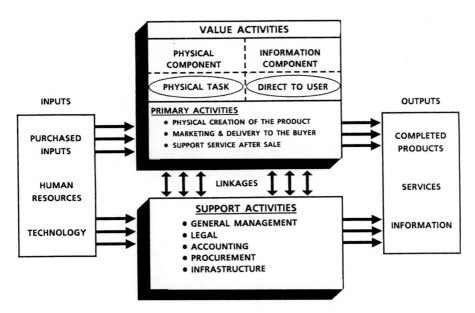

FIGURE 11.9   Value chain of organization.

Information technology can transform the value chain in any number of ways. The most obvious is that information technology is generating more data about every aspect of the company's operations. This enables management to analyze, improve, and control every facet of its activities more extensively. Computer models, simulation techniques, and artificial intelligence make it possible to evaluate different variables, scenarios, and strategies.

The impact of information technology is also being felt in the physical processing side of the company. For example, computer-controlled equipment of all types speeds up processes and increases accuracy, reliability, and quality. The equipment can also provide readouts of its performance or problems. This makes it possible to improve linkages between operations and to streamline many of the company's internal and external planning and coordination functions.

Consolidating the Infrastructure.   Organizational infrastructure provides support services, including physical facilities, utilities, logistics, procurement, and legal support. These are necessary to sustain the organization's primary activities. Although not directly involved in the actual creation of the company's product or service, these support services have an impact on the company's competitive advantage.

To achieve competitive advantage, companies must find ways to reduce, improve, and streamline all their support activities. Since supporting activities tend to be information-intense, they can be improved substantially by the utilization of information technology. For example, the use of EDI (electronic data interchange)—the direct computer-to-computer exchange of purchase orders, invoices, inventory status, bills of lading, and other standard business documents—will reduce the paper flow, eliminate layers of support organizations, and provide for better integration and control of the company's value chain activities.

Information is no longer confined to the point where it originated; hence it is possible to control parallel activities. For example, orders sent via the customer's computer to the supplier's computer will trigger a chain of activities such as adjusting inventories, issuing production orders, and generating delivery schedules and status reports. Thus information technology can tighten the infrastructure and provide for corresponding improvements in value chain activities and linkages.

Enhancing Management Control.   An organization can improve its speed and responsiveness by getting its management level as close to the value-added activities as possible. Advances in computers, database systems, and communications in general have enabled companies to shrink organizational levels and eliminate many staff and management functions. Gone are the legions of management personnel that collected, organized, analyzed, summarized, and presented information to top management and the rest of the organization. Computer workstations, CAD/CAM, LAN, fiber optics, satellite communications, electronic conferencing, office automation, and a host of other advancements in information technology now make it possible for management to create the networklike organizations. Network organizations have few management levels, and the support and value-added elements of the organization are thoroughly integrated by information technology.

To be successful in the global marketplace, network companies must manage well and have the speed and flexibility to adjust quickly to change. But this is not enough; companies will also have to be innovative. They will have to create new products and services for an ever-changing global market. And they must constantly redefine their role in the market place, that is, they must practice systematic entrepreneurship.

STRATEGIES FOR GLOBAL PROJECT MANAGEMENT

The global project manager must lead the design and implementation of processes, procedures, systems,and structures to organize work better, accelerate decision making, and improve communications and intelligence gathering. All are important and necessary to create high-performance flexible network organizations. However, the most critical task for project management is to establish, clarify, and reinforce the values, behaviors, and knowledge to create a new culture—a culture that synthesizes the strengths and differences of all the people who will be part of the global team. In this effort, project management must act as the role model and the champion of the new values and the new way of doing business.

A strategy for accomplishing this difficult task is to design and implement project activities that help to transform the frame of reference of the people assigned to the network organization.[14] Some of these activities will involve traditional team-building techniques. However, the major elements in the reframing process are the following.

1. *Establish cultural expectations.*   The culture needed to ensure success must be defined. Special project teams, involving representatives from all the participating firms, should be organized to study and decide what this culture looks like. These teams should determine what values are important, how people are to communicate and support each other, and the behaviors that will be rewarded. Once a clear charter of the organization's culture is established, it must be endorsed by top management (to ensure sustaining sponsorship) and communicated to the total organization.

2. *Design the organization.*   Organizational design objectives must be defined and implemented. Project management must lead teams to identify the processes, procedures, structures, and systems to be created, revised, or eliminated. The teams must also define the behavioral expectations for all members of the network organization. This will include methods for formal and informal communications, conflict management, decision making, problem identification and resolution, and protocols for social interaction.

These organizations will involve participants from different regions of the world, so it is important to develop national and cultural sensitivity and understanding. It is also necessary to know the important legal, ethical, moral, religious, and social codes that govern how each member must operate within his or her own organization and country. The goal of this effort is not only to provide understanding, but to harmonize values to build a team culture that produces synergy.

3. *Manage the projects.*   The global project manager's objective is to synthesize the values, beliefs, and experiences of the team. This can be done while team members are designing the organization's processes, procedures, and systems and when they are carrying out the organization's business. Project management must implement projects to make this happen. Diverse project experiences tend to create unique cultures.

4. *Create the environment.*   To deal with the harsh realities of global competition, network organizations will have to revitalize themselves constantly. Thus the global project manager will have to create an environment for learning, innovation, and entrepreneurship. The network organization will have to improve its processes, procedures, and systems constantly. In addition it will have to devise new products and services to meet the changing needs of the global marketplace.

The fundamental strategy for project management is to challenge constantly the established forms and norms. Project management must not allow the organization to become too attached to its traditions, history, and technology, and it must not allow the participants to fall victims to group think.

Creating an environment where innovation and entrepreneurship are the norm and not the exception is a key element in the cultural strategy of the global project manager.

CREATING AN ENVIRONMENT FOR INNOVATION AND ENTREPRENEURSHIP

Network organizations must construct an environment where the creative powers of the individual can be stimulated by the experience, knowledge, and resource of the total organization, and directed in a formal way to achieving business goals. In order to work well, innovation and entrepreneurship must be approached in an organized, professional manner without stifling the creative powers of the individual (Fig. 11.10). An organization that provides coordination and control based on shared goals, values, processes, and traditions, rather than rules and procedures, is a fertile environment for innovation and entrepreneurship. Some organizations have this type of environment; most do not.

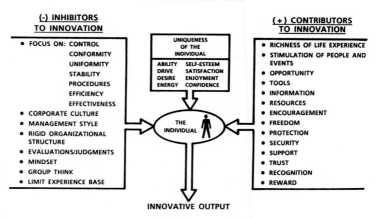

FIGURE 11.10   Inhibitors and contributors to quality and degree of innovation.

In order to create an environment that stimulates innovation and entrepreneurship, the organization must have a culture that actually enjoys the challenge of change and new ways of doing business. This may be difficult for organizations with well-entrenched cultures. However, project management can be a powerful instrument for creating the values and frame of reference for a new culture.

The Framework for Entrepreneurship

The typical corporate environment is not conducive to innovation and entrepreneurship. Unless an organization makes a specific effort to be innovative and entrepreneurial, it will tend to focus mainly on matters dealing with efficiency and control. The problem is further complicated by the fact that we tend to think of innovation and entrepreneurship as highly individualistic efforts. And to some extent they are. However, the corporate environment needs to harness individual creativity and enterprise in a constructive manner. Organizations wanting to excel in the global marketplace must have a formal approach to innovation and entrepreneurship that combines the synergy of the team with the creative abilities of the individual. As Drucker notes[15]: "Entrepreneurial businesses treat entrepreneurship as a duty. They are disciplined about it...they work at it...they practice it." The question is how should organizations gear themselves to "work at it"? How can they carry out this most important "duty"?

The framework for innovation and entrepreneurship can be created as follows:

- Establish teams specifically for innovation and entrepreneurship. These need not be full-time jobs, separate and distinct from the normal operational responsibilities of the corporation. However, they must have a specific creative, entrepreneurial mission.
- Formulate objectives, schedules, and budgets for all innovation and entrepreneurial activities. Innovation and entrepreneurship must be organized and managed like a project.
- Provide a well-defined structure for responsibility, accountability, and reporting.
- Develop and implement procedures, routines, methods, and techniques to facilitate the business of innovation and entrepreneurship. Some of these procedures and routines may have to be contrary to the norms of the organization.
- Staff teams with capable, experienced people. Assign the best people possible to give credibility to the undertaking.
- Monitor and measure the team's accomplishments against expectations. Provide resources and encouragement to those willing to undertake efforts that may be uncertain, unclear, ill defined, and fraught with risk.
- Build a culture to sustain the continued metamorphosis of the total organization.

The actions described are strategies for any well-managed project. This is why a project-management approach can be used to implement systematic innovation and entrepreneurship. The key to success is to combine the creative power of the individual and the synergism of the team.

Designing Teams for Innovation and Entrepreneurship

To thrive in the global marketplace, an organization must think and act like an entrepreneur. To bring this about, we need to involve everyone in the organization (Fig. 11.11). As Reich puts it,[16] there is always a need and place for the charismatic leader, but "if America is to win in the new global competition, we need to begin telling one another a new story in which companies compete by drawing upon the talent of all their employees, not just a few maverick inventors and dynamic CEOs."

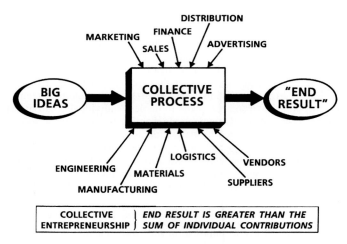

FIGURE 11.11 Collective entrepreneurship—end result is greater than sum of individual contributions.

To create a culture for innovation ad entrepreneurship it is important to launch these endeavors with a well-crafted sense of theater. Also, it is useful to perpetuate war stories, myths, and legends to sustain the new culture. It is particularly worthwhile to create excitement and generate a feeling of energy about the whole undertaking.

Organizations can commission special teams to lead innovation and entrepreneurial efforts. These special teams can create a vision of products, services, processes, procedures, and systems that can help the organization to gain competitive advantage. Furthermore, using a process of entrepreneurship, commitment, and team effort, the vision can be put into actual operation to achieve the desired results. An appropriate name for these entrepreneurial teams is VECTOR. The name is an acronym of the team's mission statement (Fig. 11.12).

A VECTOR team is empowered to capture the creative energies of the total organization, give form and substance to its visions, and accomplish an entrepreneurial undertaking. VECTOR teams are intended to carry out their mission through an interactive process based on shared goals, values, ideals, and traditions.

ESTABLISHING PRACTICES FOR INNOVATION AND ENTREPRENEURSHIP

Drucker[15] states that "organizations must be made receptive to innovation." He also says that we have to find ways to "make innovation attractive to the organization." Drucker recommends that organizations establish specific practices for entrepreneurship, including organizational structure, staffing and manning, compensation, incentives, and rewards. In short, we must create a structure that not only empowers people to be entrepreneurial, but demands it. We can use VECTOR teams to design and implement the organization structure, policies, procedures, plans, and reward systems

VECTOR TEAMS

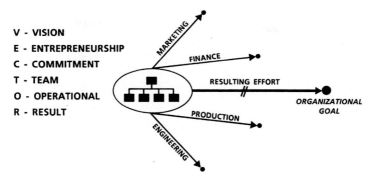

V - VISION
E - ENTREPRENEURSHIP
C - COMMITMENT
T - TEAM
O - OPERATIONAL
R - RESULT

OPERATIONAL APPROACH:

- MAKE KNOWN IDEAS, PROBLEMS, NEEDS

- LOOK FOR PARTICIPANTS

- CREATE THE CULTURE

- OBTAIN COMMITMENT

- CHALLENGE THE TEAM MEMBERS

- LEGITIMIZE THE TEAM

- MANAGE THE ENTREPRENEURIAL EFFORT

FIGURE 11.12   VECTOR teams.

to provide for systematic innovation and entrepreneurship. However, in order for VECTOR teams to be effective, the organization needs to establish a supportive environment that will foster its success.

Creating the Right Environment

To foster innovation and entrepreneurship throughout the organization it is necessary to change the perception of success in the corporate environment. In most cases advancement in the corporation is based on individual accomplishment over everything else. However, in today's highly competitive global environment, innovation and entrepreneurship cannot be left to the gifted few. The new values that have to be admired and rewarded must be those which spur the creative energies of the total organization. Individual efforts should be rewarded on the basis of the value that is added to the team effort.

Creating these new values will be very difficult for organizations that focus on the superstar. However, these organizations will have to build a "clan culture," one that is based on a collaborative decision-making process, open discussion, and an emphasis on teamwork, problem solving, and creativity.[17]

Getting the Right People

The first step in creating a VECTOR team is to select the right people. The best candidates are individuals from the ranks of project management. They have the mental attitude, experience, and environmental conditioning needed for a new undertaking. Management must select the best seasoned performers from the project-management ranks. This is important for two reasons. First, the job at hand will demand individuals who are comfortable in an environment of uncertainty and change. Second, those selected must be recognized as seasoned performers by the total organization. This gives credibility to the undertaking and will do much to convince the whole organization of management's commitment to this new undertaking.

Selecting the right people involves identifying those who share the vision and are truly innovative. VECTOR teams need people who tend to be iconoclastic. Typically these individuals know how to get things done despite the organization. We are looking for people who are totally dedicated to the success of their undertaking and who have the initiative and courage to work around the rules when necessary to get the job done. Every organization has these people.

Defining Mission and Structure

A project team is organized to accomplish a specific objective on time and within budget. In a similar manner, VECTOR teams are created to stimulate, plan, organize, lead, and direct innovation and entrepreneurship for the organization. Whereas the mission of a project team is usually definable in quantitative terms, the mission of a VECTOR team is more appropriately defined in qualitative terms. Hence VECTOR teams work to increase the capacity of the organization's resources to create wealth or to increase the consumer's value and satisfaction from the organization's products and services. In fulfilling this mission, VECTOR teams work to stimulate an idea, conceptualize a need, identify a problem or opportunity, and institute the plans and programs to translate these into enhanced organization resources, products, or services.

VECTOR teams will have to identify routinely and systematically the changing concerns, trends, values, and practices that impact the economic viability of the organization. More importantly, VECTOR teams will act as the catalyst for the creative talents, abilities, and experience of the total organization.

Using Formal and Informal Teams

There are two types of teams—formal and informal (Fig. 11.13). A formal team is recognized by management, that is, the team has a specific charter, entrepreneurial goal, budget, and schedule. This team has been commissioned by management to pursue a venture that is considered important to the future viability of the organization. Team members may have full- or part-time assignments. Informal teams are encouraged but do not have formal assignments or charters. Informal VECTOR teams work on proj-

VECTOR TEAMS

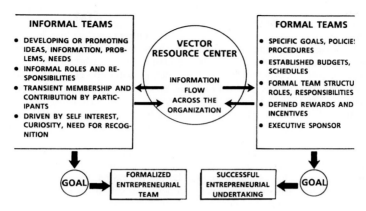

FIGURE 11.13 Types of VECTOR teams.

ects on their own time and at their own pace. The only resources provided by the company are those available through the VECTOR resource center.

This formal and informal team approach provides for maximum organizational participation at lowest cost. However, to get maximum participation, management must create an environment where the team members can experiment without fear of criticism or risk of loss of security. Thus the network organization must establish a process that encourages, rewards, and protects people who want to develop new ideas.

A PROCESS FOR INNOVATION AND ENTREPRENEURSHIP

The focal point for innovation and entrepreneurship in a network organization is a VECTOR team resource center. The resource center can be as simple as a convenient meeting place or as complex as an electronic information center. The center must be accessible to everyone in the organization.

People in the organization use the resource center to identify and explore problems, needs, and ideas. The center can provide databases and electronic bulletin boards for individuals to explore their ideas with others in the organization regardless of their physical location. The VECTOR team resource center must ensure the security of all users. It must function so that participants can present, explore, develop, and experiment with their ideas without fear of criticism or ridicule. Also, the identity of the participants and their ideas must be protected until they are ready to formally present their programs to management.

The goal of the resource center is to provide a platform for stimulating innovation and entrepreneurship apart from the mainstream activities of the organization. Individuals use the resource center to establish a common framework of interest. When the individuals feel that they have enough interest, they can move forward and organize an informal VECTOR team. Participation in a VECTOR team is voluntary.

Process Scenario

An engineer working in one of the divisions of a network organization has an idea for a new product. He thinks he has come up with something novel, but he is concerned because his idea is far removed from the company's current product line. He is also worried about management's interest and support; he wonders if his product can be produced economically and if there is a large enough market to warrant development. He has a dozen or more questions. To see if his idea is viable the engineer implements a process that the company has established to encourage innovation and entrepreneurship.

The engineer submits a brief outline of his idea to the VECTOR team electronic bulletin board (Fig. 11.14). He identifies himself only by a code number which is assigned by the resource center. He asks interested individuals to respond with comments and suggestions. Within a few days the engineer is pleased to find several responses to his proposal. Some of the inputs offer ideas for enhancing the original idea, others point out significant problem areas, still others provide information on similar projects which have been attempted before but failed. Our engineer is encouraged by what he has learned. He refines his ideas and submits a more definitive proposal to the electronic bulletin board along with a request for team participants. This is where the concept of the VECTOR team begins to take shape.

INFORMAL VECTOR TEAMS

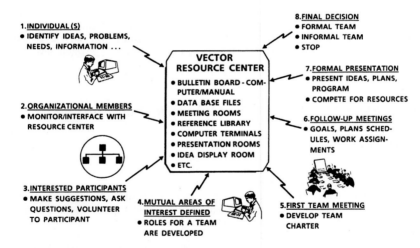

FIGURE 11.14   Informal VECTOR teams.

At some point interested individuals meet to organize their team and develop goals and strategy. Collectively they produce the plans, schedules, procedures, and ground rules appropriate to their undertaking. All team efforts are voluntary and done on the members' own time. The company contributes no support other than that provided through the VECTOR resource center. This means that the team must achieve its goals on its own time and with its own resources. The team may solicit support from other teams, utilize information gleaned from company databases or other sources, and it may even choose to work on its project at the expense of its members' more routine responsibilities. Teams are expected to be creative in finding ways to develop their ideas.

The team members work on their project until satisfied that it has developed to the point that it can be sold to management. All VECTOR teams are informal teams until they sell their ideas to management. Teams have to convince management not only that they have good ideas, but also that their ideas are better than those being proposed by other VECTOR teams.

At regular intervals the owner's group of the network organization will convene formal seminars where VECTOR teams present their ideas. Each team will make a presentation to persuade management to fund its project. Successful teams will receive formal VECTOR team status. This means that team members will be assigned full- or part-time to make their proposals a reality. Formal VECTOR teams are distinct and separate business development ventures, operating within the framework of the total network organization (Fig. 11.15). Each team has an executive sponsor who is responsible for its success.

FORMAL <u>VECTOR</u> TEAMS

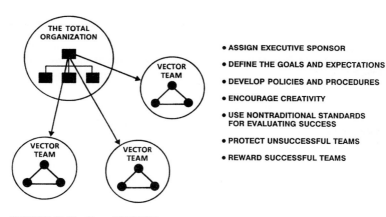

FIGURE 11.15   Formal VECTOR teams.

Formal VECTOR teams are segregated from the established routines of the corporation. The teams decide how to operate; they create their own policies, procedures, and standards. They are judged on how well they accomplish their objectives. Successful teams are rewarded by sharing in the benefits or profits of their ventures. Unsuccessful teams are protected and ensured a vital role in the organization. The executive sponsor's job is to see to it that the team's creativity, innovation, and entrepreneurship are not inhibited by fear, worry, or concern about its members' future in the organization.

THE GLOBAL PROJECT MANAGER'S CHALLENGE

The global project manager's goal is to manage the network organization in the global marketplace to obtain market leadership. To be successful, the global project manager will have to deal with issues far more complex than anything in the past. These issues include (1) technology—using it more effectively while adjusting to all its ramifications; (2) culture—creating organizational culture, work relationships, attitudes, and values for everyone in the new organization; (3) organization—building new types of organizational structures and work relationships; (4) work processes—finding better ways through technology, creativity, and innovation to get work done; and (5) perception—changing how we view the environment in which we work and live and developing a proactive mentality for responding to the threats and opportunities that this environment will present.

The global project manager will also have to find and develop a new breed of individual, a professional who has a well-developed portfolio of technical, managerial, and human relations skills. This new breed of professional will be a technologist, behavioral scientist, and pragmatic participant, all rolled into one. Team members will have to know how to utilize the power of satellite communications, electronic conferencing, computer-based workstations, information databases, decision support systems, artificial intelligence, and a host of other information technology tools. In addition, these individuals will have to have an extensive organizational and business background. They have to understand implicitly how organizations function and how specialized areas of knowledge contribute to the overall accomplishment of a major undertaking. Most important, these individuals will need to be skilled in dealing with people of different backgrounds, cultures, and nationalities, in an environment of uncertainty and change. Clearly, in today's world of the specialist it will be difficult to find the renaissance professional that will be needed for the challenges of the global marketplace.

SUMMARY

The key to survival in the global marketplace is speed, flexibility, innovation, entrepreneurship, and a strategy that enables the organization to dominate its market.

Organizations must band together in networks or communities of interest to systematically develop new ideas, products, processes, and services to meet the needs of a changing world. They must also introduce their innovations into the marketplace before the competition. In the global marketplace, network organizations will integrate the values, beliefs, and ideals of people from different lands to create an organizational culture that transcends national interest and traditions. The new network corporation will manage culture as a resource which can give competitive advantage. This new environment will provide project management ample opportunity to build on its unique management tools and techniques to create innovative processes, procedures, systems, and culture for the future. Project management is one of the few management disciplines that is well equipped to master the challenges of tomorrow.

REFERENCES

1. V. Royster, "How Stands the American Republic?," *Wall Street J.,* Apr. 21, 1988.
2. R. E. Carlyle, "Managing IS at Multinationals," *Datamation,* p. 54, Mar. 1, 1988.
3. P. Drucker, "The New World According to Drucker," *Business Month,* May 1989.
4. P. Drucker, "The Transnational Economy," *Wall Street J.,* Aug. 25, 1988.
5. J. B. Barney, "Organizational Culture: Can It Be a Source of Sustained Competitive Advantage?," *Aca. of Management Rev.,* vol. 11, no. 3, 1986.
6. G. Hofstede, "Cultural Dimensions for Project Management," *Proc. INTERNET 82* (Copenhagen, Denmark).
7. J. Tuman, Jr., "Information Technology, Organizational Culture, and Engineering Management for a Volatile World," *Engineering Management J.,* vol. 1, Mar. 1989.
8. J. J. Sherwood, "Creating Work Cultures with Competitive Advantage," *Organizational Dynamics,* Spring 1988.
9. J. Tuman, Jr., "Project Management in Different Organizational Cultures: A Tool for Creating Competitive Advantage," *Proc. INTERNET 90* (Vienna, Austria).
10. J. Tuman, Jr., "Transnational Project Management for the Global Marketplace," *Proc. INTERNET 88* (Scotland).
11. R. Hayes and R. Watts, *Corporate Revolution,* Nichols Publ., New York, 1986.
12. T. M. Hout and M. F. Blaxill, "Make Decisions Like a Fighter Pilot," *The New York Times,* Nov. 15, 1987.
13. M. E. Porter and V. E. Millar, "How Information Gives Competitive Advantage," *Harvard Business Rev.,* July–Aug. 1985.
14. D. B. Fetters and J. Tuman, Jr., "Project Management—Agent for Change: Building a New Culture for a Nuclear Engineering Organization," *1989 Proc. Project Management Institute* (Drexel Hill, Pa.), p. 22.
15. P. Drucker, *Innovation and Entrepreneurship: Practice and Principles,* Harper & Row, New York, 1985.
16. R. B. Reich, "Entrepreneurship Reconsidered: The Team as Hero," *Harvard Business Rev.,* May–June 1987.
17. M. Elmes and D. Wilemon, "Organizational Culture and Project Leader Effectiveness," *1986 Proc. Project Management Institute* (Montreal, Canada).

BIBLIOGRAPHY

Applegate, L. M., J. I. Cash, Jr., and M. D. Quinn, "Information Technology and Tomorrow's Manager," *Harvard Business Review,* Nov.–Dec. 1988.

Chase, H. W., *Issue Management: Origins of the Future,* JAP, Inc., Stamford, Conn., 1984.

Denison, D. R., *Corporate Culture and Organizational Effectiveness,* Wiley, New York, 1990.

Drucker, P. F., *Managing in Turbulent Times,* Harper & Row, New York, 1980.

Hamermesh, R. G., *Making Strategy Work, How Senior Managers Produce Results,* Wiley, New York, 1986.

Kanter, R. M., *The Change Masters, Innovation for Productivity in the American Corporation,* Simon and Schuster, New York, 1983.

Kuhn, R. L., *To Flourish among Giants, Creative Management for Mid-Size Firms,* Wiley, New York, 1985.

Peters, T., *Thriving on Chaos,* Alfred A. Knopf, New York, 1987.

Porter, M. E., *Competitive Advantage, Creating and Sustaining Superior Performance,* Free Press, New York, 1985.

Porter, M. E., "From Competitive Advantage to Corporate Strategy," *Harvard Business Rev.,* May–June 1987.

Strassmann, P. A., *Information Payoff, the Transformation of Work in the Electronic Age,* Free Press, New York, 1985.

Waterman, R. H., Jr., *The Renewal Factor: How the Best Get and Keep the Competitive Edge,* Bantam, New York, 1987.

Zuboff, S., *In the Age of the Smart Machine: The Future of Work and Power,* Basic Books, New York, 1988.

CHAPTER 12

IMPLEMENTATION OF PROJECT MANAGEMENT AT THE PHILIPS CONCERN

J. Kessel and S. Evers

Philips Project Management Training and Consultancy,
Eindhoven, The Netherlands

Sjaak Evers, from The Netherlands, received his degree in industrial engineering from the University of Eindhoven in 1978. He served as management-trainer and -consultant for the Dutch Ministry of Agriculture and Fishery from 1978 to 1982 and as an industrial engineering specialist for the Network of Innovation Consultancy Centres in Holland from 1982 to 1987. Since 1987 he has been (project-) management-trainer and -consultant for the Philips company. The implementation and the integration of project management has to do with projects and organizations, mainly within but sometimes outside Philips, usually in Holland, but sometimes in other areas.

Joop Kessel, from The Netherlands, studied economics and statistics and followed a postgraduate training for Business Consultant. He worked with Philips for 36 years, 20 years in the commercial field and 16 years as internal business consultant. During these 16 years he gathered a worldwide experience in the implementation of project management. Nowadays he works as external business consultant in the field of strategy formulation, implementation of organizational change, and project management.

INTRODUCTION

Setting the Scene

Thirteen years after the invention of the light bulb, the ground was broken for The Netherlands' first "Gloeilampenfabrieken." With this, the Philips family carved its niche into history. Flush with the profits from a thriving family tobacco trading enterprise, Anton Philips concluded that the lighting industry constituted the wave of the future and invested accordingly. Anton, a student of commerce, formed a partnership with his brother Gerard, an outstanding physicist.

The year was 1891. From the outset the brothers pursued opposite ends. Anton believed success lay in solid financial policies. Gerard contended that progress lay dependent on technical innovation. This historic rift between the fledgling research and commercial efforts of the firm, which insiders call the "Gerard and Anton syndrome," still haunts Philips.

Sphere of Influence

From one modest 4000-square-foot brick structure on the outskirts of urban Eindhoven, Philips evolved into a $30 billion multinational behemoth. The company, the third largest in Holland and currently ranked as the twenty-second largest industrial company in the world, controls 197 subsidiaries in 61 nations. With sales in 165 countries, their sphere of influence is indeed ecumenical.

Product Divisions

Philips has been grouped into nine product divisions:

1. *Lighting.* Lamps, professional luminaires and gears, automotive and signaling lamps
2. *Consumer electronics.* Video display products, video equipment, audio, personal information products, car stereo
3. *Domestic appliances and personal care.* Home comfort, kitchen appliances, personal care
4. *Components.* Passive components, magnetic products, liquid-crystal displays
5. *Semiconductors.* Discrete semiconductors, integrated circuits
6. *Information systems.* Data systems, dictation, laser magnetic storage industries, customer services
7. *Communication systems.* Business communication systems, data communication, cable transmission and network access, private mobile radio, public radio communications, radio transmission and rural telephony, cable and glass fibers
8. *Industrial electronics.* Testing and measurement, analytical, industrial automation, communication and security systems
9. *Medical systems.* Röntgen diagnostic systems, x-ray diagnostics, computer tomography, magnetic resonance imaging, ultrasound equipment, radiation therapy

National Organizations

In about 60 countries the world over Philips has national organizations, representing Philips in those countries and supporting the commercial, economic, and technical activities of the product divisions.

Organizational Structure

The main structure of Philips is a matrix of product divisions and national organizations (Fig. 12.1).

Philips matrix

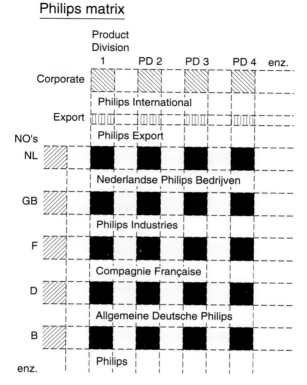

FIGURE 12.1 Philips matrix.

On May 14, 1990, Jan Timmer, head of the Consumer Electronics Division, became CEO of Philips, a year ahead of schedule. Investors were horrified when Philips announced that its operating earnings for the first quarter had plummeted to $3.25 million from $121 million a year earlier. Timmer's appointment caused their stock to rebound, but getting Philips in shape for the competitive 1990s will be much tougher.

The company has historically experienced difficulties parlaying technical breakthroughs into hard profits. It was Philips, after all, that virtually invented first the audio cassette, then the video cassette recorder, and most recently the compact-disc player, only to have the bulk of all three new markets snatched away by Japan.

In December 1990 Timmer presented the following evaluation of Philips:

Strengths

- Leading market positions in various sectors
- Deeply rooted sales organizations in 60 countries
- Key technologies within own scope
- Philips brand name

Weaknesses

- Profitability: structurally too low-level
- Therefore decreasing financial maneuverability
- Loss-giving activities too often labeled "strategic"
- Efficiency (sales per employee) too low by competitive standards
- High restructuring costs
- Vulnerable to currency changes (dollar, yen)

He presented the sectors in which Philips has leading positions (Table 12.1).

TABLE 12.1. Philips' Market Positions

Global	Europe
No. 1 in lamps	No. 1 in integrated circuits
No. 1 in color picture tubes	No. 1 in compact-disc equipment (including spinoffs)
No. 1 in shavers	No. 1 in workstations for financial industries
No. 1 in passive components	No. 1 in car radios
No. 1 in Compact Discs	No. 1 in analytical equipment
No. 1 in dictation equipment	No. 1 in medical imaging equipment
No. 1 in coffeemakers	No. 1 in public audio systems
No. 1 in x-ray spectrometry	No. 1 in simultaneous interpretation systems
First manufacturer reaching 100 million television production level	
No. 1 in color television (including Grundig)	

Timmer has undertaken drastic steps in the attempt to change corporate direction. As has been said, "People expect Philips to turn around like a sailboat, but they're working with a supertanker." The results are far from tallied, but projections are in order. Many predict that there will be few European survivors in high technology. There is no question, however, that Philips will be among them.

PROJECT MANAGEMENT AT PHILIPS

At the end of the 1960s Philips became acquainted with project management (PM). A concept called systems management (Fig. 12.2) was introduced in the product division Data Systems. This division developed large computer systems.

Systems management was originally developed for application at the U.S. Department of Defense. Its capabilities were rapidly recognized by the American industry and implemented as their standard approach to new business at that time. The system was introduced in Data Systems by a company called Computer Sciences International (C.S.I.).

When introducing the system, the management of the product division wrote:

SYSTEMS MANAGEMENT CONCEPT

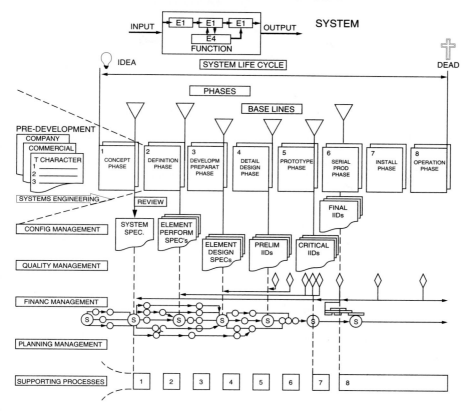

FIGURE 12.2 Systems management concept.

The need to introduce this type of system is evident; we produce a very complex product which, preceding delivery to the user, takes years of development work.

Indeed the development time for computers at that time was around 3 years. Systems management was designed to cover such a long period.

The need for such a system was created by the complexity of both the technology and the management of such developments. To quote management once more:

Today's modern business is faced with a condition of interdependence of materials, elements and people the like of which has been non-existent in the past. We can no longer work in isolated groups; we must recognize these interdependencies and operate as a team directing all our efforts towards a common goal. Everybody concerned in the development must be constantly aware of the relationship between his work and that of his colleagues in his own and other departments.

Nowadays this sounds familiar, but not so at the end of the 1960s. Philips was still a product-oriented company based on a functional (organizational) structure and struggling to adapt itself to a more market-oriented approach.

The use of systems management in Data Systems was intended to:

- Facilitate interfaces in the development activities while maintaining the functional structure
- Define a process of planning and controlling

The basic principles of systems management can still be recognized in the present PM concepts within Philips. This is not the consequence of a consistent top-down policy, but the result of an organic process that developed itself during the 1970s and 1980s.

Characterizing the organization of Philips in the 1960s can best be done as a "federative" matrix with high degrees of autonomy for the product divisions (the one side of the matrix) and the national organizations (the other side of the matrix). "Constructive friction" was a slogan at the time. It meant that you had to convince your counterparts rather than imposing decisions upon them.

Apart from the administrative system, which was centralized, every division and national organization could organize itself according to its own needs. Management of Medical Systems was in need of a PM concept and chose systems management. This did not mean that now the other divisions also had to use this concept.

PM could develop itself organically because of a number of factors.

1. A restructuring of the Philips organization took place in the 1970s as a consequence of, among others, the growing impact of the European Economic Community, which meant that trade barriers gradually disappeared. Production and development facilities could therefore be reallocated. Most of the development activities were situated in The Netherlands.

2. In the 1970s Philips was still a Dutch company. Many of the more strategic management positions—inside and outside The Netherlands—were in the hands of Dutch managers. All these managers, in the first years of their careers, followed the same series of general management training (GMT).

3. Management training departments understood at an early stage the importance of PM for the company. As a consequence, in GMT sessions explicit attention was given to this topic.

4. Many of the managers who took these training sessions were already very well schooled and task-oriented. PM emphasizes the task-oriented management style.

5. These young managers, especially those who worked in the product development field, were eager to introduce PM in their departments. They asked for the support of the management training group.

6. As a result an iterative process started. A body of knowledge was built up with regard to the various sorts of need for PM, the various sorts of concepts, and the implementation skills. This knowledge was passed on again in GMT sessions.

At the moment many departments in various product divisions make use of PM in one way or another. The extent depends on the need for such a tool and the willingness to make use of it. Some departments use it occasionally when a big project comes up; others made it their basic organizational philosophy. There still is a great variety.

The central automation department originally developed its own concept on which to base its projects. This concept differed to quite some extent from the concepts used in the product development departments. In the course of time, however, it gradually adapted its system in such a way that both the concept and the jargon now have great similarity with the more generally accepted concept.

Sometimes the jargon used is a problem. For instance, when we compare the phasing and the milestones of PM concepts used, they often look very much alike, but they have been given different names. This makes discussion on PM between departments sometimes difficult. Later in this chapter we describe a case that is based on this aspect.

From the foregoing one may assume that the implementation of PM was in general a smooth process. Unfortunately this was not the case. Philips was—and still is—a functional organization and very dedicated to that structure. Time and again it is amazing to see how soon young managers submit themselves to the culture of functional management. This is the "achilles heel" of PM. PM demands a form of matrix structure, including a shift in responsibilities and authority between functional managers and project managers—an ongoing struggle which takes a great deal of energy and goodwill. In summary, we can say that PM has a strong foothold at Philips, but this does not take away the fact that much still remains to be improved.

HOW TO IMPLEMENT PROJECT MANAGEMENT

We have built up a considerable body of knowledge on the implementation and integration of PM. The former is the subject of this section, the latter will be discussed later. Our general strategy for the implementation of PM, visualized in Table 12.2, is described in this section, and a specific case is presented. The case is about the implementation of PM in a development and manufacturing plant of the Philips product division Domestic Appliances and Personal Care (DAP).

TABLE 12.2. Implementation Strategy

Management team (MT)	Phase 1
Key personnel (KP)	Phase 2
Discussion between MT and KP	Phase 3
Conditions (manual)	Phase 4
Project start-ups	Phase 4
Progress control	Phase 5

Implementation Approach

Phase 1: The Management Team. A very important condition for successful implementation is the support of the management team (MT). Especially at the beginning it is necessary to discuss the pros and cons thoroughly. Management sometimes seems to be convinced of the necessity to introduce PM, but if you interview MT members personally, you often find that not everyone agrees. Therefore it is important to discuss in the MT two issues:

1. The context, that is, the situation of the organization in which PM has to fulfill **a** function. *Why* do we have to implement PM?

2. The PM concept itself. *What* do we have to implement?

To prepare for discussion on the first issue, we generally start with a strategic analysis of the organization. This can be done via the McKinsey 7 S model. With the help of this model we define the "ist" and the "soll" situation of the organization. Another approach is to make use of the "SWOT" analysis. In this analysis we confront external "opportunities and threats" with internal "strengths and weaknesses" from which we derive the relevant issues for the organization.

These analyses can be done by us alone or together with the MT. We prefer the latter because we have the feeling that this leads to a better understanding of the situation and will produce more commitment from the MT. This approach, however, takes more time. Discussions can last from one to three days.

After we have made "a map" of the situation of the organization, we can discuss whether the implementation of a PM approach is useful. The questions to be answered are:

- Is PM logically fitting in the situation?

- Is PM really a tool to get strength inside the organization, necessary because of opportunities outside the organization, or should other measures be taken first?

If the answer is in favor of PM (a satisfactory answer has been given to the *why* question), then the relevant questions are:

- To what extent shall we introduce PM?

- In which form will we apply PM?

(The answer to *what* we have to implement.)

Finally we discuss together the conditions and consequences for the organization. Items are:

- The matrix organization structure

- The management of capacity, change, and product programs

- The human side of change

It depends on the situation how deeply subjects are discussed.

Phase 2: The Key Personnel.   The objective of phase 1 is to achieve a joint decision of the MT on the necessity of the implementation of PM. Once this has been achieved, the next step is to involve those people who will have a major influence on the ultimate success. We call these the key personnel (KP), by which we mean (potential) project managers, project support experts, and the like.

With these people we actually have the same discussions on the pros and cons of PM, but we are more specific about the organizational conditions that have to be provided. We do this by means of workshops—mostly two workshop sessions of two or three days each.

In the first session we convey the principles of PM. As homework, in between the two blocks, the participants try to structure running projects according to the PM principles. They also compare the present working conditions with the conditions necessary for PM.

Phase 3: Discussion between MT and KP.   In the second workshop we discuss the homework findings and prepare points of discussion with the MT at the end of the second workshop. Generally these points are:

- Conditions which have to be fulfilled in the organization in order to implement PM successfully
- A ranking of the conditions
- A proposal for action plans

Such action plans differ of course according to the situation. Generally speaking, there are two kinds of action:

- To work out proposals for changes in conditions that are high on the priorities list
- To identify jobs that can serve as trials for PM

Phase 4: Continuation—Project Start-Ups or Manual. At the end of phase 3 there is in fact a milestone discussion. As a result there should be a go/no go decision about the implementation. If the decision is affirmative, arrangements should be made to draw up a plan for action. From the foregoing it will be clear that—for a successful continuation—the most crucial factors are "projects" and "conditions for projects."

The attention given to each of these factors differs of course depending on the situation. Some organizations give their attention to both aspects from the beginning. Other organizations first pay attention to projects. They start with project start-up sessions (PSUs). Other companies start to develop conditions. They appoint a working team to draft the concept of a PM manual. The decision of what to do first and what later depends on the priorities within the organization.

Priorities can be:

- *Structure*-related conditions, such as drafting the tasks, the responsibilities, and the degree of authority for the various stakeholders
- *Strategy*-related conditions, such as what can be considered as a project and what not
- *Systems*-related conditions, such as software and planning tools
- *Staff*-related conditions, such as training and development of employees
- *Style*-related conditions, such as the way management and employees communicate with each other at the beginning of a project and at decision points

A carrier for working out these conditions can be the drafting of a PM manual. It forces the organization to write down these conditions, and it has been our experience that putting things on paper really makes people think about them. Advantages of drafting a manual are that conditions have to be discussed thoroughly and the results are made visible. Also, differences with the present way of working are made clear. A disadvantage may be that later on the manual starts to lead a life of its own, and it may increase bureaucracy.

Another aspect to observe is that management sometimes thinks that once the manual has been discussed and agreed upon, PM has been implemented, whereas in fact the implementation is actually only starting. It takes at least 1 to 3 years before PM has been institutionalized in an organization.

Phase 5: Progress Control—Evaluation and Planning. With a view to the foregoing it is necessary to evaluate periodically the progress against objectives set at the beginning of the operation. The second point is that the PM concept—after its implementation—should be evaluated regularly against changing conditions. The concept should not be considered as a mechanistic tool, but as an organic instrument that should adapt itself to the changing circumstances.

Final Remark.   We described the implementation as a sequential process, without overlaps. Practice can be different, with related consequences, of course.

The DAP Plant Case

The internal management consultant of the DAP plant contacted us to discuss the implementation of PM in "his" plant. According to our implementation strategy, we organized a meeting with the MT. Problems in the innovation process with respect to throughput time, coordination of activities, capacity, and budgets were the reason why the MT was interested in PM.

We had discussions with the MT during three sessions on the *why* and *what* questions, as mentioned. To prepare for the next phase in the implementation, a list of key employees with respect to PM was composed by the MT. The appointed KP attended their two training sessions, as explained before.

In block 1 we taught them the principles of and the conditions for PM. We introduced, discussed, and exercised:

1. Project result description
2. Phasing
3. Controlling
4. Decision making
5. Conditions for the application of PM

At the end of block 1 we organized five teams to prepare the second block. We defined the following tasks for each team:

1. What are we calling a project?
   What should we call a project?
2. How are we doing the phasing of a project?
   How should we do it?
3. How are we doing the controlling of a project?
   How should we do it?
4. How are we doing the decision making because of a project?
   How should we do it?
5. How are we doing the organization control of a project?
   How should we do it?

On the basis of this homework, the group, assisted by us, developed in block 2 a draft of a manual or framework in which they described how PM could be done in their situation, including agreements with respect to tasks, responsibilities, and authorities of the project owner, project manager, departmental manager, and so on. At the end of block 2 the KP presented their drafts of the manual to the MT and discussed them with them.

As a consequence of the analyses and the discussions on the conditions for PM during block 2, and according to our suggestions, the KP proposed to the MT to install an implementation team to deliver:

1. A manual for the management of a product creation process
2. A plan for the training of other people of the plant, involved in projects

3. A proposal about software to control the projects

In the following MT meeting the MT decided to agree with the different proposals. The MT decisions have been communicated to the KP.

An internal management consultant became the chairperson of the implementation team and the plant manager became its project owner. They agreed on the implementation assignment.

The first activity of the implementation team was to organize a project start-up (PSU) workshop for the project "mini food processor." The purpose of the workshop was to generate a platform for the management of PSU projects.

Before the start of a PSU workshop the project owner appointed the project manager and the team members. The project owner and the project manager made the necessary preparations for such a session.

During the workshop the following topics were treated:

- Introduction regarding the project and explanation of the importance of the output
- Scope of the project
- Functional requirement specification
- Work breakdown structure (WBS)
- Responsibility charting
- Budget allocation on the various main activities of the WBS
- Control aspects

The results of the PSU were:

- Project owner, project manager, and project team agreed on the project.
- Participants became involved in the project; team spirit and commitment have been established.

An important aspect of these sessions was also the negotiation with the project owner on the conditions in terms of the time, money, and capacity necessary to achieve the project results.

The workshop started with a presentation by the project owner in which he gave his view on the project: the problem or reason why, what should be the result of this project. The participants, divided into different subgroups, tried to formulate in their own words what they understood as being the problem and the result of the project.

The different presentations of the subgroups were integrated by the project leader after the discussion.

As the next step the participants, divided into new subgroups, tried to define the phase plan for this project: which activities in which phase. The presentations of the different subgroups were integrated by the project team, after which they defined the control plan (time, capacity, funds, and so on) for the project. At the end of the PSU workshop the project manager, assisted by the group, presented their views of the mini food processor project to the project owner:

- Problems and results
- Phase plan
- Control plan

A strong basis had been established for a successful project.

In addition to organizing the PSU workshops, the implementation team coordinated the following activities. The team was very quick with respect to the choice of software to control the projects. The development of the manual for the product creation projects needed more time. To introduce the manual effectively to the organization, the implementation team proposed to combine its introduction with the training of other people involved in PM.

The chairperson of the implementation team explained the manual and the principles of PM to the other employees in the plant in a series of six workshops, each for 15 employees. We assisted in the first two of the six workshops. The chairperson did the other four alone.

After the manual had been developed and implemented, it was recognized as an important guide in the projects.

After 1 year the first evaluation was made on the use of the manual. The main conclusion was that the MT was on the right track, but that improvements in the content had to be made. The MT adapted its project manual on several points. A year later a second evaluation was made. Although part of the original MT had mutated, the new MT members continued working according to PM principles and improved the manual on the basis of suggestions brought forward during the evaluation. Apparently the way of working according to the PM principles had been integrated gradually.

PROJECT MANAGEMENT AND THE PHILIPS CULTURE

In the introduction we mentioned that in 1991 Philips celebrated the 100th birthday of the company. During these 100 years Philips has been applying successfully the method of routine work for at least 85 years. The efficiency of mass production, the specialization of functional departments, and the standardization of working methods were a key to success.

The necessity of meeting the demands for new and better products and services at shorter delivery times called for shorter lead times, better quality, and, above all, greater flexibility. These changes required other working methods and opened good perspectives for PM.

Culture

Culture is a much discussed topic at present, and rightly so. Organizational structures cannot be changed permanently when the culture has not been adapted adequately to the new situation. Culture sometimes is described as "the way we are doing things." We may add: "and without asking ourselves why we are doing the things the way we are doing them." Here we touch upon an essential element when discussing the possibilities of implementing aspects of PM. Functional managers within Philips (but also outside Philips, as we learned from colleagues) see PM as a set of procedures and control, handy to have in addition. They remain, however, very dedicated to their functional organization.

Still, to implement PM means to implement, at least to a certain degree, a matrix structure. The more you shift from a functional organization into a matrix organization, the more important it will be to divide responsibilities and authority between functional managers and project managers. Functional managers do not like that. They love PM as long as you do not touch their power positions.

In the Introduction we also mentioned that, especially in the past, the culture within Philips included that you had to convince your counterparts and that, consequently, it was difficult to enforce decisions upon them. Although a lot of things have since changed, these culture aspects still play their subtle roles.

Delegation

PM means delegation of decision making. Functional managers are not accustomed to that. They feel responsible for all results in their departments. Anyhow, they are accountable. Therefore they want to stay in control. Although they understand rationally that delegating decision power belongs to the game, mentally they are inclined to intervene as soon as something unusual occurs.

Organizational Change and Culture

It is often difficult to convince management that the implementation of PM is an organizational change and not merely the implementation of a set of rules and procedures. Tichy[1] explains that with each organizational change you always have to consider three aspects:

1. *Technical aspects.*   In our business the rules and procedures of PM
2. *Political aspects.*   In PM reflected, for instance, in the negotiations on the different responsibilities between functional managers and project managers
3. *Cultural aspects.*   Reflected in the integration of PM in such a way that it is accepted as a fact of life

To achieve cultural change takes time. Our experience with PM shows that it takes at least 2 to 3 years before full integration is achieved. The crucial factor that has to be changed is the style of leadership.

Style of Leadership

To explain what we mean by changing the style of leadership, it is useful to introduce the concepts of Etzioni.[2] An important characteristic of leadership is the use of power.

Etzioni distinguishes between position power and personal power. Position power is based on formal authority, while personal power is generated through follower acceptance (for example, as a consequence of expertise or personal characteristics). Leadership in functional or bureaucratic organizations is mostly based on position power. Within Philips, too, the accent is still on position power.

Autonomy and Project-Management Manual

The style of leadership is a big obstacle for successful PM, especially if one realizes that very important aspects of project management are the assignment of functions and tasks and the delegation of authority.

The drafting of a PM manual can be of help to try to achieve these conditions, not only for the project team and stakeholders, but also for functional groups, article teams, and so on. All important points have to be put on paper. If done well, the drafting of the manual is done in consultation and often in negotiation with all relevant parties involved. Every aspect or condition is made visible in text and is discussed before it is accepted.

The manual is not meant as a replacement for the former bureaucratic structure. It gives the limits within which groups can work autonomously. Once the manual is drafted and accepted, the transition period toward integration can begin. The manual can be consulted in cases of conflict and adapted if necessary.

Integration of Project Management in an Organization

Two aspects endanger the successful integration of PM during the transition period. First we already mentioned the danger to fall back on former procedures and routines of working as soon as difficulties arise in the progress of the work on the projects. Before you realize what is happening, "things are back to normal." At least that is what some functional hardliners say.

The second danger is that part of the MT is transferred in the course of the transition phase. In Philips managers move from one job to another quite often. These transfers are not always very well planned. In most cases there is no room for a profile for a new manager, neither is there time to pass on the work from the parting manager to the newcomer. Most newcomers want to give an imprint of their own on the working procedures and want to have a say in running the business right from the beginning. Especially if the new manager is not fully acquainted with PM, these interventions may disturb the transition process to quite some extent.

A transition process takes 1 to 3 years. During that time we try to hold attitude interviews. The purpose of the interviews is to find out how far the new working methods have changed the original working activities of the employees and how they react to it. Findings are discussed in plenary sessions, and actions to be taken, if necessary, are worked out.

FURTHER DEVELOPMENT OF PM CULTURE WITHIN PHILIPS— TWO CASES

In the first case we give a description of an intervention by means of attitude interviews in a development department. In the second we describe how we dealt with a merger of various PM cultures.

Case 1: Providing the Right Conditions for PM

A few years ago PM was implemented in a development department at Philips. Lately a new development manager had been appointed. After some time he invited us to discuss with him a practical problem. He missed an overview of the development activities that were going on and what was happening in his department. A good planning system was absent, so he said.

When discussing this problem he informed us that the department was still working on the basis of PM, but that he doubted that the present procedures were still up to date. The department had grown from 40 persons to over 70 and the intention was to grow further to around 110.

We proposed to hold interviews on these aspects. These revealed the following results:

- The shortcomings in the system were confirmed. The PM procedures were based on a small department that had a special culture at the time of the introduction of PM. Meanwhile the department had grown to the present size of over 70 persons, making the department more complex.

- The original staff of 40 people were accustomed to a certain style of working—a small department where everyone knew each other and where many informal horizontal contacts existed. This style could best be described as "adhocracy."[3]

- At the same time that PM was introduced many new members were hired for the department. These were young well-trained people who liked to work task-directed and liked PM. The original group, however, increasingly missed the former informal contacts, and gradually a kind of generation gap grew, which made cooperation increasingly difficult.

From the interviews it further became clear that in the present group the members had different preferences for the work that had to be done. Some liked to be project managers, to achieve something together; others liked to be functional group leaders, to see to it that the department kept in pace with the technological developments. Others wanted to remain scientific developers. Clearly, an issue for resource management: to forecast the technological changes that may take place in the work of the department and the consequences thereof for the staffing of the department.

The outcome of the interviews was discussed with all relevant members of the department in a plenary session. During that session it was decided that all members would participate in working out new concepts for working together, including proposals for improvement in the existing PM procedures.

In this case an important lesson was that the new manager rightly concluded that changes had to be made. However, he shared our view that, without the attitude interviews, he might have overlooked the interrelation problems between the various groups.

Lesson learned: Attitude interviews, to be held at regular intervals, are useful instruments to measure the climate for PM in an organization.

Case 2: Matching Different PM Procedures

After a reorganization, which included the geographic transfer of a number of departments to one and the same development and production facility, it appeared that practically all departments used PM, but partly with different procedures. To improve the effectiveness of the facility, it was decided by the MT to develop one set of procedures for all departments. The need was urgent because one of their major clients wanted the phasing of development activities to be in line with their procedures. The issue we faced was the fact that every department claimed more or less that their procedures were the best and should be taken as the guiding system—a topic for endless and fruitless quarrels.

The problem was solved by appointing a project team with representatives of all relevant departments. Their assignment was to work out a PM manual which had the approval of all parties involved.

The project was kicked off with a PSU session in which all relevant stakeholders were present. In the session the following steps were taken:

- The scope of the project was established.
- The specification was made.
- A work breakdown structure (WBS) was constructed.

The WBS was constructed in such a way that sections of the new manual could be worked out by working teams. For each of the sections working teams were appointed. The leader of each team was a member of the project team.

Departments were represented in the working teams as much as possible. Procedures on progress reporting were established in such a way that all those interested could give their comments before formal approvals were made by the project owner. These approvals were made in formal milestone sessions. Present at these sessions were:

- The functional leaders of the departments (members of the MT)
- The project owner (the chairperson of the MT)
- The project manager

The total operation of compiling a manual that had the approval of everyone took nearly half a year. Implementation of the manual is an ongoing process. On the basis of the experience gained with projects and as a consequence of changing circumstances, the manual is adapted regularly. The discussions preceding any changes contribute that the manual remains a flexible instrument instead of becoming a new bureaucratic tool.

Lessons learned:

- To take your time to explain the reasons for a change. It helps to avoid invisible resistance.
- To involve as many persons as possible in working out the new procedures. It furthers the commitment.
- To make use of plenary sessions to discuss proposals and progress.
- Technically it was not too difficult to bring the new procedures in line with the requirements of the customers. After all, the principles of PM used within Philips do not differ to such a great extent.
- The objections made at the beginning of the process had been mostly of political and cultural nature.

OUTLOOK INTO THE NINETIES

As a consequence of the multinational and multiproduct structure and of the "freedom for all" culture at Philips, different parts of Philips had started PM in their own manner and at the moment of their choosing. In some places some people in the Philips hierarchy started PM on the basis of a strong belief in the benefits of PM.

In 1990 the CEO of the Philips group, Mr. Timmer, started a centrally managed improvement process called "Operation Centurion, Movement for Change." "Quite simply, the sole objective of Operation Centurion is to improve the profitability."

Operation Centurion implies a companywide scrutiny of performance with respect to the following five issues:

1. *Portfolio of activities,* including turnaround activities of loss-making and under-performing activities
2. *Asset management,* including improvement of the productivity of human resources; increase of the rollover speed of capital
3. *Product creation process,* including improvement of the efficiency and effectiveness of the process
4. *Quality,* including quality as a state of mind and as an integral part of the company's shared values; quality as an all-pervasive companywide drive for excellence
5. *Organization and management,* including decentralization as the guiding principle, business units as the basic building blocks

An important aspect is the stricter application of the principles of accountability. It means keeping your word by following through on your promises. This is a prerequisite for successful PM.

The change strategy is perhaps the newest element at Philips, the most important aspect of Centurion. There is a great deal of attention directed toward implementation, communication, change, processes, and so on. As stated by the CEO, "Change has to start at the top: top management has to practice what it preaches. However, improvement can be achieved only if we can successfully mobilize and involve all of the company's human resources. Because motivation of people is 90% of the battle, Operation Centurion, therefore, will focus on communication and training programmes in the form of CASCADE schedules."

Philips' top management and the top managers of the product divisions and the national organizations started the so-called Centurion 1 program. Later on we got Centurion 2: product division top management together with the top managers of their business units, national organization top management together with the managers of their plants, and so on.

To accelerate the cascade, by the middle of 1991 we got the so-called town meetings: groups of about 100 employees held discussions with plant, business unit, product division, and national organization management. The Centurion 1, 2, 3, and 4 programs are a top-down process; town meetings are a bottom-up movement.

To facilitate change, so-called change leaders were trained in the second half of 1991. One of the tools we were allowed to introduce to them is PM:

- To achieve "targets," it is useful to choose the PM approach for some issues.
- To bring about "cultural change" the PM approach is useful as well. A more result-oriented attitude is necessary anyhow.

We would like to state that a more open-minded and more businesslike attitude is expected of the Philips managers and employees. A primary aim is the achievement of joint targets, and the safeguarding of positions is no longer warranted.

A product, inside-directed attitude combined with a functional structure and culture is "out." A market, client, outside, competitor-directed attitude in combination with an innovative organization and flexible mind is "in."

Another most important factor for Philips in the 1990s is to accelerate the assimilation process of innovation; to create an innovative climate and an adequate control of the product development processes. It will be obvious that PM will also play an important role in the achievement of these targets.

We foresee a development in which PM receives a more dominant place. Certainly the positive aspects of a functional organization will have to be safeguarded, but working according to PM concepts and principles will replace the functional culture in the 1990s.

REFERENCES

1. Tichy, "Managing strategic change," 1983.

2. Etzioni, "Comparative analysis of complex organizations on power, involvement and their correlates," 1961.

3. H. Mintzberg, *The structure of organizations,* Prentice-Hall, Englewood Cliffs, N.J., 1979.

CHAPTER 13
PERSONNEL MANAGEMENT IN (SMALL) PROJECT-ORIENTED COMPANIES

Georg Serentschy

Austrian Aerospace Company ORS,
Vienna, Austria

Georg Serentschy holds a Ph.D. degree in Nuclear Physics and Mathematics from the University of Vienna. He came to the Austrian Aerospace Company (ORS) after 7 years as Project Manager in software automation and the aluminum industry. Since 1982 he has been Managing Director of ORS and since 1991 President of AUSTROSPACE, the Association of the Austrian Space Industries.

INTRODUCTION

Personnel Management is the core managerial challenge for project-oriented companies (companies driven by a project-oriented business), and in particular for knowledge industries (companies in which human resources are more crucial than other production factors such as classical assets, including financial capital and equipment). The high level of qualitative requirements for the work force and the demand for quantitative flexibility of the work force are defining this challenge. The managerial behavior in a project-oriented company should therefore in general differ significantly from the standard: the requirements, goals, and tools for personnel management have to be adapted to the changed organizational climate, especially when a project-management organization is being introduced.

The main task of a company is to identify product-market combinations and to make them succeed through the competitive process in the marketplace. Project-oriented companies should therefore handle their organization and structure as the "best fit" between the project requirements (which of course have to be driven by customer requirements) and the available abilities or competences of their work force. This impacts the managerial behavior in two directions. First there is the development of the organization, which may be summarized as "organization follows projects." The second impact is on the area of personnel management. One of the lessons learned during the first years of the buildup of ORS was the necessity to develop specific ways to attract sufficiently experienced staff in quantity as well as quality. Personnel management in a project-oriented high-tech company is considered a key managerial

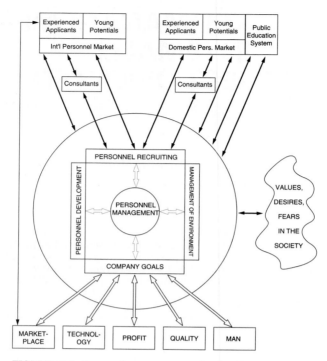

FIGURE 13.1   Personnel management as a key function.

function because it is at the focal point of all areas involved: the company and its goals, the corporate identity as it is seen from outside and inside the company, and the domestic and international personnel markets. Moreover, every company is part of society and has therefore a close interface with the perception of the technology within society. Companies have to balance their responsibilities vis-à-vis shareholders, employees, and society. Taking into account this complex interdependence, it seems evident that personnel management plays a key managerial role (Fig. 13.1).

This chapter represents a kind of personal account of a process of several years. The system described hereafter is considered an "unfinished mosaic." Some parts of the system were completed quickly, some were left in the conceptual phase, some elements have been postponed only timewise. The most important result—discovered during the writing of this chapter—was the confirmation of the process character of the development of the system. Therefore the author gratefully acknowledges the numerous contributions made by friends and colleagues during fruitful discussions. Their contributions were essential for the development described.

THE COMPANY AND ITS MARKETPLACE

ORS, the Austrian Aerospace Company, is a highly specialized enterprise in the field of design, development, and—partly—production of aerospace products and services.

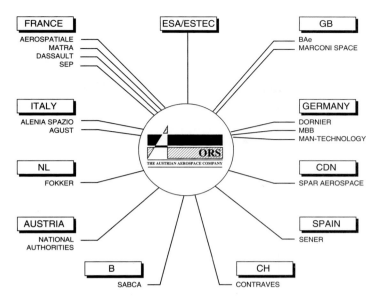

FIGURE 13.2   ORS customers.

Today ORS is the aerospace market leader in Austria because of the accumulated experience stemming from the great number of successfully completed projects, compared with all Austrian competitors. Now nearly all European aerospace companies are customers of ORS services and products (Fig. 13.2). Although the company was founded in 1983 as a joint venture between the German aerospace company Dornier GmbH (part of Daimler-Benz's Deutsche Aerospace—DASA) and Austrian Industries AG (the largest Austrian industrial group) to reap the greatest benefits from the Austrian membership in the European Space Agency (ESA), which was finally achieved in 1987, its roots reach back to the mid-1970s, when the first European manned space module Spacelab was to be developed and Austria decided, for reasons of industrial policy, to participate in this program. In retrospect, the start of ORS's corporate activities was well timed—not too early and not too late. In 1991 a work force of more than 80 specialists organized into four profit centers will achieve sales figures of approximately 12 million U.S. dollars. For the organizational chart of ORS see Fig. 13.3.

The activities of ORS are covering four major areas. The largest field, Mechanical Systems, spans a broad spectrum from mechanical ground support equipment to mechanical structures for satellites and mechanisms for the deployment of large antennas in space. The second field deals with the development and production of thermal hardware, especially the so-called multilayer insulation system. These thermal protective blankets cover satellites, space probes, and manned modules in order to protect the systems against the extreme thermal conditions of space. These multilayer insulation systems consist of a series of thin foils (0.01 to 0.1 mm) coated with a metal layer to reflect the heat radiation. The prefabricated foils are tailored, sewed, or ultrasonically welded to achieve three-dimensional shapes (somehow similar to the production of clothes or hats). This manufacturing process has to be undertaken in a clean-room environment to fulfill the cleanliness requirements for flight hardware.

The first two corporate activities can be considered generic (cross-sectional) tech-

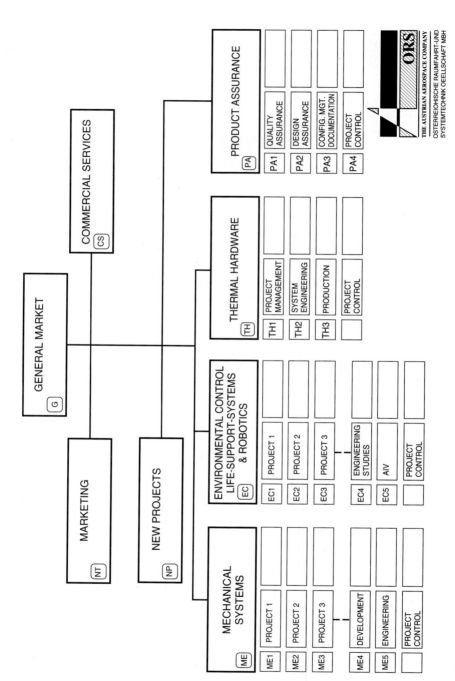

FIGURE 13.3 ORS organizational flowchart.

nologies, as these kinds of products or services are used for all types of satellites and spacecraft.

The third profit center focuses its activities on projects in the field of manned space flight and space robotics. When bringing humans into space it is imperative to provide all conditions necessary to ensure survival in the hostile environment of space. This is the aim of the so-called environmental control and life support system. In particular this means the control of the composition and condition of the breathing atmosphere and the provision of food and water and personal hygiene systems. Within this framework ORS is developing a waste management system for the European manned space flight program and a regenerative atmosphere revitalization system for future applications in the next century. Within the same long-term scenario space robotics will play a major role. ORS concentrated its activities in this area on autoadaptive end effectors (grippers) for space robotics. The activities in this profit center required reasonable support from both public and company R&D funds to accomplish the entry into the market.

The development and application of RAMS (reliability, availability, maintainability, and safety analysis) procedures for the European Space Agency represent the major part of the fourth profit center. This activity carries a remarkably high potential for diversification into other industrial (nonspace) applications.

As a company fully dedicated to project-oriented work, ORS was never confronted with the problem of introducing a project-management organization. The company was forced from the very beginning to organize and reorganize everything according to the project requirements. This approach requires a culture willing to accept permanent change. The relevant learning processes are described in this chapter.

GENERAL PERSONNEL REQUIREMENTS IN A PROJECT-ORIENTED COMPANY

Looking at the requirements for the technical work force, it seems useful to consider technical and nontechnical skills separately. The technical work force in a project-oriented-company such as ORS needs project managers, system engineers, development engineers, draftspersons, manufacturing specialists, integration and verification engineers, and specialists for product and quality assurance. Their educational backgrounds can be in aerospace, mechanical, or electrical engineering, physics, chemistry, process engineering, or a combination of engineering and business administration.

One of the key issues in a project-oriented company is to have the right people (qualitatively and quantitatively) for each project at the right time, that is, to find the optimum path between over- and undercapacity. To discuss this matter in more detail, the term "redundancy" is introduced and defined. From a purely economic standpoint it is very attractive to have as many "universal" staff members as possible (people who can accomplish "everything" in the project). In fact it turns out that there are some rare examples of "universal" engineers. These people are well suited for lateral responsibilities such as project management and product assurance. Their flexibility (exchangeability between different tasks) is expressed by a high redundancy factor (greater than 50 percent).

On the other hand these "universal" people are normally not sufficiently specialized to perform the necessary detailed engineering work (such as structural mechanics, fraction mechanics, production specialists), which is of enormous importance when it comes to the design and manufacturing of "real" space hardware. For this type of work a highly specialized engineering work force is imperative. It will be very inefficient to use such highly specialized engineers in areas outside their specialization, even if done so on a short-term basis for economic reasons. The latter type of engineers offers a lower degree of exchangeability within the project, which is expressed by a low redundancy factor (less than 50 percent).

Among development engineers there is usually a lower degree of redundancy (10 to 50 percent), while project managers, integration and verification engineers, and product assurance specialists have redundancies of 50 to 90 percent. When managing a project-oriented company, it is important to be aware of possible bottlenecks, gaps, and overcapacities because the low-redundancy part of the staff represents the major part of the total work force. For economic reasons it is advisable to run the company with a planned undercapacity of approximately 10 percent. This in turn creates the problem of an overload for the work force and requires management to keep the people sufficiently motivated.

Quantitative Requirements

The management of a project-oriented company is permanently faced with one central question: how many people are required by the projects, and which kind of specialists are needed? This section deals with the first part of the question. It is simple to answer for contracts already awarded. But there are a lot of projects close to contract signature and, normally, many projects in a more or less vague acquisition status. In a competitive commercial situation the company has to steer permanently between over- and undercapacity. Many involved in this challenge are waiting for a balanced situation. Experience shows that exactly this kind of uncertainty is one of the constituents of a project-oriented company.

To make the decisional environment for the managers not too simple, the question concerning the number of employees required is highly interconnected with other parameters:

- Different types of specialists required (with high and low redundancy factors)
- Dependence on the number of projects
- Development of the quantitative demand for each type of specialist over time

With respect to the different types of specialists required, an in-depth work-load planning procedure is essential. Software tools are available for this study. We have also developed our own tools. Regarding the dependence on the number of projects (both awarded and under acquisition), the simple application of a weighting factor (based on the estimated probability of receiving a contract) turned out to be a very well-suited approach. It has to be noted that, according to probabilistic theory, this method is successful only when a rather large number of individual projects is involved, with project values on a similar scale. If the company is dependent on one large project, this approach fails. In case the contract is awarded, the planned capacity will be much too small; if the bid is turned down, the capacity will be much too high. Regarding the time dependence of the quantitative requirements, the ideal picture is a company that depends on a high number of independent projects with a time distribution similar to bricks in a wall. Specialists with high redundancy factors (such as project managers) are required in a more or less uniform way over time. Engineers with low redundancy factors such as production specialists or verification engineers may transfer from one project to the next if there is a significant number of different projects, and the "brick-wall" picture regarding their time distribution applies.

These considerations are leading directly to the "gray area" between quantitative and qualitative requirements. How to keep the "overloaded work force"* sufficiently motivated, is one of the most difficult tasks for management. Project-oriented companies are extremely dependent on motivated specialists dedicating all their technical and personal skills to the job. It also requires the employees to start a very critical process: the development of their personalities. Management has to develop in addition an "impresariolike" competence in leading "primadonnas" and "stars." Besides standard motivational factors such as the challenge of the job and an attractive payment system, other "benefits" have to be developed: the pride to make personal contributions to a complex international program (being part of a large "project family"), as well as relevant appreciation received internally and externally and the possibility to develop one's own personality under the challenges of the job. The higher the responsibility, the more a developed personality turns out to be the most important instrument. A culture willing to support such individual development is one of the constituents of success.

Furthermore it is important to cultivate and train the nontechnical skills. The international, multilingual, and multicultural environment especially requires a facility for communication, a team spirit, and a well-developed conflict culture. In addition there is one technical competence which sometimes seems to be underestimated: the ability for developing a synoptic engineering approach to the system as a whole. The so-called systems engineer (the individual technically responsible for the entire system) carries this responsibility and needs to develop this specific ability. It especially requires a second priority for all kinds of one-dimensional technical optimization.

Qualitative Requirements: Demand for Technical and Nontechnical Skills of the Technical Work Force

Technical excellence[†] is an unquestionable must, but it offers to the project team no guarantee for success. In addition nontechnical competence is required, such as facility with communication (presenting and listening, two or, better, three languages, taking into account the European situation, technical and nontechnical terminology), team competence, and a suitable conflict culture.

Each European space project is usually driven and strongly influenced by a certain European nation. This situation is reflected by the management style visible throughout the entire project and expressed by a very specific managerial behavior of the relevant industrial prime contractor, even if each project-management action is based on the same formal procedure. There have been four European "cultural clusters" observed: the Nordic (Sweden, Norway, Denmark), Anglo (United Kingdom,

---

*In practice it turns out that in running a project-oriented company under competitive circumstances the work force (especially the key personnel) is always overloaded.

†It is not intended to give the impression of underestimating the importance of the technical skills. In fact everybody will admit immediately that an excellent education and relevant technical experience of the work force are a must for a high-tech company. But approaching this matter in a more holistic sense it turns out that "education" and "experience" mean more than gaining technical knowledge; it means also to know how to work together and to communicate with other individuals inside and outside the company, how to solve conflicts, how to present and defend one's own ideas, and how to find a synthesis with the concepts of others.

Netherlands, U.S.-influenced), Latin (France, Belgium, Italy, Spain), and Germanic (Germany, Switzerland, Austria) clusters. This situation, where one nation is represented by an industrial prime contractor and its domestic home base, is useful on the one hand, because national financial contributions may be compensated partly by "identification" and "pride" factors (beside the "real" benefit gained by the system or product to be developed). On the other hand it requires the contractor (for instance, ORS) to develop specific corporate strategies in order to cope with these challenges.

But cultural clusters are not limited to national levels. Usually they appear on the company and project levels as well. This is expressed by the classical frontiers such as hardware versus software people, Marketing versus Production, Engineers versus Scientists, or Marketing versus Administration. The development of specific company and project cultures means bridging across cultural frontiers. This goal—on both the company and the project levels—requires the development of a common language, a systematic approach, shared by all team members. International technology projects should be treated as a paradigm for the post-1992 situation (creation of the single European market) and can therefore be used as a "laboratory" for the development of future-oriented (project) managerial behavior. Maintaining the cultural differences and integrating them in a team ensures both problem-solving capacity and sufficient thrust for the necessary creative potential.

RECRUITMENT IN PROJECT-ORIENTED COMPANIES

The Challenge

The challenge for the recruitment process is closely connected with the specific situation in Austria. For the time being the aerospace industry in Austria is by far not as developed as that in the well-known high-tech areas of the world. This is closely connected with a "dry" personnel market and forces the companies active in this field to devise their specific solutions.

The situation in Austria regarding the high-tech personnel market was strongly influenced by the lack of a coherent public policy on industry, technology, and education since the end of World War II. Austria's industrial policy had been driven, over a long period, by the objective of autarky in (state-owned) basic industry and agriculture. During recent years, in parallel with the emerging unified Europe, this principle lost its importance permanently, and it will be fully obsolete in the mid-1990s when Europe is going into full integration. This concentration on basic industry led to a growing gap between science and the industrial world and to a relatively low R&D coefficient (compared with other EEC countries).

Furthermore this situation caused a steady brain drain.* A significant number of well-educated engineers and scientists left the country to work abroad on a long-term basis. In addition no migration in the other direction could be managed. This situation is one of the main reasons for the dry high-tech personnel market in Austria.

One of the major tasks for the immediate future would be for the universities to build new educational focuses more related to the aerospace industry, which in Europe

---

*"Brain drain" occurs when a relevant percentage of highly qualified graduates migrates to a foreign country for long-term employment because of a gap between educational possibilities and the demand for relevant job profiles in the domestic industry.

employs a work force of nearly 500,000 with annual sales of approximately 50 billion dollars.* In addition it is important to encourage the development of a relevant national high-tech industry to take full economic advantage of the public investments in the universities. On the political level it is important to implement an integrated industrial, technology, science, and education policy as soon as possible, as from the point of view of macroeconomics it is of very limited use to undertake research in areas where no national (in the future, after the start of the common market in Europe, one ought to say regional†) industrial potential exists to apply the knowledge gained at the university. This consideration of course only applies to technical sciences. It is based on a macroeconomic point of view and does not take into account the purely intellectual value of scientific knowledge. It is intended here to underline the necessity of a more coherent policy in this field (compared with the given situation), and not to express an extreme standpoint.

The International Personnel Market

A small or medium-sized company working in purely international markets, but based in a country with a more or less dry personnel market, is forced to develop international recruitment strategies in addition to the local strategy of the "Mental Campus Concept." This concept represents a central part of the strategies. The first goal of international recruitment was to overcome the shortages of the local personnel market. Later on it turned out that international recruitment offers also a chance to be more responsive to customers' wishes. The multicultural and multilingual European situation is a touchstone for the relevant skills of each company working on a competitive basis in this market: being somehow a reflection of the European situation on the company level gives the customer the feeling of being "at home." This is a competitive edge, and it very often turns out to be a successful "door opener." Thus we learned that by being confronted with the problem of a dry personnel market, we had the inherent chance of providing the customer with a better service. In this sense the second goal—originally a consequence of the first—became the most important driver for the international recruitment. For international companies their worldwide recruitment is a routine exercise to fulfill all personnel requirements. To keep this process alive, they use their own international networks, represented by the relevant national subsidiaries, to work on local personnel markets. A small or even a medium-sized company (if it is not participating in the activities of an international industrial group) has to develop its own network to make contact with potential employees. Our strategy includes several ways to build up the international network: international consultants and advertisements in local papers or international aerospace magazines. It turned out very helpful to get advice or support from the local Austrian trade delegate, who belongs to the Austrian Chamber of Commerce, or from local bankers. In most cases the support from both was very effective. We learned that the "active way" via the international network, consisting of consultants and other company-external persons,

---

*Commission of the European Communities, "The European Aerospace Industry, Trading Position and Figures," Brussels, Belgium, 1989.

†Within a unified Europe regional and local politics should be the successors of the current national(istic) politics. For an increased and more efficient cooperation between universities and companies local policy has to make sure that sufficient communication opportunities and an open "interactional area" are available for the parties involved to get a high return on the investments from the taxpayer's money.

works properly. But to our great surprise there is also a "passive way" to get in contact with people from abroad. We are faced with an increasing number of engineers applying for a job directly. Obviously there exists some kind of informal network among interested engineers worldwide. Some write letters to the European Space Agency asking for job opportunities, others read the different directories of aerospace companies and look for job offers in publications. It should however be emphasized that the passive way also needs some prior (active) communications: indicating to the public that the company is interested in receiving applications. In general this strategy requires an attractive corporate identity in the pure sense of the word.

We discovered the efficiency of the passive way earlier on the local level. When we started to search systematically for high potentials at the Vienna Technical University, we noticed after a while an increasing flow of incoming applications. One of the explanations for this phenomenon was the combination of the company's public relations and the experience gained by individuals in the area of overlap* between the company and the university. When the right person is found, not all the problems have been solved, however. Since for the time being Austria is not a member of the European Community, anybody without Austrian citizenship needs a working permit from the local authorities. This imposes an additional delay of several months before the new employee is allowed to start working. This legal requirement (which was created for the protection of the domestic workers in standard or low-level jobs) became a bureaucratic stumbling block for international recruitment because of the enormous time lag until the new employee becomes effective. Especially in project-oriented companies it is essential to act very quickly, which—in this case—is forbidden by law. We hope that this obstacle will be removed in the coming years when Austria becomes a member of the unified Europe.

International recruitment is an expensive program because of the high cost of communications. This is part of the "high intangible investments" which a knowledge-based company has to make.* In essence this is an important cost element, affecting the hourly rate of the company which, of course, cannot be compared with the hourly rate of a company working only locally.

The Strategies

Considering the specific situation of Austria, it turned out to be most advantageous to follow a binary recruitment strategy, represented by two parallel recruitment paths.

Networking with Local Universities.   This recruitment strategy requires specifically active networking between the industrial world and the educational world. Early (timely) recruitment of interns, hiring of working students, and support of a thesis or doctoral dissertation are elements of this strategy. This approach is furthering the ties necessary between the educational world and industry. It is the aim of this first recruitment path to attract "young and hungry" high potentials who will be able to find the right orientation in a project-oriented company.

---

*The commercial value of a small or medium-sized knowledge-based company is very difficult to determine. Equipment and machinery—even if they were expensive—represent normally only a fraction of the real value. This value is mainly represented by the expertise of the experienced key personnel. We always considered it as somehow incomplete not to include this value in the company's balance sheet. On the other hand, know-how is a volatile value and difficult to express in precise business figures.

International Recruitment.   The second path is aimed at attracting internationally experienced engineers. Their responsibility is to assure the required technical competence and experience, but also to stabilize the team.

This twofold strategy requires the company to make high intangible investments (that is, it requires a relatively large training budget and an expensive recruitment system) in order to ensure the technical and personal competence of the company in the long run. Both paths are complementary elements of the recruitment system as a whole. It took several years to develop this strategy. At the beginning the dry personnel market was considered a permanent threat to the ability of the company to fulfill its contractual obligations. It is obviously not recommended to hire people only on the basis of future contracts. Part of the job of personnel management is to know enough potential employees able to start working on a short-term notice. In addition it is important that personnel work overtime when required. As a result the German and English personnel markets were researched extensively and contacts were made with some high-tech personnel markets outside Europe. One of the main tasks was to develop an information network through personal contacts and advertising in order to acquaint potential candidates with the company. The development of our recruitment strategies evolved out of initial "naive" approaches such as advertising in local papers (assuming that Austrians working abroad would read these papers) or in university bulletins. But it was noted that these elements of standard recruitment procedures failed to a great extent. Some new approaches had to be worked out.

Short-term demand for experienced personnel is very specific to project-oriented companies. This also means a significantly higher turnover compared with non-project-oriented companies. A culture able to absorb "hire and fire" tides can be very helpful in such a company. To fulfill this short-term demand for personnel, the criterion "relevant experience/education" is considered essential. This will require a Europe-wide search via different information channels. In addition to the active search on the job market, there is an external pool of basically interested potential employees if the occasion arises. These people need to be integrated into the information network, otherwise this external pool does not work properly.

Students Play an Active Role in the Company.   One of the successful strategies in the recruitment of young potentials is based on the normal summer or vacation work of university students.* This period of summer work is an excellent opportunity for both sides to learn more about each other. We are using this period for the search of high potentials. We learned that students with technical and personal excellence will approach management on their own initiative in search for more highly qualified jobs or greater responsibility as compared with the normal "clerk-type" jobs, which are routinely given to inexperienced students during their summer vacations. Students making an excellent impression during a first period will be offered more challenging job opportunities. This step aims at different goals, such as recruitment path "one," "repair-shop" function (to learn more about questions like how does a company operate? what are my personal strengths?...), and additional capacity for peak demand. In essence the student is invited to learn more about his or her own personality in addition to gaining increasing technical competence. The next steps are the students' participation in specialized classes and the definition of their diploma work, which is needed to complete their studies. This work is very often part of one of the company's projects.

---

*The strongest motivations for students to do summer or vacation work are economic reasons. Only a small fraction of students are searching primarily for additional technical experience.

In recent years we attracted some very promising young engineers by following this strategy. These people have become part of the organization, and although they are very young, they are all most appreciated by the customers as well as by the corporation.

The Inverse Brain Drain.   One of our successful strategies was to initiate an inverse brain drain* as part of our international recruitment process. This was approached by systematic advertisements in foreign—mainly German—journals and in international aerospace magazines, indicating that the expertise of domestic engineers working abroad is being highly appreciated. We tried to pool the interests of several Austrian companies in one ad, thus offering more job opportunities and showing a broader spectrum of different industrial areas involved in this program. In subsequent interviews the most important issue was to convince applicants that working in Austria can be very attractive for them. To manage an inverse brain drain is one of the possibilities to reach quantitatively and qualitatively sufficient personnel. In the beginning we tried to address specifically Austrians working abroad. Later we changed our policy because we found that the "old" value, "we employ only (mainly) Austrians," was undermined by the upcoming new Euro value, "we are proud to offer our customers an international team." As part of the international recruitment activities we now also reach Austrians living and working abroad.

The Mental-Campus Concept.   In the context of this contribution it is obvious that cooperation with the educational system in general is a very important support in the recruitment of young high potentials. On the other hand, the company is responsible for giving enough feedback to the educational system to support the right qualification levels for the students. In general, cooperation with the universities offers many more possibilities than recruitment alone.

Increased cooperation between companies and universities is very often recommended from the point of view of industrial and science policy. In fact, from a macroeconomic standpoint the entire system (the industrial and scientific worlds) should be treated in a more holistic sense: when investing the taxpayer's money in universities and research establishments, the return on investment is much higher if there are local opportunities for applying the know-how gained at the university. Taking into account the direct interests of both partners involved, it seems important to understand that real cooperation needs a clear mutual benefit; otherwise it will never work properly.

If scientists and industrialists share the same interests in a specific field, this can help pushing forward the relevant market sector. We learned how a permanent networking process can make and keep this cooperation alive. This networking process needs to be developed permanently.

According to our experience, cooperation between companies and universities is aiming at three targets:

1. Support of the first recruitment path described
2. Preparation of project-related cooperation (subcontracts)
3. Increased (political) lobbying in the common-market sector

This cooperation needs to be driven by two factors: an increased efficiency of the allocated resources on both sides and a mutual benefit. Mutual benefit means, for

---

*"Inverse brain drain" is an organized way to convince as many former job emigrants as possible to apply for domestic jobs.

example, to accept a "grey area" between both parties: personnel shift and subcontracts in both directions are constituents of a good working relationship. One of the criteria of a good relationship is the feeling, for both sides, of living in a dynamically balanced environment.

Professors and other scientific personnel need to know the general strategies, the projects, technologies, and gaps,* and the specific personnel requirements of the company. Students are interested in knowing the company's goals and projects and the types of specific skills required. They are looking for ideas of how to specialize in their studies (such as diploma work or dissertation) and for opportunities to develop and prove their skills within concrete tasks set by the company.

On the other hand, the company is normally interested in discussing the technical and nontechnical skills required of young engineers with university faculty in order to learn more about the educational process, but also to influence the study curricula.

Intensive networking between university and industry on all levels helps to understand more about the "other side" of modern technology: industry specialists should give presentations and lectures at the university; professors and students should work within industrial projects. Financial support of students' diploma work, provided it is useful for the company, is one of the prime possibilities to create an active interface.

The Mental-Campus Concept requires people to be able to think beyond usual boundaries: university faculty should actively approach the company, offering their expertise; the company should consider the capabilities and personnel of the university as a virtual part of its own organization. To accept the "other side" as being a virtual part of one's own domain requires the willingness of all persons involved to live with frequent personnel changes: the Mental-Campus Concept needs a "grey area" between the organizations. This area represents the necessary intellectual, know-how, and spiritual overlap between both parties. We are aware that the Mental-Campus Concept sounds quite avantgarde, but we are sure that it will eventually become the future concept for implementing cooperation between science and industry. This concept was developed in the course of an individual and institutional learning process and tailored to solve a specific problem at ORS. The initial idea came from the fact that Austria is lacking an officially supported or encouraged working relationship between universities and companies. Therefore we tried to overcome this problem when we organized our own "mental campus," which is supported by a physically close location of our company to the Vienna Technical University. But after a while we realized that it may be worthwhile to develop the concept further into the "future concept" for cooperation between industry and university.

PERSONNEL DEVELOPMENT IN PROJECT-ORIENTED COMPANIES

The specific task to develop the human resources of a small project-oriented company can be described by the development of ORS. This is the story of a personal and an institutional learning process to integrate different cultures, due in part to the binary

---

*Unveiling a company's (general) strategies to university faculty is probably too risky because strategies are normally considered top secret. Nevertheless, it is recommended that strategies be communicated in a qualitative way. In addition entering into a confidential agreement (and living up to it) is advisable. Otherwise a fruitful cooperation seems impossible.

recruitment strategy described in the preceding section. ORS was started about 10 years ago as a "two-caste" system—experienced (hired) British engineers carrying most of the know-how and local "apprentices." Communication between both parties was very limited and biased. The key question at that time was: What is my hierarchical level in the organization or project team?

The next phase was the initiation of a program of inverse brain drain. The people hired through this measure formed an important part of the company's staff. In parallel, a number of young graduates were hired while we were searching systematically for "high potentials." These young but inexperienced engineers were sent to a training program at the German mother company. This training program lasted more than 3 years, and about 20 engineers graduated from it. This group, together with the engineers hired internationally, became the basis of our core team.

Ten years later the flat and flexible organization is fully visible. A relatively high proportion of foreign engineers leads the company to be more successful in the multilingual, multicultural, and, of course, technically highly demanding work. The experience found through industrial cultural exchange, the universalists, the specialists, the "young and hungry" protect team members (in part students), and the experienced engineers has been challenging. The degree of redundancy is now much higher than earlier. This redundancy is warranted by a personnel policy where an engineer is requested to contribute to one project as a system engineer while at the same time doing detailed analysis on another project. This development also brings about an important change in the understanding and appreciation of hierarchies—from formal to functional authority.

A key question under this culture is: What kind of responsibility will I be able to take over in the project? It is recommended to use this test question to check for different development phases in one's own organization. This process is not yet finished and most probably it never will be. It is a process that requires a highly developed conflict culture, and there are more than enough conflicts because of the different interests of each team member. It requires a redefinition of benefits, because the project-oriented company does not offer sufficient classical career opportunities. This redefinition includes in particular a payment system that offers a maximum of success-dependent salary components instead of a system designed to perpetuate "fixed-fee thinking." In particular, it includes different qualitative and quantitative targets and a variation formula to compromise between agreed-upon targets and actual achievements. This system can be used for both profit center managers and project managers. For engineers a premium system that depends on concrete targets to be achieved has turned out very useful. In all cases a success-dependent payment in cash representing a reasonable fraction of the basic income is still a very powerful and necessary incentive, but as a single measure it is not sufficient. It should be underlined that the project-oriented approach is the prerequisite for target definition and a viable premium system.

In addition there are "alternative" benefits: the pride for each individual to contribute personally to a successful project, to present his or her results to the (sometimes skeptical) customer, to convince the customer in the end by overcoming difficult project phases through special personal dedication to the task, and the appreciation accorded by others (inside and outside) to a successful team.

"Internationalize internally!" At ORS special emphasis was given to the process of developing the corporate culture under the above-mentioned ideas. This process includes strategies, focused on the recruitment of new staff members, to offer an international team for international customers and their projects. The main objective of this strategy is the integration (that is, primarily, not a change or assimilation) of different cultures, thus ensuring the necessary problem-solving capacity without losing the cultural differences. In a first approach, on a working level these cultural differences have been considered partly as disturbances. Subsequently lessons have been learned to use these differences as a creative potential. This example shows clearly that the imperative that introduces this paragraph is valid on both the company and the project level. The development of specific company and project cultures requires the development of a "common language" shared by all team members to bridge across cultural frontiers. This common language can be represented by a systematic approach mediated with special team and communication training.

This is part of the "repair shop" function of the company. Usually team and communication competence is not developed sufficiently by the educational system. Companies are challenged to act as repair shops to compensate for this lack. But acting in that way requires a common basis with the educational system. It needs the "overlap area" described in the next section. One of the problems is that a company normally is not paid directly when doing this type of repair work. This function should be neither in competition with the public education system nor an initiative to ask for the taxpayer's money.* It is proposed to split the relevant training budgets† available in a company between technical and nontechnical courses or workshops and to create and develop a network between universities and industry.

Each staff member has the opportunity of taking one nontechnical (such as communication or team training) and one technical course a year. To use these cultural differences as a source for creativity is very important since there are different types of projects: study projects ("software"), requiring different approaches and different types of engineers, and hardware projects. Therefore the profile required for an engineer within a project does not only depend on the engineer's functional responsibility (such as project manager, systems engineer, development engineer, draftsperson, manufacturing specialist, integration and verification engineer, specialist for product and quality assurance), it depends probably even more on the type of project. This means that for personnel development the term "qualified" carries a twofold meaning—function and project.

The project-oriented approach to performing a complex and challenging program, such as the development of a space plane, has to be seen as an international, multicultural, and multilingual environment in which the project has to be accomplished. The small subcontractor has to accept the fact that the project is driven not only technically by the prime contractor and, partly, by the customer. The customer, who is the prime contractor of the industrial team, also defines the communication culture‡ of the project and the specific ways how things are being approached and handled. In addition the customer also defines the means of communication, such as the software standards, for the project reports. This means that the small subcontractor must be able to use, for example, several different project-management software types to fulfill the requirements of different customers.

Personnel development in a project-oriented company therefore comprises basically three major constituents:

1. The opportunity for job rotation within the company (on the employee's initiative)

---

*It is highly recommended for a company that depends to a major extent on public procurement policies (as is usual in the aerospace business in Europe) not to increase this institutional dependence.

†It is advisable to focus, from time to time, a major fraction of the training budget on a specific "subject of the year" (either in the technical field or in the area of personnel development).

‡There is a significant difference in project cultures, depending on who is the lead company. Sometimes it is surprising to see that different cultural approaches lead to the same result.

2. Training opportunities both internally and externally
3. The right for each individual to be led and to be part of an international team

CONCLUSION

- The corporate culture of a (small) project-oriented company should express that personnel management is treated as a key managerial task that deals with the acquisition, guidance, and development of the most important resource of the company—the motivated specialist.

- The corporate culture should support a concept of close cooperation between the company and the universities. The Mental-Campus Concept turned out to be a useful guideline in the environment of the author's company, where no official institutional support is given to a campuslike situation.

- The corporate culture of a small project-oriented high-tech company should support the permanent willingness to adapt its own strategies, operational philosophies, and technical approaches to a permanently changing environment which in the end will turn out to be a stronger company.

- The corporate culture should support a willingness to undertake responsible experiments in a search for new solutions in all areas. These solutions have to be implemented quickly. An "80 percent solution" quickly implemented is much better than a "100 percent solution" requiring a long time for its realization.

- Each company has to develop its own success strategies. The combination of individual experience and a corporate culture supporting individual and institutional learning processes is one of the most important ingredients for success.

CHAPTER 14
PROJECT MANAGEMENT: AS A PROFESSION

John R. Adams

Western Carolina University
Cullowhee, North Carolina

John R. Adams is a Professor of Project Management and Director of Project Management Activities at Western Carolina University, Cullowhee, N.C. In this position he manages the first nationally accredited Master of Project Management Degree Program established in the United States. A Fellow of the Project Management Institute (PMI), he has served PMI for over twenty years, including over nine years as Director for Educational Services, and is currently President of the Institute (1993). Widely published in the field, Dr. Adams has also held a variety of responsible project management positions, notably in the development, production, and employment of major U.S. military weapon systems. Dr. Adams holds a Ph.D. in Business Administration from Syracuse University, as well as MBA and BS Electrical Engineering degrees.

INTRODUCTION

A Profession of project management. A dream for some, an objective for others, and totally unheard of by the vast majority of the world's managers and project managers alike, this term profession has been used widely in the project management literature since the late 1970s. At that time, a number of the officers of the Project Management Institute (PMI), supported by a small but vocal group of the Institute's members, expressed concern over the conditions that existed then, and still exist to some extent, regarding the level of professionalism within the field of project management.

The concern was not a *lack* of professionalism within the field. Most project managers are technically educated and trained before they became involved in the management of projects, and they bring their concepts of a profession with them from their various technical fields into the arena of project management. In fact, that was the basis for the concern. There was no set of professional standards specifically defined for the field of project management. What sparked the concern among this initial group of PMI members was a recognition that anyone, regardless of their working experience, their academic education, or for that matter the lack of either, could be a "project manager" simply by claiming the title. There were no standards that could be used to define a prospective project manager's qualifications.

In response to these and other concerns, PMI established a multifaceted program designed to "improve professionalism in project management" and eventually to develop a recognized profession of project management. This paper documents the progress that has been made by project managers in professionalizing their field and examines some of the implications of a project-management profession on the careers of those who choose to pursue this occupation.

PROFESSIONALISM IN PROJECT MANAGEMENT

Dr. Linn Stuckenbruck, professor of systems management at the University of Southern California (USC), past Secretary of PMI, consultant and ex-project manager on a number of multibillion-dollar utility and Federal Government projects and programs, was asked how he became a project manager. His comments are revealing[1]:

> Most of us never really plan to be project managers—it happens almost by accident. Top management is having problems with some existing product, process, project, or project manager, and immediate action is indicated. The expedient solution is to implement a new project or appoint a new project manager. In either case, volunteers are seldom requested. The new project manager is called in by the boss and told, "Congratulations, you're it—the new project manager. Now your first meeting on this problem..." Notice that the project is immediately a "problem." You are selected, of course, because you are either (1) the top technical expert available to deal with the problem, (2) an experienced and well-trained line manager knowledgeable of this type of problem, or (3) a promising youngster to be "challenged." In any event, you just "happen" to be available, an individual *IN THE RIGHT PLACE AT THE RIGHT TIME!*

Most project managers have experienced something similar to that at one time or another. In this way they became members of what has been called "the accidental profession,"[2] meaning that no real planning had gone into their becoming project managers, nor had they prepared for such a position. It can even be argued that they had not become professionals at all, for a profession is an occupation people plan for, study and prepare for, and build a career within. To the contrary, they entered the new job with no formal preparation and not much of an idea about what skills were needed, what knowledge was available, or how to go about dealing with this new challenge. Most project managers would agree that with better preparation they could have done a much better job on their first project. They would certainly have had an easier time of it and would have made a lot fewer mistakes.

For the specialized management discipline known as project management, the situation just described is the rule rather than the exception. Project management is an action-oriented field, and the individual practicing in the field are so busy managing projects that few people stop to think about *how* to manage projects. When a project manager is named, there is seldom time provided in which to study the new field, and there is certainly precious little opportunity for learning about it. As a result, each new project manager is left to learn this trade the hard way—through the "school of hard knocks," or by "trial and error." The time, money, and energy that are wasted as a result are beyond calculation.

Yet project management is essential in today's complex world of rapidly expanding technology; increasing control by federal, state, and local governments; growing pressures from environmental protection and special-interest groups; and rapidly changing economic conditions. Project management is a management technique specifically designed to deal with a high rate of change, a characteristic requirement facing most organizations in this time of high pressure for rapid change.

When project managers are asked what they do, they almost invariably recite the detailed tasks of their jobs. They coordinate work among others who must contribute to the project. They conduct meetings, make presentations, prepare reports, and develop plans. They check progress on the project to determine whether the plans are followed and whether milestones are being met. They emphasize their skill requirements as being coordinative in nature—building teams, generating commitment, and overseeing the progress toward a specified objective. Rarely, if ever, do project managers refer to their contribution to society, yet if this field is to warrant recognition as a profession, its contribution to society must be a widely recognized and paramount concern of those serving in the field. When viewed in the broadest sense, we can recognize the role of project managers in our society as a creative one. Someone performing the function of a project manager is responsible for designing and constructing every building that is used by members of our society. Project managers build the roads, design the cars, construct the buildings, develop the chemicals which keep us healthy (or which destroy our health), manage the research projects that add to our knowledge, develop the weapon systems that are used by our military forces, and otherwise create the facilities, products, and systems that will determine our society's standard of living in the future. In short, project managers manage and implement change, and they manage our society's investment in the future. It is clear that this is a significant service, but those who provide it are widely scattered throughout all of industry and government, in widely different specialty fields, and have widely varying educational backgrounds. They work on projects that range from huge to infinitesimal in scope, from exciting and unique to mundane in nature, and from unknown to world-renowned in public acclaim. Drawing those who practice project management together to provide some form of professional recognition is a formidable task at best.

A profession, by definition, is a vocation or occupation requiring advanced education and training and involving intellectual skills.[3] There are many professions in our society, including doctor, lawyer, teacher, engineer, dentist, accountant, and other such recognized and formalized occupations. These occupations are distinct from others in that they engender a level of respect from the population as a whole, and their standards of entry require education and training specific to their respective fields. Perhaps more important, professionals generally provide a service to the population as a whole, but the average recipient of that service has little opportunity to judge the qualifications of the professional. Thus the individual who wishes to make use of the services provided by professionals must rely on their professional memberships to determine their qualifications.

DEVELOPING A PROFESSION

There are five attributes that are generally identified as being common to all recognized professions.[4] These are listed here with a brief description of what each means. Please note that if these attributes are to be useful to project managers, they must be representative of all recognized professions, including law, medicine, engineering, accounting, and all other occupations generally considered to be representative of "the professions."

1. *A unique body of knowledge.* This first attribute implies principles and concept that are unique to the profession and are codified and documented so that they can be studied and learned through formal education. In most professions, the body of knowledge is taught in graduate or professional schools; for example, the specialized body of knowledge of the legal profession is taught in law schools. A degree does not necessarily qualify an individual to practice in the profession, but it does provide a means of assuring that the individual has at least been exposed to the basic principles on which the profession is based. Every profession has at least one degree that can be earned by those wishing a knowledge of the profession's principles. Some professions have many degrees, allowing higher levels of specialization by the professionals.

2. *Standards of entry.* Defined minimum standards for entry into the profession imply progression in a career; entry standards define the place from which a career path begins. All professions must have an accepted route open to the public by which a person can become a recognized member of the profession. Law, engineering, accounting, medicine, teaching, they all have entry standards. These standards typically involve formal education leading to an academic degree; several years of experience, as in an apprenticeship program or as a beginner in the profession; test score requirements, which may or may not be legally enforceable; or some combination of the three.

3. *A code of ethics.* Ethical standards, or a code of ethics, are common to most professions. Their purpose is to make explicit appropriate behavior and to provide a basis for self-policing of unethical behavior, thus avoiding or limiting the necessity of legal controls.

4. *Service orientation to the profession.* The service orientation is actually an attitude of the members of the profession, an attitude by which members are committed to bettering the profession itself. Professionals will commit their time, money, and energy to attending conventions, publishing their ideas and experiences, and generally contributing to the body of knowledge and the administration of the profession. A professional's commitment to the profession is frequently stronger than to the employer. In many cases professionals will leave their employing organization rather than violate the profession's standards or ethics of practice.

5. *A sanctioning organization.* The authenticating body or sanctioning organization has many purposes. It sets standards and acts as a self-policing agency. It promotes publications and the exchange of ideas, encourages research, develops and administers certification programs, and sponsors and accredits education programs. Through public information and recognition of professionals, such organizations provide a voice for their profession. In a word, the purpose of an authenticating body is to administer the profession.

It holds that if these attributes are common to all professions, then project management is a profession only to the extent that these characteristics are represented in its structure. To become more of a profession, these attributes must be developed and incorporated into some structure capable of sustaining and expanding the professional ties that bind qualified project managers together. Work to accomplish this task has been going on for some time under the auspices of PMI. The following is a brief review of the progress to date.

PROJECT MANAGEMENT: THE POTENTIAL

First of all, we need to recognize that project management is not engineering. It is not architecture. It is not medicine. It is not accounting. It is not research. Project management is the management of certain aspects of all these fields, and more. The key term in project management is *management,* and it is generally recognized as being the management of specialized activities designed to achieve a predetermined objective or goal. The characteristics of this profession must thus reflect these unique facets of the occupation.

1. *A unique body of knowledge.* In 1979 PMI initiated a project to document, the "Project Management Body of Knowledge" (PMBOK), the knowledge that would be considered key to any individual expecting to function as a project manager. At the time this was considered a straightforward and relatively simple task; after all, all project managers understand what they need to know, don't they? Some 13 years later PMI is preparing to publish its third iteration of this Body of Knowledge. The PMBOK now exists in an outline (work breakdown structure) form, with the specific content defined in eight different sections. The key to gaining widespread recognition and acceptance of the PMBOK has been PMI's long-term commitment to provide for revisions, modifications, and elaborations to reflect advances and changes in the field. This commitment is represented in the PMI Standards Committee, a standing committee of the Institute which periodically reviews the PMBOK, solicits comments and suggestions from the field, and makes the revisions that seem warranted. The Committee is also sponsoring the development and publication of nine handbooks, an overview, or "framework," for each major section of the PMBOK, designed to document in more detail the body of knowledge needed by practicing project managers. Agreement has been reached specifically by considering project management as the *management* of technically oriented tasks, and not the technology itself.

2. *Standards of entry.* The acceptance of defined minimum standards for entry into the profession of project management is some distance into the future, but a good start has been made in this area. PMI is using the engineering model, in which entry to the profession can be gained by education (one can be hired into an engineering job directly out of college with an engineering degree), by examination (one can obtain engineering jobs by passing the professional engineer's exam), or preferably by a combination of the two (many practitioners earn a degree, gain experience, and then go on to take the professional engineer's exam, thus gaining status in their profession). The Project-Management Professional certification exam has been offered by PMI since 1984, and there are over 2000 certified project managers today, and the number is growing rapidly. Further, since PMI instituted its program to encourage development of project-management degree programs, five master's degree programs have been established in the United States, three of them in nationally accredited universities. There are at least two master's degree project management programs in England, seven in Canada, five in Australia, and one in Europe (University of Vienna), with other programs developing rapidly. Programs accredited by PMI are reviewed to assure that they adequately cover the PMBOK.

3. *A code of ethics.* There are lots of ethics in project management, but most of them are borrowed from the technologies in which project management is used. This approach has worked reasonably well in the past on a project-by-project basis, but the technically oriented ethics appropriate to an engineering construction project, for example, may be totally inadequate for a pharmaceutical project developing a new medication. As part of its certification program, PMI has published a set of ethical standards believed to be appropriate to the project-management field regardless of the technology involved, but these need to be developed and refined. They have been rather widely accepted by the PMI membership, however, and they provide a good start in this area. Along with the ethical standards, PMI has established a standing

Ethics Committee to review and propose modifications to the standards, and to hear complaints of ethical violations. This committee can recommend corrective or punitive actions to the PMI Board of Directors in the event that ethical violations are substantiated.

4. *Service orientation to the profession.* PMI now claims nearly 10,000 members in some 68 countries, about 70 percent in the United States and another 20 percent in Canada. There were over 67 chartered chapters of PMI at last count. This number is growing rapidly, with some 24 chapters in some stage of organization. The willingness of the chapters and members to commit their own time and money to participate in PMBOK and certification reviews, advise universities and encourage establishing degree programs, write articles and books for publication, and carry on the administration of a major international professional association is simply remarkable. This service orientation is not unique to PMI. INTERNET in Europe, the Australian Institute of Project Management (AIPM), and the Engineering Advancement Association (ENAA) of Japan, to name only a few, all report high levels of interest and activity. The interest and support needed to develop our new profession seem to exist, and a large number of project managers are actively participating in this work.

5. *A sanctioning organization.* The final requirement for generating a new profession is the authenticating body that provides leadership, establishes standards, and provides a means of communication and coordination among practitioners. PMI has taken the lead in providing this service. It is the largest of the project-management professional organizations, and is represented in the largest number of countries. The International Project Management Association (INTERNET) has recently taken steps to develop a European certification program along traditional English lines. The Australian Institute of Project Management and the Engineering Advancement Association of Japan, along with INTERNET, have support and working agreements with PMI to help extend the influence of professionalism in Project Management. PMI welcomes participation, criticism, and support, and is willing to make its work available to other project-management organizations for adaptation and potential use within other countries. Other cooperating organizations include the Western Australian Project Management Association, the Associazione Nazionale di Impientisica Industriale (ANIMP) of Italy, the Construction Management Association of America (CMAA), and the Performance Management Association (PMA) of America.

There is clearly a tremendous amount of work to be done by countless numbers of people if project management is to become the recognized profession envisioned by practitioners today. Nevertheless, a start has been made, and the basis has been established within each of the five major attributes of a profession. It is necessary that project management be identified as the newest of the recognized professions. There is simply too much to be gained by the individuals concerned, by the organizations they represent, and by the society they serve to allow this opportunity to slip by unfulfilled.

PROJECT MANAGERS IN A PROFESSIONAL WORLD

It is difficult for most project managers to examine their occupation in perspective and philosophize over its contributions and its place in society. After all, most project managers are not philosophers. By education, training, experience, and inclination project managers are much more likely to be pragmatic, dynamic, objective-oriented "workaholics" than the quiet, reflective persons we think of as philosophers. Project managers are generally educated in a technology appropriate to the project they are managing, a project that is defined by a carefully worded and documented objective to be achieved. They are trained in the art of breaking that objective into its component tasks and in sequencing those tasks in the manner best calculated to achieve the objective "on time, within budget, and meeting all performance requirements and specifications." Individuals usually accept the position of project manager in the belief that they will accomplish something that is useful and worthwhile beyond the basic need to provide themselves employment. Their experience with projects quickly demonstrates the pressure, stress, conflict, and time-critical nature of the job. Because of the very nature of their work, and considering their education, training, and inclination, project managers almost invariably have an abiding preoccupation—almost an obsession—with "getting the job done." This obsession leaves very little time for philosophizing.

Of course, these characteristics are not useful only to those who choose to manage projects for a living. These characteristics can be formed among other managerial personnel in large companies, particularly among those who are advancing rapidly through the ranks. This obsession with meeting objectives can be developed and reinforced by a period of work within the project-management field, so project management has long been considered an excellent training and testing ground for line managers. For most individuals selected to be project managers, the end of a project is the chance either to return to their technical occupation or to move into more permanent "line management" positions. This is true particularly if the project has gone well, for it has provided the opportunity to demonstrate an ability to manage. In fact, most successful managers have managed one or two projects as part of a career in line management. In such a case, project management has provided a valuable training ground for line management, and this is a valid and useful service to provide an organization. It is unlikely that line management will stop availing itself of this training opportunity in the near future. Unfortunately this does not do much for developing the field of project management.

The real problem with this approach is that it provides no means for developing and identifying those individuals qualified to take responsibility for the really big and important projects; the space programs; the major transportation and regional development programs; the major drug, genetic, and medical treatment programs; the projects that will shape the future dimensions of our society. For that we need a way by which project managers can develop and grow as project managers. We need a structured method by which individuals can (1) study what project management is all about, (2) enter the project-management field at a relatively low level of responsibility to learn and to demonstrate the needed skills, (3) be "promoted" to positions and projects of ever-increasing responsibility and magnitude, and eventually (4) achieve a reputation that would warrant major project responsibility. In other words, what is needed is a means by which an individual can choose to make project management a career. Along with this choice will come the implied responsibility to help train line managers in the tasks and methods of project management. With a profession of project management, therefore, the majority of projects will still be managed by nonprofessionals— line managers with little, if any, formal project-management education managing their first or second project, demonstrating their abilities in management, and looking for promotions up the line. These are likely to be the smaller, less critical projects, however, with which management can afford to take risks.

A career path along these lines is compared in Fig. 14.1 with the proposed career path of the professional project manager. Training and advising the "interim" project manager would be a function provided by the professional project manager, providing one of his or her most valuable services to society as a whole.

Issue	Interim project manager	Career project manager
Education	*Technical* Engineering Hard science	*Project area* Management Medical Engineering Etc.
Entry to PM	Accidental—handed the job	Planned—after qualifying studies
Progression desire	*Return to technical field* Professional engineer Technical specialist Management specialist Management	*Remain in management/project management* Project specialist Project engineer Project manager
Advanced education	Technical professional engineer	Management (MBA) or project-management (MPM) certificate
View of PM	Necessary evil, one career step	Interesting, challenging, rewarding
Professional-orientation goals	Specialist Technical recognition Scientist Chief engineer	Generalist Managerial advancement Manager of project managers VP projects Management of large, key projects
Knowledge or skill required	Technical Detailed In-depth Line manager with project-management experience	General Broad in scope Professional project manager

FIGURE 14.1 Alternate PM career paths.

The professional project manager's career path, however, would be somewhat different. In the first place, entry to the profession would be attained through a program of formal education available at a wide selection of universities in several countries. The "typical" professional project manager would receive an undergraduate degree in a technical specialty, where the term technical could refer to engineering, architecture, finance, computer information systems, social welfare, theater arts, or whatever field in which the individual will be managing projects. Several years' experience in that field, with high evaluations, would prepare the individual for advancement into project management. Project-management education would be obtained at the graduate level, followed by entry into the profession as an assistant project manager, a project coordinator, a project specialist, or a functional manager responsible for providing extensive support to projects. As performance warrants, the individual would be promoted to manager of a small budget project and then on to the larger projects. At some point during this period, the individual would probably sit for exams and become a certified project manager. Contributions to the profession and personal recognition would come in the form of research, publications, and participation in the local, national, or international activities of a project-management professional organization. Again, as performance warrants, our project manager might next become an in-house project-management consultant or a general project manager of a very large, critical, society-affecting project activity. Finally, with this level of skill and experience, our project manager might become a private consultant to other organizations, or a senior-level project-management executive, a "manager of project managers," or a vice-president for projects.

ORGANIZING FOR PROFESSIONAL PROJECT MANAGERS

Recently a new category of project manager has come onto the scene—the individual who is competent in project management and considers this field to constitute a career. These individuals come to project management either through the traditional school of hard knocks and project-management certification, or through one of the new project-management degree programs which can short-circuit the process of gaining experience or knowledge of the field. Regardless of how they come to project management, the crucial difference from the traditional project manager is that they consider project management as a career, a lifelong specialty that allows them to grow through a sequence of projects of increasing responsibility to become professionals in this field, sought after for managing the large, society-impacting projects of the future.

There are a few organizations that have embraced the concept of using project management to achieve their goals. Those organizations provide innovative approaches to allow individuals to become professional project managers. These approaches provide practical guidance for other organizations that are trying to adopt similar career paths. Two such organizations that have adopted the project-management philosophy are examined hereafter. One does not have a formal structure for project management, but has all the tools available to encourage an individual with initiative to pursue a career in project management. The other company has established a formal structure for developing a career in project management (Figs. 14.2 and 14.3).

Case 1: Electric Utility

For those involved in projects within a large organization, a career-oriented view of project management requires as a minimum that the employing company have the processes in existence that allow the project manager to specialize in managing projects and be recognized as having a significant skill that contributes to accomplishing the company's goals.

One of the largest and most highly respected electrical utilities in the United States, for example, is heavily involved in project management. The company operates three nuclear, eight fossil, and several hydroelectric power-generating facilities, while distributing this power across two states to some of the fastest growing residential and

---

CASE STUDY 1

Electric Utility (EU) produces, purchases, and distributes electric power throughout two southern states in the United States, providing for the needs of both residential and commercial customers within the two-state region.

Founded: 1901 as one of the pioneer electric power producers in the country. EU pioneered the development of both hydro and nuclear methods of power production.

Capitalization: Highly capitalized in a capital-intensive industry.

Employees: Approximately 19,000.

Scope of Activities: Government-regulated monopoly covering a two-state region.

Number of Regional or District Offices: 40 to 50.

Basis of Project Activities: Projects internal to the organization, designed to develop and enhance the organization and its resources for effective and efficient operations.

FIGURE 14.2   Case study 1.

CASE STUDY 2

Environmental Engineering (EE) performs environmental studies for both initial-development and corrective-action projects, conducts environmental monitoring programs, and is developing environmental correction and clean-up capabilities.

Founded: 1988 through the consolidation of (1) an old-time, traditional engineering firm that had evolved into working on environmental issues; (2) an environmental laboratory conducting long-term environmental monitoring studies; and (3) a small environmental studies division of a much larger engineering-oriented corporation.

Capitalization: Limited (service organization, not highly capitalized).

Number of Employees: Approximately 725.

Scope of Activities: Nationwide, contracts to conduct studies, design, and implement solutions.

Number of Regional or District Offices: 13.

Basis of Project Activities: Mainly external to the organization, conducted by contract. Projects are designed to study, predict, propose solutions to, and recover from other organizations' environmental problems and concerns.

FIGURE 14.3   Case study 2.

business communities in the country. The company has received several national and regional awards for the excellence of its operations and management processes. Electric Utility (EU) is in a highly regulated, capital-intensive, mature industry that is facing the threat of deregulation and dramatically increasing competition. Hierarchically organized and centrally managed, its service area is geographically bound, and its business processes are highly regulated. It is widely accepted, however, that these regulations may be relaxed in the relatively near future as the electric utility industry becomes the next target of deregulation. One of the hallmarks of EU's current management success is an effective and continuing emphasis on the development and use of project-management philosophies and techniques as a principal method of achieving its strategic goals and objectives.

The company does not recognize a formal, documented project-management career path, and such a formally documented program would probably be too limiting for a profession that cuts across so many aspects of the company's activities. It does have all of the methods and processes in place, however, for the individual who wishes to take charge of his or her career and pursue a profession of project management within the company. There are several elements that contribute to the ability of the individual employee to build a project-management career. These include:

- *Top management's support of project management as a principal method of achieving corporate objectives.* The leadership of the company supports the concept that project management provides a method of controlling schedule, quality, and cost specifications for major projects. Project managers can, therefore, significantly contribute to achieving corporate strategic goals by meeting the objectives of those projects that provide the resources for implementing the company's strategic plans.

- *Formal project-management training by leading professional experts.* The organization has hired respected experts in the project-management field to provide realistic project-management training that exposes individuals to practical applications of project-management techniques and concepts. This training covers a 6-month period, including four days of concentrated project-management simulation training which develops an excellent overview of the discipline, followed by a 6-month program in which the participant must develop and manage his or her own project according to the methods taught during the simulation. After the initial training, individuals can develop a more in-depth project-management knowledge by taking other training programs in any of the specific areas of PMBOK.

- *A network for communication between project managers across different functional disciplines.* This enables a sharing of project-management techniques. The formal project-management training is conducted with participants representing project managers from across all functional areas of the company. The participants can therefore share techniques and information across functional and academic disciplines. This promotes the ability of a project manager to move from one functional area to another, and to learn from techniques that work well in other parts of the company.

- *Formal training taken by the project sponsor as well as the project manager.* The top management sponsors of major projects also participate in formal project-management training in order to understand the processes used by the project managers. This encourages free communication of ideas and concerns between the project manager and the sponsor.

- *Encouragement from management for individuals to move between functions within the organization.* An individual wishing to develop a career as a professional project manager is encouraged to work in a variety of areas across the company to obtain a holistic and more balanced perspective of the company's operations. This perspective is important for managing the increasing number of interdisciplinary and cross-functional projects undertaken within the company.

- *Publishing all available internal positions within the company across all departments.* By posting the positions available within the company across departments, a professional project manager can be aware of all job opportunities available within the company. This program is designed to enhance movement of personnel throughout the company for the development of a breadth of knowledge, and serves the purpose of helping the project manager identify available positions that may enhance his or her career.

In sum, these elements allow an individual who wishes to pursue the professional project-management career the latitude to do so. Of course, the individuals must demonstrate their own initiative to develop such a career, but this is a characteristic typical of project managers in general. The individual must seek out projects to work on and use the techniques learned in formal training to build a reputation for delivering projects within time, cost, and quality specifications. Such a reputation goes far in building a project-management career.

Case 2: Environmental Engineering

This case involves a newly formed private company, national in scope, that functions within the environmental protection industry. This is one of the fastest growing industries in the United States, involving both government and private funding of many relatively small engineering and research studies. The studies range from environmental impact assessments investigating specific sites for possible industrial development to remedial design, construction, and cleanup activities aimed at correcting past damages to the environment. This company provides examples of some more detailed management processes and procedures developed to implement specific working relationships among those who must accomplish the company's primary work in the newly reorganized project structure.

Environmental Engineering (EE) is in a rapidly expanding industry that is growing at the compound rate of 30 percent per year. The company was newly formed by combining previously independent organizations having widely differing locations, histories, structures, and cultures. Employees are highly committed to the new-wave philosophies of protecting and cleaning up the physical environment. The company is highly dispersed, with permanent and semipermanent offices scattered across the United States.

Newly formed and dispersed across the United States, the company had little choice but to establish a relatively flat, decentralized organization that emphasized and rewarded individual initiative. The concern here was to empower their project managers, recognized by senior management as the key company representatives in contact with clients and responsible for delivering the company's service, to interact freely across the organization, marshaling the company's resources to meet the clients' needs. The difficulty was that several cultures had been inherited when the company was formed, with each portion of the company striving to maintain its identity in the face of demands for change. Empowering the project manager was, in this case, a clear strategic necessity for senior management.

Faced with developing an entirely new company, EE defined its contracts as projects, concentrated on the sharing of responsibilities among all those who must contribute to completing the projects successfully, and developed a formalized project-manager career path. The formal career path is summarized in Fig. 14.4.

SUMMARY

The profession of Project Management is a fact. All the characteristics of a profession are in place, and major companies are beginning to take notice. Organizations such as AT&T, Digital Equipment Corporation, Procter and Gamble, and the U.S. Post Office, to name only a few, are sponsoring in-house project-management training for their

Title	Grade	Complexity or impact	Education or experience
Principal Project/ Program Manager	12 and above	One or two major projects; high penalties for nonperformance; manages interregional/division projects. Typically manages projects in excess of $5 million.	B.A. or B.S.; 12 + years relevant experience; PM training; finance and contract-law training
Senior Project Manager	11	Three or more major multidiscipline projects; high risk for nonperformance; manages interdivision projects. Typically manages projects from $3 to $6 million.	B.A. or B.S.; 3 + years relevant experience; PM training; finance and contract-law training
Project Manager	10	Three or more moderate to major multidiscipline projects; moderate risk for nonperformance. Typically manages projects from $3 to $5 million.	B.A. or B.S.; 5 + years relevant experience; 4 + years PM experience; PM training; finance and contract-law training
Associate Project Manager	9	Multiple single or limited multidiscipline projects, small to large scale; moderate risk for nonperformance. Typically manages projects up to $500,000.	B.A. or B.S.; 3–5 years relevant experience; 2 + years major project responsibility; PM training within 1 year; contract-law training within 2 years
Contract Specialist	6/7	Participates on multiple projects or major federal programs; knowledge of finance and legal aspects required.	B.A. or B.S., or equivalent; 0–3 years finance/business administration; PM training within 2 years
Project Assistant	VII	Participates in preparation of proposals, contracts, and PANS; checks invoices and prepares letters; coordinates preparation of deliverables.	A.B. or equivalent; 2 or more years of related experience

FIGURE 14.4 Project-management career track matrix.

personnel, and many have either established or are examining the possibility of project-management certification programs as a factor in their project-manager progression. It is not reasonable to expect that in the foreseeable future all projects will be managed by professional project managers. To the contrary, we can expect many if not most line managers to include project-manager experience as a required step in their own progression. One of the required tasks for the professional, then, will be to assure that these line managers learn the project processes well, that they learn to appreciate the abilities of the professional, and know when to call for their assistance. The professional, then, will be available for the large projects where risks must be minimized, ready with a proven and documented track record of performance on ever more significant projects, to take charge and make the project manager's contribution to the standard of living our children will enjoy. Our immediate task, then, is to work toward the increased recognition of our profession, and to support the continued development of our professional activities.

REFERENCES

1. L. Stuckenbruck, personal communication, Nov. 1986.
2. J. C. Davis, "The Accidental Profession," reprinted in *Project Management J.,* spec. Summer issue, p. 6, Aug. 1984.
3. *New World Dictionary,* 2d college ed., The World Publishing Co., Cleveland, Ohio, 1976.
4. L. C. Meginson, D. C. Mosley, and P. H. Pietrie, Jr., *Management Concepts and Applications,* 2d ed., Harper & Row, New York, 1986, pp. 16–17.
5. J. R. Adams, "From the Education Director's Desk," column appearing in each quarterly issue of *Project Management J.,* Dec. 1983–Mar. 1986.
6. D. Ono, "Professional Project Management: Who Needs It?," *Project Management Monthly,* vol. 1, pp. 23–29, Oct. 1991.

P · A · R · T · 4

STRATEGIC MANAGEMENT AND STRATEGIC PROJECT MANAGEMENT

On the one hand, projects are defined in order to implement company strategies; on the other hand, side projects influence and contribute to the organizational learning and to the continuous adjustment of corporate strategies.

In Chap. 15 Wendy Briner and Colin Hastings provide insight into the role that projects play in the development and execution of global strategy. Their chapter delves into why strategy fails, how projects can be used in the strategy process, and how strategic projects can be fitted into the solution of strategic problems. Using strategic management as a common denominator, these authors offer some pragmatic prescriptions on how to effectively integrate the interdependent forces of project management and strategic management.

Peter Sutter, in Chap. 16, perceives projects and project management as strategic instruments of integration in international companies. The impact of the corporate culture on the internationalization process is analyzed. Further, the role of projects in the strategic development of the value chain between a company and its clients is described.

Anthony Buono describes, in Chap. 17, how strategic alliances are becoming an increasingly important aspect of corporate project management in the international realm. Buono notes that it is becoming increasingly clear that many of the key problems and barriers in global interfirm alliances are as much organizational and behavioral as they are strategic and financial.

Rajeev M. Pandia describes, in Chap. 18, the various economic, social, political, and demographic considerations likely to be encountered in the implementation of international projects. According to the author, such projects provide valuable opportunities to all participants. He further notes that if suitable strategies are formulated—and appropriate approaches are adopted at the initial stages of such projects, and go on through implementation—many of the threats and dangers endemic to global projects can be reduced or avoided.

Hans J. Thamhain and Aaron J. Nurick provide an examination of project team development in multinational environments in Chap. 19. They summarize the characteristics of multinational team environments to provide a better understanding of the existing challenges and opportunities. Also, the authors offer a model for organizing and developing the multinational team, and suggest a framework for evaluating team organization and performance.

CHAPTER 15

THE ROLE OF PROJECTS IN THE STRATEGY PROCESS

Wendy Briner
Colin Hastings
Consultants

Doctor Colin Hastings and Ms. Wendy Briner are independent consultants and founding members of New Organisation Consulting, in London, specializing in the application of the project way of working and learning in business development, innovation, and change in organizations.

The world sometimes intervenes in our plans.
HENRY MINTZBERG[1]

Smart strategists appreciate that they cannot always be smart enough to think through everything in advance.   JAMES BRIAN QUINN[2]

WHY STRATEGY FAILS

We have been hearing in recent years from senior managers their increasing dissatisfaction with their strategy processes. Managers are irritated that they have not achieved what they expected. If we analyze these dissatisfactions, a number of reasons consistently emerge. We hear these managers saying things like:

- *"Our current strategy is ok. We've done our 5-year plan."* This may not seem a cause for dissatisfaction, but it is the very lack of dissatisfaction that is the problem. Many managers talk of the dangers of complacency, believing that everything is "under control" while they see the world changing rapidly about them, and suspect that they are not keeping on top of vital new issues. John Harvey-Jones, ex-chairman of ICI, talks about dissatisfaction being the "engine of change" in his excellent book *Making It Happen.*[3]

- *"We have no strategy. We need one but we don't know how to go about developing one."* This situation may occur where the company's board members have come up through the operational ranks of a stable organization. The process of thinking forward, thinking conceptually, and making decisions about where they want to take the business is an alien one, particularly when their focus has been short-term.

Alternatively, there may be a strong Chief Executive who believes he or she has a clear strategy which resides within his or her head! The rest of the organization cannot see this strategy; and even if they could, they might not have confidence that it is soundly based.

- *"We have a strategy but not everyone at the top agrees with it."* There may have been some kind of debate within the Board but the conclusions are neither widely nor actively supported. The debate may have been railroaded to a decision by vested interests or powerful figures, thereby overriding objections and leaving some people passively complying. The lack of "buy-in" among the top team results in resistance or undermining when it comes to action.

- *"The Board has an agreed strategy but no one else understands it or is committed to it."* Some senior managers believe that if they publish their strategy widely within the organization, the strategy will be implemented. The assumption is somehow that, if you tell people to do things, this will produce the desired results. This misconception goes even further! Some managers seem to believe that, if people are told to do things, they will necessarily be positively motivated and know exactly what they have to do differently. In practice, understanding, doing, and feeling committed are three separate aspects of implementation. If the published strategy is questioned, this is frequently perceived as unreasonable resistance. The board members forget "where they came from." They have been working for a long time with the issues, thinking and talking about them, so to *them,* the messages are clear and simple. To the first-time hearer they are difficult to understand.

- *"Everyone understands the new strategy but nothing is changing."* Some organizations put considerable effort into involving people in the development of a strategy or into explaining it. Once this is completed, managers wait for it to happen, believing that they will see rapid results. They seem to feel that the thinking is over and that it is now up to those lower down in the organization to make it happen. There seem to be a number of reasons behind this:

  1. There are unreasonable expectations about how quickly strategy can produce visible results. Significant shifts of activity are likely to take 3 to 5 years.
  2. There is a widespread assumption that implementation will be smooth—a linear progression rolling out in clear sequence. Recent research into the behavior of complex systems suggests that a series of quite pronounced ups and downs will be experienced as people try things out, partly fail, and then modify their activities until they find a method that works.[4]
  3. Managers ignore the effect of history in their organizations. New strategic initiatives which in themselves are very worthy can be blighted by people's experience of previous initiatives. This can lead to a cynical response: "Keep your head down and let this one pass." At worst, previous failures have injured people so much that this adversely colors their reactions to the new strategy.
  4. Management, who have communicated the strategy well, and have depended, for example, on levers such as reorganization, companywide training, or a computer system to implement the strategy, are still disappointed. In practice, implementation penetrates many aspects of an organization and a wider range of levers being pulled in concert is required.

- *"We know where we want to get to but not how to get there."* This is increasingly common as the markets in which businesses operate become more turbulent and uncertain. Most managers believe that once you know *what* you need to do, you automatically know *how* to do it. This works when you have had relevant previous experience. When companies operate in new environments, the skills, attitudes, sys-

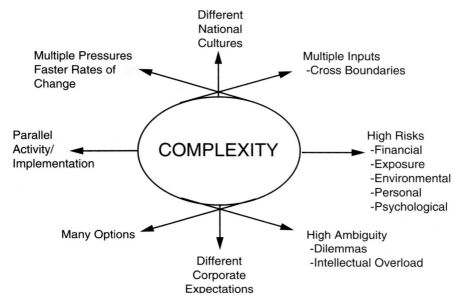

FIGURE 15.1   Sources of complexity in the corporate environment.

tems, and structures to enable them to operate successfully do not exist. These have to be learned, and this takes planning, management, and time. It is hard for managers to know what they do not know, and even if they do, it is often politically unacceptable to say so. This tends to lead managers to do what they have done before and to be surprised when little has changed despite their efforts.

- *"The Board just can't keep a handle on all the issues facing the business...they are becoming a blockage to us moving forward."* The strategy process is becoming overloaded by the sheer scale, volume, and complexity of the issues facing businesses. Figure 15.1 illustrates a range of factors that make both understanding and decision making more taxing. Boards and senior managers who feel they ought to be able to comprehend and get to grips with all the issues find themselves paralyzed, intellectually and emotionally.

- *"We've tried involving more people in the strategy process and it just doesn't work."* Some organizations have tried to involve a wider group of people in the strategy process, but they find that this can lead to opening up so many issues not related to strategy that the thread gets lost. Also the consultation process is perceived as being unrelated to the normal decision making processes and is therefore not seen to be important or genuine.

Using Projects in the Strategy Process

Moving an organization from one position or state to another involves two sorts of decisions: (1) deciding what the organization is going to do and (2) actually finding out how to do it effectively. Many of the problems highlighted above stem from the traditional role split when senior management have undertaken the thinking and the rest of the organization, the doing. We now understand that both aspects have to be integrated, and ways need to be found to link thinking and doing within a contained framework. We believe that projects as temporary organizational structures, combining thinkers and doers, different functions, and different organizational perspectives, offer a way forward.

The use of different kinds of organizational teams is suited to dealing with these kinds of problems. By pulling together a diverse range of brain power both inside and outside the organization from different functions, they can master the complexity of strategic issues. The multidisciplinary team is a vital element of the strategic management process. It recognizes that Boards alone cannot do all the thinking, nor can they provide sufficient integration. A range of different project teams, drawing on many people within the organization both directly and indirectly, creates many opportunities for contribution and involvement, thereby enhancing commitment. Project teams provide strong focus and follow through, particularly at the implementation stage. Using dedicated resources and traditional project management skills ensures that targets are achieved amidst distracting uncertainty and turbulence. The management of the overall process, with its portfolio of different kinds of projects, provides the infrastructure for managing continuous uncertainty.

For many managers, this remedy may seem to produce side effects that are worse than the original disease. Their perception is that strategy is about finding direction for the organization and that this is the Board's job. However, the problems of complexity and of gaining people's commitment suggest that the Board cannot keep this function to themselves. Widening the net to those who can contribute or are involved taps into the collective brainpower and wins commitment. Yet this process of involvement feels as if it increases the complexity. Focus seems to be diluted because the widened net may increase conflict and make decision making more difficult. So, for the Board, the process may appear out of control and a negation of their responsibility. When answers are unclear, there has to be a more extensive process of exploration, and this process may distract the organization from committing itself to successful implementation. Strategic projects also enable people to move into areas where they may not have been before. This can be exciting and threatening. Senior managers may find themselves questioning their role and their power and ability to control as issues are investigated and the full complexities exposed. Line people may wonder whether they will be taken seriously and whether they should be doing management work.

Is there a way of moving forward and finding a different process, a different mental model that can help managers to create direction for their organizations under conditions of complexity and uncertainty? Is there a way of exploring while being focused? Is there a way of involving the brainpower of the organization without getting distracted or degenerating into anarchy? We believe that the use of projects within a clearly understood framework for strategic management offers a practical process to solve some of these problems. Whereas formulating strategy was something you did every 5 years, today it is suggested that it should be done all the time. Increasingly the words *strategic management* rather than *strategy development* are used to denote the continuous process of finding direction and implementation. In the next sections we will explain the steps needed to fit projects effectively into a strategic management process and to get the best out of them.

FITTING STRATEGIC PROJECTS TO STRATEGIC PROBLEMS

Businesses do not have one strategic problem. They have a range of different strategic problems. We have come to realize that different kinds of strategic problems require various project approaches to manage them effectively. We believe that a lack of awareness of the differences in both strategic problems and project approaches may be why strategy task forces have not been universally successful. It is not just a question of "set up a project or task force and let them solve the problem!"

The Spectrum of Strategic Problems

Figure 15.2 articulates four broad types of strategic problem that we have identified and illustrates briefly how the project focus is different. All these types of strategic problem are complex, but they differ in their degrees of uncertainty and the levels of previous experience that the organization has about how to go about solving them.

Problem 1—"We know where we are going, we have experience of how to get there, but we are doing it in different conditions."   A soft drink manufacturer with a worldwide brand builds a new bottling plant in a country where they have not previously manufactured. The technology is familiar but the local cultural setting is not, although the company has been distributing through others in that market for a long time. Past experience is reliable. Technical know-how, skills, and management systems will be transferable. There may be new challenges that are risky and specific to

	Problem (1)	Problem (2)	Problem (3)	Problem (4)
Type of Problem	More of the same in different conditions. We have core experience	We know where we are We know where to go but getting there is demanding	We know where to go but we don't know how	Don't know where to go but we can't stay here
	Lots of Organizational Experience		Little Organizational Experience	
Project Focus	Actioning Solutions	Learning about and Testing Solutions	Exploring for Solutions	Searching for Direction

FIGURE 15.2   Strategic problems—a spectrum.

this situation. The project focus here is on action and review to keep on top of the specific situation. In this instance results, outputs, and deadlines are very important. Traditional project management approaches of planning and control are extremely valuable and relevant.

Problem 2—"We know where we are going, but getting there is going to be very demanding and stretching." A manufacturer of four-wheel-drive vehicles has developed an enviable reputation based on a basic rugged machine. Its engineering processes are not sophisticated; its quality is not consistent. The product is moving away from the rugged to the fashionable end of the marketplace. So it needs to be able to build a quality product to the standards of a new customer profile at a competitive price. Linking engineering, manufacturing, sales, marketing, and suppliers together enabled the company to reduce its new product development time by more than one-third, while maintaining the quality and cost standards required.

The project focus is on rapid learning, based on cycles of planned experimentation, review, and further experimentation. The "what" is known, but the "how," in terms of internal organization, processes, task organization, problem-solving mechanisms, and teamwork is very different.

Previous know-how is relevant. Detailed planning with regular progress checks is vital, with a problem solving mentality which constantly challenges "How are we doing this? How could we do it better?"

Problem 3—"We know where we want to go, but we don't know how to get there." One of the police forces undertook comprehensive internal and external surveys of their services, from which it was relatively easy to identify what needed to happen. However, getting there challenged many of the established ways of working, organizing, managing, and rewarding. So the key task was to discover workable alternatives that would enable the force to deliver the sort of service they and their customers wanted.

The project focus here is to explore and evaluate options. The direction and purpose is a given but the established means do not deliver, so new ones must be found and built. Many more people will tread into unknown areas in this process, and they will become increasingly aware of what they do not know.

Problem 4—"We don't know where to go, but we know that we cannot stay where we are." A major international insurance company which has lived on a reputation earned many years ago finds itself suddenly in turmoil. Competitive pressures have increased in the United Kingdom, and it realizes that it needs to respond urgently to the single European market. It has limited experience outside the U.K., so it is acting on little direct know-how. Both what to do and how to do it are open questions. The level of ambiguity is high.

The project focus here is on exploring for direction. There is a very high level of ambiguity, as the problem itself is unclear. The project team will have to create its own project definition and objectives because nobody else can give direction.

At any time, an organization will have a range of strategic problems in front of it. In stable markets they will tend to fall into the Problem 1 group. In this case the organization feels itself well-equipped to deal with the issues. If there are many in the Problem 4 group, there will be high uncertainty, as the sense of direction is being questioned. If a direction emerges from a Problem 4 project, then the problem will gradually move from right to left as both the "what" and "how" become clearer through focused experimentation and gradually acquired experience.

These broad groupings of strategic problems need very different types of project organization to deal with them effectively. Figure 15.3 indicates the differences in

	Problem (1)	Problem (2)	Problem (3)	Problem (4)
PROJECT FOCUS	ACTIONING SOLUTIONS	LEARNING ABOUT & TESTING SOLUTIONS	EXPLORING FOR SOLUTIONS	SEARCHING FOR DIRECTION
Key Question	How to make it happen	What do we have to do differently to get there	How do we get there	Where are we going
Project Characteristics	• Clear, strong objectives • Data heavy • Strong control of progress • Cost monitoring • Formal milestones and reviewing • Commitment taken for granted	• Clear, strong objectives • Flexibility about how to get there • Prototypes, pilots • Building commitment to the best way • Rapid application of focused learning on what we don't know	• Building commitment to the direction • Mobilizing support • Creativity • Some early judgment	• No direction • Multiple iterative cycles • Ambiguity • Focus on rapid, extensive and continuous exchange of learning and ideas • Direction emerges not imposed

Fitting the Project to the Problem

FIGURE 15.3

project characteristics. Problem 1 is more suited to a formal project structure with work breakdown structures, milestones, and controls. At the other end, Problem 4 needs to be very flexible, with only some outline planning and broad milestones. The project team for Problem 1 will be structured with roles and technical contributions well-known. For Problem 4, there will be a wide ranging search, with many and varied contributions coming both from inside and outside the organization. The team may vary, and the leader will coordinate and focus ideas and alternatives.

Senior management may be unaware of the kinds of strategic problems that they have, and therefore what sort of project team to set up, what to expect of it, and how to manage it. Such confusion can lead to disappointment with the value of the project's outputs. Many organizations face a bewildering portfolio of different projects, and the task of trying to keep a helicopter view of where each one is and how it links with others in the evolving strategic management process is not easy.

We will consider how senior management can manage this portfolio effectively in a later section. The company in an uncertain environment is like Christopher Columbus preparing to discover India. It builds, crews, and provisions different kinds of "ships" (which represent strategy projects), each one for the different kinds of conditions or purposes that it anticipates it might meet. Once those ships set sail, even if they start together, they gradually spread apart as they meet different conditions of wind and weather. In subsequent sections we will be describing some practical ways to help those navigating their organizations to bring their strategic management process speedily and safely to its destination, despite the inevitable hazards.

HOW TO SET UP PROJECTS FOR STRATEGIC MANAGEMENT

Whether a company's strategy is a journey into relatively familiar territory or one into the unknown, a considerable amount of preparation is necessary. Any sailor makes careful preparation before setting off on even a familiar local trip. This section examines some of the factors that senior management should pay attention to when preparing for and setting up projects as part of a strategic management process. Although for convenience these are presented in a series of steps, in practice they need working on in parallel. They are interrelated, and there is no logical linear order. Management need to be seen attending to these activities simultaneously so that those involved can visibly identify the parts of the strategic management process.

Step 1—Be Clear about the Different Kinds of Strategic Problems You Have Got

Although this is ultimately the responsibility of the Board, the mapping of the range of strategic issues facing the business is best done with a wide population of people with experience and different perspectives. By using a wider group you begin to "get on board" a community of people who know what is being thought about. Some of these may become project leaders in subsequent stages of the strategic management process.
   An off-site workshop for this broad community should aim to:

- Diagnose where the business is now
- Speculate on future developments
- Identify blockages that need immediate attention
- Identify clusters of strategic problems that can be addressed by project teams
- Agree where these strategic problems fit in the spectrum
- Identify people who are interested and willing to contribute to a project team
- Identify external resources
- Agree how the strategic management process is to be organized and coordinated
- Establish a champion, preferably the Chief Executive

As this is a challenging process, it is often useful to have an outside consultant helping to keep things moving.
   Once these basic elements have been put in place, then more detailed consideration and planning can go ahead for the three other steps.

Step 2—Match the Project Team to the Problem

Starting up a project team and "throwing it in at the deep end," hoping that it will somehow survive is usually counterproductive. Many come to grief, generating cynicism and disappointment which is damaging to the team, senior management, and the strategy process.

Different kinds of strategic problems or challenges will require different types of project teams. Greater tolerance for ambiguity and the ability to pull together new configurations of ideas are necessary in Problem 4–type projects, where the task is to find the direction. So volunteers need to be found from people who are interested and have a preference for investigating new things. In the case of Problem 1–type projects, it is important to have people with specific know-how and the willingness to find creative solutions within the existing frameworks. It is unlikely that one person can operate with equal effectiveness across the whole spectrum of projects, so the right combinations should be matched to the problem type. In any case, the project leader needs to be somebody with perceived weight and credibility within the organization. The ideal is somebody who appreciates the complexity and nature of the topic but who is not an expert. The leader should have the people skills not only to draw together and motivate a project team but also to establish good relationships throughout the organization. Team members should be aware of how the organization functions, who the key decision makers are, and what the underlying politics are about.

The Chief Executive or a senior manager should brief the team on the overall process, and agree with them on the contribution that their project should make to the strategic process. The impact that taking on a strategic project will have on their normal work should be agreed on and any conflicts settled.

Step 3—Prepare the Ground...Creating the Necessary Preconditions for Effective Strategic Management by Projects

As with any expedition or journey which involves a considerable number of people, it is important to consider what resources need to be put in place *before* the journey starts. The Chief Executive or his/her coordinating team asking "what if" questions will give a rough estimate. As there is little reliable data at this stage, planning will have to be on a revolving "plan/do/review" basis, adjusting with each phase as it rolls out.

In our experience there are some particularly important foundations that need to be planned and established for a project-based strategic management process to be successful.

1. Ensure that all the key players have realistic expectations of what might be involved and how long it might take. It always takes much longer than you think!

2. Establish some baseline measurements against which the progress of the overall strategic management process can be monitored. We call this *benchmarking*.

3. Begin to make realistic and pragmatic estimates of the resources that might be required.

4. Develop among management a common language for being able to talk about strategy, the strategic management process, project management, and change.

5. Identify groups that require specific preparation briefing or training to be able to carry out new roles.

6. Audit existing communication processes to assess whether they are effective, or to put in place and test out new ones to deal with the volume and variety of communication that will be needed.

7. Make visible and available relevant expertise, from inside or outside the organization.

8. Put in place mechanisms which will enable each project team's learning to be rapidly and effectively codified and disseminated to and understood by other projects and the whole organization.

Let us now look at each of these in more detail.

Realistic Expectations.   Our own experience and research by Michael Cross[5] show that one of the main reasons why major change processes fail is that expectations of results are unrealistic. There are two major areas where expectations tend to be unrealistic. First, there is usually a massive underestimation of people's time and commitment that needs to be put into a strategic management process in order for it to bear fruit. This can be due to unrealistic appraisal of the uncertainties and therefore the new approaches to be learned, as well as an underestimation of how much time it will take people just to get to understand the full complexity of the issues facing the business.

Second, if an organization has to learn significantly in terms of what it does and how it does it, this takes time. It is impossible for large numbers of people to make sizable shifts in less than 5 years. Unfortunately, management's normal expectation is that change should be completed in 1 to 2 years.

Many changes get aborted because it seems as if they have not succeeded, when in fact, the new learning is just starting to grow and establish itself. It pays dividends for senior management to adjust their expectations to avoid unnecessary setbacks and disappointment.

Success Criteria and Benchmarking.   A difficult question for any organization to answer is "How good are we at what we do?" The challenge of the strategic management processes is to define: If we were different, how would we know we were successful? There are many ways of establishing how good an organization is but external verification is the strongest basis. The comparison of aspects of one company's performance against others in a sort of league table is called benchmarking. John Harvey-Jones says in his book *Making It Happen,* "Dissatisfaction grows when we know we are threatened by the superior performance of others (other organizations, other countries, other products, other individuals)."[3]

Internal measurement criteria will require that hard and soft parameters of performance are specified as a baseline so that improvement can be tracked. Management should agree on the range of significant measurements that help them pay attention to what influences business success. In the less tangible areas, regular attitude surveys can gauge people's reactions to the company's future direction, the style of management, the culture of the organization, levels of understanding and commitment, and the effectiveness of communication about strategic issues. Measurement gives feedback that indicates targets for further improvement.

External benchmarking can be based on narrow criteria such as financial ratios or stock market indexes. New groupings like Productivity 2000 are being formed by companies in noncompetitive industries, collaborating to exchange information on wider parameters of performance.[6] Competitors' databases or trade associations' market research make possible comparisons between different organizations from a customer perspective.

Resourcing. One of the central dilemmas of an effective strategic management process is the problem of how people find the time to do this work plus their normal jobs. Without consistent dedication of time and effort to the strategic management process, it is likely to peter out.

One of the most common misconceptions is that senior management thinks that "they" (i.e., the rest of the organization) will be doing all the work, while less senior management feel that "they" (i.e., senior management) should be doing most of the work. There are a number of ways to work this dilemma through.

1. Persuade management that strategic management is real work and that it constitutes a normal part of their job on a continuous basis.

2. Dedicate resources to the strategic management process by appointing a full-time strategy process leader who acts as a permanent reminder. Often external consultants provide an additional resource to help both with the planning and the ongoing project management.

3. Get people to stop doing things that are not essential. It surprises people to find that some of what they do is unimportant.

4. Get senior managers to delegate more of their current work to lower levels to free up their own time. Alternatively, junior staff can be brought into the strategic management process to provide extra resources and valuable development opportunities.

The strategic management process does take people's time. Well-managed and -planned, it can release latent motivation and energy which otherwise remains untapped. As people become stimulated by feeling that they can contribute to the future of the organization, they somehow find the time to make it happen. It also takes money. This must be budgeted to cover consultants, workshops, training programs, conferences, travel to visit other companies and other countries and competitors, internal communications mechanisms, attitude surveys, and market research.

Common Language. When new concepts are introduced into an organization, it is assumed that they are commonly understood and assist in communication, but often this is not the case. This problem can be overcome by a senior group developing a common language for talking about the strategic process. Creating a shared language with common definitions enables clear communication and also creates a common set of ideas about what is going to be involved.

We help organizations develop this language and approach to strategic management processes through a program designed with the senior players. This program has two major components. First, it focuses on language, providing useful concepts and tools that the organization can translate into its own words and uses. The range of concepts and tools offered may include strategic analysis, SWOT analysis (strengths, weaknesses, opportunities, and threats), competitor analysis, service models, and tools used in total quality. Additionally, the nature of strategic projects is highlighted, particularly the approaches required for different project types on the strategic problem spectrum. Finally, the overall "big picture" is drawn, linking the strategic problems with an outline of the strategic management process.

Second, the program provides emotional preparation for the nature of the journey ahead. Little is ever done to prepare people for the kinds of experiences that they may face in the overall strategic management by projects process. This kind of preparation requires a very different approach from the didactic approach of most senior management training. For example, analogies, stories, and images can be used to help people explore what it might be like and how they might react. In working with senior management on their role as sponsors of strategic processes, it helps to use an image that characterizes the organic nature of the activity. Sailing or gardening images are useful, but essentially the steering group needs to find its own image to express its feelings about its experiences and hopes.

Training. Projects are not commonly used in organizations for managing internal strategic issues. People find them different from normal line management practice and unlike external customer projects that they may be more familiar with. Most organizations have little accumulated know-how and find it hard to appreciate how strategic projects are different from the relatively traditional projects. The strategy projects that start off at the "direction finding" end are unclear and hard to handle. Training to prepare people for strategic projects helps them understand how to approach different projects along the problem spectrum, particularly at the scoping stage. Clear scoping emerges through an iterative process involving work with the important stakeholders. The more open the problem, the more iterations it takes to achieve a clear scope.

Generally, training for strategic projects has four key target groups:

1. Board members or steering group
2. Senior management as sponsors of strategy projects
3. Project leaders
4. Project team members

Steering Group (Board). The steering group needs to be welded together into an effective team. It has a complex job to carry out: planning, setting in motion, coordinating, monitoring, and sustaining multiple sets of activities. Inevitably different vested interests are represented and people will have different levels of skill. Working early on with such a group will help them to become clear about: (*a*) their role, (*b*) their interactions with project and project leaders, (*c*) how to plan and integrate, (*d*) their leadership function, and (*e*) how to build their own commitment and that of others.

Sponsor Development. The role of the sponsor is to be the most senior person who steers, champions, and supports strategy projects and act as one of the conduits through which a project's findings can be fed in to the Board.[7] For many senior managers it is an alien role. It demands a sense of direction amid ambiguity. It means encouraging and investigating confusion, while making decisions and explaining why. And yet it is one of the crucial elements in the success of strategic projects. Without an effective sponsor with clout, projects frequently get stuck against organizational blockages which they do not have the power to remove.

Sponsors can be developed in several ways. For example, sponsors may be coached on a one-to-one basis as the role develops with the progression of the strategic management process, or through a workshop which brings together several sponsors to establish a common focus on what they are doing and how they are putting it into practice. There are major problems with strategic processes if the top and middle-level sponsors are giving contradictory messages to the rest of the organization. Figure 15.4 summarizes the main aspects of the sponsor's role.

Project Leader Development. The project leader's role is probably the most vulnerable in the whole strategic management process. On the basis of our book *Project Leadership,*[8] we have developed a series of flexible learning modules that can be used for project leader development in workshop form or for one-to-one coaching. The broad topic areas frequently covered are the identification of stakeholders, project scoping, clarifying success criteria, assembling and developing a team, matching planning tools to the project, developing relevant mechanisms for monitoring and review,

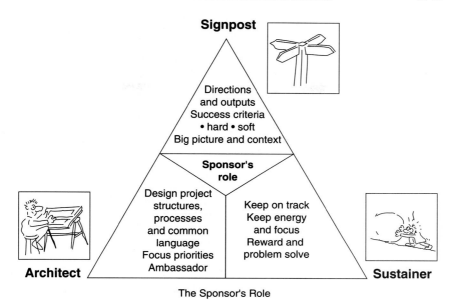

The Sponsor's Role

FIGURE 15.4   Main aspects of the sponsor's role.

internal marketing, and linking the project to the big picture. When developing project leaders for strategic projects, it is most effective to use their "live" material in the workshops.

Project Team Start-up.   Project teams when brought together do not work effectively by magic. Members come with many different agendas, anxieties, and expectations. Many strategic projects will be designed to bring together people from different levels and functions who never expected to work together. Helping these teams to get clear about their task and to understand the processes through which they can move forward on their task is the primary purpose of a start-up workshop. A second aspect is to help them to understand how they are part of the bigger strategic process and why they are important. The third aspect is to help them to develop strategies for interacting with other parts of the organization whose help they may need. And last, we work with them to understand what kinds of teamwork strategies will fit with the style of their particular organization and are likely to get results. Our experience suggests that different kinds of organizations interpret successful teamwork in quite different ways and that teams need to understand these differences in order to make the system work for them.[9]

Intensive Communications Processes.   One factor that is consistently and hugely underestimated is the amount of time that will be taken explaining things to people, reminding people of things, and listening to people. One problem highlighted when the Board develops strategy in isolation is the gulf that opens between the thinkers and the implementers (the rest of the organization). Explaining complex ideas and getting people to a point where they feel that they understand them is a lengthy task. Furthermore, understanding is still a long way from action. Building the ability and willingness to act demands detailed two-way discussions to determine appropriate ways forward. These paths then need to be tested out in action. The newly tested approaches in turn need to be introduced to others in similar situations. The strategic management processes that we are describing involve many more people in developing and testing, but it is not practical to involve everybody at the same time. So plans must be made to bring everybody up to speed with the new ways.

Senior managers may be very reluctant to communicate new half-formed thoughts as they emerge and before they are fully "baked." However, drip-feeding ideas, seeking feedback and clarification, developing them a little bit more, retesting and getting further feedback, is a more efficient way of developing ideas and builds support as you go. A cornerstone of the whole strategic management by projects process must be that the thinkers and implementers are continually brought together as frequently as possible, so that all learn and move forward as far as possible in unison.

Communication to the rest of the organization and helping them to understand what is going on should go on in parallel so that they have time to understand and accommodate the new thinking. Creating mechanisms for the dissemination takes thought, time, and investment. We advocate planning at the beginning to put communication mechanisms in place and make resources available.

An array of communication approaches can be orchestrated over a period of time, by concentrating on stimulating interaction rather than one-way communication. These might include, for instance:

- Training programs for a wide population to publicize the strategic problems facing the organization and to develop a language to discuss them effectively
- Cascading briefing groups through line management
- An exhibition or road show that visits sites, using videos or printed materials to build awareness of business problems and how they are being handled
- Senior management constantly getting out to the front line and "walking the talk"—discussing with groups of about 25 employees the portfolio of strategic projects and how they are being handled, and listening to their reactions
- Strategic project teams informally marketing the importance of their strategic problem
- A newsletter for the strategic management process, reporting on the milestones of the portfolio of strategic projects
- A strategic project office that coordinates communication activities.

Accessing Know-How. Strategic management by projects is above all else a process of organizational learning—helping others learn as you are learning what the organization should do and how it can do it. Some of this learning comes through creative problem solving inside the organization, but a very important part of it comes from accessing people with relevant expertise or experience. In a small company this may not be a problem because everybody knows everybody else. The problem will be accessing external know-how.

For the larger company, discovering the breadth and depth of the expertise that lies within it is not easy. Early workshops that seek to explain strategic problems and test participant interest can be used to identify hidden capabilities. In addition, exciting new possibilities are opening up for companies to create powerful databases that make visible and accessible the full expertise and experience of people across the organization.[10]

Informal methods of networking are also vital as channels through which expertise gets transferred. The steering group and consultants will be giving thought to creating opportunities through the communication opportunities mentioned previously for people to meet each other and to network.

Reviewing and Exchanging Learning. In the early stages of the strategic management process, some projects will be of the direction- and solution-finding types. The teams involved are reconnaissance teams exploring unknown territory and bringing back the results to help the organization understand better. In these early stages, rapid and frequent exchanges between different project groupings need to be designed to ensure that learning takes place. Good exchanges of learning speed up the learning curve of a large number of people as they acquire wider appreciation of the overall issues, challenges, and blockages facing the organization.

There are many ways in which learning exchanges can take place. Part of the steering group's role is to be a broker of information and learning between projects to ensure that wide and rapid dissemination happens. In addition, on occasion they should bring together the steering group and project leaders to present and discuss their findings, to sift significant aspects, and then to integrate the implications into the next phase of action. In the early stages this may be done informally. But later it will be important to have formal *learning conferences.* These set pieces provide milestones to pace the whole activity, and they give opportunity for larger numbers of people to be brought up to date.

SUSTAINING MOMENTUM—WHAT TO PAY ATTENTION TO

After the preparation and planning, with the project teams invisible but busy, senior management experiences a vacuum. The relatively hands-off approach of strategic management by projects leaves them feeling uneasy. What should they do?

Their prime focus is to keep up the momentum—first, by zealously promoting, supporting, reinforcing, propagating, problem solving, defending, and extending the positive ideas and activities from the project groups. Alternatively, they need to be ruthless in removing, undermining, finding ways around, and diluting blockages either to the strategic process or a project team. Specifically, however, there are six areas to which senior management should now be applying themselves.

1. *Managing a portfolio of strategic projects.* Project teams will be like sailing ships bobbing around in different places with different kinds of weather. Senior management's job is to be the communication satellite with these different ships to know where they are, perhaps replacing crews if necessary or replenishing supplies. Their job is not to skipper the ships but to provide all possible help to ensure that the ships reach harbor. In addition, project portfolio management will entail setting up a formal project "slate" (denoting what projects are on and not on), deciding when projects should be stopped, resourcing and agreeing to means to settle conflicting demands for resourcing between different projects, monitoring each project's progress using traditional project management tools, and enabling interchange between projects to keep learning moving.

2. *Personal leadership.* Once people get involved in the strategic management process they need constant reassurance that the process is for real. The level of reaffirmation and required reassurance is much more than seems reasonable to senior management. People further down the organization may continue to believe that nothing is really going to happen as a result of all this project activity.

This may be acute in the first months where there will be visible action but few tangible results. There may be more uncertainty as the organization discovers more aspects that it didn't know it didn't know. This frequently culminates in a well-established pattern, which we have called "the watershed." This takes the form of a crisis point around 12 to 18 months after inception. There may be a feeling of failure. Senior management's nerves are put to the test. This is the supreme testing point, rather like "the wall" in marathon running, when senior management have to reassert and demonstrate their own conviction, commitment, and courage to continue and to make the breakthroughs required to move the organization forward. To make progress when sailing you cannot go straight. Periods of progress are followed by periods of confusion, so leadership from the top must keep sight of the horizon, despite the turbulence.

3. *The emphasis on communication.* Constant reinforcing communication from senior management helps to keep the goals in people's minds. In the early days the emphasis is on explaining why the changes are necessary, what the causes are, and demonstrating senior managers' personal belief that it is important. Later it will shift toward providing firm support and stubborn reassurance while people battle through difficulties.

As the process unfolds, senior management must draw together the threads of possible directions and solutions as they emerge and express them. Directions become more visible as people see them more clearly, test them, and ultimately become committed to them. Senior management should communicate these emergent strategies[11] to project groups to narrow their attention from exploration toward implementation. Throughout the process, communicating small successes and rewarding them demonstrates that progress is being made in small cumulative steps.

4. *Managing project team performance.* Early on, it is hard to set specific output goals, so senior management should set teams very ambitious targets and time scales to achieve certain agreed stages. Also they should ensure that, in the exploration stages, teams search inside and outside the organization for know-how and expertise. They are brokers in identifying relevant expertise. When a project gets into trouble, senior management needs antennas to pick up early warning signals either that a project is getting stuck and needs rescuing or that it is overstretching itself and losing focus or fragmenting its energies. Linking the progress of individual projects to each other and the big picture gets the most value from the strategic process.

5. *Measuring progress.* Gathering information and feeding it back widely is crucial. At intervals the benchmarks will need to be monitored. Every 18 months senior management should be looking for measurable signs of improvement. This feedback provides data to review the strategic management process in order to develop new directions and amend previous plans; it is the prime means of keeping the organization's performance on track.

6. *Orchestrating learning through interaction.* The direction that emerges is highly dependent on the interaction between the Board, the steering group, and the project teams. The final responsibility for strategy decisions rests with the Board; it cannot be delegated. If the strategic management processes are effective, the choices will be widely known and debated to the point where they are obvious. This then requires the Board to make them official. The Board need to be well-connected to the formal and informal processes to choose the right time to consolidate what is being established.

The strategic projects approach will demand difficult decisions when initiatives must be stopped. The Board needs to think through how it will handle openly the evaluation of options so that people who have put in a tremendous amount of work do not get turned off, but understand the tradeoffs that have to be made. Senior management has a significant part in integrating progress to sustaining momentum. Energy does tend to flag, and there are dispiriting times, so reinjections of purpose and energy are required to keep momentum going until visible results produce their own satisfaction.

A PROJECT PROCESS MODEL OF STRATEGIC MANAGEMENT

So far we have described some of the considerations and activities for an organization that might want to use different kinds of projects in its strategic management process. This section describes a conceptual model which integrates these activities into six processes. Figure 15.5 summarizes the model. The model is presented as a sequence of linear phases but, in fact, these processes are going on simultaneously in parallel. They are a continuous set of processes operating on each strategic topic. Figure 15.6 gives a better feel for the model's dynamic nature; the processes of *directional feel, parallel exploration, interactive exchange, and controlled exploration* are in flux, incrementally building buy-in, clarifying directional focus, and building capability to operationalize. Each of the key processes will be described in more detail with a summary of learning points derived from our own consulting work and others' research.

The Key Project Processes

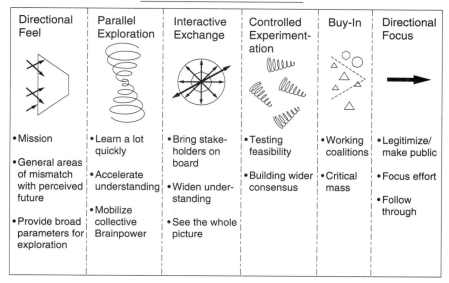

Directional Feel	Parallel Exploration	Interactive Exchange	Controlled Experimentation	Buy-In	Directional Focus
• Mission	• Learn a lot quickly	• Bring stake-holders on board	• Testing feasibility	• Working coalitions	• Legitimize/make public
• General areas of mismatch with perceived future	• Accelerate understanding	• Widen under-standing	• Building wider consensus	• Critical mass	• Focus effort
• Provide broad parameters for exploration	• Mobilize collective Brainpower	• See the whole picture			• Follow through

FIGURE 15.5   Strategic management by projects.

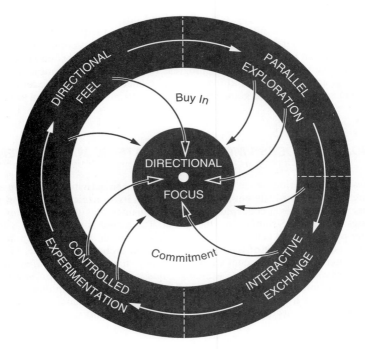

FIGURE 15.6   The dynamics of the strategic management process.

Directional Feel

Strategic Problem 4 illustrated a situation where there is need to move but where to go is not clear. Even in these circumstances there will be broad indicators based on hunch, perceived business strengths, and competitive threats or quite clear "no-go" areas. The thing about directional feel is that it provides some sense of future focuses without rigid parameters. It may say in very broad terms what the organization seeks to do but not how to do it. Many organizations express directional feel through statements of mission,[12] vision, and values or general "strategic intent."[13] Directional feel provides strategic projects with the broad territory to be mapped and explored.

Learning Points.   This stage is difficult for senior managers because they have to say that they do not know all the answers. They are asking others for help. Being open about what they are trying to do and why they want to involve people is crucial. This position is contrary to the conventional image of a decisive manager.

     All project team members need to understand senior management's thinking behind the directional feel. It is frequent for those involved in the process not to believe that senior management are asking them to participate. They think management have the answers up their sleeve and will not listen to the project group's findings.

Parallel Exploration

Parallel exploration is the process whereby project teams explore the potential strategic alternatives in parallel, working to cover as much territory as possible in a short space of time. The organization's objective is to accelerate understanding about the strategic options. Overlap between project teams is not a cause for concern; the aim is to build up a picture. There are seldom "right" answers, the purpose is to investigate the territory. James Brian Quinn[2] advocates this parallelism as a method to manage uncertainty. The Japanese use parallel teams in new product development to get the best products, either taking the best or combining aspects of all.[14]

Learning Points. The accent in parallel exploration is on the creative search for directions, not on seeking early closure on solutions. Project teams should be encouraged to look outward into the organization and to scan widely the external environment. Of paramount importance is a project team's ability to tolerate ambiguity and to not drive toward premature convergence and closure.

Interactive Exchange

The benefits of exploring the strategic territory can be realized only if the emerging understanding is widely shared and debated. Interactive exchange between project teams through major reviews of progress stimulates thinking, avoids duplication, and integrates outputs supportive to the big picture of the business's future. The flotilla of small boats are in radio communication with each other, sharing information on weather conditions and relative progress.

Learning Points. Interactive exchange does not happen by itself. It has to be planned and managed by the steering group.

The volume of information can cause overload (especially for the Board), therefore each team needs to synthesize and present its findings so that they are easily comprehensible. Synthesis does not mean reducing to obvious generalities. Early exchanges are the foundations of rapid understanding and promote the emergence of the sense of direction. Feeding progress back to the steering group and to other project teams is continuous. This encourages the important but undervalued process of "chewing over" information over a period of time, getting to understand fully its importance and implications.

Project teams may be challenging Board decisions or ways the organization has of doing things. Consultants can facilitate open and constructive communications to prevent a defensive reaction, meaning that important options are closed off. The Board's task is to sense areas of agreement that emerge and communicate them. Each iteration of exchange helps to reduce the number of options, articulate commitment, and narrow the focus of the strategic project.

Controlled Experimentation

Parallel exploration and interactive exchange processes reduce the field of strategic activity. Directions emerge and options are excluded. The nature of the problem moves from being highly ambiguous, to a concept, and then to feasibility testing. The accent moves from divergence and exploration to convergence, evaluation, and implementation. The idea is to test how things will work in practice, so people involved with implementation must be included. Detailed implementation plans with costs and human resource estimates must then be formulated. The commitment of wider groupings should now be tested. Objections are still listened to so that ways around resistance or other implementation barriers can be found. This phase develops the critical mass of people who are prepared and able to work with the strategy. The Board is looking to assemble evidence and commitment to give the formal green light to go live with the tested new strategy.

Learning Points

- Testing strategic options for workability uses the planning disciplines of project management.
- Bringing on board the wider organization requires intensive communication, the objective being to create widespread alignment and attunement[15] with the new directions. By now this is not new to the whole organization, as a critical mass is already working with the new initiatives.
- The project teams need to assess what else in the organization needs to be modified for successful implementation. Reviewing systems such as rewards, control, forecasting, and attitudes to work is necessary, as they are needed to support the emerging strategy.

Buy-in and Commitment

Buy-in is the fundamental goal of strategic management by projects. Each of the previous processes is designed to build up commitment gradually through phases of wide involvement and testing. There comes a time when senior management formally commits the *organization* to a course of action. This is the Board's clear leadership signal that the whole organization is moving forward toward the chosen objective. If achieved effectively people will already be well down the road of working to the new strategy so the gulf between planning and implementing will be minimized.

Learning Points

- You need enough commitment to form a working coalition to champion the strategy and establish the critical mass of people and resources to make it work.
- Pushing hard against those who are perceived as resisting creates subversion. A powerful approach to building commitment is to acknowledge these people's concerns and to work with them to reduce the blockages that they perceive.
- Choosing the moment to commit the organization does seem to have elements that require patience and perhaps cunning. James Brian Quinn advocates waiting until the last possible moment before committing, particularly where the environment is very dynamic.[2]

Directional Focus

Once the organization's direction has become clear and established through exploration and experimentation, the task is to embed the new strategy into the day-to-day running of the business. Gradually it becomes part of normal operational management and ceases to be anything separate. It has taken root.

However, as we have said right from the beginning, strategic management by projects is not an on-off process. As soon as some strategic problems are in the implementation stage, there will be other issues appearing. These now become topics for further project groupings, and so the cycle continues. The organization should think of itself as being in a constant state of what Ralph Stacey[16] has called "bounded instability." At any time there are many areas of certainty and areas of uncertainty. Some strategic problems will be at the direction finding end and others down the route of concrete implementation. There needs to be a perpetual tension between these opposites, for this is the energy that provides an organization with the ability to be stable and deliver while being flexible and innovative in response to environmental changes.

When Columbus set out to discover "India," little did he know what he might discover. Much of the philosophy behind strategic management using projects recognizes Mintzberg's findings that, particularly in uncertain environments, strategies emerge instead of being planned. Strategy becomes a justification of what is already happening rather than a statement of what is intended to happen.[11] This is similar to the findings of Robert Burgelman, who found that new strategies emerged through the work of internal product champions who conceived of an idea and managed to sell it by linking it to top management's view of the future of the organization.[17] Increasingly, strategy is a post hoc rationalization of what you have already done rather than a pseudo-logical statement of what you intend to do.

However, the difficulties of such a process—and particularly its differences from many conventional approaches to strategy in management—should not be underestimated. It places considerable demands on those taking part, but, at the same time, if properly planned and managed, can succeed in releasing considerable energy, helping significantly to build a common vision and to ensure successful follow-through at the implementation stage. Perhaps its most important outcome is to create what Andrew Campbell[12] calls the collective sense of mission, probably one of the strongest sources of direction and motivation an organization can hope to mobilize.

REFERENCES

1. Henry Mintzberg, quoted during talk to Strategic Planning Society Conference "Three Views of Strategy Formulation," London, February 1989.

2. James Brian Quinn, "Strategic Change: Logical Incrementalism," *Sloan Management Review,* pp. 7–21, Fall 1978.

3. John Harvey-Jones, *Making It Happen,* Collins, London, 1988.

4. Ervin Laszlo, *Evolution: The Grand Synthesis,* New Science Library, Shambala, London, 1987.

5. Michael Cross, *Developing a Multiskilled Organisation,* mimeo, Manchester Business School, U.K., 1990.

6. Private communication—Productivity 2000 is an informal association of senior production management in major European companies.

7. Wendy Briner and Bob Farrands, "The Sponsor Role in the Project Based Organisation," presentation to European Foundation for Management Development Research Conference, Palermo, 1991.

8. Wendy Briner, Mike Geddes, and Colin Hastings, *Project Leadership,* Gower, Aldershot, England, 1990.

9. Wendy Briner and Frank Tyrrell, *Leadership and Membership Competences in High Performing Organisations,* mimeo, Ashridge Management College, Berkhamsted, HP4 1NS, U.K., 1988.

10. 3E Research and Development Ltd., 44 South Street, St. Andrews, Fife, KY16 9JT, Scotland. Markets innovative consultancy and software designed for this purpose.

11. Henry Mintzberg, "Crafting Strategy," *McKinsey Quarterly*, pp. 71–90, Summer 1988.

12. Andrew Campbell and Sally Yeung, "Do You Need a Mission Statement?" *Economist*, Special Report No. 1208, 1990.

13. Gary Hamel and C. K. Prahalad, "Strategic Intent," *Harvard Business Review,* pp. 63–76, May–June 1989.

14. Ikujiro Monaka, "Redundant Overlapping Organisations: A Japanese Approach to Managing the Innovation Process," *California Management Review,* pp. 27–38, Spring 1990.

15. Roger Harrison, *Towards the Self Managing Organisation,* mimeo, Harrison Associates, Berkeley, Calif., 1986.

16. Ralph Stacey, "Is Planning Still Relevant?," talk to Strategic Planning Society Conference, Chaos Theory and Corporate Planning, London, March 1991.

17. Robert Burgelman, "Corporate Entrepreneurship and Strategic Management," *Management Science,* vol. 29, pp. 1349–1364, 1983.

CHAPTER 16

PROJECT MANAGEMENT: THE DRIVING FACTOR FOR INTEGRATION IN INTERNATIONAL COMPANIES

Peter Sutter

Digital Equipment Corporation,
Austria

Peter Sutter has held management positions with Siemens and Wang. Since 1985 he has been with Digital Equipment Corporation, Austria, first as manager and at present as director of the Digital Services Department as well as of Machinery Industry Europe.

INTRODUCTION

During the past few years change management has become the management challenge in international companies, especially in a highly competitive environment such as high-tech industries. Digital Equipment believes that project management is an integral discipline both for mastering the constant change and for developing new business relationships with its customers.

CHANGE MANAGEMENT AT DIGITAL EQUIPMENT

At present we are facing a dramatic change in the high-tech market. Customers are now expecting their vendors:

- To be global partners with global business practices
- To offer the highest price-performance ratios
- To be able to deliver a complete range of services
- To be outsourcing partners (that is, to take over data center responsibilities previously delivered by the customer's own organization)

Vendors on the other hand are faced with a tremendous acceleration in technology cycles. While in the past you had years to develop and market a product, you are now down to about 6 months.

What has Digital Equipment done to meet this market challenge?

- First they changed from a pure product vendor to a full-service company. Projects, especially project management, became integral parts of their marketing strategy.
- They introduced a corporate-wide project methodology, including the associated processes and tools, to enable them to delivery projects on an international basis.
- They introduced account management with a focus on both international accounts and large national accounts.
- They implemented systems integration centers on both international and national levels. Those centers are able to perform all project management and other skills necessary to deliver large projects.
- They integrated account management and project management principles in a new management system with a very low level of hierarchy and largely distributed responsibilities ("entrepreneurship"), as illustrated in Fig. 16.1.

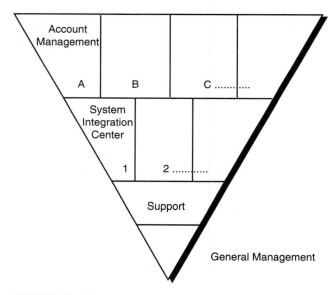

FIGURE 16.1   New management system.

- They linked together account management for international accounts and, subsequently, for accounts of similar industries (account groups). At the same time system integration centers with similar skills were joined together on a European and a worldwide basis, thus creating a "network of entrepreneurs."

THE INTERNATIONALIZATION PROCESS

Approaches to Internationalization

Central Approach. When you go international, you generally start with a central approach, that is, you define the portfolio centrally and leave the implementation up to the subsidiaries. This is usually the stage where the company founds an international division with a VP International.

Worldwide Approach. Soon you realize that there are some problems. Local markets do not follow your central strategy, and there are a lot of ideas in the subsidiaries which somehow get lost as it is very difficult to channel those ideas back to corporate headquarters.

Therefore at this stage you may decide to choose an approach of worldwide product divisions. Instead of a VP International you assign to each of your product lines' VPs the authority of worldwide business.

What you have achieved now is a better mechanism for channeling back and the possibility of picking up local ideas and marketing them worldwide. However, usually the sum of these activities no longer gives a clear marketing picture at the individual subsidiaries. You start lacking synergies, and efficiency goes down the drain.

Multinational Approach. Therefore you may decide to go the other way and implement a multinational structure. Providing just a core set of products, corporate headquarters leaves the full marketing responsibility up to the subsidiaries. What happens now is an effect of "reinventing the wheel" in all subsidiaries. Investments cannot be placed properly as they are split up into hundreds of individual pieces. This approach also makes it almost impossible to serve international customers properly.

So far we have concentrated on the dimensions "business" and "area." There is a third dimension in an international company—the functions. Initially they were established to optimize the skills in your company, but over time they have developed into the main carriers of your corporate culture. Take sales, service, purchasing, production, and finance as the usual examples, and you will see that with time these functions develop individual cultures which have a dramatic impact on all the changes caused by the approaches mentioned in the previous section.

Ghoshal[1] described the relationships between the three dimensions mentioned, as illustrated in Fig. 16.2. We can also show the dilemma of internationalization as presented in Fig. 16.3.

How can we make this global approach happen?

- We must channel input from the subsidiaries into tasks of innovation (providing multilevel and multifunctional communication channels).
- We must ensure that development efforts are linked to market needs.
- We must manage responsibility transfer (implement career paths and personnel flows and empower local management).

National subsidiaries have to change their roles from local implementors to strategic assets, as illustrated in Fig. 16.4.[1]

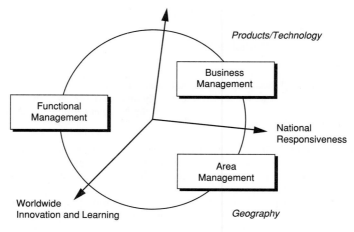

FIGURE 16.2   Global efficiency.

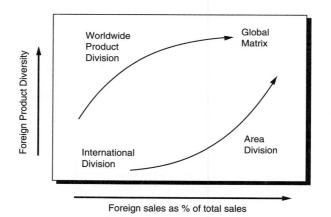

FIGURE 16.3   Dilemma of internationalization.

Impact of the Corporate Culture on the Internationalization Process

A model for corporate culture has been developed by Harrison[2]:

> What you actually see from a company, its results, its market positioning, its public relation is only the surface. All these visible external relations are based on a set of norms, behaviors, traditions, values, beliefs and identities. These basic assumptions and patterns are influenced by history, ownership, technology, objectives, environments and people. Often these patterns are more stable than formal structures and procedures. Success stories, jokes or company failures often tell a lot about them—often more than facts and figures.

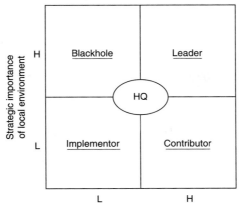

L = low
H = high
HQ = Headquarter

FIGURE 16.4   Changing role of national subsidiaries.

In order to classify different corporate cultures we focus on two aspects: (1) the degree of centralization within an organization (that is, the degree to which power is distributed), and (2) the degree of formalization within an organization (that is, the degree to which processes are defined in a formal way). Following these two aspects, we end up with a model of four different types of corporate cultures (Fig. 16.5).

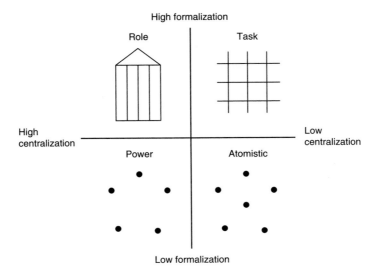

FIGURE 16.5   Four types of corporate culture.

What are the typical aspects of these different types of corporate cultures?[3]
Typical aspects of the *power culture* (central/informal) are:

- Person in center is vital
- Few rules or procedures
- Control through selection of key individuals
- Political organization, balance of influence
- Moves quickly

You find typical examples for the power culture among companies which are still run by their founders or owners, especially families. A self-explaining symbol for this type of culture is a crab. Faith in individuals and political and loyal relationships toward the person or persons in the center are determining factors.

Typical aspects of the *role culture* (central/formal) are:

- Reason, logic, rational
- Specialization into departments
- Procedures for roles, communication, dispute settlement
- Individuals exchangeable for given jobs
- Power is with the position, not with the individual
- Experts are tolerated, but not liked (line versus staff)
- Fit for stable environments
- Inflexible at turbulences
- Security for individuals

A typical example for a role culture would be an army. Bureaucracy or a Greek temple represent this culture.

Typical aspects of the *task culture* (decentral/formal) are:

- Right people at right level
- Expert power
- Team culture
- Individuals identify with organizational objectives
- Very adaptable, speed of reaction
- Mutual respect based on capacity
- More creative than specialized
- Difficult to control, less economics of scale
- Difficulties in handling hierarchies and scarce resources
- Fit for competitive markets

It is difficult to find examples representing the task culture 100 percent, but there are companies in the market which come close to this model, such as some of the large consulting firms run on a partner concept. The symbol is the net or matrix.

Typical aspects of the *atomistic culture* (decentral/informal) are:

- Constellation of, more or less, independent individuals
- Everybody does his or her own thing
- High expertise or quality (everybody a star)

- Unstable conglomerate
- Shared influence
- Psychological contracts
- Control and hierarchy almost impossible
- High creativity

You will find typical examples for this culture among smaller partner-oriented companies such as law firms. There are probably no real examples among larger companies as they have to evolve into a role or task culture along with their growth.

As in any model, there are very few examples of companies that really fit any of these models 100 percent. You will always have more or less fitting examples. Companies will unify within themselves aspects of every culture (different functions).

Now is there any connection between the foregoing and the approaches to internationalization? Usually you start with the central approach which, most likely, will produce a power culture. When you choose to go multinational, there is a tendency to go atomistic. When you go for worldwide divisions, there might be a good chance that you will end up in a role culture. Finally "global linkage" might produce a task culture.

In any case, in managing international projects, it is important to be aware of the corporate culture and also the linkage of this culture to the degree of internationalization this company stands for.

What Has All This to Do with Project Management?

Assume that company A—like most international companies—is in a situation of constantly increasing both its international sales and its international product diversity. Therefore A is in the middle of a constant change process in order to come as near to the "global matrix" as possible.

Both internally and externally (vis-à-vis customer C) project management has to be the stable factor. Clear methodologies and a clearly defined role of the project manager vis-à-vis line management are the most critical success factors in a constantly changing environment focusing on moving targets.

Once you start using project management as a methodology to implement your products and services and therefore as a marketing weapon, you immediately run into the same problems as with product internationalization.

Should you then build up central expertise centers and leave only the implementation part up to the subsidiaries (central approach)?

The central expertise centers will soon keep an expertise nobody in the subsidiary needs. Channeling back the needs is a very clumsy effort, and constantly adjusting the quality and quantity of the resources needed is a Sisyphean task. Making a decision as to whether, at a given moment, you support a project in subsidiary X or in subsidiary Y when the resources are available only once is the kind of decision everybody hates.

Should you rather set up project management as a worldwide division not linked into the other divisions (worldwide approach)?

Separating project management from the rest of the company has different dangers. You end up in an ivory tower and as you have no coherent strategy with the expertise in the company, you end up managing processes with less and less content contribution. This will at the end be a very small business ("lease a project manager").

Should you then leave everything up to the subsidiaries and maybe just institute common methodologies (multinational approach)?

Leaving the whole business up to the subsidiaries and instituting only a common methodology seems to be a reasonable approach. However, as in Ghoshal's example, you end up with hundreds of strategies and no investment strategy at all. International projects for international customers are impossible to manage out of such a structure.

It therefore seems that, as in the example of Ghoshal,[1] a globally linked strategy is the only way to go, however difficult its implementation may be.

How Could "Global Linkage" Work for Project Management?

First of all, get rid of central expertise centers and transfer them into the subsidiaries (see Fig. 16.4). Find a balance between central investments (for your most strategic segments) and national responses. Do not put all your investments into the "leader" box. Usually the resources of the leader are all absorbed by the subsidiary where the leader is located. Find a balance between "leader" and "contributor." Encourage contributors, give them investments and rewards so that they are motivated to help the "implementors." Finally get rid of the "black holes."

Now link project management as closely as possible into these expertise centers. Knowing the formal and informal links into the right level of expertise is suddenly adding content to the project manager's role and increasing the business potential dramatically.

VALUE CHAINS OF THE BUYING COMPANY AND THE SELLING COMPANY

The value-chain concept as introduced by Porter[4] is providing us with a good tool to analyze the relationships between our company A and its customer C. Let us assume that at some point in time A is building products and selling them to C (Fig. 16.6). In order to do this, A needs a certain level of support organization, including everything from the finance department, personnel, and general management to a certain level of assets. Hopefully at the end of the value chain there is some profit left over to satisfy the shareholders and to finance future growth.

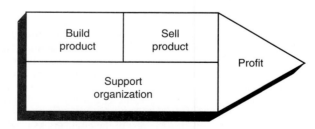

FIGURE 16.6   Value chain for company A.

In order to be able to do anything with the products of A, C has to install a value chain, as shown in Fig. 16.7. It needs a purchasing department to buy the products,

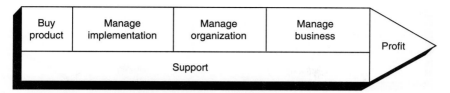

Buy product	Manage implementation	Manage organization	Manage business	
Support				Profit

FIGURE 16.7   Value chain for company C.

project management to implement them and to integrate them into an existing technical environment, and in many cases an organizational development process and associated human resource development processes in order to embed the products successfully into an existing organization or to change the organization according to the product requirements. Next there has to be a link of this implementation into C's business strategies, so that C may be managing its business more effectively or efficiently and, finally, will be achieving increased profit.

The only knowledge that A must have in order to sell its products to C is how C will be managing their implementation. The remainder of the value chain is C's business, and there is no connection to A. Technical specifications and price usually are the selling points in this type of business.

Now assume that A will be going into the business of implementation by using project management. We shall see in the next section that there is a clear pulling effect from C in order for A to go further down the value chain.

INTEGRATION OF VALUE CHAINS BY PROJECT MANAGEMENT

Integration by "Classical" Project Management

By "classical" project management we mean management of the implementation of a contract. The contract defines the deliverables clearly (functional specifications), as well as time schedules, the aspects of integration into an existing technical environment, and last but not least the costs of this implementation, in many cases at a fixed price.

What is the impact on the value chains? Company A must now have a good knowledge of the next interface in C's value chain, that is, the organization of C (Fig. 16.8).

Most classical projects that failed did so mainly because the organizational aspects had not been considered. While the solution was delivered in time, it was useless for the end user. In most cases large sums of money were spent in lawsuits which ended in settlements since the judge did not understand what the lawyers and their clients were talking about. How can this problem be overcome?

Integration by Systemic-Evolutionary Project Management

In the German-speaking literature a substantial amount of research is being published on the subject of systemic-evolutionary project management. Basically these contribu-

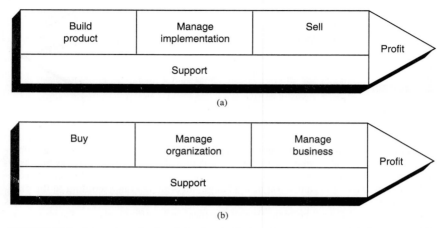

FIGURE 16.8   Integration of value chains by classical project management. (*a*) Company A; (*b*) Company C.

tors are stating that the mechanistic view of the world is being replaced more and more by a holistic-systemic view. Königswieser summarizes these views as follows[5]:

Mechanistic view	Holistic-systemic view
Hierarchy	Network
Machines	Living organism
Planning according to schedules	Visions
Manager as "doer"	Manager as "developer," "gardener"
Hard thinking, logic	Soft thinking, "psychologic"
Organization	Self-organization
Exert pressure	Let grow
Objectivity	Subjectivity
Structure	Process
Cause and effect	Mutual effect
Right and wrong	Functionality
Planned change	Balance between change and keep

Generally in projects where elements of the systemic-evolutionary approach can be incorporated, not all of the typical project constraints, such as cost and price, date, or customer satisfaction, are fixed. There is usually a certain degree of flexibility in at least one or two of these dimensions. Therefore it would be in the best interest of company A and customer C to add elements of the systemic-evolutionary approach on top of the classical implementation approach, discussing their impact on the project dimensions openly.

Now we are entering new ground. To our knowledge systemic-evolutionary project management has been either applied to internal projects or used on a consultant basis to help customers with their internal projects. An interesting development of course would be to find a contractual legal framework which makes it possible to really manage a project for a customer on these grounds. An interesting approach is presented in Bolka.[6]

The value chains would now look as illustrated in Fig. 16.9. The big difference for A is that in order to be able to manage the organizational aspects of the project for C, it must have substantial knowledge of C's business.

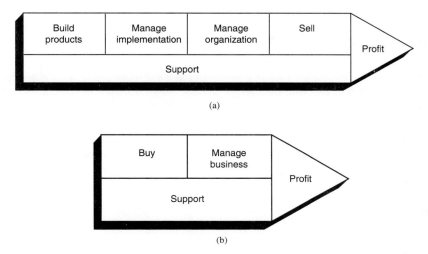

FIGURE 16.9 Integration of value chains by systemic-evolutionary project management. (*a*) Company A; (*b*) Company C.

New Contractual Relationship

The "Deal."   The deal is a very interesting example of this kind of value-chain integration. A offers the total project to C on the principles mentioned in the preceding section. A is not quoting a price for the project, but a return on the business created by the project. This could, for example, be a certain percentage on all transactions created by an automatic travel agent system.

The deal is in fact the last step prior to entering into a joint venture, which will be addressed next.

Joint Ventures.   There is a pulling effect coming from the market once you have entered the business of project management, which, after several deals, will produce joint ventures.

In a joint venture the value chains of A and C have virtually merged, as illustrated in Fig. 16.10. One interesting detail, however, is that the "sell" and "buy" parts have disappeared, two elements of the value chain which usually carry significant cost. This means that the joint venture ultimately will be more competitive than two separate companies. The interesting fact is that project management helped the relationship between A and C to grow from an initial relationship of supplier and buyer into a real business partnership.

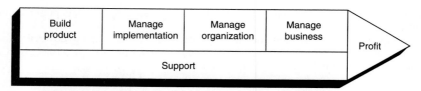

FIGURE 16.10   Value chain of joint venture.

CONCLUSION

We often tend to emphasize project management as a discipline per se. This is correct as far as methodologies and tools are concerned. When it comes to critical success factors, however, we recommend developing a multidimensional view of project management, considering the dependencies on internationalization, corporate culture, organizational development aspects, and the value chains of all parties involved. Only by addressing all these dimensions can we transform projects into more than the sum of their pieces, namely, into driving factors for integration in international companies.

REFERENCES

1. S. Ghoshal, "Managing across Borders: Strategic and Organisational Challenges," Lectures at INSEAD, Fontainebleau, France, 1989.

2. R. Harrison, in *Understanding Organisations*, C. B. Handy (ed.), Penguin, Harmondsworth, U.K., 1986.

3. Heitger and Sutter, "Project Management in Different Corporate Cultures: Success Factors for Internal Projects," in *Handbook of Management by Projects*, R. Gareis (ed.), Manz, Vienna, 1990, pp. 134–140.

4. M. E. Porter, *Competitive Advantage*, Free Press, New York, 1985, pp. 36–61.

5. R. Königswieser, *Das systemisch evolutionäre Management*, Verlag Orac, Vienna, 1990, pp. 1–9.

6. G. Bolka, "Elemente des EDV Vertrages," *EDV & Recht* (Verlag Medien und Recht, Vienna), no. 2, pp. 56–59, 1990.

CHAPTER 17

STRATEGIC ALLIANCES AND INTERNATIONAL PROJECTS: MANAGERIAL AND ORGANIZATIONAL CONSIDERATIONS*

Anthony F. Buono

Department of Management, Bentley College,
Waltham, Massachusetts

Anthony F. Buono is Professor of Management and Organizational Behavior and Chair of the Department of Management at Bentley College, Waltham, Mass. His recent books include *The Human Side of Mergers and Acquisitions: Managing Collisions between People, Cultures, and Organizations* (Jossey-Bass, 1989), *A Primer on Organizational Behavior* (2d ed., Wiley, 1990), and *Corporate Policy, Values and Social Responsibility* (Praeger, 1985).

Strategic alliances between organizations—mergers, acquisitions, joint ventures, technology sharing arrangements, and industry consortia—are becoming increasingly important aspects of corporate project-management strategy in the international realm. As the forces of global competition continue to escalate, R&D efforts grow larger and riskier, technologies become too expensive to afford alone, and the ability to dominate markets becomes increasingly difficult, these interfirm partnerships hold significant promise as part of a broader competitive strategy for U.S. firms.[1,2] Over the course of the 1990s we are likely to witness a growing number and influence of these alliances as more and more companies with a "go-it-alone" policy will be forced to reevaluate their position.[3,4] Indeed, as advanced information technologies continue to increase the interconnectedness of firms operating in a global economy, some theorists have even questioned whether the traditional notion of organization is a sufficient unit of analysis.[5] Considering the growing complexity of organization-stakeholder networks and the blurring of organizational boundaries, efforts to improve organizations are beginning to be viewed as at least partly multiorganizational in nature.

The growing use of such interfirm arrangements offers companies the opportunity

*This chapter is adapted from "Managing Strategic Alliances: Organizational and Human Resource Considerations," in *Business in the Contemporary World*, vol. 3, no. 4, 1991. I would like to thank Tim Macauley, my graduate research assistant, for his preparation of materials and research support for this article.

to expand their project options beyond existing capabilities or current product-market domains. While these alliances are created for a variety of reasons, their primary focus is on improving the competitive and financial positions of one or more of the partners. A common objective is to avoid the disadvantages and limitations faced by single organizations—for instance, resource scarcity, restricted economies of scale, undue risk—by drawing on the capacity and potential of multiple, autonomous firms.[6] Thus interfirm alliances have emerged as a way of raising complementary resources and attaining important attributes—marketing, technological, financial, managerial, raw materials—from partner firms.[2,7-9]

While these advantages suggest increased opportunities abroad for domestic firms, not all observers are favorably inclined toward such alliances with foreign partners. One criticism is that using interfirm alliances as a strategy will only ensure a company's mediocrity.[10] Although alliances with foreign firms promise a number of benefits (such as entry into new markets, hedging against risk), they exact significant costs—the need to coordinate two distinct systems and harmonize goals and objectives with an independent entity, while creating a potential rival in the marketplace and relinquishing profits. Accordingly, this perspective views alliances as short-term "transitional devices" instead of part of a stable, long-term relationship. A similar argument points to the myriad ventures involving U.S. and Japanese companies, which are suggested to work much more to the advantage of Japan than of the United States.[11] Although basic research leading to initial product design is carried out in the United States, specific design and production of the most complex parts and sophisticated assemblies are done in Japan, while final assembly and sales and marketing functions shift back to the United States. As a result, these ventures indirectly support the growth and skill development of Japanese workers while shortchanging their U.S. counterparts. Essentially, the United States has taken control of the two ends of the innovation-production process—initial research innovations and final assembly and sales—while Japan has concentrated on the complex production in between, where large numbers of workers are able to develop sophisticated technological and organizational competence.

A related problem with strategic alliances is that many, if not most, managers are unfamiliar or uncomfortable with the problems and difficulties of managing in these interorganizational arrangements. Indeed, although these types of alliances are not new per se, they are unique in both a strategic and an organizational sense. Most assessments of management emphasize fairly conventional contexts—unitary, free-standing companies, small businesses, or entrepreneurial activities in larger organizations (intrapreneurship). Strategic alliances, in contrast, are hybrid organizations with distinctive characteristics and problems: they have an "instant" presence and size; they face performance expectations that are immediate, substantial, and externally imposed; the partners often lack a common cultural domain or experience base; and their management and governance structures, which are based on cooperation and collaboration, are generally untested.[6] Thus there is an inherent fragility in these agreements, which makes them quite complex to manage.[12,13]

Finally it is important to realize that an alliance that might look good on paper does not necessarily translate into success in the marketplace. One only has to look at one of the most talked about U.S.-European joint ventures of the 1980s—the partnership between AT&T and Olivetti (the Italian office equipment manufacturer)—for an illustration of how promising ventures can collapse. Similarly, merger after merger that appeared to make good strategic sense in the 1980s is now being viewed as a disastrous mistake, and many of the most promising industry consortia have either failed outright or have not fulfilled their promise.[14-16] In fact, in far too many instances the actual performance of these alliances simply falls short of the operational, strategic, and financial goals envisioned during preformation feasibility studies and assessments. The reality is that, in spite of the growing number of such alliances, a disproportionately large percentage ends prematurely and in failure.[2,7,14,17,18]

Despite these concerns, it appears that a range of interorganizational endeavors will continue to grow in both number and influence. Indeed, current and emerging pressures and forces in the international business environment literally demand greater collaborative endeavors between organizations in existing and yet to be developed alliances. Accordingly, this chapter focuses on four basic interfirm strategic alliance types* and some fundamental organizational and human resource concerns associated with each type: (1) *mergers,* the total consolidation of two or more organizations into a single entity; (2) *acquisitions,* the purchase of one firm by another such that the parent takes control of the target; (3) *joint ventures,* cooperative agreements between two or more firms that want to achieve similar objectives which create overlapping organizations; and (4) *industry consortia* as found in the microelectronics and semiconductor industries, which combine the financial resources and technical expertise of member firms for proprietary research endeavors. The chapter concludes with a discussion of the role that managers as change agents can play in facilitating successful alliances.

COMPARATIVE ASSESSMENT OF STRATEGIC ALLIANCE TYPES

It is important to underscore that individual strategic alliances are unique entities because of the myriad ways in which firms might interact with each other as well as the potential diversity of the partners themselves.[2,14,18] Thus efforts to make one alliance work may very well result in quite different outcomes in another. Moreover, much of the evaluation of alliance effectiveness depends on the strategic intent of the partnership.[19,20] Nevertheless, due to the growing attractiveness of such strategic alliances and the myriad questions that remain about coalition dynamics, it is important to draw lessons from existing ventures that can begin to serve as a guide for thinking about managing such partnerships.

Table 17.1 summarizes a number of key organizational and human resource concerns in strategic alliances. Each alliance type presents a different set of organizational relationships and, as a result, disparate human resource and managerial considerations. The degree of interfirm consolidation in a merger or acquisition, for instance, is much greater than in either a joint venture or an industry consortia agreement. Although the actual level of consolidation and the related dynamics are influenced by the strategic intent of the combination and the degree of integration desired between the firms,[14] merger or acquisition partners do not retain the level of autonomy maintained by joint venture or industry consortia members.

To bring about this higher level of consolidation, integration efforts in mergers and acquisitions tend to be much more normative in nature, relying on culture change efforts and group socialization processes.[6] As a result, the direct effects on the individuals involved tend to be much more extreme than in either a joint venture or industry consortia. Although there are a range of cultural integration possibilities—from cultural pluralism to cultural blending[21]—in any merger or acquisition, the degree of threat perceived by the firms' human resources tends to be high. Especially in horizontal mergers and acquisitions, organizational members tend to expect reductions in

---

*These alliance types are adapted and extended from Borys and Jemison.[6]

TABLE 17.1 Organizational and Human Resource Dimensions of Four Strategic Alliance Types

	Mergers	Acquisitions	Joint ventures	Industry consortia
		Organizational		
Degree of interfirm consolidation	High ———————————————————————			Low
Risk/return potential	High ———————————————————————			Low
Primary integrative mechanisms	Normative/ cultural ————————————			Contractual/ structural
Mission strategy	Ownership ———————————————————			Collaborative competition
		Human resource		
Level of HR uncertainty	High ———————————————————————			Moderate
HR perception of threat	High ———————————————————————			Low
Recruitment needs	Low ————————————————————————			High
Potential of turnover	High ———————————————————————			Low to moderate
Compensation/remuneration requirements	Traditional —————————————————			Innovative

force as overlapping job responsibilities create a certain degree of organizational redundancy.[22] Yet even in unrelated acquisitions, the potential of job loss and significant shifts in organizational culture and practices can create a perception of threat among managers and employees.[23,24] As such, the potential of turnover of key organizational members and resistance to the combination tend to be high as well.

In a recently studied technology-guided acquisition, for instance, valued technical experts began a mass exodus following the acquisition of their chemical company (Petro) by a firm that constructed steel and nuclear power plants (SteelCo).[14] SteelCo's acquisition of Petro was part of the firm's diversification strategy, and the jobs of the technical experts, engineers, and scientists at Petro were not threatened. In fact, SteelCo had acquired the firm largely due to the expertise of these individuals. During the postacquisition period, however, as operational consolidation of basic functional areas—human resources, accounting, and finance—began to take place, many of the technical staff interpreted the terminations and changes in other sections of their company as a "sign of things to come." The result was a high level of voluntary turnover among the very people that influenced SteelCo's decision to buy the company in the first place. Thus SteelCo found itself in control of the petrochemical company, but without many of the petrochemical experts that made the company successful in the first place.

The Burroughs-Sperry combination that created UNISYS, in contrast, illustrates what can be done to create a partnership between members from two firms.[25] Despite the fact that Wall Street and industry analysts criticized Burroughs' hostile takeover of Sperry, Burroughs CEO W. Michael Blumenthal's vision of the combined firm and his efforts to directly confront the uncertainties and anxieties perceived by organizational members facilitated the consolidation of the two firms. A merger-coordination council, staffed by high-level executives from both firms, oversaw a series of task forces to study the ideal structure and operation of the consolidated company. Focused efforts at creating a unified culture, supported by value statements and guidelines for bringing that culture to life, were also undertaken: a clear vision of the new organization (the "power of 2"), top-management speeches and support, a sense of partnership between the firms, sensitization seminars, counseling sessions, merger-related newsletters, a "name the company" contest, and ongoing feedback to employees. Although many of the tensions and stresses that accompany mergers existed—politicking and infighting between members of each firm, distrust of the "other" firm's leaders and policies, fears about what the combination would mean for individual jobs and careers—over time the focused efforts to create a true partnership between members of the two firms appear to have paid off. There has not been anywhere near the level of turnover and dissatisfaction as found in the SteelCo-Petro acquisition. While UNISYS is being impacted by the general downturn in the computer industry, at this point the firm appears to have created a high level of member commitment to the venture.[26,27]

Thus in mergers and acquisitions top managers must emphasize interventions that can ameliorate some of the tension, strain, and discontent associated with the post-combination integration of two previously autonomous organizations: two-way communication channels, realistic merger previews, combination-related workshops and counseling efforts, survey feedback, transition teams, interfirm mirroring and team building, parallel organization structures that oversee and guide the combination, organizational symbols and rituals to support the integration, humanistic approaches to employee retention and dismissal, and joint evaluation and integration review.[14,28,29] It is important to realize, however, that while such efforts can minimize employee fears and anxieties, research has indicated that members of acquired organizations still maintain feelings of suspicion and, at least in the immediate postmerger period, never feel fully informed or comfortable in the combined organization.[23,30,31]

Joint ventures, in contrast, do not typically involve the same level of threat to the firms' human resources. Rarely, if ever, are these alliances associated with reductions in force or major organizational culture change efforts. They do, however, face a host of managerial and organizational difficulties. Conflicts over which partner has the most control, changes in partners' objectives, different management styles and organizational customs, a lack of trust between partners, and declining interest from either partner can easily contribute to failure.[17] Moreover, due to competing pressures and work demands, managers and employees often direct their attention toward other job responsibilities and away from the venture itself. Accordingly, one of the keys to joint venture success is based on the degree of fit between the venture's strategic purpose and the actual activities of the venture partners.[6] Yet, as research has indicated, the stated purposes of a joint venture are often much broader than the actual efforts and activities of the venture partners.[32] Moreover, while the need for consolidation in joint ventures is not as high as in a horizontal merger or acquisition, it is critical that member roles be clearly defined and sufficient motivation to ensure cooperation be present. While some normative control and related socialization efforts are important, joint ventures typically rely on more explicit contractual and structural agreements between member firms. An underlying problem, however, is that even apparently clear agreements are often interpreted quite differently by venture partners. This is one of the reasons why incorporated, stand-alone ventures, especially those without return paths to the parent organization (such as the IBM-Motorola venture that created ARDIS), are more likely to succeed than ventures without their own identities and institutions.

An example of the types of problems that can emerge in joint ventures is illustrated in a domestic alliance between two entrepreneurial computer services firms recently studied.[17] With the intent of adding valuable services to their firms without expensive and time-consuming expansion efforts, the presidents of the two companies agreed to enter into a joint venture agreement. The goal was to create a series of interfirm projects that would lead to new venture opportunities, enhance their presence in the marketplace, and improve their general ability to meet changing customer needs and demands. Expectations were so high, a merger stipulation was even part of the venture. Although there would not be any initial trading of stock, it was agreed that the joint venture would be a type of "dating period" after which a formal merger between the companies was the likely outcome.

Since both firms were relatively small, little, if any, difficulty was anticipated. Once service-level employees began working toward joint venture goals, however, a sufficient number of misunderstandings and conflicts between the companies took place that the staffs became increasingly resistant toward the venture. Moreover, members of both firms devoted their energies to their existing jobs and responsibilities instead of to venture-related projects. As a result, nothing of any substance emerged from the alliance. Within 8 months conditions between the firms had deteriorated to the point where the two presidents began to question the wisdom of their initial strategy, and all merger talks were put on indefinite hold. Approximately 1 year later the venture was officially terminated. The stated purposes of the joint venture were much broader than the actual efforts and activities of the two firms, and the resultant conflicts and interpretive disparities over the goals and objectives of the venture contributed to its failure.

Thus in joint ventures it is important that planning and interactions be viewed as part of an iterative process. Realizing the probability of cultural differences between firms even in the same industry, spelling out the goals and obligations of the partners, attempting to create and maintain a vision of the partnership, using appropriate parallel structures to oversee and support the venture, and building and mobilizing coalitions of key organizational members provide a basic foundation upon which many of the internal pitfalls that can undermine joint ventures can be dealt with in an effective and constructive manner.

Industry consortia involve somewhat different dynamics and processes. Compared to the three other strategic alliance types discussed, industry consortia entail a much lower degree of interfirm consolidation with relatively narrow strategic purposes. As a result, contractual and structural mechanisms replace culture change efforts as key integration devices. Instead of assuming ownership of the core technology, product, or service as found in a merger, acquisition, or incorporated joint venture, member firms engage in collaborative competition that involves a combination of basic research and prototype development at the consortium level (collaboration) and new product testing and development at the firm level (competition). Accordingly, it is critical that member firms develop appropriate structural changes that facilitate communication flow and information dissemination if they are to fulfill the goals and objectives of consortium membership.

An illustration of such an organizational structure is presented in Fig. 17.1. First, a program office would be established that serves in a liaison capacity between the consortium and the individual member firm. Second, a series of project development groups would be formed as a way of readying the company to make the most efficient use of the basic research emerging from the consortium. The program office would be responsible for initially assessing the technological developments, research insights, and breakthroughs produced through the consortium, and channeling that information to appropriate project development groups throughout the organization. Any training or other support necessary for the full utilization of this new knowledge or experimentation would also be set up and coordinated by this office. The project development groups, in turn, would be responsible for generating potential product offerings and prototypes, which would then be shared, as appropriate, with the consortium through the program office. This type of structure reduces the probability that pertinent information would be lost between the consortium and its individual members, and pro-

FIGURE 17.1 Technology transfer program structure.

vides a mechanism to enable subsequent requirements for new product advances to be funneled back to the consortium in a timely manner.

While the level of threat for organizational members tends to be low in industry consortia compared to mergers and acquisitions, the uncertainties and ambiguities that surround consortium research and related activities can still be moderately high. The failure of U.S. Memories, a consortium of American high-technology firms that was formed to compete with the Japanese in the development of basic memory chips, is a good example of the pitfalls inherent in such ventures.[15] The basic idea was to create a secure, stable source of memory chips and related technologies that would not be burdened by short-term earning goals and the pressures associated with individual firms, which would enable the United States to compete more effectively with the Japanese. Within less than 1 year after its formation, however, U.S. Memories was unable to raise sufficient capital, interest, and commitment from its members and initial backers. As a prominent social critic suggested, "the ability at once to share and to compete in the same industry, turned out to be alien to us."[33]

In contrast to the other alliance types, since industry consortia include direct competitors, member firms often view their participation as risky because proprietary interests might be inadvertently compromised.[34] Moreover, given the underlying apprehension that other member firms might be more successful in translating the consortium's findings and breakthroughs into new product offerings, companies and their managers are often further reluctant to devote their full energies and attention to them.

As a result, these alliances experience an array of intraconsortium conflicts due to real and perceived equity problems.

Another issue focuses on recruitment needs and retention efforts. While mergers and acquisitions typically involve some reduction in force, industry consortia usually have the reverse problem—how to attract top-quality individuals to work at the consortium. The Microelectronics Computer Technology Corporation (MCC) and its member firms, for instance, were faced with the prospect of luring scientists, engineers, and support staff to Austin, Tex. In addition to hiring outside talent, partner firms also had to draw on their internal labor markets to send their own engineers to Austin.[35] While these individuals are expected to return to their organizations when their specific research projects are complete to ensure that the technology gets fully transferred and assimilated, high-technology firms have been found to literally raid each other's engineering talent as a way of life.[36,37] The question that remains is whether these individuals will, in fact, return to their original companies. Accordingly, specific interventions which increase the probability that these individuals will come back to the member firms are an important dimension of consortium success. For example, companies with leading-edge R&D functions, such as 3M and Eastman Kodak, have enjoyed success in keeping valued scientists and engineers by setting up special "innovation banks" that are used to fund intrapreneurial ventures. The employees are provided with the opportunity to start a new venture without assuming the risks associated with entrepreneurial endeavors.[38]

A closely related concern focuses on how individuals should be rewarded for their contribution to the creation and innovation process. Appropriate reward systems should contain a mix of intrinsic and extrinsic motivators, such as autonomy and the opportunity to pursue one's ideas, recognition and promotion opportunities, and special compensation systems.[39] Dual-career ladders, where technical experts can be given promotions and salary increases without having to assume managerial responsibilities, are a good example of such systems. In most large corporations upper-level managers and executives receive the large financial bonuses while engineer-inventors tend to receive much more nominal rewards. In the video game industry, in contrast, manufacturers typically include royalties that range from 10 to 15 percent of the profits on the games, a reward system which is suggested to contribute to the significant growth and creativity in the industry.[40,41] Similarly, many small, start-up microcomputer and electronics companies attract top talent from major corporations by offering these individuals a "piece of the action." Another approach is to reward idea generators with one-time monetary awards based on the relative success and profitability of the breakthrough.[39] These strategies are beginning to force corporations to expand their profit sharing, stock options, and related programs. Unlike compensation systems in a merger or acquisition, where a valued engineer might be given a bonus to remain in the company, unless equitable compensation systems are devised, the real possibility exists that engineers and scientists at the consortium could take their breakthroughs elsewhere, forming their own start-up operations.

COMPETITIVE REALITIES OF COOPERATIVE VENTURES

The success of any cooperative venture depends in large part on what has been referred to as the "coalition game"[42]: an alliance will be effective only as long as the benefits of membership or group activity are greater than the costs of participation for individual members. While the emphasis is typically placed on the potential breakthroughs that can be gained through collaboration, it is important to realize that there are competitive realities inherent in these cooperative ventures. Thus it is just as critical to learn and assimilate the skills of your partner as it is to gain access to different products and markets.

As research has indicated, outside sources of knowledge are often crucial contributors to organizational innovation processes.[43,44] The joint ventures between General Motors (GM) and Fujitsu Fanuc and GM and Toyota, for example, are illustrative of how a firm that desires to become more knowledgeable about the application of advanced manufacturing processes joins with another already expert in their use and management.[45] To a large extent, however, the ability to evaluate and utilize that outside knowledge is a function of (1) the level of prior related knowledge—basic skills, shared language, technological acumen, technology insights—that exists in a firm and (2) the internal supports and resources provided for organizational members.[43,46] Thus if such interfirm endeavors are to be successful, it is critical that our top mangers utilize both cultural and structural supports to ensure that (1) the firm and its members learn from their venture partners, (2) knowledge is disseminated to appropriate organizational units, and (3) appropriate levels of consultation and collaboration—which go beyond our traditional patterns of management and organization—are developed.

One of the keys to successful alliance building is to ensure that the role of the alliance is tightly coupled to the firms' development, specific project plans, and overall business plans. For instance, while accelerating the product innovation process has been a managerial concern for some time,[47] it has acquired greater importance due to the competitive realities of joint ventures and interfirm consortia. Although a growing number of firms have been able to pursue research paths that would have been too costly for each to attempt on its own, through such coalitions these shared research and technological innovation efforts also contribute to a redefinition of competitive technological weapons. As the same research base becomes available to partner firms, competitive strategies begin to shift away from internal R&D endeavors to internal technology transfer efforts that disseminate breakthroughs and new technological insights to appropriate project development groups on a timely basis. Thus organizations that can translate that basic research quickly and successfully into new products will enjoy premium prices for their products, the ability to incorporate the latest technology in their goods, and the advantages associated with being the initial market entrant. This reality underscores the importance of the type of structural changes and supports illustrated in Fig. 17.1.

In a merger or acquisition, in contrast, such competitive realities are more internally focused, as when the interaction between groups in the different companies takes on a "we" versus "they" mentality.[14] Yet while these dynamics stress the importance of a joint inquiry process in organizations,[5] many U.S. firms continue to maintain inappropriate structures and internal barriers that hinder rather than promote exchanges of ideas across departments and divisions.[48]

SOCIOCULTURAL AND POLITICAL DYNAMICS IN INTERNATIONAL ALLIANCES

The types of problems and difficulties discussed thus far apply to domestic as well as international ventures. Intersocietal alliances, however, are further complicated by additional sociocultural and political tensions. As suggested earlier, strategic alliances tend to be fragile agreements, and communication and language problems make their operation even more difficult. Differences in national cultures and their reflections in managers' assumptions about organizational processes and behaviors can easily hinder understanding between alliance partners.[49] Cultural inhibitions, for instance, varying perceptions of appropriate levels of "sharing" and the importance of "running things your own way," play a crucial role and sometimes lead to unnecessary reluctance or mistaken zeal on the part of member firms.[50] When compounded by contrasts between interorganizational cultures, these societal differences precipitate misperception and dispute between alliance partners. Thus while it might seem like a simple verity, for U.S. firms the importance of having a partner with an adequate English language capability should not be overlooked.[51] An inability to communicate with one's alliance partner effectively has resulted in more than a few disasters.

Along these lines, it is important that general managers in international alliances be wary of being perceived as giving systematic preference to the well-being of any one individual partner. It is critical that these individuals maintain effective relationships with *each* of the partners. An appointment of a manager who operates in a biased manner is likely to undermine the venture's success.[52] Thus these individuals must have the vision and leadership skills to inspire confidence and motivate people drawn from different cultures, and to satisfy the often divergent interests of the parents.

Similarly, political barriers can impede the creation of international alliances, especially when wholly or jointly owned facilities in critical industries are involved. Both Japan and the United States, for example, have expressed opposition to direct foreign investment in their domestic semiconductor industries.[53] At the same time, the United States' emerging focus on support for our domestic firms has paradoxically stimulated the creation of alliances spanning national boundaries. While antitrust policy has traditionally viewed such interfirm alliances with suspicion, the legal environment has become increasingly supportive of such ventures.[54,55] Thus firms that purport to be global in nature must be increasingly aware of and conversant with the policy environment of the nations in which they are producing or marketing goods and services. As part of this process, companies should ensure that they have sufficient background— by hiring either executives knowledgeable in international business and law or experts with specific country knowledge—before considering any foreign alliance.[56]

MANAGING STRATEGIC ALLIANCES

The development of strategic alliances is rapidly becoming a major force contributing to the further politicalization of management's role. Instead of simply controlling a set of subordinates or even mobilizing a coalition of supporters within the organization, managers must be increasingly adept at juggling a shifting set of stakeholders over which they do not have direct control. As management experts point out, this idea is not new, but goes back to the early work of management theorists who recognized that an important dimension of management is to develop and mobilize a network of cooperative relationships across people, groups, departments, and divisions.[4] Indeed, the underlying theme of agenda setting, network building, and agenda implementation through that network is reflected in a growing number of depictions of the general manager's role.[57] The emergence of a broad array of strategic alliances, however, has essentially multiplied the complexity of this reality.

Interorganizational partnerships require a level of consultation and collaboration that goes well beyond typical patterns of management and organization.[2,4,14] Indeed, the dominant American management and organization paradigm—based on assumed needs for hierarchical control and organizational stability—appears to be increasingly outdated for such alliances and needs to shift toward one based more on involvement and continuous change.[58] It does appear that, in an increasing number of instances, a shift is in fact occurring, moving from a control-based management model to one based much more fully on collaboration and commitment.[33,59,60] Similarly, many practices, which just a decade or two ago would literally have been viewed as heresy, are becoming commonplace. Organizations, for example, that were and still are competitors in the marketplace are collaborating on a broad array of projects and products that go against traditional notions of public policy (for example, antitrust laws) and corporate strategy (such as shifting notions of competitive weapons). Thus while we do need to be skeptical of such alliances, especially those on an international basis, it is also clear that they are rapidly becoming a reality of our business world.

As a way of coping with these changes and emergent structures, managers need to focus on ways to mobilize and bring order to such alliances. Yet while traditional organization development–related interventions have focused on what may be referred to as "overorganized" systems, strategic alliances are more appropriately conceptualized as "underorganized" systems.[61] As suggested in this chapter, alliances are loosely coupled systems that vary across a number of important organizational and human resource dimensions, with leadership and power dispersed among autonomous organizations and sporadic member commitment to interfirm collaboration efforts. Thus efforts to facilitate such linkages should be guided by a transorganizational perspective, focusing on the dynamics associated with this higher level of social system, identifying and convening the member organizations, and designing structures and mechanisms for regulating collaborative efforts.[62,63]

Compared to the typical phases of planned change—entry, diagnosis, intervention, and evaluation—the process of change in strategic alliances assumes the need to create new systems and assimilate those involved—identification of relevant systems and appropriate representatives, convention of those members and initiation of linkage processes, and organization of the system to regularize behaviors.[61] A problem, however, is that we are still just beginning to fully comprehend and understand the nature and underlying dilemmas involved in such interfirm interactions. For instance, although reactive change is largely viewed in negative terms, especially when contrasted to proactive management efforts, given the uncertain nature of strategic alliances, significant questions linger as to whether managers can accurately predict exactly what is going to happen in such linkages.[64] Moreover, it has been suggested that the nature of the relationship between alliance members should model the change process.[61] Accordingly, it seems that being prepared to cope with the resultant processes and changes appears to be a more realistic posture than one of directly managing such dynamics.[14]

Within this context it is also important that a broad-based, long-term time frame should be used to assess alliance effectiveness and contribution. Instead of simply using traditional profitability and cash-flow measures, it is important to move toward longer-term measures that reflect the true state of the alliance—for instance, adaptiveness, innovativeness, and productivity appraisals as well as the level of morale and harmony between the partners. While such an approach may be cumbersome and highly subjective, it holds greater promise than financial resource indicators alone.[65] This broad-based view, of course, does not mean that alliances should be given unlimited time to prove their worth. It does, however, reflect the reality that the intended value that such alliances create is often wide-ranging—from knowledge acquisition to competitive protections to increased market share—and requires corresponding input and output measures.

As suggested in this chapter, it is becoming increasingly clear that many of the key problems and barriers in international interfirm alliances are as much organizational and behavioral as they are strategic and financial. Given the changes that are taking place in the business world today, it is critical that we develop an appropriate understanding and appreciation of (1) how these alliances vary across important organizational and human resource dimensions and (2) the resultant processes and dynamics inherent in each type if managers are to facilitate the ability of interfirm ventures to fulfill our expectations of them.

REFERENCES

1. F. J. Contractor and P. Lorange (eds.), *Cooperative Strategies in International Business,* Lexington Books, Lexington, Mass., 1988.

2. K. R. Harrigan, *Managing for Joint Venture Success,* Lexington Books, Lexington, Mass., 1986.

3. D. Gold and K. P. Sauvant, "The Future Role of Transnational Corporations in the World Economy," *Business in the Contemporary World,* vol. 2, no. 3, pp. 55–62, 1990.

4. R. M. Kanter, "The New Alliances: How Strategic Partnerships Are Reshaping American Business," in *Business in the Contemporary World,* H. L. Sawyer (ed.), University Press of America, Lanham, Md., 1988, pp. 59–82.

5. S. A. Mohrman and A. M. Mohrman, Jr., "The Environment as an Agent of Change," in *Large-Scale Organizational Change,* A. M. Mohrman, Jr., S. A. Mohrman, G. E. Ledford, Jr., T. G. Cummings, E. E. Lawler III, and Associates (eds.), Jossey-Bass, San Francisco, Calif., 1989, pp. 35–47.

6. B. Borys and D. B. Jemison, "Hybrid Arrangements as Strategic Alliances: Theoretical Issues in Organizational Combinations," *Academy of Management Rev.,* vol. 14, no. 2, pp. 234–249, 1989.

7. N. Alster, "Dealbusters: Why Partnerships Fail," *Electronic Business,* vol. 12, pp. 70–75, Apr. 1986.

8. A. H. Blank, "The Growth of Joint Venture Marketing," *The Bankers Mag.,* vol. 170, pp. 60–63, Mar.–Apr. 1987.

9. C. E. Schillaci, "Designing Successful Joint Ventures," *J. Business Strategy,* vol. 8, no. 2, pp. 59–63, 1987.

10. M. E. Porter, "The Competitive Advantage of Nations," *Harvard Business Rev.,* vol. 68, no. 2, pp. 73–93, 1990.

11. R. B. Reich, *Tales of a New America,* Times Books, New York, 1987.

12. K. R. Harrigan, "Joint Ventures and Global Strategies," *Columbia J. of World Business,* vol. 19, no. 2, pp. 7–16, 1984.

13. S. Waddock, "Building Successful Social Partnerships," *Sloan Management Rev.,* pp. 17–23, Summer 1988.

14. A. F. Buono and J. L. Bowditch, *The Human Side of Mergers and Acquisitions: Managing Collisions between People, Cultures, and Organizations,* Jossey-Bass, San Francisco, Calif., 1989.

15. W. J. Murphy, *R&D Cooperation among Marketplace Competitors,* Quorum Books, New York, 1991.

16. E. Richards, "Consortia: Cure-All for US?" *Boston Globe,* pp. 11–12, May 29, 1989.

17. A. F. Buono, "Managing Joint Ventures: Interfirm Tensions and Pitfalls," *SAM Advanced Management J.,* vol. 55, no. 2, pp. 28–32, 1990.

18. O. Shenkar and Y. Zeira, "International Joint Ventures: Implications for Organization Development," *Personnel Rev.,* vol. 16, no. 1, pp. 30–37, 1987.

19. J. E. Finnerty, J. E. Owers, and R. C. Rogers, "The Valuation Impact of Joint Ventures," *Management International Rev.,* vol. 26, pp. 14–27, 1986.

20. J. J. McConnell and T. J. Nantell, "Corporate Combinations and Common Stock Returns: The Case of Joint Ventures," *J. Finance,* vol. 40, no. 2, pp. 519–536, 1985.

21. American Bankers Association and Ernst & Whinney, *Implementing Mergers and Acquisitions in the Financial Services Industry,* ABA, Washington, D.C., 1985.

22. A. F. Buono, J. L. Bowditch, and J. W. Lewis, "The Cultural Dynamics of Transformation: The Case of a Bank Merger," in *Corporate Transformation,* R. H. Kilmann, T. J. Covin, and Associates (eds.), Jossey-Bass, San Francisco, Calif., 1988, pp. 497–522.

23. M. L. Marks and P. H. Mirvis, "The Merger Syndrome: Stress and Uncertainty," *Mergers and Acquisitions,* vol. 20, no. 2, pp. 50–55, 1985.

24. D. L. Schweiger and J. P. Walsh, "Mergers and Acquisitions: An Interdisciplinary View," *Research in Personnel and Human Resource Management,* vol. 8, pp. 41–107, 1990.

25. D. L. Kanter and P. H. Mirvis, *The Cynical Americans: Living and Working in an Age of Discontent and Disillusion,* Jossey-Bass, San Francisco, Calif., 1989.

26. K. P. DeMeuse, "Corporate Surgery: The Human Trauma of Mergers and Acquisitions," presented at the 95th Annual Meeting of the American Psychological Association, New York, Aug. 1987.

27. A. R. Malekzadeh and A. Nahavandi, "Making Mergers Work by Managing Cultures," *J. Business Strategy,* pp. 55–57, May/June, 1990.

28. M. L. Marks, "How to Treat the Merger Syndrome," *J. Management Consulting,* vol. 4, no. 3, pp. 42–51, 1988.

29. D. L. Schweiger and A. DeNisi, "Communication with Employees Following a Merger: A Longitudinal Field Experiment," *Academy of Management J.,* vol. 34, no. 1, pp. 110–135, 1991.

30. D. T. Bastien, "Common Patterns of Behavior and Communication in Corporate Mergers and Acquisitions," *Human Resource Management,* vol. 26, no. 1, pp. 17–33, 1987.

31. A. F. Buono, J. L. Bowditch, and J. W. Lewis, "When Cultures Collide: The Anatomy of a Merger," *Human Relations,* vol. 38, no. 5, pp. 477–500, 1985.

32. P. Pekar, "Joint Venture," *Planning Rev.,* pp. 15–19, July 1986.

33. D. Halberstam, *The Next Century,* William Morrow, New York, 1991.

34. W. M. Evan and P. Olk, "R&D Consortia: A New U.S. Organizational Form," *Sloan Management Rev.,* vol. 31, no. 3, pp. 37–46, 1990.

35. A. F. Buono and A. J. Nurick, "Emergent Forms of Technology and Human Resource Development," presented at the Annual Joint Meeting of the Institute for Management Science and the Operations Research Society of America, San Francisco, Calif., May 1984.

36. T. J. Murray, "Silicon Valley Faces up to the 'People' Crunch," *Dunn's Rev.,* pp. 60–62, July 1981.

37. E. M. Rogers and J. K. Larsen, *Silicon Valley Fever: Growth of High-Technology Culture,* Basic Books, New York, 1984.

38. L. R. Gomez-Mejia, D. B. Balkin, and G. T. Milkovich, "Rethinking Rewards for Technical Employees," *Organizational Dynamics,* vol. 18, no. 4, pp. 62–75, 1990.

39. J. R. Gailbraith, "Designing the Innovating Organization," *Organizational Dynamics,* vol. 10, no. 3, p. 19, 1982.

40. A. Deutsch, "How Employee Retention Strategies Can Aid Productivity," *J. Business Strategy,* vol. 2, no. 4, pp. 106–109, 1982.

41. N. Orkin, "Rewarding Employee Invention: Time for Change," *Harvard Business Rev.,* vol. 62, no. 1, pp. 56–57, 1984.

42. T. W. Kang, *Gaishi: The Foreign Company in Japan,* Basic Books, New York, 1990.

43. W. M. Cohen and D. A. Levinthal, "Absorptive Capacity: A New Perspective on Learning and Innovation," *Administrative Science Quart.,* vol. 35, no. 1, pp. 128–152, 1990.

44. J. M. Pennings and F. Harianto, "Innovation in an Interorganizational Context," *Managing the High Technology Firm Conference Proc.,* (University of Colorado, Boulder, 1988), pp. 228–234.

45. D. D. Davis, "Integrating Technological, Manufacturing, Marketing, and Human Resource Strategies," in *Managing Technological Innovation,* D. D. Davis and Associates (eds.), Jossey-Bass, San Francisco, Calif., 1986, pp. 256–290.

46. K. Pavitt, "Sectoral Patterns of Technical Change: Toward a Taxonomy and a Theory," *Research Policy,* vol. 13, pp. 343–373, 1984.

47. A. K. Gupta and D. L. Wilemon, "Accelerating the Development of Technology-Based New Products," *California Management Rev.,* vol. 32, no. 2, pp. 24–44, 1990.

48. M. L. Dertouzos, R. K. Lester, R. M. Solow, and the M.I.T. Commission on Industrial Productivity, *Made in America: Regaining the Productive Edge,* M.I.T. Press, Cambridge, Mass., 1989.

49. Y. L. Doz, "Technology Partnerships between Larger and Smaller Firms: Some Critical Issues," in *Cooperative Strategies in International Business,* F. J. Contractor and P. Lorange (eds.), D.C. Heath and Co., Lexington, Mass., 1988, pp. 317–338.

50. W. T. M. Koots, "Underlying Dilemmas in the Management of International Joint Ventures," in *Cooperative Strategies in International Business,* F. J. Contractor and P. Lorange (eds.), D.C. Heath and Co., Lexington, Mass., 1988, pp. 347–367.

51. J. M. Geringer, *Joint Venture Partner Selection: Strategies for Developed Countries,* Quorum Books, Westport, Ct., 1988.

52. C. A. Frayne and J. M. Geringer, "The Strategic Use of Human Resource Management Practices as Control Mechanisms in International Joint Ventures," in *Research in Personnel and Human Resources Management,* B. Staw, J. E. Beck, G. R. Ferris, and K. M. Rowland (eds.), JAI Press, Greenwich, Ct., 1990, pp. 53–69.

53. D. C. Mowery, "Collaborative Ventures between U.S. and Foreign Manufacturing Firms: An Overview," in *International Collaborative Ventures in U.S. Manufacturing,* D. C. Mowery (ed.), Ballinger Publishing, Cambridge, Mass., 1988, pp. 1–22.

54. S. S. Samuelson and T. A. Balmer, "Antitrust Revisited—Implications for Competitive Strategy," *Sloan Management Rev.,* vol. 30, no. 1, pp. 79–87, 1988.

55. S. G. Zwart, "The New Antitrust: An Aerial View of Joint Ventures and Mergers," *J. Business Strategy,* vol. 7, no. 4, pp. 68–76, 1987.

56. L. E. Nevaer, E. V. Louis, and S. A. Deck, *Strategic Corporate Alliances: A Study of the Present, a Model for the Future,* Quorum Books, Westport, Ct., 1990.

57. J. P. Kotter, *The General Manager,* Free Press, New York, 1982.

58. A. M. Mohrman, Jr., and E. E. Lawler III, "The Diffusion of QWL as a Paradigm Shift," in *The Planning of Change,* 4th ed., W. G. Bennis, K. D. Benne, and R. Chin (eds.), Holt, Rinehart & Winston, New York, 1985.

59. E. E. Lawler III, *High Involvement Management,* Jossey-Bass, San Francisco, Calif., 1986.

60. R. E. Walton, "From Control to Commitment in the Workplace," *Harvard Business Rev.,* vol. 63, no. 2, pp. 76–84, 1985.

61. L. D. Brown, "Planned Change in Underorganized Systems," in *Systems Theory for Organization Development,* T. Cummings (ed.), Wiley, New York, 1980.

62. T. G. Cummings, "Transorganizational Development," *Academy of Management Organization Development Division Newsl.,* pp. 8–10, Summer 1989.

63. T. G. Cummings, "Transorganizational Development," in *Research in Organizational Behavior,* vol. 6, B. Staw and L. Cummings (eds.), Jai Press, Greenwich, Ct., 1984.

64. A. M. Mohrman, Jr., S. A. Mohrman, G. E. Ledford, Jr., T. G. Cummings, E. E. Lawler III, and Associates, *Large-Scale Organizational Change,* Jossey-Bass, San Francisco, Calif., 1989.

65. R. Anderson, "Two Firms, One Frontier: On Assessing Joint Venture Performance," *Sloan Management Rev.,* vol. 31, no. 2, pp. 19–30, 1990.

66. W. E. Halal, "The New Management: Business and Social Institutions for the Information Age," *Business in the Contemporary World,* vol. 2, no. 2, pp. 41–54, 1990.

CHAPTER 18

INTERNATIONAL PROJECTS: OPPORTUNITIES AND THREATS

Rajeev M. Pandia

Herdillia Chemicals Ltd.

Mr. Rajeev Pandia has been closely associated with the chemical and petrochemical industry in India for the last 20 years. He currently holds the dual positions of Managing Director of Herdillia Chemicals Ltd. and President of Herdillia Unimers Ltd., its subsidiary.

INTRODUCTION

Several political, commercial, economic, and social developments in various parts of the world have led to an increasing number of international projects being implemented. The recent trend toward globalization, even in those countries which had opted to operate in isolation in the past, indicates that international projects will receive a greater thrust in the years ahead.

There are various ways in which corporations in any country can participate in international projects (taken up outside its geographic boundaries): as vendors, contractors, consultants, licensers, investors, or members of a consortium. Each role presents varying levels of opportunities and threats.

Unlike domestic ventures, international projects require exposure to an environment having several features both at the country and industry levels with which the participant is not familiar fully or even partly. These include the local culture, language, political system, economy, laws, regulations, infrastructural facilities, work force, and skills. The success or failure in this environment usually depends on the preproject country and industry studies carried out at the macro and micro levels, proper understanding and appreciation of the major differences, and appropriateness of the approaches adopted to minimize the risks and maximize the benefits.

This chapter deals with the scope of a preproject study that is essential for an international project to succeed. The major areas normally covered in this study include a country profile, project-specific regulations, industry status and characteristics, regional highlights, feasibility of the project, and infrastructural facilities.

The sources which could provide useful and relevant information necessary for this study include international banks, multinational corporations with past exposure to the particular country/region, industry associations, government departments, international and domestic business/trade journals, consultancy organizations, and consulates.

This study normally enables a corporation to take a preliminary position on, and quantify its exposure in, an international project. It is particularly useful if it has to make a selection out of several options available to it with limited manpower or financial resources, leading to the need for prioritization.

The corporation concurrently needs to address itself to several strategic issues in greater detail. These include its objectives and synergies, expected return on investment, the extent of its exposure (which could be one-time or long-term), weighing of risks in a relatively unknown environment on the one hand against broadening its basket of projects on the other, and its inherent flexibility and strengths and weaknesses in adapting to a new culture.

The approaches arising out of the preproject country/industry studies can form the basis for the formulation of an appropriate strategy matrix for the success of the selected project.

SPREAD OF GLOBALIZATION

The process of globalization, which has become more pronounced and accelerated during the past few years than it has been earlier, was very aptly described by former U.S.S.R. President Mikhail Gorbachev when he addressed the faculty members and students of Stanford University, California, in June 1990 and shared with them his "vision of a new world." Since globalization has contributed more to international projects than any other social, political, or economic phenomenon, it would be appropriate to begin with a few excerpts from his speech which also referred to the need for strategic alliances for fostering cooperation among nations:

> Perestroika was born of profound processes taking place in our society. In five years, our country has become different. But it seems to me that the whole of mankind is making the transition into a new time of life, a new age, and that, in all of the world, some kind of different system of laws and rules has begun to operate, a system to which we are not yet accustomed. Fifty years ago, the great Russian scholar Vernadskii wrote: "Humanity is one. The life of mankind in all its multiplicity has become indivisible, one. The events, which happen at any point on any continent or ocean, affect and have repercussions, great and small, in a series of other places everywhere on the face of the earth. The telegraph, telephone, radio, airplanes and balloons have embraced the entire globe. Communications have become simpler and faster. Every year the level of their organization increases, and they grow rapidly. This process of man's totally settling the biosphere is conditioned by the historical course of human thought, and is inseparably linked with the rate of communication, with advances in transportation technology, with the possibility of instantaneous thought transmission, and with the simultaneous discussion of a thought everywhere on the planet."
>
> I am delighted—after all, this was said 50 years ago—that this prediction has proved itself true. We are seeing the essential issues of international relations evolve toward ever-augmented, mass-scale, and open communications. In this connection, the role of today's creative, constructive policies is sharply increased. But the price of mistakes also grows, the mistakes which, to no small degree, stem from adherence to old dogmas, accustomed rule of action, old ways of thinking. All people must change so that a new type of process

might become the reality throughout the world. The alpha and omega of the new world order is tolerance. Without tolerance and a respectful attitude towards one's partner, no understanding of the other's care and anxieties can be expected. This supposes the culmination of an approach...that is new and in keeping with the times. It requires cooperation and even alliance-building.

I am convinced that we are on the threshold of revising our concept of alliance-building. Until now, alliances were built on a selective, in essence, discriminating basis, on the opposition of one party against another. They separated countries and peoples to a significantly larger extent than they united them....But the time is approaching when the very principle of alliance-building should change. And its point is unity, a unity that enables us to achieve living conditions worthy of humankind as well as the preservation of the environment, unity in the war against hunger, disease and ignorance.

Apart from the importance of globalization, and the need for strategic alliances, the references to tolerance (toward other cultures, operating styles, and methodologies), as also to avoidance of dogmas and of conventional approaches which have outlived their utility are as relevant to international projects as they are to global statesmanship.

THE NEED FOR INTERNATIONAL PROJECTS

Those projects in which assistance or involvement is sought or obtained from one or more countries outside the nation in which they are implemented could be termed *international projects*.

The association of other countries with these projects could be in one or more of the following areas:

- Project evaluation
- Technical know-how
- Patent license
- Basic engineering
- Detailed engineering
- Supply of equipment
- Procurement services
- Third-party inspection
- Insurance services
- Freight-forwarding assistance
- Supply of labor (with varying degrees of skills)
- Construction services
- Project management expertise
- Training of personnel
- Commissioning assistance
- Financial services including syndication of loans
- Joint ventures/portfolio investments
- Project audits

- Turnkey projects
- Member of a consortium
- Management/operation of the facility after commissioning
- Marketing and buy-back arrangements.

In order to identify the advantages and opportunities that international projects present both to the host country and to all the participating nations, it is necessary to study the reasons for the phenomenal growth in the number and sizes of international projects implemented during the last two decades, a trend which is likely to be further accelerated by the process of globalization referred to earlier:

- Certain developed countries, such as the United States, which grew at an extremely rapid rate around the middle of the century, reached a level of saturation in certain industrial segments. In order to sustain their individual growth rates, their engineering/construction companies, like the manufacturers of industrial goods, had to turn to markets and projects outside their national boundaries.

- Concurrently, countries outside North America, Western Europe, and Japan, which were trailing behind in basic infrastructure, utilities and industrial facilities, began to work aggressively to reduce the gap in development, aided by political, social, or economic factors.

- Those countries, such as China, the Soviet Union, and those in Eastern Europe, which had followed the system of centrally controlled economies for many years, went through a process of major restructuring and began their experiment with market economies. They opened their territories to the developed countries, expecting them to bring in technologies, equipment, capital, and services to improve the standard of living of their people. These have suddenly provided tremendous project opportunities to Western countries.

- Another current example of changes in political thinking contributing to opportunities for international projects is South Africa, which, with its recent transformation, is becoming the focal point of considerable interest for companies in the United States.

- The logistics and local availability of mineral or agricultural resources have also contributed substantially to international projects. For instance, the discovery of crude oil and natural gas in West Asia led to the involvement of a large number of developed and developing countries from various parts of the world in local projects varying from oil exploration and production to refining and petrochemicals. The boost that these developments provided to the West Asian economies, in turn, led to implementation of a large number of projects, such as airports, roads, utilities, infrastructural and trading facilities, office buildings, and hotels. Similar developments have also been witnessed in countries which have large resources of other minerals (such as metal ores) or agricultural products, some examples being the development of titanium-based industries in Australia (which is rich in these metallic ores) or those based on natural rubber and vegetable oils in Malaysia.

- Economic factors, such as the dramatic changes witnessed in the relative values of currencies, the resulting trade imbalances, formation of trading blocs among nations, protectionist tendencies in the form of quotas and tariffs, and shortage of hard currencies, have given rise to the need for fresh thinking on flexibility in project locations. The joint ventures set up by Japan in the United States for the production of automobiles when the yen became increasingly strong against the U.S. dollar, the high levels of tariffs introduced by some countries in Asia to promote local production and thereby to reduce imports, and the reduced imports by the Eastern bloc countries in view of shortage of hard currency, resulting in rerouting of imported products through soft-currency trading partners, are some instances of how economic factors have contributed to international projects.

- Following the continuing and unprecedented evolution in scientific knowledge and research, it has become increasingly expensive to develop new products or technologies—partly due to the short lives of technologies because of obsolescence and partly due to relatively short or inadequate protection granted for intellectual property rights. This is particularly relevant to areas like electronics and pharmaceuticals.

    It is estimated that it costs over $100 million to develop a new drug from the test tube to the drug store and yet, due to the long process of U.S. Food and Drug Administration (FDA) approvals, the useful life of a patent is often too short to justify developmental expenditure of such a magnitude. For instance, the patents of all of 1983's 200 top-selling drugs have already expired, making it legally possible for other companies to produce the same drugs without having incurred similar developmental expenditure. This has given rise to international projects of two types—joint ventures in R&D to share the developmental costs among various companies/countries and joint ventures for local production of drugs in various regions in order to maximize revenues during the useful life of patents. The growth of multinational drug companies is an indication of the spread of international projects in this industry.

    Another example of the need to spread costs, risks, and manufacturing facilities is the aircraft industry in which the consortium approach has been increasingly common due to commercial compulsions.

- Demographic trends indicate that the population growth rates in several newly industrialized or developing countries are significantly higher than elsewhere. These trends, combined with high savings rates, have created large pockets of urban and semiurban middle class with substantial disposable income, and these people, in turn, have created a high demand potential for several consumer goods, justifying local production facilities.

- On the other hand, some countries, such as Sweden, are faced with low population and growth rates, entailing a shortage of industrial labor, necessitating promotion of joint venture projects in countries such as South Korea, Thailand, and India, where abundant skilled labor is available at relatively low costs.

- In order to attract foreign investment/technology, create local employment opportunities, and export products, some newly industrialized countries have formulated highly attractive schemes for corporations desirous of setting up new industrial facilities, with several incentives including tax exemptions and low-interest loans/subsidies. These improve the profitability of projects significantly and can serve to disperse industries from other countries.

- With the growing concern on environmental impact of industries, some countries have had to formulate very stringent domestic ecological standards because of already unacceptable levels of pollution. Corporations in these countries have found it advantageous to shift or implement some projects having an adverse ecological impact in other regions where the standards are more easily attainable.

A study of the above factors would show that the trend toward international projects will continue in the years to come and that they will present even greater opportunities than they did earlier.

In the past, the response of American, Western European, and Japanese corporations to opportunities of participating in international projects (as licensers, consultants, vendors, contractors, or investors) was varied. Some kept away from them, a few succumbed to their dangers, and many others reaped dividends. However, in view of the lasting opportunities that international projects will present in the future, there is need for detailed reappraisal of approaches that these corporations followed in the past and to formulate long-term strategies based on a global and comprehensive review of the dangers and dividends of international projects.

OPPORTUNITIES AND BENEFITS

Many of the advantages that can accrue to a country or corporation through participation in international projects are apparent from the factors outlined earlier. However, in view of the preconceived and strong negative views and reservations that some corporations have about projects outside their countries, it it necessary to highlight the benefits arising from international projects:

- These projects provide protection against domestic recession, stagnation, and competition and assist corporations in optimal utilization of resources and production capacity.
- They provide a hedge against inflation, currency fluctuations, tariffs, quotas, and similar adverse economic factors and widen the basket of risks and rewards to reduce cyclical vulnerability.
- They provide access to large international markets.
- International projects provide countries or companies access to mineral, natural, and agricultural resources in the host countries.
- Exposure to an international business and project environment can considerably broaden the perceptions and sensitivities of a company's personnel at various levels, thereby contributing to the development of human resources.
- An optimal mix of local and foreign workers, materials, and equipment and the benefits of fiscal and financial concessions can reduce costs and increase productivity and efficiency.
- The presence of a corporation in a particular country, even if it is achieved through entry in a small way, can lead to larger and recurring future business opportunities in that region.
- Successful implementation of projects in an international environment can enhance the image of a company and elevate its status from the domestic to a transnational level.

DANGERS AND THREATS

While international projects present several opportunities and advantages, they also involve risks and threats beyond those characteristic of domestic projects, and these vary depending on the country or region and nature and duration of association of a corporation in a specific project. Some of these dangers are listed below with examples:

- *Political turbulence/warfare.* The political developments in Iraq and Iran a decade ago and more recently in Kuwait and Iraq have led to damage to installations, freezing of assets, uncertainty about safety of personnel, and varying force-majeure situations, which have had adverse effects on companies having large stakes in those regions.

- *Economic downturns.* The recent financial difficulties faced by a few Latin American countries, leading to hyperinflation, currency devaluations, and debt traps have entailed uncertainty about profits, repatriation, and performance of joint venture projects. While it is easier to foresee these possibilities in domestic projects and initiate corrective steps, the speed of response could be slower in international situations.

- *Sudden changes in laws relating to taxation or repatriation.* Drastic changes of this nature usually result form major internal political upheavals, and it is not uncommon that in small underdeveloped countries, political revolutions or coups entail the installation of governments following totally different philosophies and ideologies on foreign investments and contracts, which, in turn, could lead to nationalization, a freeze on repatriation, and high levels of taxation, thus adversely affecting the operation of foreign companies.

- *Delays due to government controls.* Corporations based in the United States, which are accustomed to a minimum level of government/regulatory interference with particular reference to project approvals, and those in Japan, which receive positive support from government departments [such as the Ministry of International Trade and Industry (MITI)], are not accustomed to inordinate delays which result from multiplicity of statutory approvals that are required in semicontrolled or fully controlled socialistic economies. The involvement of several departments, disagreements among them, lack of a specific time frame, subjective interpretation of laws and regulations, and an outdated bureaucratic regulatory apparatus (often operating under the influence of the political system) can be highly frustrating and agonizing, especially for corporations from Office of Economic Cooperation Development (OECD) countries.

- *Barriers due to cultures or languages.* Though it is normally expected that operation in foreign lands could entail communication problems due to language and cultural barriers, the magnitude of these is often realized only when work is commenced on projects. These can, on the one hand, affect project documentation, interpretation of finer points in contracts and training of personnel. On the other, differing cultural norms (such as local restrictions on dress in some West Asian countries), business etiquette, climate, and lack of social infrastructure conforming to Western standards can come in the way of attracting suitable professionals who would be willing to be sent as expatriates for medium or long durations.

  The levels of industrial infrastructure (including transport and communications), technological sophistication, computerization, quality control and consciousness, and commitment to occupational safety and environmental protection also differ in various countries, and it requires considerable effort and tolerance on the part of all project participants to understand and appreciate their respective approaches. It is also necessary to realize that while addressing themselves to cost-quality tradeoffs, unlike their counterparts in OECD countries, corporations in developing nations tend to lean toward lower costs, partly due to endemic shortage of resources and partly on account of their focus on a domestic market which is itself highly cost-conscious.

- *Protection of intellectual property.* A few countries and companies are known to be less particular about the protection of intellectual property, either on account of their weak patent laws or due to their cavalier attitude toward business/trade secrets. Once valuable know-how is transferred to them, its improper use could entail creation of unexpected and unwanted competition, dilution of a strong technological position, and unfair business practices.

- *Industrial unrest.* The labor laws in socialistic countries often lean toward labor, and this ideological trend has encouraged labor in some countries to resort to irresponsible practices and militant trade unionism, entailing heavy loss of worker-hours. Strikes and lockouts in these countries lead to greater loss of production on a comparable basis than they do in Japan, the United States, or even the United Kingdom. These, in turn, can adversely affect project schedules (since unrest even among a contractor's personnel or in an important vendor's factory or in an ancillary unit could delay a large project). Contingency planning for each major item could, on the other hand, prove very expensive.

- *Inadequate prior assessment.* Participation in an international project in any capacity requires extensive studies because of the several unknown but influential parameters and the inadequacy of data in contrast to readily accessible information on domestic projects. There are many instances of corporations having plunged into international projects without a proper assessment of the country/region/industry-specific opportunities and threats or of the project's technical/commercial/financial feasibility, resulting in inadequate safeguards or preparations and an inappropriate strategy mix.

In view of the increasing importance of international projects, it is imperative for corporations desirous of spreading their wings beyond their national boundaries to consider approaches and strategies which could reduce the threats and dangers and enhance the opportunities or dividends.

PREPROJECT COUNTRY STUDY

Some of the major risks encountered in international projects are endemic to the host country—its culture, economy, political system, infrastructure, and laws.

It is therefore highly desirable that any company which contemplates participation in a project abroad should undertake a preproject country study to gauge the extent of risks. The depth of the study could vary depending on the nature of its association—it could be very significant for long-term involvement through a joint venture or as the leader of a consortium and considerably less if the involvement is one-time and of a shorter duration, as for a vendor.

In case a company is faced with selection from several projects being implemented in many different nations, the country studies can also be used to prioritize its interests among the possible candidates.

The broad areas which a country study should cover include economic indicators, political stability, legal and fiscal framework, business environment, and infrastructure. The scope for each of these typical areas and their significance are highlighted here.

Economic Indicators

- The gross domestic product (GDP) of a country indicates the size of its economy. The past growth rates of GDP over a few years could provide an indication of the dynamism of the country and the possibility of new projects materializing. It is necessary to study the size of GDP and its growth rates in conjunction, since for very small economies, in view of the low base, the growth rates could, at times, appear to be very high. The range of industrial products manufactured in the country also provides a fair indication of its development.

- The rate of inflation can provide a basis for judging the management of a country's economy. Since projects involve long gestation periods, unless a safeguard is built into the compensation structure, high inflation rates could prove highly detrimental to a company's profitability.

- The variation in the value of a country's currency with respect to other world currencies is another parameter which also requires careful scrutiny, for the reasons mentioned above, and could help a company in its financial negotiations.

- The capital markets—their buoyancy, size, variation in the index, and membership—as well as the country's banking system could provide an indication of the local fund raising capacity.

- The balance of trade and external borrowings resorted to in order to fill a shortfall could affect the attitude of a country to import/export regulations, foreign investment, foreign debt, and revaluation of its currencies. In cases of acute shortfalls, these could lead to a debt trap, which was observed in some Latin American countries recently and had an adverse effect on several foreign investors.

- A parameter of particular concern to a foreign investor, licenser, or contractor is the track record of a country related to repatriation of dividends, servicing of loans, and payment of compensation in accordance with the agreements entered into with them. Any country which has a steady track record could be considered reasonably safe, especially if its management of economy is sound, as could be judged by the earlier parameters.

Political Stability

The internal and external political stability in a country can be gauged from its history and record in the following contexts:

- The relationship with neighboring countries could affect a nation's economy, defense, and implementation of projects. A reference was made earlier to the effect of warfare on projects in the Iran-Iraq-Kuwait region. A country's trade, defense, and similar pacts with other nations can also provide an indication of its political ties and affiliations.

- Internal strife, revolutions, turmoil, and unrest, especially when they are frequent, widespread, or for causes which are still live, could lead to force-majeure situations and affect the stay of the expatriate personnel sent by a company for projects in the host country. The recent turmoil in China which had opened its economy and which, until then, seemed to be a very promising prospect to several U.S. corporations, is an example of the need for a careful study of factors affecting the political stability in a country.

- The impact of political instability can be significant on a country's laws and attitudes, especially to external trade, international projects, and foreign investment. In this context, changes in a country's government may not necessarily be as significant as those in its basic political structure, ideology, or philosophy, since the former might not disturb continuity in major policy directions whereas the latter could entail major swings.

- In some countries the federal-state relations differ from state to state, especially if the party controlling the government in a state is politically distant. It is then not uncommon that a specific state does not receive the same level of federal support as the others, and this could affect the performance of projects in that state.

Legal and Fiscal Framework

International projects are often subject to the laws of the host country and hence these need careful scrutiny.

- Following the Bhopal gas tragedy, the legal liability of a foreign company (as a licenser, investor, consultant, or contractor) in case of a disaster has come under sharp focus. The provisions in this respect vary from country to country and need careful scrutiny in order to ensure that proper safeguards are built in at the contractual stage.

- With increasing consciousness about the ecological impact of projects, federal, state, and local governments in several countries (varying from Germany to Taiwan) subject pollution-prone projects to extensive scrutiny and the environmental laws and approval procedures can affect the project schedules by several years.

- It is not uncommon that disputes arise between the contracting parties in a project and these are often settled in accordance with the local laws. This is another reason for a quick appraisal of the relevant legal provisions.

- The tax provisions, including deduction of tax at source, double tax avoidance treaties, divestment provisions, taxation of expatriates, and tariffs on imported equipment or materials also vary significantly and could seriously affect the after-tax cash flow of a company.

- While any corporation desirous of participating in a joint venture would normally do so with a long-term association in view, it would understandably like to keep its options open in case withdrawal from the venture or country becomes desirable at a later stage. In this context, it is necessary to study the laws and regulations related to exit (including sale or transfer of shares, repatriation of the proceeds, and liabilities at the time of exit).

Business Environment

The success of a project depends considerably on the general business and industrial environment since its cost, duration, and quality could be seriously affected by regulatory delays, nonavailability of materials or skills, labor unrest, and productivity. The following facets of the industrial climate could therefore form part of the country study:

- The local availability of skills under various categories could be particularly relevant to the project cost, since local labor is often more economical and easy to deploy. In some cases, despite the availability of skills, the levels of productivity, familiarity with sophisticated equipment or techniques, attitudes toward work, and language or cultural restrictions could create differences or barriers, many of which could be resolved by a proper understanding and suitable provisions in the project schedule for relevant training.

- Federal, state, or local regulations and approval procedures could entail very significant delays which could, in turn, completely alter the schedule and success of a project. It is therefore very necessary to have an early appreciation of the nature and types of approvals required, the time frame for these consents, their sequence, and the attitude of the bureaucracy or regulatory authorities.

- U.S. corporations, which are accustomed to free and unrestricted availability of construction materials and equipment at prices governed entirely by the market forces of demand and supply, often do not visualize shortages of basic materials, such as cement or steel, restriction on their imports or high tariffs, regulated prices, and problems of logistics. The entire area of materials management under local conditions thus acquires significance in the context of international projects.

- Problems of business ethics and integrity tend to arise in an economy of shortages or one which is overregulated, and if the extent of these problems could be gauged at an early stage, it could facilitate formulation of suitable strategies or tactics.

Infrastructure and Logistics

Infrastructural inadequacies could be as common in some newly industrialized countries as shortage of materials and therefore require some attention at the early stage of a project.

A large country in Asia had planned a chemical project and had reached an advanced stage of negotiations with a U.S. corporation for technology and engineering services. It was at that stage, after many worker-months had been spent on preproject activities, that an infrastructural study revealed acute shortage of water in the region identified for the project. Since the requirement of water for the project was unusually high, it had to be abandoned.

Communication is another facet of possible infrastructural inadequacies, and U.S. corporations, accustomed to immediate communication through facsimile transmission and electronic mail, need to be acquainted fully with the level prevalent in the host country and get reconciled to it.

Narrow roads, inadequate jetties, small bridges, and short transport vehicles may necessitate site fabrication of large equipment instead of shop fabrication and could therefore have a significant effect on the procurement planning.

The other areas of infrastructure and logistics which need scrutiny are availability and supply reliability of power and fuel, erection equipment, and a railroad network.

In order to ensure that expatriate personnel and their families can live under conditions which are reasonably close to those to which they are accustomed, it is also necessary to study the available social infrastructure, including housing, medical care, clubs and recreational facilities, and shopping arcades, with a view to assessing the need for their augmentation in case the number of expatriates is to be large and the duration of their stay extensive.

INDUSTRY STUDY

Along with the country study referred to earlier, it is also desirable to get an overview of the particular industry to which the project belongs in the host country. Some of the facets which could be covered in this study are

- Present demand for the product and installed capacity to meet it (both at the national and regional levels)
- Sizes of individual projects
- Past growth in demand and capacity and factors influencing them
- Projections for future demand potential and the other competing projects contemplated for meeting the anticipated increase
- Competing products and their effect on demand projections
- Market shares
- Price elasticity of demand
- External trade (imports/exports) in the specific product
- Prices, variations, and taxes
- Availability of raw materials
- Marketing channels and efforts required for market development
- Industry-specific statutory approvals required for the project

It is becoming increasingly normal for technology licensers and major engineering/construction companies based in OECD countries to seek information on the above lines from potential licensees, clients, and consultants in other countries before making a preliminary decision on their interest in a specific project, since it helps them in identifying serious projects with a fair chance of being implemented before they decide to commit many worker-hours on preparation of bids for offering their services.

SOURCES OF INFORMATION

The country, regional, and industry studies referred to above could be carried out in-house by a corporation (through its strategic planning, international, marketing, or legal department or its regional offices). In the absence of these services being available internally, they could be entrusted to consultancy organizations specializing in international economic/industry studies.

The information needed for these studies could be obtained from publications and through discussions with the concerned international organizations or those based in the country/region being studied. The following sources could be useful in this context:

- Embassies and consulates
- International financial institutions such as the World Bank and Asian Development Bank
- International country credit rating agencies (such as Business International, Standard & Poor's, or Moody's investors services)
- Stock exchange publications

- Official government guidelines and trade/industry statistics
- Country promotional literature
- Chambers of commerce and industry
- Trade/industry associations
- International professional societies (e.g., Licensing Executives Society/Project Management Institute)
- Trade publications
- International trade fairs
- Foreign companies operating in that particular country
- Foreign and local banks
- Local and international consultancy organizations
- Law and accountancy firms (especially those with international affiliations)
- Government ministries or departments which are responsible for industry, foreign trade, foreign investment, or the specific type of business to which the project relates
- Local companies operating in the same area of business and their annual reports/brochures
- Multinational corporations which have had extensive business/trade dealings with the particular country and region.

For those corporations which participate in a large number of international projects and for which revenues from foreign businesses account for a significant part of the total income, it is even desirable to maintain in-house data banks on countries of interest and to update them periodically.

PROJECT FEASIBILITY STUDY

Once the preliminary country, regional, and industry studies indicate that the conditions in a particular country and in the industry to which a project belongs appear to be worthy of deeper interest for possible association (in absolute terms or in comparison with other available alternatives), the next step, especially for a corporation seeking financial participation, income sharing, or turnkey supplies or a long-term involvement, is usually to undertake a project feasibility study.

Though the nature of this study does not differ from that of the ones that the corporation would have undertaken for an internal or domestic project, the variables could vary considerably both in absolute terms and in the levels of probability that could be assigned to them. These studies are therefore often combined with sensitivity studies to observe the effect of changes in key variables on the project profitability.

The key areas considered in the project feasibility studies are

- Technical arrangements (for technology, engineering, construction, project management, commissioning, and operation) with particular reference to capabilities of local/international companies selected or shortlisted (listed for final consideration) for each part of the project, assignment of responsibilities, handling of interface areas, adequacy of contractual arrangements, and safeguards.
- Sources of supply of equipment and construction materials (local and imported) and their past record, capability, reliability, and quality assurance procedures.

- Local promoters of the project, their resources, resourcefulness, management expertise, past achievements, and standing in the industry. The local promoters have an extremely important role to play in the smooth implementation of a project and its subsequent operation, and hence it is highly desirable that their background is assessed in great detail to ensure their suitability and capability for the successful execution of the project.
- Schedule of implementation—the important activities and milestones, extent of realism in the assumptions made, and critical activities and contingency plans.
- Short- and long-term availability of raw materials and utilities and their quality and prices.
- Logistics.
- Standards, codes, and units to be followed.
- Environmental and safety standards to be complied with.
- Availability of skilled personnel and their rates.
- Need for expatriate support and facilities envisaged for them.
- Project cost, broken down into various segments, degree of their accuracy, impact of inflation during the project span, and margins for contingencies.
- Financing plans, extent of project finance which has been committed, major terms of loan agreements (especially those which can affect the participating foreign corporations) and average cost of capital.
- Market potential (along the lines of the industry study referred to earlier but with a greater degree of depth and accuracy).
- Profitability projections under local conditions (with the basis for assumptions) and the return on investment. (The sensitivity studies referred to earlier could be carried out concurrently with these projections.)
- Profitability under international conditions. (This is particularly applicable to those countries which operate as sheltered economies and in which projects are supported by high levels of tariffs. It is the belief of the World Bank and other international development finance institutions that, in the longer time frame, each project must be capable of being viable without the props of tariffs and quotas, which could be withdrawn later. Hence it is desirable to carry out an analysis of this nature.)
- Statutory approvals required, their status, major conditions stipulated, their sequence, time frame, extent of subjectivity, and difficulties likely to be encountered in obtaining them.

IDENTIFICATION OF PROJECT OPPORTUNITIES

Just as country and industry studies assist a corporation in aggressively seeking, short-listing, or rejecting a project opportunity, the project feasibility study described above assists it in quantifying its revenue earning potential. The studies also enable it to decide on the type, extent, and length of exposure it should consider for a specific project, with appreciation of the dangers and dividends of the project, including those endemic to the host country.

In order to formulate strategies for successful participation in an international project, it is necessary to first define, at the level of a company's top management, its philosophy and commitment in respect of international projects and specific countries or regions. These could cover:

- Initial short-term and limited participation to get acquainted with the country
- Medium- or long-term involvement through long-duration projects
- Investment in projects through joint ventures
- Use of a country as a base for exploitation of future identified business opportunities in that region.

The outcome of this strategic exercise could assist in assignment of weights and the prioritization of resource allocation for international projects.

The preproject studies referred to earlier can then be used to form a matrix of problems and opportunities. In order to quantify the outcome of the study, each of the areas covered by it can be given a numerical rating on a scale from 0 to 10 on the basis of certain predefined criteria.

Since the importance of each area of the study might not be comparable and would depend largely on the type of involvement of a corporation and the kind of international project, these can be used as the basis for assigning weights in order of importance to each of the study areas. For instance, in a joint development project for computer software, the availability of construction materials or equipment would have little relevance and hence could attract a zero weight.

The weighted average rating for a country and an international project that would result from this exercise could be extremely useful both to the top management team of the corporation in making major strategic decisions and, later, to the project management team in formulating project-specific strategies to reduce the inherent dangers. It could reduce subjectivity and also the effect of prejudices or preconceived ideas, which are currently not uncommon in several corporations.

Once an international project has been shortlisted or identified for being pursued aggressively, it is desirable for the foreign corporation to get a more detailed appreciation of some of the relevant parameters of the country study. This could be done partly by identifying the local country affiliates of its law and accountancy firms. An appreciation of the reasons for a country's laws, regulations, and procedures can be extremely helpful not only in complying with them but also to find solutions to problems which could arise in case strict compliance is not possible for genuine reasons.

An appreciation of a country's legal and regulatory framework needs to be accompanied by that of its cultural practices in order to harmonize with them and to promote healthier interpersonal relationships.

It is also necessary to gauge the optimistic, pessimistic, and realistic time frame for those project activities (such as regulatory approvals and equipment supplies) which are not strictly within the control of the project management team. This evaluation could avoid slippage and lengthy explanations at a later stage.

The selection of the local partner assumes great significance for the reasons cited earlier. Whenever a corporation has several options available, it should not be averse to devoting considerable resources in researching the credentials of the likely candidates.

DOCUMENTATION

In an era of paperless offices and standardization of contract documentation, there is a general tendency to reduce the paperwork involved in projects. However, in the case of international projects, even at the cost of the time spent on building up lengthy contracts, it is desirable to spell out certain clauses which become necessary either due to the involvement of more than one country in a project or the differing interpretations and laws which could create midstream interpretation problems. It is therefore essential for all the project participants to visualize the likely peculiar situations that could arise at various stages of the project and to address themselves to those possibilities at the outset.

A few typical clauses and provisions which are endemic to international projects are outlined below:

- *Jurisdiction.* In cases of disputes arising between or among project participants, the legal framework which would apply needs to be clearly spelled out. While this is often a subject of negotiations, in some cases the governments of host countries insist that their laws should apply.

- *Arbitration.* In addition to the applicable law, the machinery for resolution of disputes also needs to be identified. This is usually achieved through arbitration to be conducted in a neutral country (which is specified) following the procedures laid down by the International Chamber of Commerce.

- *Patent/trademark infringement.* The supply of know-how for setting up a manufacturing facility, equipment, or certain services and the marketing of products could entail violation of third-party patents or trademarks in one or more countries. In the event of such a contingency, the responsibilities, roles, and liabilities of each party as well as the indemnities need to be clearly defined.

- *Units, codes, and standards.* With the United States continuing to follow the British system of units and an increasing number of nations switching over to the metric one, the system to be followed in all technical drawings and documents needs to be identified. In some cases it has been found beneficial for all sides concerned to follow both the systems concurrently by having two sets of numbers. A similar understanding needs to be reached on design codes and standards applicable to equipment and facilities, since several countries have their own counterparts of ASTM/ASME/OSHA/UL/SAE standards and guidelines. The same type of consideration needs to be given to the language of official communication and translations, wherever needed.

- *Conformity with local guidelines.* In case the foreign suppliers, licensers, and contractors are not in a position to follow each of the applicable local codes, standards, and guidelines, it is normal for them to seek waiver of any legal liability for nonconformity with them, thereby shifting the onus of their compliance to the project owners or their local representatives or consultants.

- *Hospitality for personnel.* During the duration of the project (and even thereafter), the personnel of the foreign participants are required to travel to and stay in the host country (and those of the project owners to travel to the countries of the project participants). The contracts in such cases normally specify the details of hospitality, such as the cost to be borne, the class of travel, type of accommodation to be provided, per diem rates and their computation, insurance covers, indemnities in respect of accidents, medical care, and the right of the project owners to reject those persons who, in their opinion, are found or considered to be unsuitable for the specific assignment.

- *Warranties.* In the case of complex equipment or manufacturing facilities, it is normally stipulated by the suppliers that the warranty would be valid only if the equipment and facilities are installed and commissioned in the presence of their representatives. Though this is usually easy to achieve in the suppliers' own country, they are often not in a position to depute their experts to distant lands and hence need to reconsider or modify a condition of this nature. It is also necessary to take into account, at the stage of designing and for purposes of performance warranties, the differences in ambient conditions (such as temperature, humidity, and altitude), utilities (electric voltage and frequency), and quality of raw materials and inputs and to provide for them in order to avoid subsequent disagreements.

In cases where several project participants are involved, questions often arise about the overall guarantees in respect of capacity of the production facility, quality of products and consumption of inputs. While the company providing the know-how or technology license is normally expected to provide these guarantees, it pleads its inability in view of the role of the other participants involved in design, engineering, fabrication, construction, inspection, and supervision, over which it has little control. This issue can be resolved by the adoption of one of several approaches: a consortium approach, back-to-back agreements, the right of the technology licenser to review or inspect critical drawings and equipment, or the appointment of an independent organization to determine individual responsibilities in case of defaults or failure to meet the guarantees.

In addition to the general clauses highlighted above, certain specific clauses apply to transfer of technology/engineering services and to supply of equipment in the case of international projects. Some examples of these are provided here:

Transfer of Technology/Engineering Services

- *Contents of technical documentation.* In spite of the perceived understanding reached at the time of negotiations, it is not uncommon to come across surprises when the technical documentation is compiled and transferred by the licenser/engineering company to the project owner. This is on account of differences in perceptions about individual roles. For instance, while in the United States, in the case of process equipment and instrumentation, only the minimum information is provided to vendors and they are expected to carry out detailed engineering and prepare fabrication drawings, in several countries vendors expect complete details from engineering consultants. Unless issues such as these are addressed at the outset by reviewing typical packages and breakdown of individual roles and recording the understandings so reached, gaps are left in the overall project arrangements which, in turn, adversely affect project schedules and costs.

- *Change notices.* There is a greater tendency on the part of both sides of international projects to make changes in the contents of the know-how/engineering documentation (after it is compiled) as each gets increasingly familiar with the other side's or country's resources, conditions, and capabilities. There is, therefore, a need for outlining a formal process of change notices, including naming of coordinators on both sides, and to stipulate a time limit for freezing the contents of the documentation.

- *Compensation for midstream abandonment.* Some of the uncertainties of international projects have been pointed out earlier. In order to ensure that the foreign participants are adequately compensated for their time and efforts as well as for lost opportunities, a procedure for working out compensation payable to them in case a project is abandoned midstream could be worked out at an appropriate stage of commercial negotiations.

A list of major clauses in a typical agreement for transfer of technology/engineering services as applicable to an international project is given in Table 18.1.

Equipment Supplies

- *Letters of credit.* Payments for equipment and, in some cases, services, are usually made in the form of irrevocable letters of credit in international transactions. Before these are established, it is desirable for both sides to agree on certain specific terms, such as the need for confirmation of the letter of credit by a bank of the foreign supplier's choice (as an additional safeguard), preconditions for drawing payment against this instrument, terminology, validity period for shipment and for negotiation, and the cost of subsequent amendments, at the instance of either side.

- *Delivery period.* The delivery period specified by suppliers in some countries presupposes drawing approval by the customer (project owner) within a specified period. While facsimile transmission and courier services have reduced the time for international mail, often adequate time is not provided and the role of the customer is not made clear at the order stage, particularly since this practice is not followed in all countries. Unless a realistic estimate is made of these factors, which could cause delays, the project schedule could have a setback.

- *Shipping procedures.* Prior agreement is also required on the type of export packaging and containerization, basis of supply (FOB/FOR/CIF), selection of shipping services, type of transit insurance (warehouse to warehouse or warehouse to port of delivery), transshipment, partial shipment, and export documentation (such as important license, certificate of origin, and certificates such as material safety data sheets for hazardous cargo).

APPROACHES FOR SUCCESS

Detailed documentation contributes significantly to clarity of roles and thus to the success of an international project, since at the time of negotiations and drafting, several issues which otherwise have been overlooked get crystallized. There are several other factors which, based on an analysis of past international projects, could contribute significantly to the success or failure of a project. Some of these are

- *Balanced view of the host country.* There is a general tendency at various levels in the corporate headquarters of large corporations (including multinational corporations) to underestimate the technical, economic, and managerial capabilities of developing countries (such as those in Africa, Asia, and even Eastern Europe) and to form judgments on the basis of some past unpleasant experiences which may not be relevant, applicable, or contemporary. The extent of subjectivity and ignorance shown by even senior executives of some multinational organizations about countries outside the OECD is often astounding. For instance, the Vice President in charge of an important division of a large corporation, when approached for a joint venture and technology transfer arrangement for a synthetic rubber project in a developing country in South Asia, was amazed when he was told that the country produced its own automobiles and tires and that it was not only self-sufficient in

TABLE 18.1. Major Clauses in International Technology Transfer/Engineering Agreements

Recitals and broad objectives

Definitions (product, plant, effective date of agreement, capacity, know-how, technical documentation, trial runs, patent rights, commercial production)

Scope of the licenser's work and its schedule with explanatory lists in the form of addenda:
Patent rights/licenses
Know-how and basic engineering package
Supervision/review of detailed engineering
Procurement of proprietary equipment
Technical assistance during engineering/construction
Assistance in start-up/commissioning
Evaluation of raw materials/products
Training of personnel
Continuing exchange of improvement/developments

Compensation
Front-end fees
Equity stock
Running royalty
Tax treatment
Payment for additional services on per diem basis and reimbursement of expenses
Schedule and mode of payments

Responsibilities of the owner
Compensation of the licenser as per the agreed schedule
Detailed engineering/plant construction
Procurement
Compliance with codes/specifications
Maintenance of records
Arrangements for start-up and trial runs

Standards, codes, language, copies of documentation

Performance guarantees and penalties/compensation to cover
Plant capacity
Product quality
Consumption of raw materials and utilities
Performance of the licenser's services within the time schedule

Scale of penalties

Performance test and plant acceptance

Trademark rights

Indemnity against patent infringements

Secrecy

Period of agreement

Assignment

Default and termination of agreement

Force majeure

Jurisdiction and arbitration

Notices/amendments

this regard but even exported these products. His impression of the country and region probably still dated back to the age of animal-drawn carts! It was only after extensive briefing on the country and its status of industrialization that the corporation changed its initial negative decision completely and became an aggressive contender for the project.

This type of attitude has many undesirable ramifications—it leads to several lost opportunities, a half-hearted approach to newer opportunities, and an impression of superiority (in the absence of the realization that pockets of excellence and talent exist in every part of the world). In an era of globalization, such attitudes are clearly outdated and could cause incalculable psychological harm during negotiations for participation in an international project, even before an agreement is reached. Moreover, these attitudes and prejudices percolate down the different layers of an organization and reflect in the behavioral patterns of other personnel of project teams. It is in this context that former U.S.S.R. President Gorbachev's references to "tolerance and a respectful attitude towards one's partner" acquire particular significance.

- *Involvement of senior management personnel.* Since international projects present challenges which differ from those to which corporations are normally accustomed, and as some of these could have political, diplomatic, or long-term implications, it is necessary that members of the top and senior management team take greater interest in the strategic aspects of projects abroad, depending on the extent of the corporation's involvement, than they would have done for domestic ventures and that, even after these projects are kicked off, they establish channels of communication (such as appropriately designed management information systems) through which they can track the progress of projects and identify the need for their timely intervention.

- *Quality of project personnel.* There is often a tendency to include or depute relatively junior, less experienced or qualified personnel from a corporation in an international project team and to avail of the better talent for domestic ventures. This is done partly after an evaluation of the corporation's respective stakes and partly on account of the reluctance of several senior employees to be relocated (for short or medium durations) to unfamiliar countries. (The second of these factors could be handled in some cases by proper briefing of the concerned employees and their families on the country concerned, the facilities available there, and the likely influence of international stints on their future career progression.)

Since international projects require persons of greater maturity, experience, and expertise, as they have to deal with more variables and from greater distances from their corporate headquarters than for domestic situations, this tendency, wherever it exists, needs to be reviewed. The project owners usually are in a position to evaluate the quality of personnel assigned to their projects, and, if they are not fully satisfied, the association of the participants and partners begins on an avoidable unsavory note.

A reference was made here to the need for proper country briefing to a corporation's employees. This is also relevant to health-related factors in view of the common tendency among people in OECD countries to believe that third-world countries present grave health hazards which cannot be fought in spite of the several vaccinations and generally mandatory medications that they are subjected to before, during, and even after each visit. It is necessary to realize that most of the developing countries have hotels and facilities which follow international standards in hygiene and that with appropriate precautions—for which the project owners' teams usually provide adequate guidance—the chances of contracting diseases are slim. Psychological fears about diseases can themselves contribute to discomfort, and hence it helps to shed them after a reasonable, dispassionate survey of health- and hygiene-related factors.

- *Drafting of documentation.* The experience of some corporations shows that it saves considerable executive time and expense if drafts or structures of agreements (for technology transfer, engineering, consultancy, and other services) are first prepared by the respective operating personnel of the project participants and owners, keeping the involvement of attorneys to the minimum at that stage, since otherwise attorneys have a tendency to provide for several hypothetical situations, especially in an international context, without appreciating their respective probabilities or importance. Once the commercial drafts or heads of agreements are compiled, they could then be vetted by the legal and finance departments for fine tuning and to avoid any serious lapses.

  It is also highly desirable for foreign corporations to use the services of firms of attorneys and accountants based in the host country—ideally these could be the local affiliates of their own legal and financial advisers. This is particularly relevant for those host countries whose legal and taxation systems are significantly different and hence beyond the full comprehension of the foreign corporation's advisers without extensive studies, correspondence, visits, discussions, delays, and expenses.

- *Flexibility of approach.* While dealing with a relatively unfamiliar region, it helps to adopt a flexible approach in both commercial and project situations. For instance, there are certain local laws, regulations, practices, and requirements which are almost mandatory and hence difficult to modify or bypass. Similarly, the specifications of locally available equipment, materials, utilities, talents, and other inputs could vary by differing degrees from the ones to which foreign project teams are accustomed. In such cases, mature judgment and a degree of innovativeness or risk taking could help resolve deadlocks which are otherwise reached by expatriates taking a very rigid stand and insisting on exact reproduction of their own standards, pleading lack of familiarity with any deviation as the reason. In many cases the flexibility can be achieved without compromising on the project quality and it contributes substantially to savings in project costs and time frames. In this context, it is also helpful to reach decisions on all important issues jointly with the owner's project team rather than imposing unilateral decisions on it.

- *Integration of project phases.* The different phases of a project, from conception to commissioning, can be classified into three groups of activities: the front-end (comprising identification, formulation, and evaluation), execution (detailed planning, design and engineering, procurement, and construction/erection) and termination/operation (completion and postcompletion activities). For domestic projects, the implementation of the different phases is carried out in a fragmented manner, with different teams handling each phase, like a jetliner being passed on from one ground control tower to another. This approach has several disadvantages, particularly in international projects, with long gestation periods and extensive financial involvement. It disturbs continuity, affects interaction with clients and other project participants, and prevents a "helicopter view" of the project from being taken. The approaches that can be adopted to promote and improve integration among project phases include (1) selection of a project manager from the very early stage for the entire span and his involvement in all the phases, (2) appointment of a core group (or master task force), (3) increased involvement of the senior management team, (4) creation of a new function for effective coordination and to ensure smooth transition/interface, and (5) review of management information systems (MIS), project documentation, and communication procedures.

- *Realistic time frame.* Management teams of several U.S. corporations tend to be concerned with the short-term bottom line and get increasingly concerned if a joint venture does not show quick profits. (A U.S. multinational had promoted a joint venture in a third-world country. The project was implemented smoothly, but since the product was somewhat new to the country and there was competition from similar products which were already in use, the venture recorded losses during the first 2 years. While it was apparent that it was headed in the right direction and would shortly turn the corner once the market was developed aggressively, the corporate headquarters lost patience and sold it to a European multinational. The condition of the venture changed soon thereafter and later became highly profitable.)

Finally, the following "five commandments" suggested by a reputable independent research, consultancy, and publication group, whose mission is to serve corporations doing business across borders, merit serious consideration:

1. Choose the right local partner.
2. Put sufficient stress on developing good relations with the (host) government.
3. Remember that getting in is half the battle. (This commandment relates to scouting for new global opportunities and developing appropriate strategies for overcoming entry barriers.)
4. Be flexible.
5. Think long-term.

SAFEGUARDS FOR PROJECT OWNERS

A significant part of the above discussion has centered around the foreign participants in an international project. It would therefore be relevant to also deal with the safeguards and precautions that the project owners in the host country should consider for the success of a project:

- *Extensive background work.* In order to obtain state-of-the-art technology and services, on the best achievable terms, the owner's project team should make a thorough review of possible options, technologies, sources, commercial terms, and the project's inherent strengths by studying the available literature and information from various sources and, if required, utilizing the services of local or international consultancy organizations. It could also avail itself of the help extended by international financial institutions and affiliates of the United Nations or approach other companies which have implemented international projects for guidance.

- *Global bidding.* In order to negotiate from a position of strength, the project promoters need to cast their net wide and to identify as many potential international project participants as is reasonably possible. In some cases, they need to resort to hard selling of their country and project in order to interest some of them adequately to bid. A properly designed invitation to bid can be of considerable assistance in eliciting all the relevant information and in compiling it for subsequent evaluation. A thorough, professional evaluation of the global bids, after placing all of them on a common denominator, cannot only improve the commercial success of a project but also improve the owner's level of awareness in respect of the various finer aspects of the project for possible improvements and optimization of resource utilization.

- *Restrictive clauses.* Once a company has gauged its strengths and limitations in commercial negotiations, it can try to avoid certain limiting clauses (such as procurement of equipment only from nominated single sources—a tendency which is not uncommon among Japanese corporations), restrictions on exports of products, and use of only nominated consultancy organizations for engineering and construction services, since these clauses have an adverse commercial impact.

  It should also be possible to insist on a favored nation clause, which would enable the project owner to obtain the best terms ever granted by a foreign corporation.

  It is also advisable to study model agreements [such as the one for transfer of technology compiled by the World Intellectual Property Organisation (WIPO)] to ensure that appropriate clauses are incorporated in the project documentation.

- *Financial arrangements.* International financial transactions are becoming increasingly complex, with wide and frequent fluctuations in currency and interest rates. Concepts such as currency/interest swaps, loan syndication, leasing, and suppliers' credits could contribute significantly to reduction in the overall cost of a project. It is therefore necessary for the project promoters to seek suitable advice from merchant bankers on available international financial options in order to work out an ideal package.

- *Preference for joint ventures.* In order to ensure lasting and deeper interest of the foreign participants and as an emotional attachment, it is desirable to invite foreign companies to take an equity position in the project and thus to become partners. Moreover, whenever there is a choice between up-front payment and running royalty or deferred payment, it is advisable to opt for the latter for similar reasons.

It needs to be appreciated by all the partners in a project that, for fruitful long-term association, all the agreements should be fair, balanced, and equitable to the extent possible.

CONCLUSION

Various economic, social, political, and demographic considerations have led to an increasing number of international projects being implemented. These projects can provide valuable opportunities to all the participants. If suitable strategies are formulated and appropriate approaches adopted at the initial stages and through their implementation, many of the threats and dangers endemic to global projects can be reduced or avoided.

CHAPTER 19
PROJECT TEAM DEVELOPMENT IN MULTINATIONAL ENVIRONMENTS

Hans J. Thamhain

Bentley College,
Waltham, Massachusetts

Aaron J. Nurick

Bentley College,
Waltham, Massachusetts

Hans J. Thamhain is an Associate Professor of Management at Bentley College in Waltham, Massachusetts. He received master's degrees in Engineering and Business Administration, and a doctorate in Management from Syracuse University. Dr. Thamhain has held engineering and management positions with GTE, General Electric, and Westinghouse. He is internationally known for his research on engineering team building and project management. Dr. Thamhain is a frequent speaker at major conferences, has written over sixty research papers and four books on engineering and project management, and is consulted in all phases of technology management.

Dr. Nurick, an organizational psychologist with clinical training, is a Professor of Management at Bentley College. His published work includes articles in *Psychological Bulletin, Human Relations, Human Resource Management, The Organizational Behavior Teaching Review, Business Horizons,* and *Human Resource Planning.* He is also the author of *Participation in Organizational Change: The TVA Experiment* (Praeger, 1985). Dr. Nurick's research has been presented at the Academy of Management, The Institute of Management Science and Operations Research Society of America (TIMS/ORSA), the Eastern Academy of Management, the Organizational Behavior Teaching Society, and various regional and local groups. He is a recipient of Bentley College's Excellence in Teaching award for his courses on interpersonal behavior in management and organizational behavior. His current research includes the study of organizational transitions and the application of psychodynamic theory to organizational life. He has also served as a consultant to a variety of public and private organizations with a focus on behavioral issues associated with organizational change.

Building effective task teams has always been a challenge to project leaders and a crucial factor toward effective resource utilization, timing, and ultimate project success. The basic process has been known for centuries.[6,11,23,47,58] However, it became more involved and managerially challenging as organizational and technical complexities increased and the external business environment became more international.[4,15,20,48] It is especially this globalization of business activities which is most elusive and challenging to managers, but potentially also most rewarding.[8,17,38]

With the growing interdependence of the world economy, international competition continues to intensify. Resources, work skills, and technology are highly mobile. The companies which will survive and prosper in this amalgamated global marketplace will be those that can innovatively integrate their resources.[3,22,36,38,48] They will be able to develop, produce, and market products and services more cost-effectively and provide higher quality and responsiveness to their customers.

The objective of this chapter is threefold. First, we will summarize the characteristics of multinational team environments to provide a better understanding of the existing challenges and opportunities. Second, we offer a model for organizing and developing the multinational team. We also suggest a framework for evaluating team organizations and their performance. Finally, specific suggestions are made for effectively managing these teams.

THE MULTINATIONAL ORGANIZATIONAL ENVIRONMENT

Project teams are complex, both in their tasks and behavioral dimensions.[4,23,29,47] Considerable efforts must usually be made by the project leader to organize the initially loosely assembled group of people into a goal-oriented unified project team, to assure proper linkages among the various task groups, and to integrate the work into desired outputs. To achieve such performance, even within a geographically confined region, requires companies to extend their managerial approach beyond traditional, departmentalized thinking to an interdisciplinary orientation.[7,24,56,58] In addition, for projects that span international boundaries, new alliances are being formed among companies and new concepts of technology transfer evolve, such as concurrent engineering and process action teams.[10,20,43] These more dynamic and intricate conditions also call for more sophisticated management skills. As shown in Table 19.1, these skills are critically important for building multinational project teams with the ability to integrate across organizational, national, cultural, and political boundaries. They are also crucial for unifying and focusing the team efforts toward a set of common objectives and goals. With these new challenges, new forms of leadership have evolved which focus on the manager's role as a facilitator with emphasis on individual accountability, cooperation, team decision making and commitment by all members to the agreed-on results.[21,41,47]

WHAT WE KNOW ABOUT WORK TEAMS

With the continuous and rapid changes of their work environments, managers are adopting their style of leadership on an ongoing basis. Yet, in spite of emerging new

TABLE 19.1   Potential Problems of Managing Multinational Team Efforts and Their Influencing Factors

Problems	Influencing factors
People-oriented:	
Communications	
Bureaucracy	
Conflict	Differences in:
Risk, uncertainty	Cultures, values
Power struggle	Languages
Mistrust	Standards, norms, measures
Indecision	Distances
Leadership	Time zones
	Government regulations
Work-oriented:	Employment laws
Task definition and delegation	Personal norms and values
Innovation and creativity	Work interests
Technology transfer	Education and skill levels
Integration	Business environment
Problem/error detection	
Quality	
Changeability	

Difficulties relate to the challenges of unifying the team behind common goals and of integrating the project according to plan.

management practices, cross-functional dependencies, and eroding sources of traditional power, and team-centered work processes, a considerable body of knowledge exists on the characteristics and behavior of work teams.[4,19,23,25,44,56,58] This established body of knowledge forms an important and solid basis for guiding managers toward adopting their style to our contemporary, more demanding work environment.[1]* It also forms the basis for new management research, theory development, and tools and techniques such as those presented in this chapter.

In fact, work teams have long been considered an effective device to enhance organizational effectiveness.[16] Since the discovery of the importance of social phenomena in the classic Hawthorne studies,[39] management theorists and practitioners have tried to enhance group identity and cohesion in the workplace. Indeed, much of the "human relations movement" that occurred in the decades following Hawthorne is based on a group concept. McGregor's Theory Y,[30] for example, spells out the criteria for an effective work group, and Likert[26] called his highest form of management the *participative group,* or System 4. In today's more complex and technologically sophisticated environment, the group has reemerged in importance in the form of project teams. The management practices apply principles of interpersonal and group dynamics to create and guide project teams successfully.

---

*Bracketed numbers refer to Endnotes.

Team Building for Project Management

With the acceptance of project management as a formal way of managing multidisciplinary activities in the 1960s, the project team became more or less a fixture of the work environment. Increasing technological complexity and specialization gave rise to the greater flexibility provided by organizational matrix structures. Such structures, organized around a product or project, demonstrate greater adaptability to change and an emphasis on common goals. A major challenge for a manager in a project environment is integrating the talents of diverse individuals with different professional orientations toward a larger task. We define *team building* as "the process of taking a collection of individuals with different needs, backgrounds and expertise and transforming them by various methods into an integrated, effective work unit."[53]

While the purpose of creating an effective team is clear, the *process* of developing a team is more difficult to determine. Effective project teams are characterized by both task and relationship factors.[47] The task factors include such things as timely performance within budget, concern for quality, and technical results, while the relationship issues center around capacity to solve conflicts, trust, and communication effectiveness. Beckhard[9] suggests that often team leaders place more emphasis on the task issues such as improving work and solving problems while those that work at team development (e.g., consultants) place a greater emphasis on the inner workings of the group and the relationships among members. As a consequence, the emphasis of team development efforts changes, depending on the viewpoint of those involved.

The technical aspects of team development are more clearly delineated, easier to measure, and thus more directly addressed. Hardaker and Ward,[18] for example, describe a technique used at IBM, known as Process Quality Management (PQM), that focuses on understanding the mission, spelling out goals, and developing specific lists of activities directed toward critical success factors. While such exercises can be useful for understanding the task, they do not address some of the common problems of misunderstanding that arise from the inner workings of multidisciplinary groups. Indeed, such barriers as differing outlooks, priorities, and interests, role conflicts, and power struggles can undermine the group process and quickly derail the task. Yet, these issues are the most difficult to see and require a leader with the necessary sensitivity to effectively confront them.[44,45,49,57]

CHARACTERISTICS OF HIGH-PERFORMING PROJECT TEAMS

The characteristics of a project team and its ultimate performance depend on many factors which are related to both people and structural issues. Obviously, each organization has its own way to measure and express performance of a project team. However, in spite of the existing cultural and philosophical differences, there seems to be a general agreement among members on certain factors which are included in the characteristics of a successful project team.

The results of recent field studies[48,53,54] suggest a simple framework, for organizing the complex array of performance-related variables into four specific categories as shown in Fig. 19.1.

Task-related variables affect specific task outcome such as the ability to produce quality results on time and within budget, innovative performance, and willingness to change.

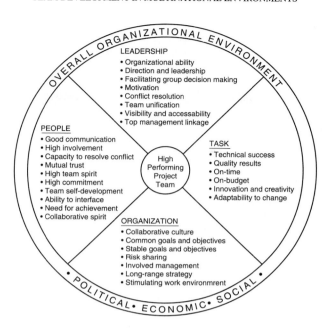

FIGURE 19.1 Variables influencing the project team performance.

People-related variables affect the inner workings of the team and include good communications, high involvement, capacity to resolve conflict, mutual trust, and commitment to project objectives.

Leadership variables are associated with the various leadership positions within the project team. These positions can be created formally, such as the appointment of project managers and task leaders *or* emerge dynamically within the work process as a result of individually developed power bases such as expertise, trust, respect, credibility, friendship, and empathy. Typical leadership characteristics include the ability to organize and direct the task, facilitate group decision making, motivate, assist in conflict and problem resolutions, and foster a work environment which satisfies the professional and personal needs of the team members.[46,50]

Organizational variables such as overall organizational climate, command-control-authority structure, policies, procedures, regulations, and regional cultures, values, and economic conditions. All of these variables are likely to be interrelated in a complex, intricate form.[55]

It is interesting to note that managers describing the characteristics of an effective, high-performing project team, focus not only on task-related skills for producing technical results on time and on budget but also stress especially the people and leadership-related qualities as shown in Fig. 19.1.

The significance of grouping and categorizing team performance variables is in three areas:

1. It provides a model for determining the factors which are critical to high team performance in a particular project environment.

2. It provides a framework for diagnosing and stimulating team-building activities.
3. The team performance variable might be useful in benchmarking the team's characteristics against the "norm" of high-performing teams, hence providing the basis for self-assessment and continuous improvement.

Taken together, within an integrated team, members enjoy their group association and derive much of their personal and professional satisfaction from the interaction with their team members. Specifically, some of the more important characteristics of such a truly integrated team are

- Satisfaction of individual needs
- Shared interests
- Strong sense of belonging
- Pride and enjoyment in group activity
- Commitment to team objectives
- High trust, low conflict
- Comfortable with interdependence and change
- High degree of group interaction
- Strong performance norms and result orientation

Creating a climate and culture that produces such team characteristics is conducive to high performance and involves multifaceted management challenges which increase with the complexities of the project and its global dimensions.

BARRIERS TO HIGH TEAM PERFORMANCE

As functioning groups, project teams are subject to all of the phenomena known as group dynamics. As a highly visible and focused work group, the project team often takes on a special significance and is accorded high status with commensurate expectations of performance. Although these groups bring significant energy and perspective to a task, the possibilities of malfunctions are great.[47,53] A myth is that the assembly of talented and committed individuals automatically results in synergy and renders such a team impervious to many of the barriers commonly found in a project team environment.[56] These barriers, while natural and predictable, take on additional facets in global project situations which are exposed to the various challenges shown in Table 19.1. Understanding these barriers, their potential causes, and influencing factors is an important prerequisite for managing them effectively and hence facilitate a work environment where team members can focus their energy on desired results. The most common barriers to effective team performance are discussed below in the context of international project environments.

Different Points of View

The purpose of a project team is to harness divergent skills and talents to accomplish project objectives. Having drawn upon various departments or perhaps even different organizations, there is the strong likelihood that team members will naturally see the world from their own unique point of view. There is a tendency to stereotype and devalue "other" views. Such tendencies are heightened when the project involves work groups from different countries with different "work cultures," norms, values, needs, and interests. Further, these barriers are often particularly strong in highly technical project situations where members speak their own codes and languages. In addition, there may be historical conflict among organizational units. In such a case representatives from these units will more than likely carry their prejudices into the team and potentially subvert attempts to create common objectives. Often these judgments are not readily known until the team actually begins its work and conflicts start developing.

Role Conflict

Project or matrix organizations are not only the product of ambiguity; they create ambiguity as well. Team members are actually in multiple roles and often report to different leaders, possibly creating conflicting loyalties. As "boundary role persons,"[1] they often do not know which constituency to satisfy. The "home" group or department has a set of expectations that might be at variance with the project team organization's. For example, a department may be run in a very mechanistic, hierarchical fashion while the project team may be more democratic and participatory. Team members might also experience time conflicts due to multiple task assignments which overlay and compete with traditional job responsibilities. The pull from these conflicting forces can either be exhilarating or a source of considerable tension for individual team members.

Implicit Power Struggles

While role conflict often occurs in a horizontal dimension (i.e., across units within the same division or across geographic and culture regions), conflict can also occur vertically as different authority levels are represented on the team. Individuals who occupy powerful positions elsewhere can try to recreate—or be expected to exercise—that influence in the group. Often such attempts to impose ideas or to exert leadership over the group are met with resistance, especially from others in similar positions. There can be subtle attempts to undermine potentially productive ideas with the implicit goal of winning the day rather than looking for what is best for the team. There is also the possibility that lower-status individuals are being ignored, thus eliminating a potentially valuable resource.

An example of such power struggles occurred in a *quality of work life project team* in an engineering organization.[34] The team was set up as a collaborative employee-management group designed to devise ways to improve the quality of work life in one division of a utility company. The membership of this representative group was changed halfway through the project to include more top managers. When the managers came aboard, they continued in the role of "manager" rather than "team member." Subsequently, the weekly meetings became more like typical staff meetings rather than creative problem-solving sessions. Although there was considerable resistance, the differences were pushed under the table, as the staff people did not wish to confront their superiors. There was also considerable posturing among the top managers in an effort to demonstrate their influence, although none would directly attempt to take the leadership position. While some struggle for power is inevitable in a diverse group, it must be managed to minimize potentially destructive consequences.

Group Think

This phenomenon of groups was identified by Irving Janis (1972) as a detriment to the decision-making process. It refers to the tendency for a highly cohesive group to develop a sense of detachment and elitism. It can particularly afflict groups that work on special projects. In an effort to maintain cohesion, the group creates shared illusions of invulnerability and unanimity. There is a reluctance to examine different points of view, as these are seen as dangerous to the group's existence. As a result, group members may censor their opinions as the group rationalizes the inherent quality and morality of its decisions. Because many project teams typically are labeled as special and often work under time pressure, they are particularly prone to the dangers of *group think.*

KEEPING THE TEAM FOCUSED

The key to continuous team development and effective project management is keeping the team focused. Field studies on multidisciplinary work groups show consistently and measurably that, to be effective, the project leader must not only recognize the potential drivers and barriers of high team performance, but also must know when in the life cycle of the project they are most likely to occur.[53,54,56] Team leaders can take preventive actions early in the project life cycle and foster a work environment that is conducive to team building as an ongoing process. A crucial component of such a process is the sense of ownership and commitment of the team members. Team members must become stakeholders in the project, buying into the goals and objectives of the project, and willing to focus their efforts on the desired results.

Specific management insight has been gained from studies by Gemmill, Thamhain, and Wilemon[49,55,57] into work group dynamics of project teams. These studies clearly show significant correlations[2] and interdependencies among work-environmental factors and team performance. They indicate that high team performance involves four primary factors: (*a*) managerial leadership, (*b*) job content, (*c*) personal goals and objectives, and (*d*) work environment and organizational support. The actual correlation of 60 influence factors to the project team characteristics and performance provided some interesting insight into the strength and effect of these factors. One of the important findings was that only 12 of the 60 influence factors which were examined were found to be statistically significant.[48,54] Other factors seem to be much less important to high team performance. Listed below are the 12 factors, classified as drivers, which associate with project team performance most strongly:

1. Professionally interesting and stimulating work
2. Recognition of accomplishment
3. Clear project objectives and directions
4. Sufficient resources
5. Experienced management personnel

6. Proper technical direction and leadership
7. Mutual trust, respect, low conflict
8. Qualified project team personnel
9. Involved, supportive upper management
10. Professional growth potential
11. Job security
12. Stable goals and priorities

It is interesting to note that these factors not only correlated favorably to the direct measures of high project team performance, such as technical success and on-time/on-budget performance, but also were positively associated with other indirect measures of team performance such as commitment, effective communications, creativity, quality, change orientation, and needs for achievement. These are especially important characteristics for high team performance in a multicultural, multinational environment where management control is weak through traditional chain-of-command channels but relies more on the norms and desires established by local teams and their individual members. What we find consistently is that successful project leaders pay attention to the human side. They seem to be effective in fostering a work environment conducive to innovative creative work, where people find the assignments challenging, leading to recognition and professional growth. Such a professionally stimulating environment also seems to lower communication barriers and conflict, and enhances the desire of personnel to succeed. Further, this seems to increase organizational awareness as well as the ability to respond to changing project requirements.

In addition, an effective team has good leadership. Team managers understand the task, the people, the organization, and all the factors crucial to success. They are action-oriented, provide the needed resources, properly direct the implementation of the project plan, and help in the identification and resolution of problems in their early stages.

Management and team leaders can help a great deal in keeping the project team focused. They must communicate and update organizational objectives and relate them to the project and its specific activities in various functional areas and geographic regions. Management can help in developing priorities by communicating the project parameters and organizational needs, and by establishing a clear project focus. While operationally the project might have to be fine-tuned to changing environments and evolving solutions, the top-down mission and project objectives should remain stable. Project team members need this stability to plan and organize their work toward unified results. This focus is also necessary for establishing benchmarks and integrating innovative activities across all disciplines. Moreover, establishing this clear goal focus stimulates interest in the project and unifies the team, ultimately helping to refuel the commitment to established project objectives in such crucial areas as technical performance, timing, and budgets.

Table 19.2 shows some of the critical components that characterize a focused team. Effective team leaders monitor their team environments for early warning signs of potential problems and changing performance levels.

Benchmarking the Project Team

Table 19.2 can also be used for examining the organizational environment and comparing it to other well-performing operations. Further, Table 19.2 can provide an

TABLE 19.2   Characteristics of a Focused Project Team: "Benchmark Your Team Against These Measures"

Factors characterizing the project team and its members	Rating by			
	Project leader	Team members	Support group	Senior managers
1. Clear project objectives				
2. Clear business mission and objectives				
3. Clear project results				
4. Clear reporting relationships, command and control channels				
5. Clear organizational interfaces				
6. Clear technical requirements				
7. Clear time and budget requirements				
8. Commitment to technical requirements				
9. Commitment to schedule and budget				
10. Clear role definition				
11. No role conflict or power struggle				
12. Project has high visibility				
13. Project has high priority				
14. Project has measurable milestones and performance feedback				
15. Involved and supportive management				
16. Frequent recognition of accomplishments				
17. Good communications with support groups				
18. Highly involved team members find work interesting and enjoyable				
19. Mutual trust, respect, low personal conflict				
20. Ability to handle change and conflict				
21. Ability to solve problems				
22. Innovative behavior				
23. Collaborative culture				
24. High need for achievement				
25. Members reinforce team concept				

important reference point for a team self-assessment or managerial audit. In this context, Table 19.2 can be used as a start-up data base for a *force field analysis*.[3] Here, team members diagnose what helps or hinders them in attaining desired performance. It is a simple, yet powerful technique which can help a project manager and team to identify those forces which help drive their projects toward success. The techniques also can help identify the barriers or restraining forces which may keep a team from attaining its goal, hence causing project failure.

Furthermore, Table 19.2 can be used for benchmarking the project environment and its leadership. That is, the typology of Table 19.2 can assist in comparing established management practices with those of other internal operations, external company practices, or global experiences. Table 19.2 can become the focus of a five-step continuous improvement process which involves: (1) defining *what* should be benchmarked and to *how* it should be compared; (2) analyzing team performance and establishing operational norms and performance targets; (3) communicating these targets to all organizational levels and developing action plans; (4) implementing these action plans; and (5) fine tuning and integrating the new practices with the total business process.

If used properly, either the force field analysis or the benchmarking process, or any combination of the two, can be powerful tools for diagnosing the need for change and implementing it. The personal involvement of the team members during the situational assessment and action plan development usually leads to buy-in and ultimate commitment to the necessary change process.

Taken together, the effective project team leader is usually a social architect who understands the interaction of organizational and behavioral variables and can foster a climate of active participation and minimal dysfunctional conflict. This requires carefully developed skills in leadership, administration, and organization and technical expertise. It further requires the project team leader to work effectively with upper management to assure organizational visibility, resource availability, focus, and overall support for the project activities and programs throughout their life cycles.

ORGANIZING MULTINATIONAL TEAMS

As competition pressures and the global business environment force organizations to become leaner, flatter, less bureaucratic and more entrepreneurial, project team building takes on new dimensions. Managers must learn to operate in a different environment. With realities of diminishing managerial power in the hands of a few and the declining effectiveness of traditional command and control channels, multinational project teams must rely on alternative forms of communication and control through networks, special linkages, alliances, partnerships, and joint ventures.

In addition, new project teams must often be organized to implement parallel multidisciplinary processes. Whether we call it design/build approach, concurrent engineering, or multifunctional team design, the objective is to organize a project team that can simultaneously execute many of the phases. This requires a great deal of cross-functional involvement and cooperation. But it also leads to shorter product or process development cycles and more predictable manufacturability, reliability, and service.

Furthermore, the multinational organizational environment and its pluralistic culture adds yet another challenge to project team building. The new team members are usually selected from hierarchically organized functional resource departments located in different geographic regions led by strong individuals who often foster internal competition rather than cooperation. Further, many of the highly innovative and creative contributors are highly individualistic and often admit their aversion to cooperation and team work. The challenge to the project manager is to integrate these individuals into a team that can produce innovative results in a systematic, coordinated, and integrated way, consistent with the established project plan. Many of the problems that occur during the formation of the new project team or during its life cycle are normal and often predictable. However, they present barriers to effective team performance. They must be quickly identified and dealt with. The following list presents typical problems which occur during project team formation and start-up.

- Confusion
- Responsibilities unclear
- Channels of authority unclear
- Work distribution load uneven
- Assignment unclear
- Communication channels unclear
- Overall project goals unclear
- Mistrust
- Personal objectives unrelated to project
- Measures of personal performance unclear
- Commitment to project plan lacking
- Team spirit lacking
- Project direction and leadership insufficient
- Power struggle and conflict

It used to be that a competent manager could build a project organization single-handed. This is no longer the case, since project team building has become a distributed process that involves many leaders at different organizational networks. These team leaders must understand the dynamics of their work groups and the organization as a whole in order to integrate their resources into a unified, focused project team. Certain steps, taken early in the project life cycles can help the project leader in identifying the specific problems and dealing with them effectively. These steps are summarized in Table 19.3, which also provides suggestions for preventive measures for eliminating the potential of these problems.

The Process of Organizing the New Team

Specific suggestions are made here for organizing a new project team. While the process is consistent with other planning practices, it must also include personal involvement of the people early at the planning phase to ensure buy-in and commitment to the plan. Further, the planning process must lead to a clear identification and understanding of the goals, resources, and activities that are critical to project success, and the measures needed to track and control the activities. This lays the foundation for organizing a multifunctional team that is focused on desired results.

Step 1: Recruit Key Players

After the project and basic objectives of the mission have been delineated, the core project management team should be recruited. Often the process starts with the appointment of the project leader, who then in turn seeks out the most qualified task leaders or systems managers who will report to him or her in the project team organization. These senior team leaders repeat the staffing cycle until a working-level team is in place; however, project organization efforts are often too intricate and dynamic to follow such a stratified process. Therefore it is not uncommon for project staffing to be conducted iteratively and with much less formality. Yet, regardless of the dynamics

TABLE 19.3 Criteria for Organizing Effective Project Teams

Criteria	Leadership actions
The assignment should be clear	Although the overall task assignment, its scope, and objectives might have been discussed during the initial sign-on of the person to the project, it takes additional effort and involvement for new team members to feel comfortable with the assignment. The thorough understanding of the task requirements comes usually with the intense personal involvement of the new members with the project team. Such involvement can be enhanced by assigning the new member to an action-oriented task that requires team involvement and creates visibility, such as a requirements analysis, an interface specification, or producibility study. In addition, any committee-type activity, presentation, or data gathering will help to involve the new team member. It also will enable the team member to better understand the specific task and his or her role in the overall team effort.
New team members must feel professionally comfortable	The initial anxieties, in the absence of trust and confidence, are serious barriers to team performance. New team members should be properly introduced to the group and their roles, strengths, and criticality to the project explained. Providing opportunities for early results allows the leader to give recognition for professional accomplishments which will stimulate the individual's desire for the project work and build confidence, trust, and credibility within the group.
Team organization should be clear	Project team structures often are considered very "organic" and inconsistent with formal chain-of-command principles. However, individual task responsibility, accountability, and organizational interface relations should be clearly explained to all team members. A simple work breakdown structure or task matrix, together with some discussion, can facilitate a clear understanding of the team structure, even with a highly unconventional format.
Critical factors for success should be clear	Defining the factors, activities, and resources which are critical to project success requires usually an intensive 3- to 8-hour session which is initially conducted for each task group and ultimately involves all key players of the whole project team. The critical factors evolve from the overall project objectives and mission. These may include performance objectives, quality, time-to-market, benchmarks, permits, social economic factors, and critical facilities.
Work interfaces should be clear	Modern projects involve multiple accountabilities and work interfaces, not only between major functional organizations, but virtually from any task team to many project support or line organizations. To ensure the necessary cross-departmental collaboration and shared commitment each task team should learn the critical links by tracing their work flow and support requirements.

TABLE 19.3   Criteria for Organizing Effective Project Teams (*Continued*)

Criteria	Leadership actions
Work input/output must be defined	As with the *work interfaces,* each task team should define its principle work outputs to and inputs from other organizations. This also includes establishing (1) proper working relations with these interface organizations and (2) an agreement with all responsible individuals on the validity of the inputs and outputs. Management tools such as the work breakdown structure, task matrix, phased activity chart, and N chart can be helpful in defining and articulating these input/output items.
Individual ability must match requirements	Team members must have the capacity to perform on their assignment. The sign-on interview is an important tool for project managers and team leaders to assess individual capabilities and to match individual talents with the job requirements. In addition, the team composition and mix of people can produce synergism if done skillfully.
Team members must have incentives	People need motivators to be self-directed, especially if the work requires change, innovation, risk taking, commitment and leadership. Some of the strongest incentives for team member participation are intrinsic to the work itself. The desire to be part of a professionally interesting and exciting project, the potential for further skill and career development, the thrill of anticipating future accomplishments, recognition, and visibility can be very strong motivators toward joining the project team and helping to organize it for a successful start-up. Project managers and team leaders should identify the attractive features of the project to their people, point out its importance to the company, its management, clients, and society. In addition, traditional incentives for joining the project team, such as salary actions and promotional opportunities, should not be overlooked.
Communcation channels should be established	Management should facilitate communcations among the various work groups of the project and its support organizations. Communications links come in many forms. Regularly scheduled meetings are recommended as soon as the new project team is being formed. These meetings are particularly important where team members are geographically separated and do not see each other on a day-to-day basis.
Provide a proper team environment	It is crucial for management to provide the proper environment for the project to function effectively. Here the project leader needs to tell management at the onset of the program what resources are needed. The project manager's relationship with senior management support is critically affected by his or her credibility and the visibility and priority of the project.

TABLE 19.3   Criteria for Organizing Effective Project Teams (*Continued*)

Criteria	Leadership actions
Manage	Especially during the initial stages of team formation, it is important for the project leader to keep a close eye on the team and its activities to detect problems early and to correct them. The project manager can also influence the climate of the work environment by his or her own actions. The manager's concern for project team members, ability to integrate personal goals and needs of project personnel with the project objectives, and the ability to create personal enthusiasm for the work itself can foster a climate which is high on motivation, work involvement, and resulting project performance.

involved, most of the senior team members should be on board before the team organization process can unfold smoothly.

Step 2: Structure the Organization

Clearly defined and communicated responsibilities and organizational relationships are crucial for successfully building a new project organization. The tools for systematically describing the project organization come, in fact, from conventional management practices:

Charter the Program or Project Organization.   The charter clearly describes the business mission and scope, broad responsibilities, authorities, organizational structure, interfaces, and reporting relationship of the program organization. The charter should be revised for each new program. For small projects a simple statement may be sufficient, while larger programs usually require a formal charter supported by standardized company policies on project management.

Project Organization Chart.   Regardless of the specific organizational structure and the terminology used, a simple organization chart should define the major reporting and authority relationships. These relationships can be further clarified in a policy directive.

Responsibility Matrix.   This chart defines the interdisciplinary task responsibilities. It shows who is responsible for what. The responsibility matrix not only covers activities within the project organization, but also defines the functional relationship with support units, subcontractors, and committees. In a simpler format, a task roster can be used to list project tasks and corresponding responsible personnel.

Job Description.   If not already in existence, a job description should be developed for all key project personnel, such as program managers, system project managers, hardware project managers, task managers, project engineers, and plan coordinators. Job descriptions usually are generic and hence portable from one project to the next. Job descriptions are modular building blocks which form the framework for staffing a project organization. A job description includes the reporting relationship responsibilities, duties, and typical qualifications.

Step 3: Define the Project

The work, timing, resources, and responsibilities of a project must be defined, at least in principle, before staffing can begin. Regardless of how vague and preliminary these project segments are at the beginning, the initial description will help in recruiting the appropriate personnel and eliciting commitment to the preestablished parameters of technical performance, schedule, and budget. The core team should be formed prior to finalizing the project plan and contractual arrangements. This will provide the project management team with the opportunity to participate in tradeoff discussions and customer negotiations, leading to technical confidence and commitment of all parties involved.

Step 4: Staff the Project

Staffing involves people, resources, commitments, anxieties, and conflict. Often, pressures on the project manager to produce lead to hasty recruiting practices before the basic project work to be performed has been properly defined. This can result in personnel poorly matched to the job requirements as well as conflict, low morale, suboptimum decision making, and, in the end, poor project performance. The comment of a project section leader who was pressed into quick staffing actions is indicative of these potential problems: "How can you interview task managers when you cannot show them what the job involves and how their responsibilities tie in with the rest of the project?"

Therefore, project leaders should start to interview candidates only after the project organization and the tasks are defined in principle. These interviews should always be conducted one-to-one. We suggest a five-step, often interrelated, interview process, shown in Table 19.4.[4]

Team Building as an Ongoing Process

While proper attention to team building is crucial during the early phases of a project, it is a never-ending process.[35] The project manager is continually monitoring team performance to see what corrective action may be needed to prevent or correct problems. Several barometers provide good clues of potential team dysfunction. *First,* noticeable changes in performance levels for the team and/or for individual team members always should be followed up. Such changes can be symptomatic of more serious problems, such as conflict, lack of work integration, communication problems, and unclear objectives. *Second,* the project leader and team members want to be aware of changing energy levels in various team members. This, too, may signal more serious problems such as anxieties, stress, conflict, and burnout. Sometimes changing the work pace, taking time off, or setting short-term targets can serve as a means to reenergize team members. More serious cases, however, may call for more drastic measures, such as reappraising project objectives and/or the means to achieve them. *Third,* verbal and nonverbal clues from team members may be a source of information and team functioning. It is important to hear their needs and concerns (verbal clues) and to

TABLE 19.4   The Interview Process

Informing the candidate about the assignments
  What are the objectives for the project?
  Who will be involved and why?
  What are the structures of the project organization and its interfaces?
  What is the importance of the project to the overall organization or work unit, including short-
  or long-range impact?
  Why was the team member selected and assigned to the project?
  What role will the team member perform?
  What are the team member's specific responsibilities and expectations?
  What rewards might be forthcoming if the project is completed successfully?
  Candidly, what problems and constraints are likely to be encountered?
  What are the "rules of the road" that will be followed in managing the project, such as regular
  status review meetings?
  What challenges and recognition is the project likely to provide?
  Why is the team concept important to success and how should it work?

Determining skills and expertise
  Probe related experience; expand from resumé.
  Probe candidate's aptitude relevant to your project environment: technology involved, engineering tools and techniques, markets and customer involvement, and product applications.
  Probe into the program management skills needed. Use current project examples: "How
  would you handle this situation...?" Probe leadership, technical expertise, planning and control, administrative skills, and so on.

Determining interests and team compatibility
  What are the professional interests and objectives of this candidate?
  How does the candidate manage and work with others?
  How does the candidate feel about sharing authority, working for two bosses, or dealing with
  personnel across functional lines with little or no formal authority?
  What suggestions does the candidate have for achieving success?

Persuading to join project team
  Explain specific rewards for joining the team, such as financial, professional growth, recognition, visibility, work challenge, and potential for advancement.

Negotiating terms and commitments
  Check candidate's willingness to join team.
  Negotiate conditions for joining: salary; hired, signed, or transferred; performance reviews
  and criteria.
  Elicit candidate's commitment to established project objectives and modus operandi.
  Secure final agreement.

observe how they act in carrying out their responsibilities (nonverbal clues). *Finally,
irregular or disruptive behavior of one team member toward another can be a signal
that a problem within the team warrants action.*

It is highly recommended that project leaders hold regular meetings to evaluate
overall team performance and deal with team functioning problems. The focus of
these meetings can be directed toward "What are we doing well as a team" and "What
areas need our team's attention?" This approach often brings positive surprises in that
the total team will be informed on progress in diverse project areas, such as a breakthrough in the technology department, a subsystem schedule met ahead of the original
target, or a positive change in the client's behavior toward the project. After the positive issues have been discussed, attention should be devoted toward areas needing

team attention. The purpose of this part of the review session is to focus on real or potential problem areas. The meeting leader should ask each team member for observations on these issues. Then an open discussion should be held to ascertain how significant the problems really are and how they impact overall team performance. Next, approaches should be worked out on how to best handle these problems. Finally, a plan for follow-up should be developed.

The process should result in a better overall performance and promote a feeling of team participation and high morale. Over the life of a project, the problems encountered by the team are likely to change, and as one set of problems is identified and solved, new ones will emerge and must be handled like any other continuous improvement program.

Our own work with clients, as well as field research,[31,44,47,48] clearly shows that there are several indicators of effective and ineffective teams. At any point in the life of the team, the project manager should be aware of these drivers and barriers toward team effectiveness. By benchmarking the team organization against desired characteristics, as suggested in Table 19.2, the team leader can identify early warning signals of performance problems and deal with them in their early stages.

RECOMMENDATIONS FOR EFFECTIVE TEAM MANAGEMENT

As more companies compete on a global scale, their project operations have become vastly more complex. The recommendations advanced here reflect the realities of this new environment where project managers have to cross organizational, national, and cultural boundaries and work with people over whom they have little or no formal control. Alliances and collaborative ventures have forced project managers to focus more on cross-boundary relationships, negotiations, delegation, and commitment, rather than on establishing formal command and control systems.

Taken together, the project leader must foster an environment where team members can work together across organizational and national boundaries in a flatter and leaner company which is more flexible and responsive to quality and time-to-market forces. To be effective in such a team environment, the leader must create an ambience where people are professionally satisfied, are involved, have mutual trust, and can communicate well with each other. The more effective the project leader is in stimulating the drivers of effective team performance (Table 19.2), the more effective the manager can be in developing team membership and the higher the quality and candor in sharing ideas and approaches.

Furthermore, the greater the team spirit, trust, and quality of information exchange among team members, the more likely the team will be able to develop effective decision-making processes, make individual and group commitment, focus on problem solving, and develop self-forcing, self-correcting project controls. These are the characteristics of an effective and productive project team.

Managing the Process

In this final section, we summarize our discussion and the established criteria for effective team management in the format of specific recommendations. These recommendations should help project leaders and managers responsible for the integration of multidisciplinary tasks in their complex efforts of building high-performing project teams in a multinational environment.

1. *Barriers.* Project managers must understand the various barriers to team development and build a work environment conducive to the team's motivational needs. Specifically, management should watch out for the following barriers: (*a*) unclear objectives, (*b*) insufficient resources and unclear funding, (*c*) role conflict and power struggle, (*d*) uninvolved and unsupportive management, (*e*) poor job security, (*f*) shifting goals and priorities, and (*g*) culture-related tension and differences in norms and values.

2. *The project objectives* and their importance to the organization should be clear to all personnel who get involved with the project. Senior management can help develop a "priority image" and communicate the basic project parameters and management guidelines.

3. *Management commitment.* Project managers must continuously update and involve their managements, in both the host and support organizations, to refuel their interests and commitments to the new project. Breaking the project into smaller phases and being able to produce short-range results frequently seem to be important to this refueling process.

4. *Image building.* Building a favorable image for the project in terms of high priority, interesting work, importance to the organization, high visibility, and potential for professional rewards is crucial to the ability to attract and hold high-quality people. It is also a pervasive process which fosters a climate of active participation at all levels; it helps to unify the new project team and minimizes dysfunctional conflict.

5. *Leadership positions* should be carefully defined and staffed at the beginning of a new program. Key project personnel selection is the joint responsibility of the project manager and functional management. The credibility of project leaders among team members, with senior management, and with the program sponsor is crucial to the leader's ability to manage the multidisciplinary activities effectively across functional lines. One-on-one interviews are recommended for explaining the scope and project requirements, as well as the management philosophy, organizational structure, and rewards.

6. *Effective planning* early in the project life cycle will have a favorable impact on the work environment and team effectiveness. This is especially so because project managers have to integrate various tasks across many functional lines and geographic regions. Proper project planning, however, means more than just generating the required pieces of paper. It requires the participation of the entire project team, including support departments, subcontractors, and management. These planning activities, which can be performed in a special project phase such as Requirements Analysis, Product Feasibility Assessment, or Product/Project Definition, usually have a number of side benefits—stimulating interest, understanding requirements, developing commitment, and unifying the team—besides generating a comprehensive road map for the upcoming program.

7. *Involvement.* One of the side benefits of proper project planning is the involvement of personnel at all organizational levels. Project managers should drive such an involvement, at least with their key personnel, especially during the project definition phases. This involvement will lead to a better understanding of the task requirements, stimulate interest, help unify the team, and ultimately lead to commitment to the project plan regarding technical performance, timing, and budgets.

8. *Project staffing.* All project assignments should be negotiated individually between the task leader and each prospective team member. Where dual-reporting relationships are involved, staffing should be conducted jointly. The assignment interview should include a clear discussion of the specific task, the outcome, tim-

ing, responsibilities, reporting relation, potential rewards, and importance of the project to the company. Task assignments should be made only if the candidate's ability and interest is a reasonable match to the position requirements.

9. *Team structure.* Management must define the basic team structure and operating concepts early during the project formation phase. The project plan, task matrix, project charter, and policy are the principal tools.

10. *Team-building sessions* should be conducted by the project manager throughout the project life cycle. An especially intense effort might be needed during the team formation stage. Self-assessment procedures such as the benchmarking process suggested in Table 19.2 can help in focusing these team sessions.

11. *Team commitment.* Project managers should determine lack of team member commitment early in the life of the project and attempt to correct related problems. Since anxiety, mistrust, confusion, insecurity, and low interest in the project activities are often the major reasons for lack of commitment, managers should try to determine why those perceptions exist and rectify them. Conflict with other team members may be another reason for lack of commitment. It is important for the project leader to intervene and mediate the conflict quickly.

12. *Senior management support.* It is critically important for senior management to provide the proper environment for the project team to function effectively. An understanding and agreement on the project scope and resource and time requirements are crucial prerequisites for establishing and maintaining top management support. Further, the project manager's ability to maintain upper management support involvement is critically important to the leaders' credibility and the priority image of the project.

13. *Problem avoidance.* Project leaders should focus their efforts on problem avoidance. That is, the project leader, through experience, should recognize potential problems and conflicts at their onset and deal with them before they become big and their resolutions consume a large amount of time and effort.

14. *Personal drive and leadership.* Project managers can influence the work environment by their own actions. Concern for the team members, the ability to integrate personal needs of the team members with the goals of the project, and the ability to create personal enthusiasm for the project itself, can foster a climate of high motivation, work involvement, open communication, and ultimately high project performance.

A FINAL NOTE

In summary, effective team building is a critical determinant of project success. With an increasing scale of multinational projects and global business operations, team building takes on additional dimensions and challenges of organizing and unifying the task groups across industrial, technological, and geographic boundaries. As one of the program leader's prime responsibilities, team building involves a whole spectrum of management skills to identify, commit, and integrate the various personnel from different functional organizations into a single task group. In many project-oriented organizations, team building is a shared responsibility between the functional managers and the project manager, who often reports to a different organization with a different superior.

To be effective, the project manager must provide an atmosphere conducive to teamwork. Four major considerations are involved in the integration of people from various disciplines into an effective team: (*a*) creating a professionally stimulating work environment, (b) providing good program leadership, (c) providing qualified personnel, and (d) providing a technically and organizationally stable environment. Further, the project leader must foster an environment where the new project team members are professionally satisfied, involved, and have mutual trust.

The complexity of the observations and recommendations discussed in this chapter clearly indicates the magnitude of challenges involved in managing multinational team efforts. However, mastering these challenges may also be most rewarding in terms of potential gains from increased productivity for individuals, organizations, and society as a whole.

Over the next decade we anticipate important developments in team building which will lead to higher performance levels, increased morale, and a pervasive commitment to final results. Success will require long-term changes in managerial thinking and leadership style, and will depend on the managers' ability to adapt to the increasingly global business topology.

REFERENCES

1. J. S. Adams, "The structure and dynamics of behavior in organizational boundary roles," in *Handbook of Industrial and Organizational Psychology,* M.D. Dunnette (ed.), Rand McNally, Chicago, 1976, pp. 1175–1199.

2. J. R. Adams and N. S. Kirchof, "A Training Technique for Developing Project Managers," *Project Management Quarterly,* March 1983.

3. M. Ames and D. Heide, "The Keys to Successful Management Development in the 1990's," *Journal of Management Development (UK),* vol. 10, no. 2, 1991.

4. J. J. Aquilino, "Multi-skilled Work Teams: Productivity Benefits," *California Management Review,* Summer 1977.

5. J. D. Aram and C. P. Morgan, "Role of Project Team Collaboration in R&D Performance," *Management Science,* June 1976.

6. K. H. Baler, "The Hows and Whys of Team Building," *Engineering Management Review,* December 1985.

7. G. Barczak and D. L. Wilemon, "Communications Patterns of New Product Development Team Leaders," *IEEE Transactions on Engineering Management,* vol. 38, no. 2, May 1991.

8. C. F. Barnum and D. R. Gaster, "Global Leadership," *Executive Excellence,* vol. 8, no. 6, June 1991.

9. R. Beckhard, "Optimizing team-building efforts," *Journal of Contemporary Business,* vol. 1, no. 3, 1971, pp. 23–32.

10. R. P. Bergstrom and G. S. Vasilash, "Inside the IBM PS/1," *Production,* vol. 102, no. 11, November 1990.

11. R. Carzo, Jr., "Some Effects of Organization Structure on Group Effectiveness," *Administrative Science Quarterly,* March 1963.

12. W. J. Conover, *Practical Nonparametric Statistics,* Wiley, New York, 1971.

13. A. Dentschman, "The Trouble with MBA's," *Fortune,* vol. 124, no. 3, July 29, 1991.

14. B. A. Diliddo, P. C. James, and H. J. Dietrich, "Managing R&D Creatively: B.F. Goodrich's Approach," *Management Review,* July 1981.

15. P. F. Drucker, "Learning from Foreign Management," *The Wall Street Journal,* June 4, 1980.

16. W. G. Dyer, *Team Building: Issues and Alternatives,* Addison-Wesley, Reading, Mass., 1977.

17. W. E. Fulmer, "Arthur Anderson: Training for Global Impact," *Journal of Management Development (UK),* vol. 10, no. 3, 1991.

18. M. Hardaker and B. K. Ward, "How to make a team work," *Harvard Business Review,* vol. 65, no. 6, 1987, pp. 112–120.

19. P. R. Harris, "Building a High-Performance Team," *Training Development Journal,* April 1986.

20. I. Janis, *Victims of Groupthink,* Houghton Mifflin, Boston, 1972.

21. H. R. Jessup, "New Roles of Leadership," *Training and Development Journal,* vol. 44, no. 11, November 1990.

22. A. Johne and P. Snelson, "Successful Product Innovation in UK and US Firms," *European Journal of Marketing (UK),* vol. 24, no. 12, 1990.

23. F. E. Katz, "Explaining Informal Work Groups in Complex Organizations," *Administrative Science Quarterly,* no. 10, 1965.

24. T. Kidder, *The Soul of a New Machine,* Avon, New York, 1982.

25. K. Lewin, "Frontiers in Group Dynamics," *Human Relations,* vol. 1, no. 1, 1947. Also see Lewin's *Field Theory in Social Science,* Harper, New York, 1951.

26. R. Likert, *New Patterns of Management,* McGraw-Hill, New York, 1961.

27. R. Likert, "Improving Cost-Performance with Cross-Functional Teams," *Management Review,* March 1976.

28. M. MacCoby, "Rethinking Competitiveness," *Research-Technology Management,* vol. 34, no. 3, May/June 1991.

29. J. K. McCollum and D. J. Sherman, "The Effects of Matrix Organization Size and Number of Project Assignments and Performance," *IEEE Transactions on Engineering Management,* vol. 38, no. 1, February 1991.

30. D. McGregor, *The Human Side of Enterprise,* McGraw-Hill, New York, 1960.

31. A. J. Nurick, "Facilitating Effective Work Teams," *SAM Advanced Management Journal,* vol. 57, no. 2, Spring 1992.

32. A. J. Nurick, J. B. Kamm, J. C. Shuman, and J. A. Seeger, "Entrepreneurial Teams in New Venture Creation: A Research Agenda," *Entrepreneurship Theory and Practice,* vol. 14, no. 4, 1990.

33. A. J. Nurick and A. N. Hoffman, "Organizational Implications of Physician Motivation," *Business Horizons,* vol. 34, no. 3, 1991.

34. A. J. Nurick, *Participation in Organizational Change,* Praeger, New York, 1985.

35. J. Pastor and R. Getchman, "Nurturing the Team Work Culture," *Supervisory Management,* vol. 36, no. 4, April 1991.

36. J. B. Quinn, "Technological Innovation, Entrepreneurship, and Strategy," *Sloan Management Review,* Spring 1979.

37. R. M. Ranftl, "R&D Productivity," technical report, Hughes Aircraft Company, 1978.

38. H. A. Riker and M. F. Roetter, "The New Business Topography," *Prism* (an Arthur D. Little Publication), fourth quarter 1990.

39. F. Roethlisberger and W. Dickerson, *Management and the Worker,* Harvard, Cambridge, Mass., 1939.

40. S. Siegel, *Nonparametric Statistics,* McGraw-Hill, New York, 1956.

41. J. Spicer, "Building Teamwork by Recognizing Corporate Cultures," *Trustee,* vol. 44, no. 6, June 1991.

42. R. M. Strozier, "Teaming Up," *World,* vol. 25, no. 1, 1991.

43. H. J. Thamhain, *Engineering Management: Managing Effectively in Technology-based Organizations,* Wiley, New York, 1992.

44. H. J. Thamhain, "Effective Leadership Style for Managing Project Teams," Chap. 22 in *Handbook of Program and Project Management,* P. C. Dinsmore (ed.), AMACOM, New York, 1992.

45. H. J. Thamhain, "Developing Engineering Program Management Skills," Chap. 22 in *Handbook: Management of R&D and Engineering,* D. Kocaoglu (ed.), Wiley, New York, 1992.

46. H. J. Thamhain, "Skill Developments for Project Managers," *Project Management Journal,* vol. 22, no. 3, September 1991.

47. H. J. Thamhain, "Developing Project Teams," Chap. 21 in *Project Management,* David I. Cleland (ed.), McGraw-Hill, New York, 1990.

48. H. J. Thamhain, "Managing technologically innovative team efforts towards new product success," *Journal of Product Innovation Management,* vol. 7, no. 1, March 1990.

49. H. J. Thamhain and D. L. Wilemon, "Leadership, conflict and project management effectiveness," *Executive Bookshelf on Generating Technological Innovations, Sloan Management Review,* Fall 1987.

50. H. J. Thamhain, "Managing Engineers Effectively," *IEEE Transactions on Engineering Management,* August 1983.

51. H. J. Thamhain, "Building a High Performance Technical Marketing Team," *Proceedings of the American Marketing Association Conference,* Chicago, August 1986.

52. H. J. Thamhain and D. Wilemon, "Skill Requirements of Engineering Program Manager," *Proceedings of the 26th Engineering Management Conference,* 1978.

53. H. J. Thamhain, "Anatomy of a High Performing New Product Team," *Conv. Rec. 16th Ann. Symp. Project Management Institute,* 1983.

54. H. J. Thamhain, "Building High Performing Engineering Project Teams," *IEEE Transactions on Engineering Management,* vol. 34, no. 3, August 1987.

55. H. J. Thamhain and G. R. Gemmill, "Influence Styles of Project Managers: Some Project Performance Correlates," *Academy of Management Journal,* June 1974.

56. D. L. Wilemon and H. J. Thamhain, "Team Building in Project Management," *Project Management Quarterly,* June 1983.

57. D. L. Wilemon et al., "Managing Conflict on Project Teams," *Management Journal,* 1974.

58. J. H. Zenger and D. E. Miller, "Building Effective Teams," *Personnel,* March 1974.

ENDNOTES

[1] This section draws in part from the writings by Aaron Nurick, "Facilitating Effective Work Teams," *SAM Advanced Management Journal,* vol. 58, no. 2, Winter 1993, which discusses the dynamics of work teams and their underlying formal concepts in detail.

[2] The Kendall-Tau rank-order correlation was used to measure the association between these variables. Statistical significance was defined at a confidence level of 95 percent or better.

[3] The force field analysis is a concept for managing change. It was influenced by the pioneering research of Kurt Lewin.[25] His process of inducing change also applies to changing attitudes and behavior in individuals and small groups. Lewin's theory is based on the notion of countervailing forces. In the management of a project, at any given point in its life of the project, there will be certain driving forces propelling the project toward success and certain restraining forces which may work against the project. In a steady state there is a balance between the driving forces pushing for change (or success) and the restraining forces which resist change. Lewin believed that change was the result of a shifting of the counter-balanced forces. In other words, if the driving forces can be increased or if restraining

forces be minimized, change is likely to occur. These processes may occur separately or together. The formal study of these forces is known as *force field analysis.*

[4] For a more detailed description of the interview process see Hans J. Thamhain, "Project Team Development," in Chap. 18 of *Project Management,* David I. Cleland (ed.), McGraw-Hill, New York, 1990. This section draws in part from the writings of Chap. 18, including Table 4, which is reprinted with permission.

PROJECT MANAGEMENT TO CONCENTRATE RESOURCES IN ENGINEERING AND MANUFACTURING

Simultaneous engineering and shared manufacturing contribute to successful project performances. David I. Cleland provides in Chap. 20 a review of one of the newer and more important applications of project management, namely, in the context of simultaneous engineering (concurrent engineering). He describes how the use of simultaneous engineering can improve global competitiveness through providing a process for the use of project management in the design of global products, and the organizational processes required to bring such products to earlier commercialization.

Shared manufacturing, the subject of Chap. 21, by Shriram Dharwadkar, Bopaya Bidanda, and David I. Cleland, describes how one community using project management conceptualized, designed, and developed a Manufacturing Assistance Center (MAC) to provide assistance to small and mid-sized manufacturing companies in the metalworking field. In this chapter analogies are drawn between the new product or services development process and the development of shared manufacturing centers. The model offered by the authors for the development of a MAC has applicability to enhance greater global competitiveness through the application of project-management systems to deal with the global change in manufacturing systems technology.

Francis Webster discusses in Chap. 22 some of the key considerations and issues to be considered in competing in global markets through the use of manufacturing technology. The demand for integration of the manufacturing strategy and the organizational business strategy is emphasized.

CHAPTER 20

SIMULTANEOUS ENGINEERING: KEY TO GLOBAL COMPETITIVENESS

David I. Cleland

University of Pittsburgh,
Pittsburgh, Pennsylvania

David I. Cleland is the author/editor of 25 books and many articles in the fields of project management, engineering management, and strategic planning. He is a Fellow of the Project Management Institute and was appointed to the Ernest E. Roth Professorship at the School of Engineering in recognition of outstanding productivity as a senior member of the faculty. He is currently a Professor of Engineering Management at the University of Pittsburgh.

In this chapter the concept and process of simultaneous engineering will be described. Simultaneous engineering also goes by other names: concurrent engineering, design for manufacturing, design for assembly, and parallel release. The keys to simultaneous engineering are teamwork and communication among the different specialists who will be working on the project team to develop both the *product* and its supporting *processes*. Simultaneous engineering has become a key strategy in a more efficient and effective commercialization of products for both domestic and global markets. Product-process design teams used in the context of simultaneous engineering represent a relatively new and strategically important application of proven project (program) management concepts and processes. Indeed it represents further growth in contemporary organizations in the use of teams as an important alternative to traditional organizational designs.

Product-process design projects are strategic decisions for the enterprise. When a new product is selected for development, or when an ongoing product or process development project is reviewed, the opportunity exists to make a strategic decision. Projects are designed to create something that does not currently exist—which, if developed and implemented, provides for the resulting products and services to enter into the operational inventory of a "bundle of values" to be provided to customers.

Simultaneous engineering is a systematic approach to the concurrent design of products and their associated processes such as manufacturing, marketing, quality, facility engineering, sales, and after-sales service. The strategy of concurrent engineering is to have the product and process developers consider all of the product's life-cycle considerations in the development of future products and services, from conception through disposal, including cost, schedule, technical performance, and user needs.

To remain competitive in the coming decade, business organizations are going to have to rethink the strategies for their entire enterprises. One of the key competitive factors will be how effectively the organizations are able to manage strategically the incremental advancement of the state of the art in their products and in the supporting processes required to develop and deliver new products to their customers, including after-sales service. The rethinking of the manner in which new products and processes are developed will require that new relationships be established with virtually all of the organizational functions: manufacturing, marketing, procurement, quality, sales, after-sales service, finance, and personnel. As these relationships are redefined, the organizational design of the companies will require modification to implement the changes that will have to be made. Teams are taking on added significance in this modification and in the strategic management of organizations.

TEAMS, TEAMS, TEAMS, ETC.

Reich says that the new hero of the business entrepreneurship is the team. Through a team individual skills are integrated into a team effort and result. This group ability to innovate becomes greater than the sum of its parts.[1]

Teams are an idea whose time has come. No matter what business you are in, teams will be coming more and more in the future. The empowerment of people through participation on teams can do much to improve quality and productivity, and, in the process, it can add to the satisfaction of the team members.

Teams are controversial. This is really an understatement. Teams force managers to give up some of their managerial prerogatives such as planning and control. If teams are working properly, they do a lot in the management of themselves. Many managers believe that "since they are appointed to be in charge, they really are in charge!" However, some teams operate as if there were no boss required; indeed the former boss now has become a facilitator—one who makes the work of the individual and the team easier. Teams do their own planning, set schedules, agree on cost and profit objectives, and may know *and* set salaries. Teams participate in hiring new members, set performance standards for the individuals and the team, and even recommend the release or reassignment of people who do not come up to the team's performance standards. Teams order supplies and recommend or even approve obtaining new equipment. Teams work with vendors and customers, develop and implement total quality management programs, and in some cases even participate in the development of long-range strategies for the enterprise. So—who needs a "boss?" What will happen to the "traditional" boss? Perhaps that type of boss will become as obsolete as mechanical cash registers and manual typewriters.

Teams develop the ability to ask the key questions about what is going on in the organization, why something needs improvement, what does not make sense, and they see things that are often invisible to the traditional boss whose "supervision" is no longer "super" because of the complexity of modern organizations and technology. Teams have shown their ability to get around the inevitable bureaucratic obstacles that exist in organizations and are able to break through the walls of functional parochialism characteristic of most organizations.

The use of teams has changed much in organizations. At 3M cross-functional teams have facilitated the creativity and innovative process at the company. At Aetna

Life & Casualty Company teams of workers have contributed to the reduction of the ratio of managers to workers from 1:7 to 1:30. That is a real saving.

But it is not all roses in the use of teams. Empowering the team carries with it the responsibility of the organizational managers to keep the team on track in doing what it is supposed to do, and not get off on activities that are not related to the team's objectives. How should members of the team be rewarded? What degree of authority should the team have? How should the conflict that comes up in the team's work be resolved? What promotion opportunities can be provided to the team members? These are some of the questions that have to be answered when a manager elects to appoint teams to help in doing the work of the organization.

When should teams be used? The most important reason is when there is a high level of interdependence among several people in doing their work, as is becoming evident in product-process development strategies. Simple production line operations may not be suitable for teamwork. But if there is complexity in the jobs, and the people have to work together in an informal team mode, then the use of formal teams could make sense. Most teams represent to one degree or another a form of cross functionalism. Some examples of the use of teams include the following:

- At Rubbermaid, a cross-functional team composed of engineers, marketers, and designers work together to develop a new product.

- At 3M, 10 self-managed teams composed of 8 to 10 people from different functions were put together to have responsibility for the day-to-day development of new products.

- The use of self-managed teams at General Mills resulted in increases in productivity by as much as 40 percent compared to the company's traditional factories. One reason for the increased productivity was the need for fewer middle managers.

- Texas Instruments' chip factories in Texas use a hierarchy of teams that, like a shadow government, work within the existing hierarchy. In the hierarchy of teams at TI, a steering team consisting of the plant manager and his staff of manufacturing, finance, engineering, and human resources set strategy and participate in approving large projects. Beneath the steering team three other types of teams exist: corrective-action teams, quality-improvement teams, and effectiveness teams. TI's quality-improvement teams work on long-term projects, such as streamlining the manufacturing process.

The use of hierarchy of teams at TI suggests that without care the teams can become just another hierarchy. Care must be taken to keep this from happening and to be flexible in the use of teams. Teams can be hung up. One of the most common problems is the failure of the team members to have empathy for the feelings and needs of their coworkers. Because teams are self-managed, there are fewer middle managers and consequently fewer promotion opportunities. Part of the care to be given to a team is to keep in touch with it and its work to see that the team's purposes are attained and that its deliberations are not becoming bureaucratic or self-serving. Work on a team can be fun and rewarding. Usually the team members are able to learn more about their work and about the team because of the inherent feedback that is found on teams as the members work together and come and go in their work through the organization.

Teams are not a quick fix. It takes planning for their use, time to get them suitably organized, and patience in the facilitation of an environment where they can work together effectively. The change to a team-driven organization is a long-term process. It requires truly strategic patience on the part of the managers and professionals.[2]

The movement to more team-based organizational designs is caused in part by the growing intensity of global competition.

THE GLOBAL CONTEXT

Simultaneous engineering has been driven in part by the need to develop strategic alliances with global partners so that complex products developed and produced by complex organizational processes involving suppliers and customers can be done increasingly on a global basis. Global competition has become magnified by the emergence of new players from across the globe—design-engineering and manufacturing cycles have had to become more compressed. Cost limitations, limited windows of opportunity in the global marketplace, and the need to develop global partnerships have simply negated the traditional way of doing business. Technology sharing has become a prerequisite for succeeding in global marketplaces.

The competitive difference in the commercialization of products in leading companies in the United States, Japan, and Europe was studied by McKinsey & Company. It was found that leading companies in these countries were able to commercialize two to three times the number of new products and processes as did their competitors of comparable size. It was also found that these companies were able to bring their products to market in less than half the time.[3]

The Japanese, to whom a lot of credit goes for the development of the simultaneous engineering process, are enlarging that process through four key areas:

- Intelligent manufacturing systems
- Intelligent CAD
- The human-machine interface
- Maintenance involving an integration of reliability, leading to a maintainable life cycle for the product

In the context of simultaneous engineering, intelligent manufacturing is perhaps the most significant subject area along with special attention to analyzing the human role in engineering.[4]

In Japan, auto production strategies pioneered by inventor Taiichi Ohno include teamwork, communication, efficient use of resources, elimination of waste, and continual improvement. The Ohno system includes the factory, the research lab, the design center, as well as dealers and showrooms. Product development under the Ohno system has engineers, planners, and market researchers working closely together. Designs are finalized early in the simultaneous engineering process, which reduces costly changes and delays. Engineering time is cut in half, and the new models emerge quickly.[5]

An appreciation of the growing importance of simultaneous engineering can be gained by continuing a review of how contemporary companies are engaged in the process.

COMPANY PRODUCT-PROCESS DESIGN STRATEGIES

At the Boeing Commercial Airplanes New Airplane Division, the chief project engineer noted[6]:

> For our product definition process concurrent engineering has two distinct elements. The first involves simultaneous design of all aspects of the airplane. Previously the air-

frame (structure) was designed early to support production requirements and then the airplane systems followed on a different schedule. The result was that significant changes in the airframe were required to accommodate the systems. This downstream process is very disruptive and expensive. The second element of our concurrent engineering involves simultaneous design and manufacturing engineering. Manufacturing planning and tool design will be accomplished with the engineering design. This will allow the manufacturing department to provide valuable producibility input to the design engineer. The expected benefit is a significant reduction in the recurring costs required to fabricate parts and assemble the airplane.

State-of-the-art technology is a fundamental source of competitive advantage. Speed is becoming more of a competitive factor in the marketplace. Concurrent computing through the use of workstations to automate professional work groups gets the work done sooner. For example, a car manufacturer uses concurrent computing to allow a group of engineers in one part of the country to design a car door or transmission that fits a car that is being assembled in another part of the country. To meet and exceed competition technology, information has to be leveraged by professional "knowledge workers."

IBM set out to develop a new-generation computer in $2\frac{1}{2}$ years instead of the 5 years it had taken to develop the previous model. A product-process design team was formed that included customers—an unusual move by a company that is as security conscious as IBM. Customers told the team that they wanted a simple new machine as well as a product that would be easy to install, use, and maintain. The Rochester unit responded with an electronic customer support system, which offers constant remote testing and on-line hardware and software assistance. Marketing representatives call customers three months after the product is shipped to find out whether they are satisfied. This strategy reflected IBM's "market-driven quality" aimed at pleasing the customer. The commitment of senior IBM people to total quality management is keynoted through the philosophy of John Akers, CEO, who announced that he wants IBM not merely to satisfy its customers but also to delight them. IBM admits that the measurement of delight is going to be challenging.[7]

It used to take Kodak an average of 3 years to move a product development from conception to commercialization. Today it takes 18 months to 2 years. The company's disposable camera was commercialized in less than 14 months. Kodak has emphasized the need to listen more closely to its customers. Manufacturing employees visit customers—such as the professional users of film in Hollywood—to find out the customers' needs. In the past the marketing department sold what manufacturing gave them. Listening more closely to the customers is a big cultural change for Kodak. Previously the company developed technology and dumped it on the market. Today Kodak is a global player.[8]

Success in the development of the Ford Taurus automobile using simultaneous engineering helped to cause Ford executives to rethink the way that the company develops and builds cars. A single team of 50 line executives representing all of the functional groups is studying how to make product development speedier and more efficient by breaking down the barriers between Ford's huge functional organizations: design, product engineering, factory engineering, and sales and marketing.

Another, more radical, approach at Ford is the use of a "skunk works" team of 20 young designers, engineers, and product planners working under one roof to help cut costs by speeding communication and reducing confusion. Ford hopes to cut 20 percent from the cost of developing a new model—now a minimum of $1 billion. Additional strategies are being pursued to reduce white-collar costs through the elimination of staff positions, increased span of control, and the independence and empowerment of people in the lower echelons of the company.[9]

Lead time for new airliner projects takes 3 to 4 years with an up-front investment of $5 billion or so for the 375–400-seat long-range aircraft. Companies such as Boeing must develop new products for future demand if it is not to lose business to competitors in case a new surge of aircraft orders develops in the future. But how many of today's older jets will be retired? No one knows the answer to that question. Some of the older aircraft are kept in service, primarily to handle peak hours. Keeping costs under control during the product development process is critical. To do this, Boeing uses simultaneous engineering techniques. By using a three-dimensional electronic prototype with massive computing power, the design and manufacturing engineers are able to determine whether the design can be manufactured and at low cost. The savings from the reduction of engineering changes compared to the old development process can be enormous. For example, on an earlier Boeing aircraft, the 767, there were six times as many drawings as for the original design in order to correct various flaws. And note: the 767 was a design that came in at design weight, on time, and on budget.[10]

At Motorola the search for one "best factory" proved fruitless. Rather the company built a factory by creating islands of excellence found from around the world and added the strength of Motorola's existing excellence in manufacturing. The "best factory" was built by a multidisciplinary team made up of product designers, process developers, tooling designers, software specialists, and financial experts. The factory was designed to build a new pocket pager, dubbed the Bravo. Some of the key strategies used in the concurrent development of the pager and the building of the factory included the following:

- Initial development of a master schedule for the development of the Bravo line. In establishing this time plan, the traditional schedule of 3 to 5 years for such a development effort was shortened to 18 months. It was in effect the scheduling of an invention. Scheduling philosophies from the Apollo space program were utilized in planning for the new plant and the new product lines of pagers.

- Great changes in quality expectations were raised when the goal of 6σ quality improvement of the company's operations was planned to include customer service. This meant that a little over three product defects were allowed per million, an extremely difficult goal to achieve. Considerable help in the meeting of such stringent quality goals was achieved by doing the job right the first time through the simultaneous design of the product and the facility.

- New supplier-customer relationships through strategic alliances were forged through an Early Supplier Involvement program in which the suppliers had almost as much to say about the development effort as anyone else. In the Early Supplier Involvement program everything was out on the table at the beginning and the suppliers were expected to participate in the concurrent design of products, facility, and manufacturing processes.

- The factory became a true CIM production facility within the Boynton Beach, Florida, facility. The degree of automation was so high that almost no human intervention from the time that an order was entered until it was shipped became standard practice. This included the successful integration of the functions of manufacturing, asset management, support systems, inventory management, product tracking, bill of material updating, material reservation, and material ordering. The best the plant has done from the time a customer order was placed until the first robot began to work on it has been 16 min. This compares to weeks in the traditional approach.

Impressive as the results have been with the concurrent product-facility development process, there were problems to be resolved, as would be expected in such a leap forward of technology. The first major problem came out of the serious underestimation of the level of effort that was required to complete the software controls. A second key problem was a failure to understand how critical the relationship with component suppliers was going to be. Quality problems arose in the incoming components. When the suppliers were visited it was discovered that the problems started with poor quality in the original equipment manufacturer's (OEM) supplier blueprints, which were antiquated or contradictory. This problem was resolved by communicating in very simple terms what was wanted. A traditional adverse relationship with the vendors had to be changed through a lot of fence mending and meeting together in a true OEM-supplier partnership. By having a concurrent design team from both firms working together, the quality levels attained were extraordinary. The list of suppliers was winnowed down to 22 to supply the 134 components that were required. Experience on this project to concurrently develop the product and the factory was sufficiently favorable to bring about the need for passing on the lessons learned to other development projects in Motorola.[11]

At Chrysler some fundamental ways of doing business are changing. Product development has been totally reorganized so that most of the company's 6000 engineers are assigned to one of four vehicle development teams, rather than to a specialty such as transmissions.

Chrysler has also sought the assistance of its suppliers in helping to support a massive cost-reduction program. Outside vendors have contributed more than 2000 suggestions for saving costs. About one-quarter of these suggestions have been accepted to date, the rest is being studied. Parts suppliers now package shipments in smaller containers so that Chrysler does not have to repack these parts before sending them on to dealers. Automated carousels have been installed at parts depots, reducing employee hours by as much as 87 percent.

Technological innovations which have been commercialized reshape the competitive factors in the marketplace. Boeing did this with its development project "Dash 80," which was a $16 million bid to capture what was believed to be a huge potential market for commercial and military jet aircraft. The resulting 707 aircraft of Boeing and its descendants opened the way for commercial jet aviation in the world. Boeing may be doing it again. Its 77 7 commercial jetliner may reshape jetliner competition well into the next century. Boeing works closely in designing the plane with three domestic and five foreign customers. The 777 has a whole new cabin width and 10 abreast seating. It has a range of 4800 miles, easily movable galleys, and lavatories for easy configuring. This new aircraft also offers folding wing tips to allow for parking at smaller gates.[12]

General Motors Corporation recently announced an organizational restructuring at its huge Chevrolet-Pontiac-Canada group into a new product-centered structure. This new design resembles Japanese car development systems and parallels the organization that was put together at GM's other big North American car-making group, Buick-Oldsmobile-Cadillac. The old Chevrolet-Pontiac-Canada organizational structure was a matrix system, where personnel involved in designing new cars were divided along functional lines—car body engineers under one vice president, manufacturing and purchasing employees under another. Managers in charge of developing new cars borrowed people from central staff operations and had little ultimate control over their pay, promotions, or transfers. Project managers had lots of responsibility, some resources, and even less authority. Now Chevrolet-Pontiac-Canada will assign engineering, finance, and purchasing people to one of six vehicle-design teams. Each team will have responsibility to develop a specific line of cars and will be managed by a platform, or chassis, program manager. All personnel on the team will report to the platform manager, and the platform manager will be accountable for the success of that line of products. Faster new vehicle development should be possible under this new organizational design: the engineers who will figure out how to build a car work alongside the people designing it from the first day.[13]

At Ford the use of a simultaneous engineering team to reduce the development time on the new Mustang model to 3 years, comparable with the way Japan does it, is being undertaken. To facilitate better communication among the team, the Mustang team members have left their home departments and moved into the same building. By having the team colocated, the people can work next to each other and talk things over during the development process. And is this not the way that Henry Ford developed his early cars? Ford executives have set some extraordinary performance goals—reduce North American salary costs by 20 percent and shorten vehicle development times by a year. To get a picture of the process, Ford product planners have plotted a 30-ft-long chart showing the product-development process. Their goal in using the chart: find where the process can be compressed and money saved. Chrysler is pushing the use of teams to improve efficiency and effectiveness. The use of product-development teams has cut off 2 years from the previous design schedules. Under Chrysler's old 6000-person bureaucracy constant fights over priorities, resource utilization, and decision authority made it difficult to do even one new product-development project a year. Now, by using teams and pushing authority downhill, better decisions are being made, and the decisions are being made faster.[14] Simultaneous engineering did not just happen—it grew out of traditional product development.

TRADITIONAL PRODUCT DEVELOPMENT

Traditional product development used a sequential process, which meant that each function performed its work in relative isolation. When the work was completed in a functional department, it was "thrown over the wall" to the next function. This sequential process passed responsibility for quality and customer satisfaction to the manufacturing plants, which were too far downstream to do more than react to problems that had been "designed into" the product. The result was long lead times, higher costs, lower-quality products, and increased customer dissatisfaction.

Under traditional vertical product and process development, friction is easily generated among the different people working in their functional disciplines. Each function has its own schedule, work load, budget, and way of operating. Problems that develop in the design are easily pushed up the line for resolution—leading to delay in decisions and loss of effort while the product and process developers are awaiting resolution of the problem or conflict. If a design engineer introduces a concept that poses problems for manufacturing, evaluation of that concept by manufacturing engineers is frequently delayed until the product has been thrown over the walls to manufacturing.

The people who design and develop the products and the processes, and the people who interact with the customer are in separate organizations. Such is the ingrained shortcoming of the traditional hierarchical form of organization. Technology in both product and process must be understood from the customers' and the suppliers' perspectives.

The problem with traditional product-process design is that the technologists who develop the product systems and the people who have the responsibility to interact with the customer come from two different functional entities in the company. Simultaneous engineering provides the opportunity to bring about an effective marriage of the technology systems people and the marketing and sales people who have a deep understanding of the customer. Improved product-process design strategies are an outgrowth of major changes in competitiveness.

COMPETITIVE CHANGES: NEW STAGES—NEW PLAYERS

Some of the forthcoming changes that should heighten the interest in simultaneous management are discussed in the following paragraphs.

There is a shift in the wealth of the world from natural resources to the ability to acquire and manage ideas and knowlege. In the past we have seen the dominance of entrepreneurial capitalism giving way to professional managerial capitalism. Today we see control going to those managers and professionals who control ideas and knowledge—intellectual capitalism if you wish.

Value through the command of distinctive property rights through patents and copyrights can vanish seemingly overnight. Consider how the fax has replaced the telex, or how the computer integrated into appropriate information systems has spelled out problems for many middle managers whose traditional responsibilities had been to transfer, monitor, and interpret information as the information passed up through the organizational hierarchy.

The challenge of going beyond quality and speed to the marketplace in providing greater customer satisfaction has to be addressed. Some companies have taken important steps in defining their products as having a natural tie with after-sales service. These companies think of every product that is bought or sold as consisting of a package of capabilities that provides service to the customers. Lexus is taking the lead in making sure that the relationship with the customer does not end at the showroom door. More and more companies are offering a 24-hour customer help line.

Customers and suppliers are being brought into the OEM's business through participation on simultaneous engineering teams, strategic planning task forces, and in general working more closely with the OEM to provide better products and services. Suppliers are becoming part of the win-win alliance with OEMs, where they get the security of a long-term relationship and customers get the opportunity to participate in the upstream activities of the suppliers. Purchasing agents are working more and more closely with a few select suppliers—reducing procurement costs and the cost of managing the suppliers.

The trend away from mass production toward specialized high-value work makes the old "command and control" management model obsolete. People expect more opportunity to participate in the affairs of the organizations to which they belong. Team management is becoming a way of life in more and more organizations. Management is becoming more "consensus and consent" than "command and control." Many managers who are imbued with the old traditional model of management are finding it difficult to adjust to the more participative environment found in more and more organizations today, and likely to increase in the future. The old style of management will not survive in today's organization.

Information has become a strategic resource. Knowing what information is required to manage an organization successfully, and how to manage that information, has become more critical in modern organizations. Computer technology, when combined with enlightened managers using relevant information, can make the manager's job easier and facilitate the restructuring and downsizing of the enterprise. Telecommunications through satellites, cable, and the more traditional means of communicating information has become a vital resource that must be managed strategically in order to remain competitive in the global environment.

The most effective way to transfer technology is through "people links"—the face-to-face conversation among people who have some vested interest in the technology. Attendance at professional meetings, trade shows, and visits to organizations using the needed elements of technology are important ways to get the most out of these people links. The focused opportunity for the use of people links on a product design team can lead to a synergy in sharing ideas, concerns, expert opinion, problems, opportunities, and strategies in the development work under way.[15] Simultaneous engineering is an idea whose time has come. It requires careful implementation.

THE IMPLEMENTATION PROCESS

The first step in implementing concurrent engineering is an in-depth look at how products and processes are currently being developed in the company. A project team can do such a review with support and direction from senior management. A review can reveal sobering information. One company that launched a "presimultaneous engineering audit" found that its production cycle was only 90 days—almost 180 days were lost in the paperwork wending its way through the traditional approval process in the company. Another company found that the design manager and staff had put an incredible abundance of procedural rules on how design was to be carried out, formally reviewed, and finally approved by the senior design engineer. During all of this time almost 60 percent of the design decisions waited for approval, often sitting on the design manager's desk or on some of the design staff members' desks waiting until people responsible for reviewing and approving the design could get to the paperwork. In another company the design documentation was often in a "hold" position until additional information was forthcoming.

Computer and information technology has facilitated the simultaneous engineering process. But computers are not absolutely necessary to bring about the integration of product-process actions. In small companies where the traditional functional walls do not exist, and where the business is conducted on an informal basis, integration can come about. But in larger companies the integration of product-process technologies requires better management, getting the people together, and managing and encouraging the members of the product-process design team to sit and work together. Through greater participation, a better integration is probable.

Simultaneous engineering is possible in large organizations because of the power of information and computer technology. Closer relationships are forged among the product-process team members which blur the traditional distinction between organizational disciplines. As the team members work together in a synergistic manner, strategies and solutions are envisioned and made possible, which in the traditional organizational design were not possible as the work of engineering design was handed over to manufacturing design and on to the marketing people to sell the product once it had been produced. The team members are able to see through and beyond the blurred boundaries of organizational disciplines and see the product and the processes as a whole related to a larger whole. As customers and suppliers are brought into the team, the parochial divisions of organizational disciplines no longer inhibit thinking across boundaries of knowledge. What matters the most for successful simultaneous engineering is the ability of the team to think imaginatively together about the integration of product-process technologies.

The team must be empowered and self-managed to make decisions free of the requirement to refer decisions up through the line organization for resolution. Such referral can waste a lot of time and can lead to poor decisions because the manager who is the decision maker is too far removed from the information centered around the decision. Then, too, there are limitations on the manager's ability to make a credible decision because of the lack of specialized knowledge within the decision context. Usually under these circumstances the manager will talk with the principals involved in the decision and balance their recommendations in rendering a final judgment. All of these deliberations and meetings take time. It is much better to empower the team members to make the decision since they are more likely to be the best ones qualified to do so. Of course, if the team members are in conflict over what the decision should be, the manager can enter the picture and resolve the impasse.

Feedback is an important element in monitoring and evaluating the efficacy of organizations. When a product-process design team is used, much of the feedback for the evaluation of the design activities comes during both ongoing and periodic design reviews carried out by the team. If a formal review process is used, having all members of the product design team sign off on the design, every step of the process can be valuable. Such a design may seem to be a free-for-all process, but if properly managed, the benefit can be a better design and a better feeling on the part of the individual members of the product design team who see that their judgment counts in approving the design. When vendors and customers are on the product design team, a much more direct path to both customer satisfaction and vendor effectiveness is available. To work successfully, concurrent engineering requires that all members of the team fully appreciate the impact that they have on other members of the team and on the outcome of the product-process design actions. Indeed, concurrent engineering will be effective only when the members of the design team are able to share information and knowledge.[16]

The composition of teams often breaks the tradition of organizational design in the firm. At Cincinnati Milacron a team was formed to design and develop a plastics injection-molding machine. For inspiration, all 11 members of the team read Sun Tzu, *The Art of War*. To make the machine competitive with the best on the market, they would have to reduce costs 40 percent, increase functionality (speed and operating time, for example) 40 percent, and reduce the development time to half. One of the first things the team did was to analyze competitors' teams and talk with customers. Team members had their offices moved together and reported directly to the vice president of plastics machinery, thereby eliminating several layers of middle management. The team used single sourcing. Parts were standardized. "Tear down the walls" became a rallying cry for the team.[17]

Experience in implementing simultaneous engineering at the Cadillac Motor Division of General Motors Corporation indicates that there are generally five roles that organizational leaders must demonstrate in order to achieve the cultural change that allows design and manufacturing engineering to work effectively together. These five roles are at the focus of the change process from the traditional way of doing product-process engineering. The roles include*:

1. Gaining the knowledge required to understand the simultaneous engineering means and how other companies have gained and applied this knowledge.

2. After having gained the knowledge, developing and articulating the vision held for simultaneous engineering so that the members of the organization can understand and embrace the vision of how things will be done in the future.

---

*Paraphrased from Roberts.[18]

3. Organization of teams and development of a culture that supports and rewards the teamwork needed to make simultaneous engineering work.

4. Enabling and empowering the employees to work in an environment in which teamwork and innovation become the accepted way of doing things. In this process training plays an important role, as does the development of a participative management style and a culture where people are kept informed about customers, competitors, organizational strategies, and how the individual and the teams can contribute to the planning and execution of simultaneous engineering activities.

5. Monitoring the simultaneous engineering process to see whether the planned successes are being attained.

The implementation of simultaneous engineering requires a change in attitude.

BEHAVIORAL CHANGE

In working with change in organizations the leader must set an example for the change and also work with the people undergoing the personal and organizational processes involved in the change. Change comes about slowly over a long period of time and requires careful planning, consideration of what could go wrong in the process, and patience in dealing with the people and their frustrations that inevitably come into play as people try to adjust to the new environment. Experience has shown that trying to move the change faster than the people are able to cope with will cause problems. People will always be concerned with how the change will affect their personal and professional lives, and where they will fit into the organization when the change is under way and when the change will be completed.

Sometimes the change can be facilitated by bringing in an outside party to work with the people and the organization undergoing the change. Consultants can help in acting as go-betweens and as a more objective group to facilitate the change. Inside "change agents" can be viewed with suspicion as having a vested and selfish interest in the change to serve their own interests rather than the interests of the organization or the people undergoing the change. An important principle in working with people to change: do not compromise on the need for the change or the performance standards for the people undergoing the change. Involve all people at all levels in the change, getting their involvement as much as possible in accepting the reason for the change, as well as the processes that must be carried out to bring the change to fruition.

The organizational design of the simultaneous engineering team shows little resemblance to the traditional pyramid organizational chart. Authority and responsibility are not partitioned off into neat boundaries isolated from other organizational units typical of the "command and control" organization. The organization chart—if one is to be drawn—consists of interlocking circles of expertise with arrows and feedback loops showing the pathways of the work and decision points. For example, the organization chart for the Taurus-Sable product development at Ford was a circle with the core product design team in the middle of working groups which branched out in all directions. The configuration of the product design organizational chart helps to communicate the message of settling issues at the lowest possible working level.[19]

An important part of change is the role expected of the company's customers and suppliers.

CUSTOMERS AND SUPPLIERS

Customers

Listening to customers is a prerequisite to providing quality products through quality processes. Attitudes in dealing with customers, promptness in responding to complaints, and follow-through on remedial action—indeed all of the attention given to the customer from the first notice of problems until the customer is satisfied—are important parts of a total quality management strategy. Attitudes of the people, such as a willingness to see the problem or complaint through the customer's eyes and then follow through quickly and effectively, are very important. The speed of customer service reflects on how the customer perceives the quality of the product or service that is provided. To see the quality issue from the customer's viewpoint in the simultaneous engineering process is to also visualize the long chain of events and processes that precede the delivery of the product or service to the customer. Total quality management theory views quality as a chain of events and processes from product or service conceptualization through the delivery and the providing of after-sales service to the customer. Customer dissatisfaction or poor quality in any of the events or processes in this chain has the potential to adversely affect the quality and image of the company. Some companies view the chain of events and processes linked through "internal customers" as getting everyone in the company working together through a quality philosophy and strategy while keeping the ultimate customer in mind. Some companies esteem the customer so much that an empty chair is provided for the customer at all organizational meetings.

Increasingly, companies are concentrating on customer satisfaction throughout the total ownership experience. They provide toll-free telephone service to discuss product characteristics and problems, offer advice on how to fix minor problems, and provide guidance on how to find a service center for more extensive repairs if required. Usually these toll-free consumer relations centers operate 7 days a week, 24 hours a day. Some companies have design and manufacturing people spend time at these consumer relations centers to hear first-hand some of the problems that the customers are calling about.

Design by itself will not ensure that a customer is satisfied with a product; efficient manufacturing will not ensure that a product is profitable. Cost, repeatability, quality, and total customer satisfaction are important as well. Total and strategic customer satisfaction requires an integrated effort of design, manufacturing, procurement, total quality management, marketing, after-sales service, as well as other support elements of the business in providing customers with the products and services needed to make them happy.

By having the customer representatives serve on the product-process design teams, all of the advantages of customer "closeness" can be realized. The customer can bring guidance on what is desired in the design characteristics of the product as well as providing insight into the product's manufacture and its after-sales service requirement. A large industrial equipment manufacturer has customers serve on all of its product-process development teams. Customer representatives are usually colocated with the team and participate actively on all of the team's deliberations. During the design review of the product the customers' opinions are greatly valued, as these customers can pass judgment on the operability and serviceability of the equipment.

Suppliers

By having equipment and parts suppliers on the product design team, assembly and component cost and technology issues can also influence the product-process design. When marketing, sales, facility engineers, and after-sales people are included on the team as well, the total life-cycle implications of the product and the processes can be considered simultaneously through the intellectual focus of the team members. By having suppliers on the product design teams, the OEM has an added resource to draw on in the development of current products and processes as well as in the development of future technological strategies for competing in the marketplace. In the past the suppliers would bid on the specification for a system or component developed by the OEM. When the supplier is a member of the team engaged in the development of specifications for the products to be sold to the OEM, an additional resource is available to the OEM. Then, too, who knows the most about the technology embodied in components and systems? Most certainly the suppliers whose business is the design, manufacture, and marketing of components and systems. Since the product design team has a mission shared by all of the members, design changes proposed from any source can be evaluated quickly. Design inefficiencies are not pushed up the organizational hierarchy—they are addressed immediately by the team with a much greater probability of resolution.

The advantages of simultaneous engineering are that products can be commercialized sooner, at a lower cost, and with higher quality. There are additional advantages as well.

THE ADVANTAGES

In the simultaneous organization, speed is the competitive factor. Speed kills the competition. When you get products to the market sooner, the market share increases because customers like to get their orders filled sooner. Inventories of finished goods shrink because they are not necessary to ensure quick delivery. Costs fall. Employees become more satisfied because they see more success in the company; they see a more responsive company which is more successful and is beating out the competition. Even quality improves, doing it just once forces you to do it right the first time.

Studies have shown that over 80 percent of the market share in a new product category goes to the first two companies that get their products to the market. Other studies have shown that a 20 percent cost overrun in the design stage of a product will result in 8 percent reduced profits over the life of the product. A 6-month overrun in time during the design stage today will result in a 34 percent loss over the life of the product.[20]

Is getting to the marketplace sooner than the competitors worth the effort? The answer is a resounding yes. It puts competitors on the defensive, reduces costs, commercializes products sooner, improves quality (because you have only one chance to do it right), and benefits from the combined expertise of focused teamwork. The placement of responsibility as far down in the organization as possible helps to develop the team members and frees the senior managers of the enterprise to spend more time on pondering future strategies for the enterprise. In summary, what are the advantages of concurrent management?

- Development of relevant information on a timely basis improves the chances that the people who have a need for information upon which to base decisions have that information available when needed.

- It emphasizes the use of teams, which become the medium through which information is processed and evaluated on a simultaneous basis by the experts.
- Team members do not have to stand around and wait for managers to make the decisions. Since they know the most about the work and the decisions that need to be made, they can make them without waiting for the managers.
- Authority in the organization depends more on the expertise, knowledge, and collective judgment of the team members than on a formal position that a manager holds.
- Productivity is improved.
- Quality is improved.
- Costs are reduced. (Elimination of overhead and waste creates a more efficient and effective operation.)
- The time span to commercialization is shortened.
- Team members provide a "self-audit."
- Team members educate each other.

The advantages of simultaneous engineering can be taken and translated into other overall organizational advantages. Bringing a project team to focus to carry out the simultaneous engineering task can be rewarding, even spectacular[21]:

> A company will save time in a business climate that heavily rewards quick response to market changes; will reduce costs by reducing surprises and rework, by getting things done right the first time, by standardizing product features and manufacturing processes, and by reducing the charges associated with tooling, fixturing procurement, and the storage of parts; and very likely will create a degree of cooperation and level of enthusiasm that has been missing from the manufacturing business for a long time.

The use of simultaneous engineering sets in motion a series of "systems effects." One of these systems effects is the general improvement in other managerial and technical processes in the enterprise. After an electronics manufacturer had successfully initiated simultaneous engineering, additional efficiencies were realized in other activities of the company. Although some of these improvements were not the direct result of the successful simultaneous engineering strategies, the participative culture fostered by simultaneous engineering facilitated a more innovative environment than had existed previously. As additional product-process development work was carried out in the company, there was a ground swell of new ideas and concepts put forth by the product design team members and other people who began to realize that the empowerment of the team members opened up many additional opportunities for the people to suggest new ways of doing things. Senior management perceived that additional change could be introduced into the organization from the example being set by the product design teams. Additional project teams were formed to provide focus points for additional innovation in the company. The result: over a period of 3 years the following innovative strategies were implemented:

- Manufacturing processes involving short-cycle production using just-in-time pull techniques to reduce work-in-process inventory and simplification of materials flow
- Reduction of manufacturing process steps and reduction of parts assembly

- Standardization and reduction of required parts and subassemblies
- Initiation of a rigorous supplier qualification to include the review of supplier process control and quality plans
- Establishment of several initial "demonstration" self-managed production teams (The success of these initial teams led to the development of plans to phase in such teams throughout all the manufacturing plants of the company.)

Teamwork done through the process of simultaneous engineering has proven valuable in improving quality. When a product-process design team works together in a synergistic mode of operation, many quality issues come up during the planning stages that can be resolved before the product goes into production. Also, experience on such teams has resulted in the elimination of problems that would have required engineering changes after the product was in production.

CONCLUSION

In simultaneous engineering a team looks concurrently at the customer needs, product design, manufacturing processes, methods for producing the product, facilities, machines, tooling, where the product will be made, how it will be supported by suppliers, how after-sales service will be carried out, how the product will be marketed, and how the transfer of technology will be carried out to the next generation of products and processes. Reduction in parts count, simplicity in fixturing requirements, and the ease of assembly are additional important factors considered by the product-process design team.

Building the plans for the product-process design team, allocating resources to support the team's objectives, dealing with the inevitable demands on the resources available to the team, and the allocation of human labor to support the team's mission are all part of the team's work. Management of the complexity of the team requires planning and budgeting for the team's effort. It requires organizing the team's human resources *and* obtaining the quality and quantity of people needed to make the team work fruitfully. Finally, the organization has to set up the systems for monitoring, evaluating, and controlling the work that is being carried out by the team.

The payoff in the use of product-process design teams can be high: getting competitive products and services to the customers sooner, improved cash flow, better satisfied customers, greater profitability, and—most important—an enhanced ability to survive and grow in an increasingly competitive global marketplace. Indeed, today's company's chances of long-term survival without using such teams is in doubt.

REFERENCES

1. R. Reich, "Entrepreneurship Reconsidered: The Team as Hero," *Harvard Business Rev.,* pp. 77–83, May–June 1987.
2. B. Dumaine, "Who Needs a Boss?," *Fortune,* pp. 52–60, May 7, 1990.
3. M. Nevens, G. L. Summe, and B. Uttal, "Commercializing Technology: What the Best Companies Do," *Harvard Business Rev.,* pp. 154–163, May–June 1990.

4. W. Kuo and J. P. Hsu, "Update: Simultaneous Engineering Design in Japan," *Industrial Engineering,* pp. 23–26, Oct. 1990.

5. A. Taylor III, "The Lessons from Japan's Carmakers," *Fortune,* pp. 165–168, Oct. 22, 1990.

6. M. R. Johnson, Boeing Commercial Airplanes, personal communication, Aug. 27, 1990.

7. F. Rose, "Now Quality Means Service Too," *Fortune,* pp. 97–110, Apr. 22, 1991.

8. S. Lubov, "Aim, Focus and Shout," *Forbes,* pp. 67–70, Nov. 26, 1990.

9. A. Taylor III, "The Odd Eclipse of a Star CEO," *Fortune,* pp. 87–96, Feb. 11, 1991.

10. H. Banks, "Running Ahead But Running Scared," *Forbes,* pp. 38–40, May 13, 1991.

11. R. R. Schreiber, "The CIM Caper," *Manufacturing Engineering,* pp. 85–89, Apr. 1989.

12. *The Wall Street J.,* Oct. 29, 1990.

13. J. B. White, "GM to Revamp an Unprofitable Car Operation," *The Wall Street J.,* June 10, 1991.

14. J. B. White and B. A. Stertz, "Crisis Is Galvanizing Detroit's Big Three," *The Wall Street J.,* May 2, 1991.

15. S. Ashley, "The Battle to Build Better Products," *Mechanical Engineering,* pp. 34–38, Nov. 1990.

16. J. H. Mayer, "Concurrent Engineering Lures Designers over the Wall," *Computer Design/News Edition,* Spec. Rep., pp. 19–23, Jan. 14, 1991.

17. P. Nulty, "The Soul of an Old Machine," *Fortune,* pp. 67–72, May 21, 1990.

18. R. S. Roberts, "Simultaneous Engineering and DFM at Cadillac," *Concurrent Engineering,* General Motors Corp., pp. 29–36, 1990.

19. J. L. Bower and T. M. Hout, "Fast-Cycle Capability for Competitive Power," *Harvard Business Rev.,* pp. 110–118, Nov.–Dec. 1988.

20. D. Brazier and M. Leonard, "Concurrent Engineering: Participating in Better Designs," *Mechanical Engineering,* pp. 52–53, Jan. 1990.

21. J. M. Martin, "The Final Piece to the Puzzle," *Manufacturing Engineering,* p. 51, Sept. 1988.

CHAPTER 21
SHARED MANUFACTURING ASSISTANCE CENTER PROJECT: A NEW PRODUCT DEVELOPMENT APPROACH

Shriram R. Dharwadkar
South Carolina State University
Bopaya Bidanda and David I. Cleland
University of Pittsburgh

Shriram R. Dharwadkar is currently an Associate Professor in Industrial Engineering Technology at South Carolina State University. He has M.S. degrees in Petroleum Engineering and Industrial Engineering from the University of Alabama, and is currently a candidate for Ph.D. in Industrial Engineering at the University of Pittsburgh. He is a coauthor of the book *Shared Manufacturing: A Global Perspective* to be published by McGraw-Hill.[*] He is a member of IIE, PMI, and ORSA/TIMS.

Dr. Bopaya Bidanda is an Associate Professor of Industrial Engineering at the University of Pittsburgh. His teaching and research interests include Computer Integrated Manufacturing Systems, Project Engineering, Computer Graphics, and Machine Vision. He has published in a variety of international and national journals and has worked for a number of years in the aerospace industry in the area of Manufacturing and Production Planning & Control. He is a senior member of IIE and SME.

Dr. David I. Cleland is the author/editor of 25 books and many articles in the fields of project management, engineering management, and strategic planning. He is a Fellow of the Project Management Institute and was appointed to the Ernest E. Roth Professorship in the School of Engineering in recognition of outstanding productivity as a senior member of the Faculty. He is currently Professor of Engineering Management at the University of Pittsburgh.

*Certain portions of this chapter have been paraphrased from this 1993 publicaiton.

In the current global economy, small and midsized manufacturers are competing increasingly in the international market. Innovative strategies are needed to help the manufacturing community be competitive in the global market. Shared Manufacturing Assistance Centers (SMACs), based on the concept of shared manufacturing systems, are suggested as a possible approach. Analogies are drawn between the new product/service development process and the development of SMACs. The project life cycle approach is demonstrated with the help of a case study in a large metropolitan area in the northeastern part of the United States. This chapter identifies globally applicable work packages that need to be accomplished during the different stages of a project life cycle in the design and development of these innovative centers.

INTRODUCTION

Manufacturing organizations around the world are facing increased competition with respect to cost, quality, and time to market. Traditionally industrialized nations have lost much of their competitive edge in manufacturing productivity and quality to emerging manufacturing nations in recent years. The reasons behind this decline are diverse and are ongoing topics of discussion and research. While the United States has lost a significant portion of its market share, some European countries have not only been able to keep pace with the Pacific Rim countries, but also increase their market share. Manufacturers all over the world are now required more than ever before to strategically plan for global competition.

The development of strategies by *large* manufacturers to compete more effectively in the global marketplace is under way in many manufacturing enterprises. Resource availability for these enterprises is usually not a problem. Reallocation of existing resources through restructuring, cost cutting, and elimination of products and services that do not provide adequate contribution to profitability are challenges facing these large companies. *Small* and *midsized* manufacturers have a more serious challenge. Finding adequate resources to learn and adopt new technical and management strategies to compete effectively in the global marketplace is difficult and often impossible for small enterprises.

This chapter details some approaches by European countries to maintain their market share. It also develops a generic strategy for the establishment of Shared Manufacturing Assistance Centers (SMACs) based on the concept of shared manufacturing. Results of an exploratory study in a large metropolitan area in the northeastern part of the United States show that such centers are vital to manufacturing competitiveness, especially in the United States. It is also shown here that the process of planning and implementation for these centers is analogous to a new product development (NPD) process.

Shared Manufacturing Systems

Shared manufacturing is defined as *sharing of modern manufacturing technologies, facilities, equipment, and management systems by different manufacturers with similar needs*. A review of the global manufacturing strategies leads one to the idea that the use of shared manufacturing centers in one form or another is a growing part of the improvement of global manufacturing productivity and quality. Manufacturing organizations in several Pacific Rim countries have brought about the use of shared manufacturing strategies through the government sponsorship of productivity and quality improvement programs. Large-scale productivity and quality awareness campaigns have been sponsored by governments toward the transfer of technology from other nations into the Pacific Rim competitors. To help the small companies, the Japanese government spends about $500 million a year supporting 185 technology extension centers across the country. On the other hand, 23 state governments in the United States spend less than $50 million a year supporting 27 technology extension centers.[1] There is no doubt that these efforts have improved the ability of the Japanese to compete in global markets. We believe that the United States needs to provide funding and support to set up a national network of such shared manufacturing centers.

The concept of SMACs is based on "flexible manufacturing networks," an idea introduced by Professors Piore and Sabel.[2] This concept was developed by these authors in Europe where small companies with complementary and overlapping capabilities operate in a cooperative alliance to compete better in the marketplace.

To take full advantage of new technologies and to keep pace with global competitors, companies must invest heavily in capital-intensive equipment that incorporates new and sometimes untested technologies—both at the process and at the system level. While this strategy is certainly effective over the long range, it is difficult to implement in small and midsized companies that do not have adequate financial resources. The situation is further exacerbated because the cost of capital investment in the United States is estimated as 2 to 3 times greater than in many other countries.[3]

The concept of shared manufacturing stems from the need for companies to evaluate new processes, new technologies, new designs, and prototypes before incorporating them into their own facilities. This can be implemented by a community-sponsored shared state-of-the-art manufacturing facility. The purpose of such a facility is to enable small companies to evaluate new process technologies, new product designs, and logistics/management systems before incorporating the advancements into their own facilities and company strategies. In addition, these facilities can also provide the buffer capacity needed by small manufacturers. The creation of SMACs capable of performing as a common community-shared manufacturing facility can provide the resources for companies that cannot afford the financial and technological risk of developing their own competitive product and process strategies and capabilities.[4]

Organizations such as Centre Technique des Industries Mecaniques (CETIM) in France and the Fraunhofer Institute in Germany have successfully used shared manufacturing philosophies to aid small and midsized firms. The following brief review of three successful centers—two European and one U.S. institute—highlight the need for these centers around the world.

Centre Technique des Industries Mecaniques. CETIM is a government-supported organization administered by a Board of Directors consisting of managers from member companies, technical personnel, and government experts. In addition, manufacturing companies in France contribute a small part of their profits toward the operating costs of CETIM. Currently, CETIM represents over 8000 French manufacturing companies involved in durable goods manufacturing. The majority of member companies are small and midsized, with 97.5 percent employing fewer than 500 people

CETIM was formed in 1966 to help French manufacturing industries remain competitive; it is now a mature and proven organization. With CETIM, companies have an answer to their needs of collective research or of specific personalized assistance, in the form of a partner in technology, thanks to a combination of qualified personnel and up-to-date equipment, for which no one individual company could assume the necessary financial responsibility.[5]

CETIM now has a staff of 630 personnel, including 370 engineers, with extensive manufacturing facilities and laboratories at three different locations. CETIM utilizes personnel with skills in diverse areas such as product design, materials, product technology, manufacturing, and production quality. Its mission also comprises of the following three complementary activities:

Clearinghouse for Information.   CETIM collects scientific and technical information from different sources within and outside France. This information is made available to member companies in the form of congresses, seminars, and exhibitions.

Research and Development.   These projects (which include product and process development) are undertaken either at the request of a specific company or are the result of the collective-interests needs of a set of member companies. The extensive manufacturing facilities of CETIM allow much of the research to be conducted on-site.

Technology Transfer to Industry.   The knowledge contained in, and acquired by, CETIM is transferred to French manufacturing industries through training sessions, publications, and audiovisual information, in-house reference libraries and data banks, and demonstrations.[5]

Fraunhofer Institute for Manufacturing Engineering and Automation.   The primary focus of the Fraunhofer Institute in Germany is in helping solve "organizational and technological problems in the manufacturing sector of industrial undertakings."[6] The Institute has extensive facilities (about 43,000 ft^2) and special laboratories for a diverse range of research projects in manufacturing systems, including industrial image processing, industrial robot development and utilization, coating technology, computer networks, interactive simulation graphics, and knowledge processing.

Though the Fraunhofer Institute is research-oriented, it still maintains strong industry ties. Research projects are mostly funded by industrial companies or by public research programs, although state and private sector joint financing is now on the increase. The Fraunhofer Institute undertakes industrial research and development projects, and all results of the research are published only with permission of the client. The Institute also provides assistance in the implementation of research results into the production work flow on the shop floor. Thus, the difficulties that arise in transfer of technology from research to production floor, particularly with small and midsized companies, can be jointly overcome. Small and midsized companies [turnover less than 500 million German marks (DM)] can utilize reserved funds from the federal and state governments to subsidize these research and development projects. These subsidies range from 40 to 60 percent of the total project funding.[6]

Shared Manufacturing Projects in the United States.   The vision of shared manufacturing projects is also being realized in the United States, although on a limited basis. According to Deborah L. Wince-Smith, former assistant secretary for technology policy at the U.S. Department of Commerce, a handful of such facilities are up and running and a number of others are in the planning stage. The National Institute of Flexible Manufacturing (NIFM), established in 1989, is one of the first such centers operating in the United States. A brief description of this center is provided below.

National Institute of Flexible Manufacturing (NIFM).   The NIFM facility is located in Meadville, Pennsylvania, within 100 miles of three large metropolitan areas. The specific objectives of the NIFM are: (1) creation of an education and training facility, (2) creation of a factory that will produce a family of parts and will provide for hands-on training in a not-for-profit mode, (3) time-sharing of equipment to local manufacturers, and (4) creation of a cadre of state-of-the-art technology services that would be for sale to regional private sector manufacturers.

The tooling and machining industry, with approximately 150 tool and die shops, is the largest industrial sector in the area. Providing services to manufacturing firms within a 40-mile radius of the facility is considered critical to the success of the center. Important industries in the area include transportation equipment (SIC 37),* fabricated metal products (SIC 34), nonelectrical machinery (SIC 35), and rubber and miscellaneous plastic products (SIC 30). There is a strong metalworking market with 29.4 percent of the total work force represented in the fabricated metals and nonelectrical machinery industries. The vast majority of the firms are small, employing less than 500 employees.

A survey was recently conducted to gain a better understanding of the market for the center. Questionnaires were mailed to 847 manufacturing firms in 14 counties in northwestern Pennsylvania. The objective of the survey was to assess the current use of "modern techniques and technologies" and to measure the interest of the area firms in the center. The response rate for the survey was 28.8 percent. The majority of the companies responding were small; over half are from metal-related industries with $500,000 in gross sales and fewer than 25 employees.

Only 20 percent of the firms responding to the survey reported use of computer numerically controlled machines in their operations. The nature of services of interest to the respondents is summarized below.

	Percent of respondents expressing interest	
Type of assistance	All organizations	Organizations with high plant utilization
Management training	29.3	44.3
Operator training	31.8	38.2
Time sharing	24.3	36.2
Consulting assistance	23.4	44.6
Personal visit	25.9	37.1

Counterintuitively, a higher interest was reported from organizations operating over 75 percent of capacity. This is probably because the busier organizations recognize a greater need for outside assistance to further improve their performance.

NIFM offers a variety of manufacturing systems services to area manufacturers, such as leasing time on advanced machines, machine operator training, assisting customers in seeking new business, product and process development, management training, and demonstration of advanced technology.

Although NIFM initiated its activities in 1988, the first machine was installed in early 1989, and the official dedication of the facility took place in mid-1990. Eight companies leased machine time at NIFM in the first year of its operation. Since then there has been a steady increase in the number of firms seeking manufacturing assistance from the center. To date, more than 30 companies have leased machine time, and more than 100 people have received education and training. The center has been operating at a profit for the last 2 years. Some of the lessons learned at NIFM and other centers currently in operation point to the need for a systematic approach for the planning and design of SMACs.

---

*Standard Industrial Classification (SIC) is an industrial classification system developed by industry experts under the guidance of the Office of Management and Budget of the U.S. government. Under this system there are 20 major groups (two-digit SIC), 143 industry groups (three-digit SIC), and approximately 450 industries (four-digit SIC) under durable goods manufacturing.

Management of SMAC Projects.   The interest in Shared Manufacturing Assistance Centers (SMACs) is on the increase around the world, and it is estimated that there could be more than 200 such centers operating worldwide by the end of this decade.[7] Unfortunately, there is only limited information in the literature about the design and implementation of SMACs. In view of the forecasted growth of such centers, there is a considerable interest in the planning and implementation of shared manufacturing assistance projects. In the next section, we use the life cycle phases in the new product development process as suggested by Cooper to show how such SMACs can be implemented within such a framework.[8]

A NEW PRODUCT PROJECT

The success of SMAC projects depends on how well critical project parameters such as cost, quality, and technical performance are attained. We see SMAC projects as going through several different life cycle phases characterized by distinct work packages and formal decision points to decide whether to continue with the project or terminate it, depending on the performance in previous phases. Different authors identify three to seven project life cycle phases with various terminologies to describe each phase.[8-10] There is a general consensus however, on the utility of the project life cycle approach and recognition of the fact that each phase is characterized by distinct work packages requiring varying levels of effort. Adams and Brandt describe four project phases: conceptual, planning, execution, and termination phase.[8] Cooper details the activities necessary for new product development and divides them into seven distinct stages.[9] These are (1) recognition of need, (2) preliminary assessment, (3) concept development, (4) development, (5) testing, (6) trial, and (7) launch. The project is terminated and responsibilities transferred to line authorities once the product is launched. Each stage in Cooper's model consists of several activities and a decision system in place to evaluate further development decisions at each stage. Cooper's model was found to be appropriate for implementation of SMAC because of its recognition of the life cycle phases and emphasis on performance evaluation at the end of each phase.

CASE STUDY OF A SMAC PROJECT: A LIFE CYCLE APPROACH

A case study that uses a modified version of Cooper's model for the development of a SMAC is depicted in Fig. 21.1. The case study illustrates the various stages and the activities conducted at each stage in the development of the SMAC in a large metropolitan area in the northeastern part of the United States.

Recognition of Need

The principle motivation for the project came from two primary sources: the needs of the local manufacturing establishments and the education curriculum at a large area university. In many interactions with local small and midsized manufacturing estab-

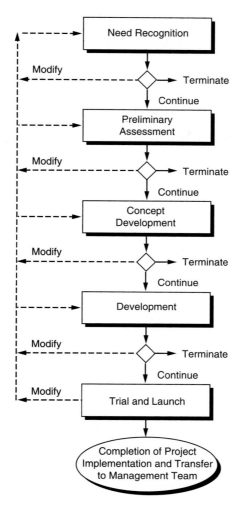

FIGURE 21.1   Approach for the development of SMAC.

lishments, it became clear to the authors that many needed a greater degree of technical and managerial assistance than was currently available to them because of their limited resources.

Concurrently, it became clear that the existing curriculum at the university, while considered progressive, was not adequately supported by manufacturing facilities in which to conduct classes and do research in an automated factory environment.

Funding to conduct the technical and marketing feasibility study to determine the need for a SMAC and the willingness of local companies to support such a facility came from the local government Industrial Development Authority. In addition, modest funding was also obtained from professional/trade organizations, dedicated to the area's economic development. The study area consisted of the seven-county region contiguous to the proposed location of the SMAC.

Preliminary Assessment

The major thrust of this stage was to collect published information to assess the existing manufacturing profile of the study area. This enabled the project team to identify the target market segment and develop an appropriate product mix for the SMAC.

A nonintrusive evaluation of the area demographics was conducted by studying literature published by the federal, state, local, and other economic development agencies. Publications of the area Chamber of Commerce served as a primary source of information. Another important source of demographic information included a statewide industrial directory published by a private publisher. The available demographic information was used to assess the employment and manufacturing profile within the study area.

The seven-county region and its communities involved in the study had experienced economic decline, population loss, employment decline, and high unemployment rates since 1980. The basic cause of these dislocations was the decline of steel and heavy manufacturing industries, which have been traditionally the most significant industrial sectors of the region.

In the period since 1980, total manufacturing employment in all seven counties declined sharply, with countywide declines ranging from 26 to 74 percent. A total of 125,875 manufacturing jobs were lost since 1980, representing a percentage loss of 45.6 percent. The largest of these declines occurred in the primary metals industries, followed by the fabricated metals and industrial machinery sectors. Unemployment rates increased in all counties in the period between 1980 and 1985, the primary period of structural decline. Highest unemployment rates were recorded at 14.6 percent.

From a detailed evaluation of the area demographics, the following observations were made:

1. A large fraction (63 percent) of the manufacturing sector was involved in the manufacturing of durable goods. Over half were in the areas of Primary Metals (SIC 33), Fabricated Metals (SIC 34), and Industrial Machinery (SIC 35). A profile of the manufacturing establishments in the study area is provided in Fig. 21.2.

2. The manufacturing employment in the study area was estimated at 181,235, with 73 percent of the employment coming from durable goods manufacturing. Over 60 percent of durable goods manufacturing was in the sectors represented by SIC codes 33, 34, and 35.

3. More than 80 percent of these manufacturing establishments could be considered small, employing less than 200 employees.

4. The manufacturing employment in the study area decreased significantly between 1980 and 1989.

5. Average size of the manufacturing establishments declined from 95 to 53 employees between 1985 and 1989. The most significant reduction was experienced by the Primary Metals sector (SIC 33), with an average decline of over 100 employees per firm.

6. Unemployment level in six of the seven counties in the study area was above the national average.

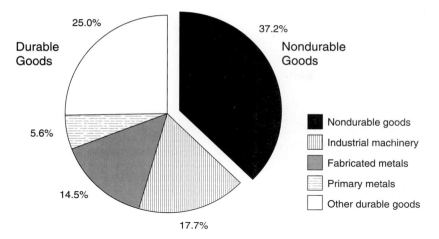

FIGURE 21.2   Profile of manufacturing establishments in the study area.[4,11]

The abundance of general economic data on the Greater Pittsburgh area provided a backdrop and the basis for the next stage of the project—concept development.

Concept Development—Market Research

The objective of this phase was to better understand the users of the project, market segments that were viable, and specific needs of those market segments. The concept development started with a market research effort to seek answers to these questions.

The first step in the market research effort began with the development of a database, which included information about small and midsized manufacturing companies within the study area. Each record in the database included the following fields: company name, address, telephone number, contact person, manufactured product, number of employees, and classification of industry sector in the Standard Industrial Classification code. The study[11] dealt only with durable goods manufacturing with SIC codes 31 to 39. This database was used to conduct a survey of area manufacturers.

The objectives of the survey were to identify the major forces driving the need for improved manufacturing and to determine what strategies should be used in the design and configuration of a SMAC for the study area. The vehicle for this survey was a three-page questionnaire which was mailed to 527 small and midsized firms in the study area. The survey was developed from the understanding gained from a study of the existing industrial demographics for the Pittsburgh community.

Survey of Area Manufacturers.   A survey that addressed 12 areas designed to assess characteristics of the manufacturers, their needs, and their interest in participating in the proposed center was conducted. The questionnaire elicited the following information:

Age and size of firm

Age of equipment

Type of equipment used

End product

Type of process flow

Type of process

Type of market

Product price

Technologies currently in use

Proposed technologies

Challenges faced by the firms

Interest level in proposed center

A total of 527 questionnaires were mailed out with one follow-up letter for nonrespondents. The overall rate of return was 60 percent. Data from the 314 returned questionnaires were entered into a database, and a manufacturing needs analysis was completed. Almost 80 percent of the respondents expressed an interest for the idea of a Manufacturing Assistance Center. A profile of the favorable responses is discussed below[11]:

1. Businesses at least 15 years old: 69 percent
2. Firms employing fewer than 250 employees: 86 percent; firms employing fewer than 100 employees: 69 percent
3. Firms using more than 15-year-old equipment: 41 percent; firms using at least 5-year-old equipment: 95 percent
4. Firms belonging to primary metals, fabricated metals, and industrial machinery segments (SIC 33–35): 64 percent
5. Firms using a job shop or batch flow: 61 percent; most reported a combination of processes such as assembly, fabrication, and machining

Figure 21.3 shows that technologies of interest included such capabilities as personal computers on the factory floor, computer-aided engineering, robots, flexible manufacturing cells/systems, and metalworking laser equipment. Respondents identified their major challenges as[4,11]:

- Quality and cost
- Adapting new technology
- Foreign competition
- Hiring and maintaining skilled work force
- Engineering and manufacturing cycle time
- Continuous improvement

Respondents in the survey felt that they needed a SMAC to provide the opportunity for them to learn and observe capabilities such as manufacturing and quality management technologies. Employee and managerial training was also a high-priority need. A complete list of the desired services at a SMAC are detailed in Fig. 21.4.

The strong favorable response to our market survey indicated a need and desire by the area manufacturers in the establishment of a SMAC. Our survey also provided suf-

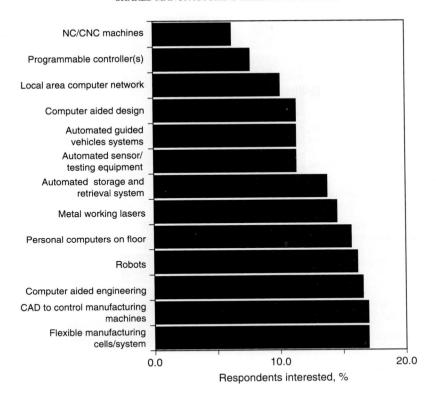

FIGURE 21.3   Technologies of interest to local manufacturers.[4,11]

ficient information on the nature of technologies and services of interest to local manufacturers. Based on this information, a preliminary configuration was developed for the SMAC in the study area. Ten of the favorable respondents were further interviewed to assess the risks involved in the development of the SMAC. Specific questions were asked to seek their likes, dislikes, preferences, reasons for possible discontent, improvement in the design, advice on major considerations, etc. From this valuable input, an improved SMAC concept emerged. This concept could be further refined and validated through focus group processes.

Development and Testing

The market research carried out in the study area provided the basis for a determination of the market need as well as the specific assistance required by the survey respondents. From the database developed from the market research, a preliminary configuration of the SMAC was developed. The set of equipment most suitable for the SMAC was identified as: computer numerical control (CNC) machine tools for traditional machining operations, wire electric discharge machining (EDM) machines, laser

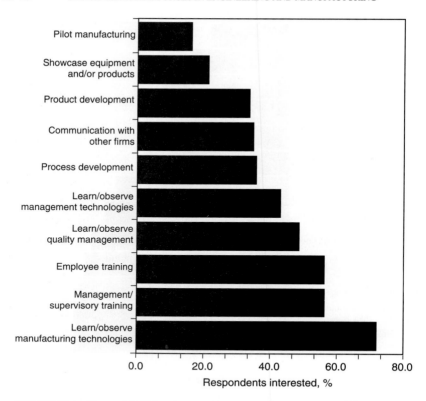

FIGURE 21.4   Nature of SMAC assistance of interest to local manufacturers.[4,11]

metalworking machine tools, automated material handling equipment, robots, metrology equipment, and computer-aided design (CAD) workstations with solid-modeling software.[4,11]

Human resource development was identified to be an integral part of the SMAC, since the manufacturers in the study area indicated a strong need for education and training. Minilecture modules taught by faculty and industry personnel would allow small and midsized manufacturers to learn and evaluate new technologies such as:

- Metalworking lasers
- Just-in-time manufacturing
- Materials resource planning
- Group technology
- Total quality management
- Computer-integrated manufacturing
- CAD modeling
- Simultaneous engineering

The equipment at the SMAC would complement these lectures and give manufacturers first-hand experience in using the equipment and technologies. To meet the needs of manufacturers, the SMAC should focus on the following capabilities[4,11]:

- Pilot manufacturing
- Product and process development
- Employee and supervisory training
- Study new manufacturing processes and technologies
- Study new management techniques
- Study new quality management technologies
- Showcase new equipment and/or products

An important part of the development and testing phase was the location of funding to design, built, test, and start up the SMAC. The local university in its Applied Research Center has existing facilities that could be retrofitted. Federal and state government agencies will be contacted for funding, as well as several large corporations and foundations. Machine-tool builders and the federal Department of Defense surplus machine-tool inventory will also be contacted for equipment to outfit the SMAC.

Additional work packages to be carried during this phase include:

- Design, engineering, construction, and start-up
- Final machine-tool and system specification
- Preparation of request for proposal (RFP) for design, engineering, and construction of SMAC
- Review of proposals and negotiation and award of contracts
- Procurement and installation of the production equipment
- Staffing and training of the plant managers and professionals
- Procurement of initial special tooling
- Procurement of initial inventory of materials
- Design, testing, and implementation of the management system for the plant
- Initial start-up of the plant
- Development of a strategy for the use of the SMAC to include:
  1. Identification of the initial users
  2. Negotiation and consummation of cooperative working agreements with the users
  3. Scheduling of the users
  4. Design and development of supporting technical data and administrative documentation to support user needs
  5. Pilot testing of the user strategy for the SMAC

Trial and Launch

Previous experiences have indicated that there are long-term and evolutionary projects requiring at least a year of trial for a shakedown of both the plant and the strategy that is used to operate the plant. System integration and troubleshooting of the plant need to be done with significant input from the users. In particular, the interfaces with local small and midsized manufacturers will be an important part of the trial and launch phase. Some of the key work packages involved in these phases of the project include:

- Pilot operation of the SMAC
- Final negotiation and execution of working agreements with local users
- Verification of the SMAC performance
- Evaluation of the efficacy of the SMAC by local manufacturers
- Integration of the technical and management systems

SMAC ORGANIZATIONAL DESIGN

Once the SMAC is established, its successful management depends on a suitable organizational design and a clear definition of authority-responsibility-accountability relationships. A *functional structure* is characterized by extensive policies and procedures, centralized authority, and high specialization. A *project structure* is on the other end of the continuum, with respect to policies and procedures, authority, and specialization. The functional structure emphasizes efficiency and production, while the project structure emphasizes flexibility and adaptability.

A *matrix structure* results from a balanced compromise between a functional structure and a project structure. It is a structure that results from superimposing a project structure on a functional structure. The successful management of SMACs requires efficient management of human and nonhuman resources without compromising flexibility. SMACs will handle a variety of projects involving a number of small organizations with diverse interests. The matrix type of organization structure shown in Fig. 21.5 is considered appropriate for the SMAC. Project teams with appropriate members from different functional groups will be formed to work on individual customer projects.

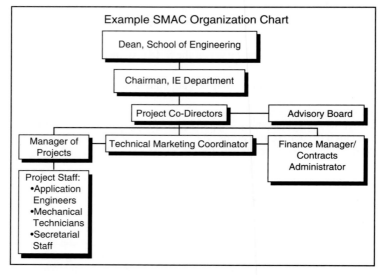

FIGURE 21.5   Example organization structure for a SMAC.

CONCLUSIONS

In our exploratory study described in this chapter, several strategic issues became apparent during the conceptualization of the project. These issues were

1. The acceptability of such a shared manufacturing facility in the community by potential users
2. The willingness of the small manufacturers to enter into a strategic alliance for the use of the plant and to share proportionately the costs of operating the facility
3. The availability of equipment from government and other sources
4. Adequate funding support from the State in the design, development, and start-up of the center
5. The effective use of the plant by university faculty
6. Maintaining the confidentiality of the work being done by the individual users of the SMAC so as to protect proprietary information

We have advanced an approach that can help small and medium-sized manufacturers regain global competitiveness by forming strategic alliances in the form of community-sponsored projects as Shared Manufacturing Assistance Centers. We have briefly described some of these centers currently operating in Europe and the United States. Further, we have shown that the planning and implementation of these centers follows a methodology similar to that of a new product development project.

REFERENCES

1. C. Farrell et al., "Industrial Policy," *Business Week,* April 6, 1992, pp. 70–75.
2. M. J. Piore and C. F. Sabel, *The Second Industrial Divide:Possibilities for Prosperity,* Basic Books, New York, 1984.
3. S. W. Sanderson, "The Vision of Shared Manufacturing," *Across the Board,* December 1987.
4. B. Bidanda, D. I. Cleland, and S. R. Dharwadkar, "Framework for the Design of Manufacturing Assistance Centers," working paper, Industrial Engineering Department, University of Pittsburgh, December 1991.
5. Editeur Conseil, *CETIM—Your Partner in Technology,* Paris.
6. Fraunhofer Institute for Manufacturing Engineering and Automation, "Survey of Working Areas and Persons Responsible," 1988.
7. E. Brandt, "A Vision of Shared Manufacturing" (interview with Deborah L. Wince-Smith), *Mechanical Engineering,* December 1990, pp. 52–55.
8. D. I. Cleland and W. King (eds.), *Project Management Handbook,* "Behavioral Implications of the Project Life Cycle," J. R. Adams and S. E. Brandt, VNR, New York, 1983.
9. R. G. Cooper, "A Process Model for Industrial New Product Development," *IEEE Transactions in Engineering Management,* vol. EM-30, no. 1, February 1983, pp. 2–11.
10. R. D. Archibald, *Managing High-Technology Programs and Projects,* Wiley, New York, 1976.
11. D. I. Cleland et al., *U-PARC Flexible Manufacturing Facility,* Final Report for EDA Grant 01-29—03001, Dept. of Industrial Engineering, University of Pittsburgh, September 1990.

CHAPTER 22
WE DON'T DO PROJECTS!

Francis M. Webster, Jr.

Editor in Chief,
Project Management Institute

Dr. Webster is editor in chief for the Project Management Institute (PMI), responsible for editorial content and publication of the *Project Management Journal* and *PMNETwork,* the technical journal and professional magazine of PMI. He recently retired from Western Carolina University as associate professor of management in the School of Business, where he specialized in project management courses and concepts.

INTRODUCTION

Perhaps the applicant for the housekeeping position can state "I don't do windows!" and get the job. Today's executive who says "We don't do projects!" is in trouble. Business literature is replete with articles promoting flatter organizations, the team approach to management, concurrent engineering, reduced time to market, and empowering employees. All of these are synonymous with the application of, or are best accomplished through the use of, modern project management.

Indeed, it has been suggested that most executives spend more time on work being performed in the project mode than in any other mode. In general, executives are concerned about the future products and capabilities of the organization, i.e., the corporate strategy. Corporate strategy is implemented through projects. Successful implementation of corporate strategy is therefore dependent on successful project management. Thus, the successful executive must not only comprehend but must be skillful in managing projects and project managers.

There have been two developments in management in the last two decades which have received intense publicity: quality management and production scheduling. Quality management has resulted in major changes in operating methods, reallocation of responsibilities from management to operators, and substantial new training requirements. New approaches to production scheduling have moved through material requirements planning (MRP) to the current emphasis on just-in-time (JIT) concepts. These have required major changes in attitudes toward inventory, efficiency, utilization, and application of engineered time standards, and created even more training requirements.

Although receiving less publicity, another "revolution" has been growing in importance: planning, scheduling, and control of project work. This new concept in management dates back to the mid-1950s and has been characterized by varying degrees of interest and understanding. Today, it is the basis for a major portion of economic activity: the management of change. It has developed from being a very specialized technique to a management philosophy supported by a range of techniques, concepts, and theory. Indeed, project management has progressed to the point of being recognized as a distinct career and profession. Project managers are served by two major professional societies—the Project Management Institute (PMI), primarily in North and South America, and the International Project Management Association (INTERNET), primarily in Europe. Among other things, these societies have staged meetings, conducted programs of education, certified Project Management Professionals, and accredited academic programs in project management.

Many people have adopted a stereotype of project management as being applicable to very large projects, primarily in construction and aerospace. While it has been very valuable in these industries, it has also been used extensively in the pharmaceutical, automotive, utility, communications, and many other industries and in product development, information systems, and many other functional areas. It has been used on athletic events including the Olympic Games and for movie production. More recently, network planning has found its way into the executive suite as a convenient way in which to express corporate strategic plans, including alternative methods to achieve corporate objectives.

Indeed, it is difficult to imagine an industry or functional area of business for which modern project management would not prove useful. Even in the most repetitive of industries and functions there is change that must be managed. Wherever there is change to be managed there is an opportunity to accomplish that change more efficiently and effectively through the use of modern project management.

And size does not negate this proposition! Size only impacts the extent to which the principles and techniques are used. Organizations are well-served to standardize many practices of project management but differentiate the intensity of application as a function of size, importance of the project to the organization, and the criticality of the resources to be employed.

International Projects

The benefits of modern project management are especially relevant on international projects. Some of those which have been reported in PMI's *PMNETwork* illustrate this very vividly.

Transmanche-Link (TML), a consortium of five U.K. and five French contractors, used modern project scheduling software and the principles of modern project management in constructing the English Channel Tunnel. "Both road and rail users will be able to take advantage of...time-saving advantages of the Fixed Link (the Chunnel) in the near future. They will enjoy this thanks to the organisational and management techniques which have allowed its creation, despite the vast size and scope of the project, in just seven years." (Ref. 1, p. 15.) In addition to the major contractors, components of the Chunnel system, as well as equipment used in its construction, involved companies and nations from around the world.

On May 11, 1990, a $70 million tin concentrator plant was dedicated in Portugal for Sociedade Mineira de Neves-Corvo. It was completed $4 million under budget and 3 months ahead of the 15-month schedule. This was "A remarkable achievement considering the tight schedule, complexity of the flowsheets, that equipment and material were imported from 13 countries on four continents and that only 5 percent of engineering had been completed before construction started." (Ref. 2, p. 9.)

In South Africa, Sasol, a South African petrochemical company, began production of polypropylene in February 1990, just 22 months after commissioning the

R541.7 million project. The completed facility was designed to produce 120,000 ton/annum of polypropylene, fed by a 150,000 ton/annum polymer-grade propylene plant. The best schedule record previously was 30 months for a facility of comparable capacities. This outstanding performance "resulted from breaking new ground in project execution methods which will definitely influence the handling of future major projects." (Ref. 3, pp. 9, 10.) The economic benefits of this project include employment for 250 people and "a massive R200 million/year saving in foreign currency for South Africa." In addition to realizing these savings as much as 8 months early, the early completion enabled Sasol to reach the market sooner and enhance their market share.

Somewhat less international in scope was the development of the Endicott Oil Field production facility, the first offshore Arctic oil field. While many of the facilities were produced on site using contractors experienced in that milieu, the major process unit was built in Louisiana, moved by a 100- by 400-foot barge down the Mississippi, through the Panama Canal and the Bering Strait, and off-loaded onto the manmade Main Production Island. This had to be accomplished within a 6-week window when the Arctic Ocean was not frozen. Not only was this accomplished but it was done at a substantial savings of both time and cost.[4]

While basically a United States effort, the Voyager 2 deep space project involved three major agencies—National Science Foundation, National Radio Astronomy Observatory, and National Aeronautics and Space Administration—and facilities in the United States, Spain, and Australia. It had a window of opportunity when the relevant planets would be in proper juxtaposition which *required* performance to schedule. The mission has been a major success to date and is expected to continue to return valuable data well into the 21st century as it speeds out of the sun's magnetic influence and into interstellar space.[5]

WHAT ARE PROJECTS?

General

There are many ways to think about projects. They can be described in many ways and they can be defined differently. Perhaps that is why there has been so much ambiguity about and lack of appreciation for project management.

One way projects can be described is that they are ubiquitous; they are everywhere; everybody does them. If they are so common, then why all the fuss? Very simply, better ways of managing projects have been and are being developed. Those organizations which take the lead in implementing these capabilities will consistently perform their projects better and be more competitive in general.

Another way to describe projects is that they are the change efforts of society. The pace of change, in whatever dimension, has been increasing at an ever faster rate. Effectively and efficiently managing change efforts is the only way organizations can survive in this modern world.

Yet another way to describe projects is by example. Most such descriptions start with things such as the Pyramids, the Great Wall of China, and other undertakings of ancient history. These were major construction projects and, indeed, construction is inherently a project-oriented industry. A modern construction project which rivals all others is the English Channel Tunnel, a $12 billion effort scheduled to be opened officially in 1993.

There are other project-oriented industries, not the least of which is the pharmaceutical industry. The search for new drugs has led to a remarkably high level of health and life expectancy. The aerospace industry is noted by its accomplishments, not only in space but also for the technological developments which have changed the way we live and work.

But not all projects are of such great magnitude. Remodeling or redecorating a house is certainly a project. A community fund-raising campaign is a project. A political campaign is a project. Developing a new product, developing the advertising program to promote that product, and training the sales and support staff to effectively move and service the product are all projects. Responding to an Environmental Protection Agency (EPA) complaint is a project, particularly if the complaint is substantial. Indeed, it is possible that most executives spend more of their time planning and monitoring "changes" in their organization, i.e., projects, than they do in maintaining the status quo.

All of these descriptions focus on a few key notions. Projects involve change, the creation of something new or different, and they have a beginning and an ending. Indeed, these are the characteristics of a project which are embodied in the definition of "project" as found in the *Project Management Body of Knowledge* (PMBOK), published by the Project Management Institute.

> **Project:** Any undertaking with a defined starting point and defined objectives by which completion is identified. In practice, most projects depend on finite or limited resources by which the objectives are to be accomplished.[6]

This definition, while useful to project managers, may not, for others, be sufficient to distinguish projects from other undertakings. Understanding alternative modes in which work is performed followed by a discussion of some of the characteristics of these modes may give a clearer perspective.

A Taxonomy of Work Efforts

It is helpful in understanding a concept to recognize its comparative others. For projects this requires a taxonomy of the modes in which work efforts are accomplished. There are five basic modes—craft, project, job shop, progressive line, and continuous flow. While most organizations perform some work in several of these modes, generally one mode is dominant in the core technology of the organization. All of these modes can be characterized as processes composed of one or more technologies/operations. *Technologies* in this sense does not imply just engineering or manufacturing technologies but includes all sorts of office technologies, including the copier as well as the computer, and the "technologies" involved in producing an advertising or political campaign, designing a training program or a curriculum, or producing a movie. The following are definitions and discussions of the five modes.

> *Craft*—a process composed of a collection of one or more technologies/operations involving homogeneous human resources, generally a single person, producing a narrow range of products/services.

This is best characterized by the single artist/craftsman producing one unit of product at a time. Other examples are a single cook preparing a meal to order or a doctor examining a patient in the doctor's office.

> *Job shop*—a process composed of a loosely coordinated collection of heterogeneous technologies/operations to create a wide range of products/services where the technologies are located in groups by function and the time required at each workstation is varied.

This is best characterized by the manufacturing plant in which equipment is located or grouped into departments by type or function and the product/service is performed by moving the unit being worked on from one department to another in a nonuniform manner. It is also the mode of operation of most kitchens and the one frequently used for physical examinations performed in hospitals.

> *Progressive line*—a process composed of a tightly coordinated collection of heterogeneous technologies/operations to produce a large quantity of a limited range of products/services in which the technologies are located serially, the operator is directly involved in the work on the product, and the time allotted at each workstation is the same.

The automotive assembly line is the stereotypical example, with the product moving from station to station in a cycle time of approximately 60 seconds. Since this mode is used for both assembly and disassembly, the general term *progressive line* is more appropriate.

Progressive line is also the typical mode of serving for cafeterias and the mode in which physical examinations are given to large groups of people such as for the military. Note that manufacturing cells and *kanban* operations fit in this category.

Progressive-line mode may be used within a project. One example is a project to construct 740 houses in a development. The houses were in fact erected in the progressive line mode with multiple crews, each crew performing a very specific task on each house. On this line the *crews* moved from house to house with a cycle time of approximately 1 day.

> *Continuous flow*—a process composed of a tightly coordinated collection of technologies/operations which are applied uniformly over time and to all the many units of a very narrow range of products/services, and in which the role of the operator is primarily to monitor and adjust the processes.

Petroleum refineries are the most popular example of this mode. In addition, based on an examination of the characteristics of this mode, electric generating stations, water as well as sewage treatment facilities, and automatic transfer lines such as used in producing engine blocks and transmission housings are examples of this mode.

Understanding the economics of these modes, as shown in Fig. 22.1, reveals a fundamental driving force for attempting to move from craft mode as far as possible toward continuous-flow mode. For a given type of work, the craft mode generally requires the least capital investment or fixed costs but the highest variable cost per unit, while continuous mode requires large capital investments or fixed costs and very low variable costs per unit. The other modes tend to be arrayed between these two extremes. Thus, regardless of the major mode for a given undertaking, there should always be a search for subsets of the work to be moved to the more economical mode. This was done, for example, for the 80,000 seats in the Pontiac (Michigan) Silverdome stadium which were installed in the progressive-line mode. It was done in the English Channel Tunnel project where the digging, moving of tailings, and pump-

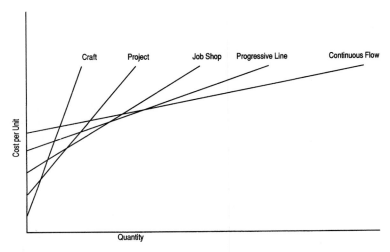

FIGURE 22.1   Breakeven curves for the five modes for arranging work.

ing of slurry to the tailings pit were all done in the continuous-flow mode. As a matter of fact, all modes can be observed on that project. At a simpler level, printed programs for an athletic event such as a swim meet have been assembled in a progressive-line mode while the overall effort to conduct the swim meet itself was a project.

PROJECT MANAGEMENT CONCEPTS

The Concept of Projects

Projects and project management may be one of the more misunderstood concepts in management. This is due in part to opinions developed in the early days of these new concepts and in part to the fact that projects have been a way of life throughout history.

The first developments in modern project management planning and scheduling were sponsored by duPont and the U.S. Navy, followed closely by the U.S. Department of Defense (DOD) and the National Aeronautics and Space Administration (NASA). From these efforts, the impression was gained that the techniques were applicable in construction and maintenance (in the case of duPont) and to very large scale defense and space systems development. This impression has been promulgated by authors of textbooks in production/operations management, management science, and operations research. With the advent of a variety of microcomputer-based project scheduling software and the availability and convenience of use of these packages, projects of all sizes have benefited from their application.

In the early years of modern project management, the cost of application of project scheduling software was substantial for two reasons: difficulty of use and the need for training of users. The first systems were *not* user friendly. Considerable computer expertise was required and the methods for these techniques were inherently labor-intensive. Today, both of these deficiencies have been overcome. User friendliness has made it possible for novice users to effectively use significant features of a package with only a couple of hours of introduction, generally from the user's manual, help messages, or tutoring programs. Training in the use of project management techniques has become widespread and many persons with project management responsibilities find it easy to plan their project in an on-line mode, with considerable benefit coming from a better understanding of the project and more accurate communication of their desired approach for the execution of the project.

Project efforts have been described in the earliest recorded history. The change efforts of society are conducted by projects. Most managers and engineers have been involved in, and many have managed, projects. The degree and scope of responsibility assigned to these individuals has typically been rather limited, often with responsibility shifting from one individual to another as the project progresses through its various phases. Today, it is more common for an individual to be given responsibility for managing the project from its inception to its closure.

Thus, many changes have taken place in the management of projects. While project management seemed to be almost entirely the application of network-based techniques, today the use of these techniques for planning, scheduling, and control can be considered at most 10 percent of modern project management. Behavioral considerations, contract management, risk management, and other concepts have been recognized to be of far greater significance. Perhaps the appropriate view is that planning and scheduling techniques have improved in both usability and usefulness to such an extent as to permit the project manager, and the project management team, to perform the planning and control functions with much less time and effort, thereby affording them the time to also perform the other functions far better.

Characteristics of Projects

In its Project Management Body of Knowledge,[6] PMI defines a *project* as any undertaking with a defined starting point and defined objectives by which completion is identified. This is indeed broad in scope, incorporating construction and defense/space systems but also including new product and/or process design and implementation and even cost reduction and methods improvement projects.

Projects are composed of activities, usually nonrepetitive, operating on an interrelated set of items which inherently have technologically determined relationships. One activity must be completed before another can begin. These relationships are described in the network diagram. This is further discussed below.

Projects involve multiple resources, both human and nonhuman, which require close coordination. Generally there is a variety of resources, each with its own unique technologies, skills, and traits. This leads to an inherent characteristic of projects—conflict. There is conflict *between* resources as to concepts, theory, techniques, etc. There is conflict *for* resources as to quantity, timing, and specific assignments. And there are other sources of conflict in projects. Thus, a project manager must be skilled in managing such conflict.

The "project" is not synonymous with the "product of the project." The word *project* is often used ambiguously, sometimes referring to the project and sometimes referring to the product of the project. This is not a trivial distinction, as both entities have characteristics unique to themselves. Some of the concepts apply to both. For example, the life cycle of a project includes the conceptual, development, implementation, and termination phases. The life cycle of a capital facility includes feasibility studies and acquisition (a project), operating it, major repairs or refurbishing (typically done as projects), and dismantling (often a project, if done at all). The project cost of creating the capital facility is generally a relatively small proportion of the life cycle cost of that product.

Similarly, there is a life cycle of product development consisting of basic research, product research, design, and production (Fig. 22.2). The first three are really projects. The objective of the production phase is typically to prolong the useful life of the product and requires a thorough understanding of the phases of the product life cycle. Some of these efforts focus on continued product improvement (redesign projects), improving production efficiency (process improvement projects), and/or developing promotional programs (advertising or merchandising projects).

Managerial emphasis is on timely accomplishment of the project as compared to the managerial emphasis in other modes of work. Most projects require the investment of considerable sums of money prior to enjoyment of the benefits of the resulting product. Interest on these funds is a major reason for emphasis on time. Being first in the market often determines long-term market position, thus creating time pressure. Finally, a need exists for the resulting product of the project, otherwise the project would not have been authorized.

Thus, time is of the essence. This time pressure, combined with coordination of multiple resources, explains why most project management systems have emphasized time management.

An Alternative Definition

The above logic leads to an alternative definition.

> **Project:** a temporary process composed of a constantly changing collection of technologies/operations involving the close coordination of heterogeneous resources to produce one or a few units of a unique product/service.[7]

Given this, it is appropriate to examine the characteristics of the process. Consider the following:

> The essential characteristic of the process by which a project is performed is the progressive elaboration of requirements/specifications.[7]

A project is initiated by a person (perhaps a member of an organization) recognizing a problem or opportunity about which some action is to be taken. That person, alone or in concert, develops an initial concept of the action to be taken in the form of a product, be it a product for sale, a new facility, or an advertising campaign. Much work needs to be accomplished to take this meager concept to the reality of the product of the project. This work, though often not conceived as such, is accomplished by instituting a project.

The general concept is expanded into a more detailed statement of requirements. These are examined for feasibility—market, technical, legal, organizational, political, etc.—resulting in further refinement of the specifications. These are then the basis for general design, the products of which become the basis for detail design. The detail designs are followed by production designs, tooling, production instructions, etc., each stage producing an elaboration on the specifications of the prior stage. Eventually, the

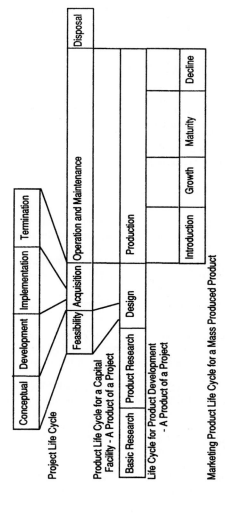

FIGURE 22.2 Comparison of project and product life cycles.

product of the project takes shape, is tested, and is ready for operation. At this stage, give or take a few details, the project is completed.

Relation to Modern Quality Management

This characterization of projects permits the adaptation of modern quality management concepts into the management of projects. Modern quality management begins with a new definition:

> *Quality:* Conformance to requirements/specifications.

The relevance of conformance can be made no more clear than as presented in the first nine pages of Chap. 4 in *Quality is Free.*[8] To begin, there must be a clear understanding of the customer's/client's requirements in a product. These then become the specifications for the next stage. This process continues throughout the project until the product of the project is completed.

Another important concept is *zero defects* or:

> Do the right thing right the first time, every time.

Repeated experience has shown that more attention to doing jobs right the first time eliminates so much rework/scrap costs as to fully support the proposition that "quality is free."

When these concepts are combined with another essential concept of modern quality management, then a new perspective is available on the project as a process. This is

> The customer is the next person/operation in the process.

Thus, as the progressive elaboration of specifications proceeds, the "customer" is the next engineer, the tool builder, the ad layout person, and so on. If the product going to them has no defects, they can perform their tasks in the most efficient manner—and do it right the first time.

The customer's requirements change over time, capabilities improve over time, and new information is discovered over time. Thus, the product or service must change over time. This leads to another concept, namely:

> Continuous improvement

One aspect of continuous improvement is essential in modern quality management for reducing defects, that is, reducing process variability. Another aspect of continuous improvement, however, is required to maintain competitiveness in a rapidly changing world. This, and the general concept of modern quality management—conformance to requirements—relates back to an earlier management theory, the marketing concept. Simply stated, the marketing concept states that all business planning and behavior relates back to understanding the customer's needs. Thus, the concepts come full circle, enabling the organization to better meet the customer's needs in the least costly manner.

Given the concept of a project as a process is essential for the application of process control, and more specifically, statistical process control, to the management of projects. One approach to this was presented in "Validating Technical Project Plans."[9] It involves procedures for reviewing the plans for a project in much the same manner as design or construction review teams analyze the design of a product for such things as structural integrity or constructability.

Another approach was outlined in "Reliability/Maturity Index."[10] This concept identified events in a project network which marked the completion of activities that either measured or contributed to the reliability of the product of the project. Other concepts/techniques were identified in "Responsibility for Quality in a Project,"[11] including failure mode and effect analysis, fault tree analysis, and stratified random sampling. Still others involve the development and application of methods/technologies which permit the operator on a project to perform inspections on the product immediately on completion or even during the performance of an operation.

SOME FUNDAMENTALS OF NETWORKS

Introduction

Many of the techniques for projects rely on a network diagram as the common language for planning and scheduling. Most of these are critical path techniques (CPTs) and have two things in common; they involve the calculation of early and late times and slack for each activity in the project, and from these derive the critical path, that sequence of activities requiring the greatest amount of time to complete and therefore determining the least amount of time in which the project can be performed. The early calculations determine the earliest times for the performance of each activity, sometimes called *forward scheduling*. The late calculations determine the latest possible times for each activity, sometimes called *backward scheduling*. The explanation of the calculations of start and finish times and slack can be found in many texts. The concept of the network diagramming language is so basic as to merit discussion here.

A *network diagram* is a schematic display of the sequential and logical relationship of the activities which compose a project. Generally, these technological relationships are very difficult to violate. For example, if getting one's self dressed is considered a project, it just does not make sense to put your shoes on before your socks. Whether to put on both socks and then both shoes or to complete the right foot before the left foot is generally a question of preference. In modern project management, a network diagram is used to portray these sequences. Figure 22.3 illustrates the use of networks to describe alternative ways of putting on socks (RSock and LSock) and shoes (RShoe and LShoe).

Figure 22.3C does not imply that both socks are put on simultaneously but, rather, it represents flexibility to determine the actual sequence based on other criteria. It is important for planners to focus on the technological relationships to prevent implicitly scheduling a project before really understanding the alternatives available. Figure 22.3D would be nonsensical in most instances for not only does it imply putting the sock on over the shoe but also putting both socks on over the left shoe.

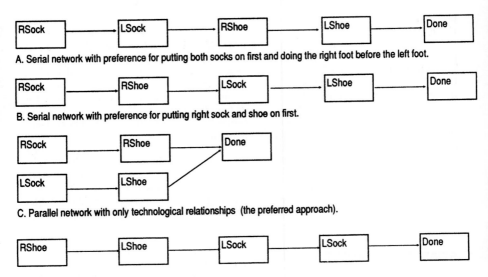

A. Serial network with preference for putting both socks on first and doing the right foot before the left foot.

B. Serial network with preference for putting right sock and shoe on first.

C. Parallel network with only technological relationships (the preferred approach).

D. A nonsensical network.

FIGURE 22.3 Example network diagrams in activity-on-node notation.

Project Network Diagram Conventions

For several reasons there has developed a great deal of ambiguity in the language of networking. PERT networks, CPM networks, and precedence diagrams are sometimes referred to as if they are unique and sometimes as if they were identical. It is helpful to develop a clear distinction which can be described by a three-dimensional matrix with the dimensions being graphic convention, focus convention, and identification convention as shown in Fig. 22.4.

Graphic Convention. Project networks can be graphically portrayed in either activity-on-the-node (AON) or activity-on-the-arrow (AOA) notation. The examples in Fig. 22.3 are shown in AON notation. An example of AOA notation, based on Fig. 22.3C, is shown in Fig. 22.4.

The original development of both PERT and CPM was based on AOA notation. At about the same time, John Fondahl, at Stanford University, was developing a comparable technique using AON.[12,13] Another development effort in Europe also used AON. In the United States, the impetus provided by the Department of Defense for use of PERT resulted in AOA being the notation learned by most users. It was not until the microcomputer versions of critical path techniques appeared that AON started becoming really popular in the United States.

There are a number of reasons for preferring AON, including ease of use, the ability to separate planning from scheduling, and the ability to maintain the integrity of identification of activities as they are subdivided when adding more detail. Probably the greatest benefit, however, is that it is exactly the same notation used in drawing process flowcharts, computer program flowcharts, and precedence (balloon) diagrams

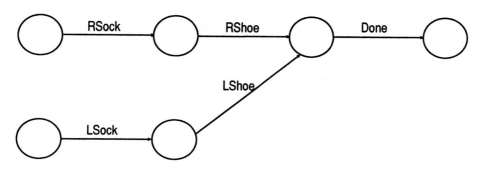

FIGURE 22.4 Activity-on-arrow notation for the example in Fig. 22.3C.

in assembly line balancing. Thus, AON requires less learning time and is less likely to result in errors from switching from one graphical mode to the other.

This does not mean that the AOA notation has no usefulness. After planning a project using AON, it is very useful to check the logic of the plan by plotting an AOA diagram on a time scale to see the temporal relationships between activities. Often it is possible to recognize that an activity is scheduled at what seems to be a totally illogical time. This can easily happen in a complex project because a precedence relation was inadvertently not specified or a relationship was specified which was really not necessary.

Most CPT software packages today have the ability to plot the project in either AOA or AON modes.

Focus Convention. PERT and CPM were developed for completely different purposes and therefore focused on the project in a different manner. PERT, as its full name implies (Program Evaluation and Review Technique), was developed to assess the probability that a specific event could take place as scheduled. Thus, the focus was on the event. CPM (critical path method) was developed to improve the planning and control of the work on a project and therefore focused on the activities or work involved. Indeed, the original development at duPont resulted in the Critical Path Planning and Scheduling (CPPS) technique, which focused on the time-cost tradeoff inherent in scheduling the project in a shorter total duration. (This is discussed in more detail later in this chapter.)

Thus, from the beginning of these two techniques, there was a difference in focus. Each has its place in modern project management. Top executives, in planning corporate strategy and approving projects, are primarily concerned with "When is it going to be completed?" Generally, they have too many other details on their minds to consider much more than the major milestones of the various projects ongoing in the organization. Indeed, there has been an axiom that no more than three milestones on a project should be reported at a briefing of top executives lest they become confused from information overload. Recent interest in "prudency," especially in utility rate cases, has led top executives to take more interest in the major projects in their organization.

Identification Convention. Both PERT and CPM, as originally developed, used the same method of identifying activities in a network, the *i-j* convention; i.e., events were given identifiers, usually numbers, and an activity was denoted by the number of the predecessor event followed by the number of the successor event, such as 110-130. The systems developed using AON notation used what is commonly referred to as *precedence* notation. Early on, these systems permitted the use of alphanumeric codes to uniquely identify activities (instead of events). Thus, an activity can be given the code name ABCDE or AESS. The relationship between two or more activities can be specified by simply listing the activity of concern followed by its predecessor(s) such as AESS − ABCDE, 34567, CM12P. An alternative also available permits listing the followers of an activity, and today some systems allow both follower and predecessor notation in the same project.

In the past, some knowledgeable persons argued for a nonmeaningful code which lead to the likes of ABCDE. More recently, it has become accepted practice to use a meaningful code such as AESS, which might stand for "Erect Structural Steel in area A." Recently developed software provides several characters for identifying activities, making it quite easy to develop a standardized coding structure for activities commonly performed in an organization. This provides at least two benefits: ease of recalling the names of frequently used activities and the ability to compare like activities across projects to improve estimating and performance measurement.

Overlap

In addition to the above conventions, a new logical relationship is now available. The original planning logic permitted only one basic logic between activities: the follower of an activity could not start until the preceding activity was finished. This proved quite frustrating in accurately portraying relationships where the follower could clearly start before the predecessor was finished. Thus, four alternative logics were introduced and are popular today. They are

Finish-to-start with overlap/delay. For [B − A(FS −3 days)], activity B can be started 3 days before activity A is finished. For [B − A(FS +3 days)] activity B cannot start until 3 days after A is finished. The former is useful to indicate that detail drawings can be started 3 days before the layout drawings are completed. The latter would be convenient for indicating that the forms cannot be removed from a concrete wall until 3 days after it was poured.

Finish-to-finish with overlap/delay. [B − A(FF +3)] is convenient to indicate that it is alright to start activity B before A is finished; just do not expect to finish it until 3 days after B is finished.

Start-to-start with overlap/delay. [B − A(SS +3)] is an alternative way to state that work can start on detail drawings 3 days after layout drawings have started.

Start-to-finish with overlap/delay. [B − A(SF +3)] is perhaps somewhat less useful, but the logic is available in many systems today.

While these overlap capabilities are very useful in many situations, they can be overused. A network incorporating many of these relationships can be very confusing and thus diminish its ability to communicate. Manual calculations are much more involved, thus reducing the ability to analyze the network. Finally, it is easy to incorporate logic which has unintended consequences. Indeed, in the extreme, it has been shown that it is possible to construct a network which has a finish time which is before the start time. With these thoughts in mind, it should be easy to remember to use these relationships with discretion.

Estimating

Estimating in projects is somewhat more involved and less precise than most estimating done by industrial engineers. For one thing, the unique nature of projects limits the usefulness of past data, makes time study of the job being done less applicable, and in general reduces the cost/benefit ratio of engineered time standards.

In addition, it is often necessary to differentiate between the work content of an activity and the actual duration which will be required. For one thing, it is generally not certain exactly who will perform the activity. One person may be very skilled at doing the activity but the person who ends up doing it may be much less skilled. Often, one person works on more than one activity at a time, thus the work content must be divided by less than a full day to determine the actual duration. Much project work is performed with little supervision, resulting in varying degrees of efficiency on the same activity. Communications of exactly what is expected as the work product of the activity may be incomplete. In addition, some activities may be pushing the state of the art of the technology and, therefore, be dependent on repeated trials, or even serendipity, to find the solution to a difficult problem.

The above tends to focus on estimating durations. Inherent in an estimate is an assumption of the methods/technologies that will be used. In addition, there are resource types to select, including not only human resources but the supplies and equipment required. There is uncertainty involved in all of these. From the above, cost estimates must be derived which could, if too high, result in the project not being approved or, if too low, result in poor performance to budget, schedule, and/or technical requirements.

PROJECT MANAGEMENT PLANNING AND SCHEDULING TECHNIQUES: A TAXONOMY

To establish the framework for discussion of the techniques available for managing projects, it is desirable to have a taxonomy of techniques. There are two classes: network-based techniques and non-network-based techniques.

Non-Network-Based Techniques

Work Breakdown Structure (WBS).   One of the most fundamental tools in project management is the Work Breakdown Structure (WBS). It is the tool by which a project, no matter how large, is subdivided into progressively smaller and smaller elements until the degree of detail is compatible with the needs of planning the project. A WBS is similar in appearance to an organization chart, but it is quite different. The WBS identifies the work which must be performed in order to complete the project (Fig. 22.5). It aids in defining and managing the scope of the project. The top level, 0, is the total project. Level 1 typically identifies deliverables which, in total, represent all elements of the project. Further levels subdivide the work by a number of criteria such as components, responsibility or trade, and geographical location. Indeed, level 1 can be subdivided by any of these criteria unless specified by the client. For example,

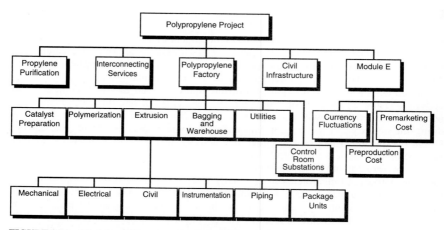

FIGURE 22.5   Work breakdown structure for a polypropylene plant.

the U.S. Department of Defense requires level 1 to be by deliverables to ensure comparability of like elements across programs and vendors.

The WBS is an important tool in controlling the work content of the project. All work that is required by the scope statement in the contract for the project must be identified in the WBS. Conversely, no work should be included unless it is in the scope statement. This then becomes the baseline for controlling changes to the scope and negotiating for appropriate budget or charges. Not only is this important to ensure proper recognition of costs for the changes and appropriate charges to the client, it is vital in managing the project. Ineffective management of scope generally results in excessive costs (often leading to reduction of profits if not incurrence of losses on the project) or to unsatisfied clients due to failure to deliver what was expected.

One of the anomalies of project management is that, while it is the management of change, it is changes to the project which are most likely to cause problems in the management of the project.

The WBS is also important in clearly establishing the strategy for the project. The manner in which the project is subdivided determines what problems will have to be dealt with by the project manager as well as each lower level member of the project team. Problems which are clearly within the responsibility assigned to the person responsible for a WBS element should be solved by that person. Problems which span two or more WBS elements require the involvement of the person who is responsible for all of those elements. One of the most important tasks of the project manager is managing the interfaces between WBS elements.

There are some basic principles which must be followed in developing a WBS, including:

Uniqueness. A unit of work should appear at one, and only one, place in the WBS.

Summative. The work content of a WBS element is the sum of the WBS elements immediately below and related to that WBS element.

Unity of Responsibility. A WBS element is the responsibility of one, and only one, individual.

Consistency. The WBS must be consistent with the way in which the work is actually going to be performed; i.e., it should serve the project team first and other purposes only if practicable.

Motivation by involvement. Project team members should be involved in developing the WBS (as well as the network plan), not only to ensure consistency, but more to ensure buy-in to the project plan.

Documentation. Each WBS element must be documented to ensure accurate understanding of the scope of work included, and not included, in that element.

Flexibility. The WBS must be a flexible tool to accommodate the changes which are inevitable in a project while properly maintaining control of the work content in the project as per the scope statement in the contract.

Careful attention to developing a WBS according to these principles will reward the project manager and the organization with many benefits throughout the project.

Gantt Chart. The Gantt chart was developed by Henry Gantt about 1915 as a means for scheduling work in a job shop. It has become the most easily understood portrayal of a scheduled plan for a variety of types of work. While it was used extensively prior to computer methods for scheduling projects, it was often demeaned by the comment, "It conceals more than it reveals." Manual means for preparing Gantt charts were so time-consuming and expensive as to make the schedule close to "cast in bronze." As a result, the tendency was to delay redrawing it until the project was well off schedule.

Modern project management software programs prepare Gantt charts rapidly and inexpensively with a wide variety of graphical formats. They are effective for communicating with clients, executives, middle managers, and line managers. Indeed, they are preferred by line managers who generally want to know what they are supposed to do and when.

Gantt charts do have deficiencies. They typically do not show the relationship between activities in the network. A special version, the connected Gantt chart, overcomes this but in so doing becomes more complicated. In general, it is best to use the simple Gantt chart for most purposes, bringing out the connected version or even the network diagram for more complex problem solving.

Network-Based Techniques

Critical Path Method. For purposes of discussion, CPM will be used to refer to the simplest form of CPT involving the precedence relationships between activities, a single estimate of each activity's duration, and the calculation of early and late start and completion times and total slack. This distinction is important because of the ambiguity which exists between practitioners and the academic literature. If one refers to the typical textbook on management science/operations research, production/operations management, or industrial engineering, CPM will probably be found to refer to the time-cost tradeoff algorithm. If one speaks with practitioners in the field, CPM will probably be used to refer to the simplest system as defined above. We prefer the vernacular of the practitioner, as presented above.

Program Evaluation and Review Technique. PERT was developed by a team representing the Naval Weapons Research Laboratory and Booz-Allen consultants to evaluate the status of the Polaris Missile Weapons System.[14] Admiral Rayborn credited PERT with taking a year out of that development project. Perhaps the more appropriate attribution is that it gave the project manager and team information which enabled them to reduce the project duration by a year.

PERT was originally developed to permit the expression of the uncertainty associated with moving from one event to another. Considerable mathematics was used in deriving the specific form of the distribution and the practical simplification leading to three time estimates: a = optimistic or shortest reasonable time, m = most likely time, and b = pessimistic or longest reasonable time. The original PERT relationship assumed that a and b occurred about one time in a hundred each. For those interested, several alternative relationships have been developed based on arguments pertaining to the relative frequency of these times. The three time estimates were used to calculate a mean and standard deviation for each activity as follows:

$$t_e = \frac{a + 4m + b}{6}$$

where t_e is the expected value of the duration of the activity, as in expected value theory in probability and statistics, and

$$\sigma = \frac{b - a}{6}$$

where σ is the standard deviation of the distribution of activity durations.

The general logic of the calculations is identical to CPM except for the calculation of t_e for the duration and the calculation of the standard deviation of the ending event or activity. While the theory of PERT assumes a beta distribution for individual activities, i.e., one that may be skewed to either the right or the left, the central limit theorem nevertheless allows the addition of the activity distributions on the critical path with the resulting distribution assumed to approximate the normal curve of probability theory. Thus, the mean of the ending activity or event is equal to the sum of the means of the activities on the longest path leading to that activity or event. Similarly, the variance of the ending activity or event is the sum of the variances (standard deviations squared) of the activities on that same path. The square root of the summed variances is the standard deviation of the completion time for the ending activity or event. Generally, this is done only for the latest activity or event in the project.

The resulting T_e is the expected completion time for the last activity or event on the critical path. There is a 50 percent chance it will take more time than is represented by T_e to complete the project but also a 50 percent chance that it will take less time.

The standard deviation for the critical path is a measure of the uncertainty associated with that activity or event. It permits the determination of the probability of completing the project on or before (or after) any specified time. Thus, for a project with a T_e of 13 and a standard deviation of 1, it can be said that the probability of completing the project in 12 days or less is 16 percent. Similarly, the probability of completing the project in 14 days or less is 84 percent. The probability of completing in 15 days or less is 97.5 percent. The probability is nearly 100 percent that the project will be done in 16 days or less.

Such information can be of considerable value. If, for example, a major meeting is dependent on completing the project discussed in the illustrations, it would be desirable to schedule it no earlier than the 17th day. In more complex projects, it is not uncommon for several paths to converge just prior to a meeting or other high-cost/high-visibility event. The more converging paths there are, the greater the probability that at least one activity will be late. Using some simple probability calculations, that probability can be calculated for alternative dates and the meeting scheduled accordingly.

There have been several papers written discrediting the uncertainty calculations of PERT. Some suggest that the optimistic and pessimistic times are not properly represented by 1 in 100 chances. Perhaps they ought to represent 1 in 20 chances, etc. The user can select one of as many as four alternatives in this regard in some software.

More serious is the error that is introduced by basing the entire calculation on only those activities which happen to be on the critical path in that particular calculation. Often, a less critical path can get into difficulty and take longer than the expected time for the critical path. The best solution to this is to use one of the software packages which employ Monte Carlo simulation to more adequately develop the distribution of the ending activity or event by including the implications of all paths through the network. This capability has only been commercially available for a few years now, but its use is growing as risk analysis becomes a pressing concern.

Resource Management.   Implicit in the above discussion has been the assumption that whatever resources are required in any given time period will be available. This is a tenuous assumption at best. Further, the discussion assumed that there is no limit on the number of resources which can be working on a given entity or in a given space at any one time. Such constraints can be incorporated in the network plan, but this is generally not a wise thing to do, as it may unnecessarily constrain other possible solutions.

Until resource implications are considered, the calculations discussed to this point merely provide time frames within which activities *may* be scheduled. There are two general classes of problems in this area, time-constrained and resource-constrained scheduling.

If time is of the essence and whatever resources required can be obtained, it is still desirable to bring the minimum necessary units of resource onto the project and avoid frequent hiring and firing. One experience by the author in using a program of this type resulted in a recommendation for the hiring of additional skilled trades—for example, 20 die setters and 10 electricians. Meanwhile, on the basis of historical approaches to the same problem, the personnel department was attempting to fill requisitions for 10 die setters and 20 electricians. After examining our analysis, a quick revision was made to the hiring plans to match the project scheduling software's recommendation. Thus, savings can be substantial in avoiding both overhiring and hiring in the wrong pattern and incurring the resulting delays of the project.

There are many software programs which can handle this problem, albeit with varying degrees of sophistication. Research has been conducted on scheduling using alternative decision rules to determine how they perform relative to different scheduling objectives. Nevertheless, most schedules derived in this manner can be improved on by experienced project managers.[15,16]

If resources are limited, the preferred approach is resource-constrained scheduling. There are many project-scheduling software programs which can solve the resource-constrained problem. There are two basic approaches: assigning resources to activities in accordance with one or more scheduling rules and letting the finish time for the project go where it may, or performing successive tries at the time-constrained resource-leveling procedure, lengthening the time available for the project at each try.

There are two types of resources which require different approaches to the problem. These approaches can be characterized as homogeneous versus heterogeneous, common versus unique, or trade versus individual. The first of the resources, software for scheduling trades, assumes that all units of resource in a class have basically the same capabilities; i.e., a carpenter is a carpenter is a carpenter. Most scheduling packages available today are of this type.

The other type of program, for scheduling individuals, assumes that each individual has unique capabilities. The author recognized this in attempts to design software for scheduling drafters and engineers in an engineering/drafting operation. One engineer could do bumpers very efficiently but was very slow on quarter panels (fenders). Another had comparable abilities of another mix. For lack of a better term, this was dubbed the "drafting room scheduling problem." It can be characterized as a three-dimensional assignment problem, where the resources and the activities form the usual dimensions but the problem is compounded by the seemingly random availability of the resources, as well as the availability of the activities to which to assign the resources. It is unclear whether this problem has been well-solved to date in commercially available software. A similar problem occurs in many architect offices and software shops where each person is assigned several activities and works on more than one at a time. Thus, the program should be able to monitor the hours allocated to each activity and the total commitment of the individual's time. Several packages have been developed to handle this problem; however, they tend to be tailored to the particular profession as opposed to being more general-purpose.

While there have been analytical solutions proposed for these problems, the analytical solutions tend to be limited to small problems of little practical significance. It is likely that expert systems concepts and even artificial intelligence will provide a new class of solutions for these problems in the not too distant future.

Critical Path Planning and Scheduling or Time/Cost Tradeoff.   Critical path planning and scheduling was the technique that was developed as a result of the pioneering effort by Kelley (Sperry Univac) and Walker (duPont) concurrent with the development of PERT.[17,18] The original solution technique was linear programming and therefore the number of activities was severely limited at the time. The problem, as originally characterized, was to determine the optimum project duration, given that each activity in the network could be performed in a range of durations, one of which was least costly. This was desirable because, once a new process plant was authorized to be built, or an existing plant was shut down for maintenance, every day the plant was not available a loss was incurred due to the inability to sell the product which it might have produced were it on stream. The problem is akin to the classic inventory problem where one cost—carrying cost—decreases with smaller order quantities, while another cost—ordering cost—increases. On the project, there may be opportunity costs—lost revenue from product which could have been sold if it had been available or charges for interest such as that on the construction loan. Such costs decrease with shorter project duration, while other costs of performing the project increase with shorter project duration. Thus, the objective is to determine the project duration for which the sum of these costs is the least.

To shorten the time for the project, the duration of the critical path must be reduced. It is realistic in practice to analyze the critical activities because there will generally be only one or a few paths which require significantly more time than all the rest. These may consist of as little as 10 percent or less of the activities in the network. Further, some of these may not be amenable to time reduction while others may be very simple to reduce. For example, on a project to design a new engineering releasing system, a substantial reduction in negative slack was achieved when it was recognized that all interchange of documents was planned as activities handled through the inexpensive intracompany mail, each requiring 3 days. Those intracompany mail activities having negative slack were changed to hand-carry delivery, thereby reducing the project duration by some 100 days.

Cost Management. Earned value analysis is the modern concept of what was originally known as PERT/COST, a system which was touted in the early 1960s, but which was found to be too rigid in practice. Earned value is primarily a client-oriented cost control system in that it aids in ensuring that the client pays only for work that is done and materials actually delivered. It essentially recognizes costs at the time when the check is drawn to pay for the work or materials.

Earned value is very similar in concept to a standard cost approach for volume production operations, permitting analyses of cost differences due to schedule variance and spending variance. Because project work is essentially one-time in nature, the measurements must be based on estimated costs for each activity or work package.

Earned value analysis has three major cost measurements: Budgeted Cost of Work Scheduled (BCWS), Budgeted Cost of Work Performed (BCWP), and Actual Cost of Work Performed (ACWP). These are measured at the activity or work package level and summed progressively through the work breakdown structure to provide an overall measure for the project as a whole. Thus, managerial analysis can begin by determining if there is a problem for the project and progress downward through the WBS to determine the contributing causes of a problem.

Cost management begins with a budget which, when approved, becomes the baseline for all further control. The first measure for an activity, group of activities, or the project as a whole is the BCWS. It is the amount which should have been expended for work scheduled during the reporting period. It is accumulated with that for prior periods to obtain the amount which should have been expended to the reporting date. The cumulative cost curve to the end of the project is generally in the shape of an S and is called the *S curve*.

Not all work is performed to schedule, so the next measure is the BCWP. This indicates the number of dollars which should have been expended during the reporting period, or cumulatively to the reporting date, in performing the work that was actually performed. The difference, BCWS − BCWP, is a measure of schedule variance, very comparable to the volume variance in a standard cost system.

The work which is performed may not cost exactly what it was estimated to have cost. Thus, the third measure is ACWP. The difference, BCWP − ACWP, is a measure of the spending variance, very comparable to the rate variance in a standard cost system.

From these measures, then, the competent project manager can determine if actual expenditures are basically consistent with the actual progress on the project and determine what, if anything, can be done about it. Because of the pyramidal structure of the accounting system built into the project management software, a variance at the project level can be traced through the many levels of detail to the specific activities contributing to the problem. Thus, it focuses the attention of the project manager on those areas most in need of diagnosis and direction.

Using various methods, based on different assumptions, a projection can be made of the Estimate At Completion (EAC) cost. The estimated completion date is more accurately projected from the time/schedule calculations.

The earned value approach has been adopted by the Department of Defense (DOD), National Aeronautics and Space Administration (NASA), Department of Energy (DOE), and other U.S. federal agencies. The DOD version is called *Cost/Schedule Control Systems Criteria* (C/SCSC) and is well-documented in Fleming.[19] Contractors performing work for the above agencies may be required to maintain cost and schedule records and controls and report in accordance with C/SCSC. It should be noted that certification of compliance with C/SCSC is based on the contractor's total management system, of which a part may be a standard project management schedule and cost system. While the commercially available project management planning and scheduling software package is not certified as such, it may be appropriate to determine if the package has been an integral part of a C/SCSC certified contractor's management system.

Project Strategic Analysis.   Two techniques developed in the early 1960s offer capabilities not found in the more familiar project management software packages. Because of the additional logic available in these, it is expected that they will become more popular as more deliberate strategic planning of projects develops and even more as project management concepts are used in corporate strategic planning.

Decision CPM (DCPM).   The techniques discussed above have some constraints which limit their flexibility in practical application. For example, there is no provision for identifying alternative methods of performing the same task or meeting the same requirement. The only way to accomplish this is to incorporate one alternative in plan A and the other in plan B. If the number of alternatives is appreciably greater than two *and* there is interaction between alternatives, the task of evaluating these can be formidable.

DCPM, a concept first published in 1967,[20] relaxes this constraint on identifying alternative methods and allows alternative task sets to be defined, i.e., mutually exclusive alternative methods for performing the task, one of which is ultimately to be selected. A task set is preceded by a decision node. The decision nodes are represented by a triangle while the tasks in a task set are portrayed by a circle, as are all ordinary tasks (or activities); i.e., activity-on-node graphic convention is used. The first task set might be identified as T_1 in the decision node (triangle) with the jobs belonging to the set being identified as $T_{1,1}$, $T_{1,2}$, and $T_{1,3}$. Based on some objective function, e.g., minimum cost or minimum time, a decision can be made between the tasks in a task set. A sample problem is discussed in Wiest and Levy,[21] in which there are two task sets composed of three alternatives each and one composed of two alternatives. It points out that there are 18 different ways of choosing from these and thus 18 different ways of performing the project. Clearly, these 18 alternatives could be portrayed as 18 different project networks.

The possible solution techniques include complete enumeration, a branch-and-bound algorithm, or a heuristic approach. While these techniques were documented in the early literature, the author is not currently aware of any commercially available software for solving this type of problem. It is discussed here due to the expectation that it will become available within the life span of this edition of the handbook. For more details on this technique see either Wiest and Levy[21] or Moder, Phillips, and Davis.[22]

Graphical Evaluation and Review Technique (GERT).   With the exception of PERT, the techniques discussed above are *deterministic* in nature, i.e., they do not involve probabilistic elements. All of the above techniques, except DCPM, assume that all activities identified in the plan will be performed. None of the above techniques allow loops in the network, i.e., an arrow which goes backward in the topology of the network, creating a path which eventually comes back to the same place in the network.

GERT portrays activities in the activity-on-arrow graphical notation and nodes are either conventional deterministic events or decision nodes. Decision nodes have both an input side and an output side. The input side determines the conditions under which a job can be released, allowing subsequent activities to proceed. For example, two or more alternatives may be planned to solve a particularly difficult problem. The first alternative finished might release subsequent activities and may, if properly coded, signal the completion of work on the preceding alternative activities. Four different logical alternatives are available on the input side, including conditions relevant to the repeat of the node in the event of looping back. Output sides can be deterministic or probabilistic. If deterministic, the logic is the same as in the above techniques. If probabilistic, one of the emanating jobs is released, depending on the probabilities assigned.

After being inculcated with a no-loops philosophy, it may not be obvious to those familiar with PERT/CPM why a loop would be useful in a network plan. Many times a part or assembly must go through a design-build-test sequence. If it fails on the test, it may have to repeat that same sequence. For example, automotive exterior lamp assemblies must pass stringent tests at the federal level and by some states, in addition to tests in the company's own laboratories. Any of those tests may fail, resulting in a return to the design-build-test sequence.

GERT is also discussed in Wiest and Levy[21] and Moder, Phillips, and Davis.[22] For more details of P-GERT, contact Pritsker & Associates, Inc., West Lafayette, Indiana.

Corporate Strategic Planning. The techniques of project planning and scheduling have not been widely used in the executive suite. Reports of interest in these capabilities for strategic planning are becoming more and more frequent. The most obvious application is to determine feasible target dates for implementing a strategic plan, often requiring the completion of several related projects. The next most obvious is to incorporate cost estimates to test the financial feasibility of the planned strategic move. Risks can be assessed on these two dimensions using the multiple time estimates to indicate the degree of uncertainty associated with individual activities and, with the proper logic, the impact on costs. At still another level, project networks might be used to test the feasibility of a strategic plan subject to limitations on the physical and human resources available. This might indicate the degree of need for additional resources.

These capabilities exist in many commercially available project management software packages today. The next capability that is likely to be introduced into the executive suite is the either/or logic of DCPM/GERTS. This will permit the definition of alternative plans to reach the same objective and more sophisticated analysis of the potential consequences and risks.

SOME OPPORTUNITIES AND CAVEATS

Opportunities

Project management is the management of change. Change is increasingly the norm in business, industry, and government. The efficient management of change is essential to survival, and modern project management is the way to achieve that efficiency.

Many organizations that operate primarily in the project mode have adopted project management as a way of life. An excellent example of this is AT&T Business Communication Systems Division which equips its project managers with laptop computers loaded with a standard set of software useful in managing projects.[23] Modern project management is used in the construction of the English Channel Tunnel[1] and in developing new pharmaceuticals.[24] It is used in developing high-energy physics research capabilities at Sandia Laboratories[25] and on the Superconducting Super Collider.[26] It is used to get new products to the market more quickly and to manage the Voyager 2 trip to Neptune.[5] It is even used in running the Olympic and Pan American games.[27]

It is becoming more popular for managers having responsibility for many small projects to simply keep track of their status and to understand the implications of both budget and resource availability. It is in this area where one of the great opportunities to contribute to the success of an organization exists. Modern project management can, and should, be applied at the plant level to manage all of the non-volume-production work being performed in the plant. It should then be expanded, where appropriate, to include all such project activity in a multiplant organization. The author had such an application working long before either time-sharing or microcomputer approaches to project management were available, which now make this a much more feasible application. Other than the slow turnaround available on early computers, the application was well-received. It included personnel studies as well as new product and facility preparations.

Caveats

Care must be taken in applying modern project management in an organization not familiar with its application. It can drive a project team too hard and develop resistance. One plant manager favored allowing supervisors to spend money in work orders rather than risk having the unspent funds resulting from better project management returned to corporate. Care must be taken to ensure that benefits being enjoyed today are not lost with the introduction of new management tools.

Managers must understand that modern project management tools are more precise than previous tools. They cannot be used in the same manner as the tools of the past. *If they are misused, project team members will ensure that they are not accurate.* One of the ways in which these tools can be misused is to put too much pressure on the team members to perform. People who are pushed to the extreme by such a system will avoid using it if at all possible. If the system is used properly, the same people can support effective team building and help projects to be completed in conformance with technical requirements, on time, and, often, under budget.

Finally, care must be exercised in using the tools to ensure that the most appropriate means are used to improve performance. The following list, in a preferred sequence based on a general concept of the cost/benefit ratios, shows ways to improve performance on projects.[28]

Managerial attention and involvement shows interest and concern, but not to the point of interfering with those who have to accomplish the tasks.

Expediting ensures that obstacles to progress are removed and resources are available when required.

Improved methods reduce the work content, time required to complete a task, the resources used, and the resources wasted.

Reassigning resources can move the most efficient resources to the activities which are most critical.

Reallocating resources can move resources from noncritical activities, lengthening their duration, to critical activities, shortening their durations.

Overlapping activities can allow work to proceed more quickly but with some risk, unless extra attention is given to adequacy of communications.

Defining activities in greater detail subdivides the work so more people can work on the project simultaneously.

Deleting certain activities is often feasible after careful review with the client of what is really essential.

Changing the technology applied is similar to methods improvement, but may be more risky if the new technology is unfamiliar to the performing organization.

Subcontracting or buying the goods or services often results in economies as well as time savings and improved quality, but with some loss in control and dependability.

Applying additional resources, using the CPPS approach to time-cost tradeoff, can reduce the total time required, but note that this is 11th in this list.

Changing target dates on the total project, while still delivering essential parts of the project at dates acceptable to the client, can minimize perceived poor performance.

Changing the scope of the project can permit timely delivery of essential elements and possibly avoid major failure on the project as originally conceived.

Changing the person responsible often results in major delays while the new person becomes acquainted with the project and the team *and* the team gets acquainted with the new person.

Changing the management information system can lead to better performance but generally only after all participants have an opportunity to become familiar with it and it is adjusted to the needs of the organization.

Changing the organization structure often seems to be a quick fix but often it takes considerable time for the project team members to begin to function as a team.

Modern project management is not a panacea, but used wisely it offers many opportunities for improving the performance of an organization's project work and, as a result, leads to better performance in other parts of the organization because projects are more likely to be completed on time and under budget.

SUMMARY

Project management has grown from an early conception that it was simply planning and scheduling, then to include cost management, and now to a comprehensive responsibility for seeing that all aspects of the project are carefully integrated and managed. One view of this progression is that the early focus on scheduling and cost management solved a major problem at the time—inadequate control of time and cost. The success in dealing with these two vital areas has substantially improved the accuracy and utility of these functions and reduced the time required for these managerial tasks, making more time available to manage the other elements of project success.

Project management experience has become recognized as one of the very important career steps on the ladder to general management. The project manager is the general manager of that project. The normal progression for successful project managers is on to larger, more challenging assignments.

The progressive manager can bring substantial benefits to an organization through project management. Purchase an inexpensive project management software package. Start with a small project in the department. Move to a larger project with another department which is receptive to change. By this time you should have a much better idea what is required in project management software for your organization. Then move up to a still larger project involving several departments. By this time, the experience gained and the successes recorded should build the acceptance to apply these concepts plantwide, if not corporatewide.

Good luck!

REFERENCES

1. J. K. Lemley, "The Channel Tunnel: Creating a Modern Wonder-of-the World," *PMNETwork,* vol. VI, no. 5, July 1992, pp. 8–21.

2. S. K. Bubna and J. J. Anderson, "PM Meets the Challenge on Somincor's Neves-Corvo Tin Project," *PMNETwork,* vol. VI, no. 3, April 1992, pp. 9–22.

3. G. J. van Zyl, "Sasol Market Share Enhanced: Record Breaking Polypropylene Project," *PMNETwork,* vol. V, no. 8, November 1991, pp. 8–21.

4. P. F. Flones, "Endicott Oil Field: First Offshore Arctic Oil Field Begins Production," *Project Management Journal,* vol. XVIII, no. 5, December 1987, pp. 41–50.

5. K. Bartos and W. D. Brundage, "The Voyager 2–Neptune Encounter, A Management Challenge: Speeding Toward Your Deadline at 42,000 Miles Per Hour...Without Brakes," *PMNETwork,* vol. III, no. 4, May 1989, pp. 7–23.

6. Project Management Institute, "Project Management Body of Knowledge," *PMNETwork,* August 1987.

7. F. M. Webster, Jr., "Integrating PM and QM," *PMNETwork,* vol. V, no. 3, April 1991, pp. 24–32.

8. P. B. Crosby, *Quality is Free,* McGraw-Hill, New York, 1979.

9. H. J. Thamhain, "Validating Technical Project Plans," *Project Management Journal,* vol. XX, no. 4, December 1989, pp. 43–50.

10. D. G. Malcolm, "Reliability Maturity Index (RMI)—An Extension of PERT into Reliability Management," *The Journal of Industrial Engineering,* January–February 1963, pp. 3–12.

11. T. Kloppenborg and F. M. Webster, Jr., "Responsibility for Quality in a Project," *PMNETwork,* vol. IV, no. 2, February 1990, pp. 25–28.

12. J. W. Fondahl, *A Non-computer Approach to the Critical Path Method for the Construction Industry,* 1st ed., Department of Civil Engineering, Stanford University, Stanford, Calif., 1961.

13. J. W. Fondahl, "Precedence Diagramming Methods: Origins and Early Development," *Project Management Journal,* vol. XVIII, no. 2, June 1987, pp. 33–36.

14. D. G. Malcolm, J. H. Roseboom, C. E. Clark, and W. Fazar, "Applications of a Technique for R&D Program Evaluation (PERT)," *Operations Research,* vol. 7, no. 5, 1959, pp. 646–669.

15. E. W. Davis and J. H. Patterson, "A Comparison of Heuristic and Optimum Solutions in Resource-Constrained Project Scheduling," *Management Science,* vol. 21, no. 8, April 1975, pp. 944–955.

16. E. W. Davis and J. H. Patterson, "Resource-Based Project Scheduling: Which Rules Perform the Best?," *Project Management Quarterly,* vol. 6, no. 4, December 1975, pp. 25–31.

17. J. E. Kelley, Jr., and M. R. Walker, "Critical-Path Planning and Scheduling," *Proceedings of the Eastern Joint Computer Conference,* Boston, Dec. 1–3, 1959, pp. 160–173.

18. J. E. Kelley, Jr., and M. R. Walker, "The Origins of CPM: A Personal History," *PMNETwork,* vol. III, no. 2, February 1989, pp. 7–22.

19. Q. W. Fleming, *Cost/Schedule Control Systems Criteria: The Management Guide to C/SCSC,* Probus, Chicago, 1988.

20. W. Crowston and G. L. Thompson, "Decision CPM: A Method for Simultaneous Planning, Scheduling and Control of Projects," *Operations Research,* vol. 15, no. 3, May–June 1967, pp. 407–426.

21. J. D. Wiest and F. K. Levy, *A Management Guide to PERT/CPM,* 2d ed., Prentice-Hall, Englewood Cliffs, N.J., 1977.

22. J. J. Moder, C. R. Phillips, and E. W. Davis, *Project Management with CPM, PERT and Precedence Diagramming,* 3d ed., Van Nostrand Reinhold, New York, 1983.

23. D. Ono, "Implementing Project Management in AT&T's Business Communications System," *PMNETwork,* vol. IV, no. 7, October 1990, pp. 9–31.

24. J. Engelhart, M. Malkin, and R. Rhodes, "From the Laboratory to the Pharmacy: Therapeutic Drug Development at Merck Sharp & Dohme Research Laboratories," *PMNETwork,* vol. III, no. 6, August 1989, pp. 11–28.

25. G. R. Barr, J. P. Furaus, and C. G. Shirley, "Particle Accelerator Research and Development at Sandia National Laboratories," *Project Management Journal,* vol. XIX, no. 1, February 1988, pp. 29–47.

26. N. Baggett, et al., "The Super Conducting Supercollider," *PMNETwork,* vol. IV, no. 8, November 1990, pp. 6–40.

27. R. G. Holland, L. Peters, and R. M. Shortridge, "The Story Behind the Games: The XV Olympic Winter Games and The X Pan American Games," *PMNETwork,* vol. III, no. 8, November 1989, pp. 5–25.

28. F. M. Webster, Jr., "Ways to Improve Performance on Projects," *Project Management Quarterly,* September 1981, pp. 21–26.

P · A · R · T · 6

CROSS-CULTURAL PROJECT MANAGEMENT

In projects, cooperation takes place across personal, corporate, and national cultures. Cultural management is a specific management qualification required in projects.

In Chap. 23, C. Michael Farr compares the project management approaches of selected international allies with management tendencies in the United States. The acquisition process and organizational structures are reviewed, and the individual differences and human factors that affect daily management at the working level are considered. The chapter concludes with a look at the impact of cultural differences on management styles.

Adnan Ehshassi presents in Chap. 24 the management styles found in multicultural construction projects in the Middle East. His chapter is based on major research studies which derived and elaborated the main dimensions of leadership. The author develops insight into leadership styles and culture, and discusses the productivity of multicultural work groups.

Ilona Kickbush and Agis D. Tsouros present, in Chap. 25, the networking approach by which "project cities" are linked with each other in order to put health on the agenda of decision makers in European cities.

N. J. Smith presents in Chap. 26 insight into construction project management in Japan and the United Kingdom. The chapter investigates the current and potential roles for project management in construction in these two countries based on a consideration of traditional and cultural practices. An assessment of potential international markets is outlined, and the associated cultural implications for project management are examined. Finally, the implications of market factors for project management training and development in Japan and the United Kingdom are outlined.

In Chap. 27, Jeffrey K. Pinto and Dennis P. Slevin describe the development and use of the project implication profile (PIP) to serve two purposes: first, to allow the project manager to assess the behavioral side of the project management process; and, second, to allow the project managers the opportunity to focus some of their attention on the strategic issues of project development.

CHAPTER 23

INTERNATIONAL PROJECT MANAGEMENT: DO OTHER COUNTRIES DO IT BETTER?

C. Michael Farr

*Systems Management Division,
Air Force Institute of Technology*

Lieutenant Colonel Farr is a graduate of the U.S. Air Force Academy and holds a Ph.D. in Operations Management from the University of North Carolina at Chapel Hill. He began his career as a Minuteman III Combat Crew member where he gained experience as a user of military systems. He has been associated with the defense acquisition community since 1980, and is currently Chief of the Systems Management Division at the Air Force Institute of Technology.

INTRODUCTION

In February 1989, Clifton Berry noted that European governments buy their weapons differently than the United States, and posed the question"…but do they do it better?"[1] This question has been considered by several authors and has also resulted in studies by the General Accounting Office (GAO), the Rand Corporation, and the House Armed Services Committee. While many dimensions of project management have been examined, a recurring theme has been the centralized, civilian-controlled acquisition process employed by many European countries.

Several authors have pondered whether the United States should adopt the European model of a "civilian superagency," and Congress has been no exception. In the wake of cost overruns and allegations of fraud, a reform-minded U.S. Congress repeatedly pressured the Department of Defense (DOD) to change the weapons acquisition process during the 1980s. During 1988, Senator William Roth and House Representative Dennis Hertel sponsored separate but similar legislation to create a civilian acquisition corps.[2] Neither bill passed, but interest in the European model still exists.

Despite Congressional interest, none of the previous studies have recommended U.S. adoption of the European centralized system. In fact, a 1989 report by the House Armed Services Committee[3] was quite clear in its recommendation *against* emulating the European system. However, several of the studies did find significant merit in various management approaches taken by U.S. allies.

Therefore, this chapter compares the project management approaches of selected international allies with management tendencies in the United States. The intent is to identify ideas and approaches that might be useful to U.S. project managers. This review begins with an overall look at acquisition processes and organizational structures and a look at "the system" and the environment in which it operates. The chapter then changes its focus more toward individual differences and human factors that affect daily management at the working level. Finally, the chapter concludes with a look at the impact of cultural differences on management styles.

COMPARISON OF ACQUISITION PROCESSES

This section first examines the debate over centralized procurement systems and whether they might offer significant advantages to the United States. The section then highlights other fundamental differences in the way various countries acquire weapon systems.

Centralized Procurement Systems: A Model to Emulate?

While there are some notable differences among the more centralized acquisition establishments found in Europe, they do have several common characteristics that stand in contrast to the U.S. approach. The pivotal difference in centralized systems is that the military services do *not* individually manage the weapons acquisition process. Instead, there is a highly civilian acquisition corps that centrally manages research, development, and acquisition on behalf of the military services. There are comparatively few high-level political appointees, and the programs are managed by experienced, highly trained civilians with a clearly defined career in acquisition management.

By comparison, the military services play a dominant role in acquiring U.S. weapon systems. Also, most U.S. project managers are military officers and usually bring considerable operational experience to their job. While many people consider this operational experience to be a strength of the U.S. process, a directly related concern is that there is no guarantee that these managers have appropriate acquisition management education or experience. (However, it should be noted that the Defense Workforce Improvement Act of 1990 has resulted in initiatives by each military service to require specified education and experience for acquisition managers.)

The pressing question is whether centralized systems work better than the U.S. approach. Advocates of the centralized model believe that such an approach might eliminate duplication of weapons development programs among the military services, curb the potential for parochial and counterproductive competition among the services, and achieve greater efficiency in the acquisition process. "The expected outcome would be lower costs and more effective weapon systems for the same defense dollars."[4] While several qualitative studies have suggested that this might be so, quantitative comparisons of actual program outcomes suggest otherwise.

For example, in their analysis of attack and fighter aircraft programs, Gansler and Henning found that schedule and performance outcomes on U.S. projects were better than those on similar European projects.[5] Similarly, Berry[6] quoted the conclusion of a 1986 Rand report, "Weapon program outcomes in Europe are generally less satisfactory than those in the United States, especially in terms of schedule length and slippage during the development phase." Other authors also agree that, at best, the evidence that centralized procurement systems produce better project outcomes is inconclusive. However, there are other issues that led the House Armed Services Committee to conclude in 1989 that U.S. adoption of a centralized system was inadvisable.

In addition to the 1989 report by the armed services committee, Gansler and Henning (previously cited) and the General Accounting Office[7] have all noted at least two serious problems with U.S. pursuit of a centralized acquisition system. First, the number of U.S. acquisition personnel and the size of the acquisition budget exceed well over 10 times that of the European country most committed to defense expenditures, the United Kingdom. The ratio exceeds 15 or 20 times that of other European countries. The acquisition infrastructure of any single U.S. military service is 3 or 4 times larger than the entire acquisition establishment of any European country. Even a consolidated U.S. agency would be massive compared to its European counterparts. "Given the existing data, and without a great deal of additional study, it is not possible to determine whether such a large organization could work."[8]

So, while previous studies question whether any efficiencies or other advantages would be gained from centralization, another serious obstacle has also been identified. The European centralized models imply more than just a central procurement agency. Any attempt to successfully emulate them would require a realignment of the entire defense establishment. For example, strong central coordination of military requirements and the defense budget would also be necessary. The United Kingdom has previously found that military control of operational requirements did not work effectively with a central procurement agency.

The 1989 House Committee concluded that deeply rooted foundations of culture, national objectives, acquisition policies, and weapon system performance goals would have to change in order for the United States to actually copy the European model. U.S. commitment to worldwide mission requirements, attitudes about technology and risk taking, its history of detailed legislative oversight, and the frequently arms-length (some would say adversarial) relationship between government and industry are all factors that would affect the success of a centralized defense establishment. The consensus of previous research has therefore expressed at least caution if not outright opposition to U.S. adoption of a centralized procurement process.

Other Fundamental Differences

However, there has also been a strong consensus that certain European policies and practices make a lot of sense irrespective of a centralized procurement system. Notably, some of these ideas are aligned with recommendations made by the Packard Commission. This section identifies some of these other differences in acquisition philosophy and highlights those that, if adopted by the United States, would improve its weapons acquisition process.

Interface of the Political and Acquisition Processes.   It is no secret that the acquisition process has been subjected to extensive legislative oversight by the U.S. Congress. The process is intensely scrutinized, is highly legalistic and regulated, and receives an abundance of "free" advice. In his article "The Joint Chiefs of Congress," Hiatt links this detailed oversight to detrimental effects on the cost and schedule of several military projects.[9]

In addition to the voluminous and ever-changing array of legislation, U.S. acquisition programs are also subjected to a line item review of the budget. Funding profiles have often been erratic, and U.S. industry finds it difficult to plan beyond one or two years into the future with any confidence.

Finally, there are many high-level political appointees involved in the U.S. acquisition process. By comparison, Europeans achieve much greater program stability through their parliamentary process. Legislative oversight is much more limited, with the exception of Germany there is no line item review of defense budgets, and there are very few political appointees involved in the weapons acquisition business. The result is an acquisition process that is somewhat insulated from politics.

If adopted in the United States, the less legalistic and less political European approach could facilitate a more stable budget environment and better long-range planning by industry.

Government-Industry Relationship.   Former Assistant Secretary of Defense Jacques Gansler, among others, has noted that the United States is perhaps the only country in the world that fails to treat its defense industry as a valued national resource. The U.S. government promotes competition in the defense marketplace as if it were the same as any commercial market. In fact, there is only one buyer, there are significant barriers to entry and exit from the defense market, and the quantities purchased are frequently quite low. Yet, R & D investments are typically high and the buyer sets unique specifications and requirements.

Governments around the world tend to work more cooperatively with their defense industries in several respects. First, Russell and Fischer note that in France "defense companies have been rationalized."[10] For example, there is one helicopter company in France compared to five in the United States. A 1986 GAO report (cited earlier) identified the number of prime producers per product type (aircraft, tanks, ships, etc.) in various countries around the world. Not surprisingly, the United States typically has more producers per product type than other countries. Given the U.S. government's insistence on promoting competition for a relatively small production volume, this larger number of producers has not always been economically sound. In fact, in a world where defense spending has cycled upward and downward repeatedly, Gansler has noted that "The U.S. result is often an unhealthy defense industry—with considerable labor instability, program uncertainty and high-cost products."[11]

Another concern related to the large number of suppliers and forced competition is highlighted by examining Japanese practices regarding suppliers. Japanese success with just-in-time inventory systems and quality derives in part from their preference for long-term, stable relationships with a limited number of carefully chosen suppliers. Close, cooperative working relationships, improved quality, and a stable planning horizon have been achieved by the Japanese with this approach.

Finally, some additional comments about the arms-length, legalistic, and often adversarial philosophy of the U.S. government are appropriate. Other governments work closely with their industries in developing and achieving technological and industrial goals and in keeping the defense industry economically viable. In his 1989 book *Affording Defense,* Gansler noted that the Japanese Ministry of International Trade and Industry (MITI) is one example of a government organization steering a country toward a specifically planned industrial structure.[12] He also noted that MITI thrives on close working relationships among universities, the government, and industry, and that great emphasis is placed on compatibility between military and civilian technology.

With the exception of recent changes in the United Kingdom,[13] other countries promote competition much less aggressively than the United States. Whenever domestic quantities are too small to economically support more than one source, these governments assist defense companies with export sales and improving their international competitiveness. They contend that, even with a single supplier, the need to provide a good price on the export market is in itself a reasonable form of competition.[14]

Education and Experience of Acquisition Personnel. In its 1989 report, the House Armed Services Committee identified three European practices it considered particularly important for the United States. One of these was "professionalism and training of acquisition personnel." The personnel system within the French Delegation General for Armaments has been cited by numerous authors as a particularly good model.

An elite group of approximately 1000 armament engineers comprise the top leadership. These individuals receive an intensive, 7-year program of higher education and also gain hands-on experience in both R & D as well as manufacturing facilities. Entry into the program is highly competitive; candidates must pass a rigorous qualifying exam. Armament engineers hold military officer status and spend a portion of their early training with operational military units.

However, they have a separate but clearly defined career path that promotes them on the basis of engineering and management skills and compensates them on a par with their worth in the private sector. French program managers always come from this group of armament engineers. They operate in a much smaller bureaucracy with fewer layers of reporting and shorter chains of command, and are granted greater authority. They usually serve in a program management assignment for at least 5 years, which provides greater continuity and program stability.

The result of this system is appealing: program managers who understand the military but have no parochial interest in duplicating weapons, who have extensive education and experience before assuming senior leadership roles, and who stay in the job long enough to encourage planning and decision making for the best long-term results.

Cost-Performance Tradeoff. Russell and Fischer[15] noted that R & D expenditures in the United States are 10 times greater than in France. They further observed that U.S. equipment is "generally better, but not 10 times better." A related observation, and perhaps a partial explanation, is that the United States starts many more projects than it has money to produce. Further, many of these multiple development projects are aimed at the same operational requirement.[16] Much of this duplicate R & D never produces any payback through production or export sales. Worse yet, the United States has withdrawn from several international partnerships because one of its "other" programs turned out to be a higher priority. By comparison, development does not usually begin in Europe until there is a reasonable expectation that the system will be funded for production.

The Gansler and Henning study revealed that both Europe and the United States generally achieve their program goals. Given the United States' focus on maximum performance, U.S. equipment does generally perform better and is fielded a little faster. However, Europeans have also been successful in achieving lower costs and risks by settling for "acceptable performance." The study supported this conclusion through an analysis of costs incurred per unit of performance, which revealed that U.S. and European outcomes were essentially equivalent when viewed in those terms.

Noting these results in his recent book, *Affording Defense,* Gansler calls for the United States to examine program affordability more aggressively and to consider a more balanced tradeoff between cost and performance. Given the dramatic changes around the world and the associated drawdown in defense spending, these changes appear more essential than ever before.

Technology, Risk, and Mission Requirements. We have already noted that U.S. acquisition has traditionally relied on aggressive advances in technology to counter a known Soviet advantage in weapon quantities. While the issue is frequently debated, we have also accepted increased costs and risks in supporting this strategy. In the worldwide defense role that the United States has set for itself, it has also implicitly accepted operational performance requirements that exceed those of other countries. This is true in two respects. First, the United States requires advanced technology across the board in every weapons category, whereas other countries are more selective in their pursuit of advanced military technology. Second, the United States chooses to be capable of operating those weapons in any environment or location around the globe.

In the absence of a political decision to take less defense responsibility for the rest of the world, it would seem that the present situation is inflexible. However, Gansler and others have suggested that a more reasonable cost-performance balance is possible. To achieve this goal, a more rational and less duplicative requirements process is an absolute necessity.

Meeting this need does not require that the entire procurement process be centralized. However, it probably *is* necessary to have a strong, independent organization with the authority to guide the requirements process through the tough choices needed to eliminate duplication and conform to the cost realities of the 1990s.

THE HUMAN FACTOR: INDIVIDUAL DIFFERENCES

This part of the chapter looks more at the working level of acquisition management, that part of the business where some people would say "the rubber hits the road." First, fundamental differences in management philosophy and individual management tendencies are examined, especially in light of the differing environments that project managers from different countries operate within. Then, the effects of cultural differences on management styles are examined.

Because of the acquisition environment, U.S. managers may not be completely free to adopt other practices that seem worthwhile. However, there may be some practices that U.S managers would want to consider. That possibility aside, there is another reason for including this part of the chapter. Many U.S. managers find themselves involved in cooperative development and production ventures with other countries, and it is a great help to at least understand how your management counterparts tend to think.

Management Differences

Project managers (PMs) operate in a substantially different acquisition environment overseas. These differences exert considerable influence on management styles, and Owen Gadeken has found that they even influence perceptions of training needs.[17] These environmental differences also partially explain why many U.S. management practices differ from those of the country's allies. Accordingly, some of the key environmental differences are recounted here.

U.S. managers work in much larger organizations than their allied counterparts, and perform much more detailed oversight of contractors in an environment characterized by lack of trust. Further, U.S. PMs are subjected to much more external oversight and are granted less authority and freedom in the day-to-day management of their projects. In *The Defense Management Challenge,* Fox offered observations from DOD managers, industry, and the 1986 Packard Commission regarding PM authority:[18]

> The program manager finds that, far from being the manager of the program, he is merely one of the participants who can influence it.

Fox goes on to describe a PM's vulnerability to a host of "special interest advocates" who levy demands on the PM, yet the system allows the PM little discretion to use judgment to balance these sometimes conflicting requirements.[19]

> In DOD there is nobody really in charge as is a contractor's program manager. Instead, we have many interest groups that can influence the system. There are so many checks and balances that decisions are very slow. Many can say no and very, very few can say yes.

Finally, U.S. PMs face many more layers of management and a political system that holds them accountable at each step of the way. The more parliamentary processes in other parts of the world interfere with PMs much less during the acquisition process and judge the results *afterward.*

In this environment, Gadeken's research revealed that U.S. PMs do not personally perform many of the analytical, hands-on management tasks typically undertaken by allied managers. Instead, U.S. managers are forced to focus on leading and coordinating the efforts of large groups of specialists. A complaint of U.S. PMs is that they sometimes lack sufficient control over these "functional" experts.

Allied PMs, who experience much less oversight themselves and who generally have close working relationships with industry, exercise less oversight of their contractors and are more comfortable in delegating certain responsibilities to industry. Gadeken's research, which focused on differences between the United States and the United Kingdom, indicated that PMs in the United Kingdom perceived much lower need for personal training in the areas of budgeting, test and evaluation, and production because they trust industry to manage these aspects of their programs.

The continual review of defense acquisition programs, which sometimes becomes both political and public, produces more uncertainty about whether U.S. projects will continue once under way. U.S. PMs, who are usually military officers with a strong action orientation and a brief tenure in their position, frequently develop a strong program advocacy and a comparatively short-term planning horizon in response to this environment. While the response is understandable and even predictable given the circumstances that often prevail, these practices have been challenged by various experts as suboptimal.

Cultural Differences

This section does not extol the virtues of copying other cultures. However, given the frequently international nature of business in the commercial as well as the defense sector, it is vitally important to be aware of cultural tendencies which may be different than our own and which may exert significant influence on our business transactions.[20]

Richard Hodgetts, Professor of Business at Florida International University, has defined culture in the following manner:

> ...acquired knowledge that people use to interpret experience and to generate social behavior....This knowledge forms values, creates attitudes, and influences behavior.

Culture can affect technology transfer, managerial attitudes, managerial ideology, and even business-government relations. Perhaps most important, culture affects how people think and behave.[21]

During 1991, Anthony Amadeo conducted an award-winning study of cultural dimensions of international business.[22] Among other issues, Amadeo examined the effect of cultural differences on management practices on international armaments projects. The remainder of the section summarizes the key findings.

While many Americans approach decision making as an exact science, other countries often use a completely different style which can prove frustrating. As compared with many international counterparts, Americans prefer quick factual decisions made by an individual who has been delegated the authority. Much of this tendency derives from American culture and its emphasis on time, individuality, and goal orientation. In contrast, other cultures place more importance on group decisions. Patience and flexibility are two mandatory characteristics for success in the international environment.

Two more management areas which are significantly affected by culture are the treatment and use of lawyers and written contracts. American ideas about the use of lawyers and contracts are not always consistent with those of other countries. Americans generally require the use of contracts with each aspect of the agreement stipulated in writing. If it is not in writing, then there is not an agreement. Also, lawyers are accepted members of a business team and are often used as a means to resolve disagreements. In fact, lawyers are often very influential members of the U.S. negotiation process. This is usually not the case internationally. Much more time is spent building relationships in other countries in an attempt to reduce the importance of written contracts and lawyers. Agreements in other countries may range from a simple handshake to a formal document. In some cases, a verbal agreement can be more important than a contract. In some cultures, a contract is merely an indication of work that is intended to be completed.[23] The implication of bringing a lawyer to an overseas business meeting is generally one of mistrust.

The significance of these two areas provides some insights into another important aspect of cultural dimensions, communications. It is essential that both parties fully understand exactly what agreements are being made. In many cases, this must be accomplished without the use of complex contracts and lawyers.

Amadeo's analysis indicated that differing management practices/styles do create obstacles that program managers must overcome. The biggest problem associated with these differences was a general slowing of the entire process.

The analysis also revealed that developing personal relationships is somewhat more important in the international environment; however, they were not considered a major problem area. Respondents were split over whether lawyers should be actively involved in international transactions or negotiations.

Also, several factors were identified that were considered important in determining the success of international transactions. The most important of these factors were preparedness and patience of the U.S. team. International dealings were generally considered to be more time-consuming; international partners lack the same sense of urgency to "get the job done and move on" that typically characterizes U.S. managers.

Based on his research findings, Amadeo drew several conclusions concerning the effects of cultural factors on successfully negotiating and managing international programs. I have summarized those that are most relevant to program management practices:

1. Several cultural factors can greatly influence the success of international negotiations. Factors associated with different managerial styles, negotiation tactics, legal

systems, financial processes, etc. presented the greatest problems. While it is impossible to completely resolve these differences, it is possible to lessen them with knowledge, understanding, and preparation. Surprisingly few problems were associated with any perception that there were deficiencies in the international counterpart team. U.S. team deficiencies such as insufficient planning, insufficient authority, and lack of experience were believed to affect success to a greater extent. The international negotiator directly controls these factors and can greatly enhance the chances of success by effectively managing them.

2. It is important to adjust your international negotiating style according to the culture with which you are dealing. An international negotiator must not only be aware of cultural differences which exist, but also individual differences. The international negotiator must be more flexible and patient than in U.S. negotiations. The international negotiation process is more time-consuming, with more emphasis on establishing personal relationships through informal conversations and entertainment.

  Small, internationally experienced negotiating teams make this goal easier to achieve. This may require the use of a local culture expert or experienced international business traveler to accompany the team. The use of a core international negotiating team with additional technical experts partially addresses these concerns. This team should also have sufficient authority to make decisions locally without excessive calling or faxing back and forth to the United States.

3. As part of the planning and preparation for an international negotiation, the negotiator must take the time to study the counterpart's culture. While the importance of this cannot be overstated, there is a definite lack of cultural understanding by U.S. business people. The successful U.S. businessperson cannot simply transfer the same knowledge, tactics, and techniques overseas and expect to be successful. The importance of this cultural preparation must be emphasized through international training courses and literature.

4. As indicated, personal relationships are more important in the international arena. It is important to create an atmosphere of trust throughout the entire program. This trust can be built through social contact, honesty, patience, and understanding of different cultures/customs. The U.S. businessperson must plan for additional time to be spent building these relationships. International transactions should not be approached with the typical U.S. attitude of "time is money." U.S. managers must also recognize that informal social events are often far more productive than the officially scheduled meetings.

5. While lawyers have a role in the international environment, U.S. teams traditionally place too much emphasis on their participation in the process. Internationally, lawyers signify a feeling of mistrust which greatly hampers the development of personal relationships. Therefore, Amadeo concluded that lawyers should review U.S. positions, but should not usually accompany U.S. teams on international travel. If it is necessary for lawyers to be a member of the team, their role during face-to-face transactions with international counterparts should not be a prominent one. Additionally, honesty and forthright explanations concerning U.S. business practices can go a long way in dismissing any feelings of mistrust.

6. Several management and/or organizational factors have important influences on the success of international transactions. These factors are directly controlled by the program manager and therefore should be considered before venturing into an international program. Like factors which affect the success of negotiations, these factors (such as preparedness, patience, and technical expertise of the U.S. team, familiarity with the counterpart's business practices and customs, and personal ties

built through the years) are important in determining the success of international transactions. Preparedness and proper planning are the keys to success. A large part of this planning is the understanding of your counterpart's culture and customs. Proper planning and understanding of differences can overcome most of the cultural barriers to successful programs.

SUMMARY

On the basis of the strong consensus found in previous studies, this chapter does not advocate copying the European centralized procurement process. However, also on the basis of strong consensus, there are some international practices that deserve serious consideration by the United States, and as noted at the outset, some of these practices coincide with recommendations of the 1986 Packard Commission. Unfortunately, as one moves from the political environment through the DOD hierarchy and on down to the working level of project management, it becomes increasingly difficult for managers to implement new ideas based on their own judgment. Certainly program managers can become more aware of the potential for adopting good ideas from overseas and can achieve varying degrees of success in making some of those changes. However, full realization of the potential benefits will require cooperation between Congress and DOD that has not been forthcoming in the past. On a brighter note, however, it has been said that necessity is the mother of invention. In the face of dramatic changes around the world, the opportunity has probably never been greater for taking positive steps toward resolving some of the historical problems with the defense acquisition business.

ENDNOTES

1. F. C. Berry, Jr., "Defense Procurement, European Style," *Air Force Magazine,* February 1989, pp. 74–77.

2. Ibid., p. 75.

3. "A Review of Defense Acquisition in France and Great Britain," Report of the Subcommittee on Investigations, House of Representatives, Committee on Armed Services, 101st Congress, 1st Session, U.S. Government Printing Office, Washington, Aug. 16, 1989.

4. J. S. Gansler and C. P. Henning, "European Acquisition and the U.S.," *Defense and Diplomacy,* 1989.

5. Ibid.

6. Berry, op. cit., p. 77.

7. GAO, United States General Accounting Office, "Weapons Acquisition: Processes of Selected Foreign Governments," GAO/NSIAD-86-51FS, Washington, D.C., February 1986.

8. Reference 3, p. 10.

9. F. Hiatt and R. Atkinson, "The Joint Chiefs of Congress," *The Washington Post,* National Weekly Edition, 12 Aug. 1985.

10. T. B. Russell and C. K. Fischer, "How to do Business with the French Delegation General for Armaments," Army Materiel Command Representative—France, Office of Defense Cooperation, New York, May 1989.

11. Gansler, op. cit., p. 35.

12. J. S. Gansler, *Affording Defense,* MIT Press, Cambridge, Mass., 1989, p. 311.

13. The United Kingdom has become very aggressive in promoting competition within its defense industry, one of the only countries in the world with policies similar to those in the United States. A major difference is that the United Kingdom is more comfortable whenever the competition has an international origin.

14. This sentiment was expressed during personal interviews conducted by the author during the summers of 1990 and 1991 with defense officials in Rome, Bonn, Paris, The Hague, London, and Brussels.

15. Russell and Fischer, op. cit., p. 6.

16. C. M. Farr, "Managing International Cooperative Projects: Rx for Success," Chap. 6, *Global Arms Production: Policy Dilemmas for the 1990s,* E. B. Kapstein (ed.), Center for International Affairs, Harvard University, 1992.

17. O. C. Gadeken, "Through the Looking Glass: Comparisons of US and UK PMs," *Program Manager,* Defense Systems Management College, Fort Belvoir, Va., November–December 1991, pp. 22–26.

18. J. R. Fox, *The Defense Management Challenge: Weapons Acquisition,* Harvard Business School Press, Boston, 1988.

19. These special interest advocates represent legitimate concerns related to issues such as small and minority business utilization, competition, preference for domestic sources, maintainability, reliability, and producibility. The difficulty is that, by comparison to allied counterparts, U.S. PMs have little freedom to exercise personal judgment in balancing these concerns.

20. The research fellows program at the Defense Systems Management College produced studies 2 years in a row, 1990 and again in 1991, on international armaments cooperation. The first report was *Europe 1992: Catalyst for Change in Defense Acquisition* by Cole, Hochberg, and Therrien. The second was *International Cooperation: The Next Generation* by Johnson, Engel, and Atkinson.

21. R. M. Hodgetts and F. Luthans, *International Management,* McGraw-Hill, New York, 1991, p. 35.

22. A. L. Amadeo, *Cultural Dimensions of International Business,* Master of Science Thesis, Air Force Institute of Technology, Dayton, Ohio, September 1991.

23. L. Copeland and L. Griggs, *Going International: How to Make Friends and Deal Effectively in the Global Marketplace,* Random House, New York, 1986, p. 94.

CHAPTER 24

THE MANAGEMENT STYLE OF MULTICULTURAL CONSTRUCTION MANAGERS IN THE MIDDLE EAST

Adnan Enshassi

Al-Fatah University,
Tripoli, Libya

Dr. Enshassi is a lecturer in the Civil Engineering Department at Al-Fatah University, Tripoli, Libya. He qualified as a civil engineer from Suez Canal University, Egypt, and was subsequently employed, in private practice, for a major construction organization as a project manager in the Middle East. He is the Middle East consultant for several German and British business organizations. He is a member of the Chartered Institute of Building, an Associate of the British Institute of Management, a member of the Institute of Management Specialists, and a founder fellow of Professional Business and Technical Management.

INTRODUCTION

In this chapter, the research studies which derived and elaborated the dimensions of leadership are to be reviewed; these are early approaches of leadership studies, normative approaches, and contingency theories. Furthermore, the relationship between culture and management style generally, and in the Middle East particularly, will be examined. Then the issue of leadership in the construction industry will be underlined.

The subject of leadership styles has been thoroughly studied in the field of organizational behavior and industrial psychology as a key factor in the improvement of industrial performance. In the construction industry, very few studies have explored the supervising leadership orientation of construction site managers and the managerial style which they adopt when managing a work force on site, and even fewer such studies have considered closely the site manager's style practiced when managing a multicultural work force on construction fields, and the association between such styles, the manager's effectiveness, and construction project performance.

Empirical data are provided which cover United Arab Emirates, Oman, Kuwait, Bahrain, Saudi Arabia, and Libya from a sample focusing on both international and local construction firms. The chapter proposes some implications for the study of leadership style, effectiveness and productivity in multinational organizations.

THE NATURE OF THE CONSTRUCTION INDUSTRY

Several scholars pointed out that there are distinct differences between features of the construction and manufacturing industries.[1-4] Such differences suggest that the process of leadership style on construction projects is not the same as in the manufacturing industry.

The construction industry has been defined in the Standard Industrial Classification (cited in Ref. 2), as follows:

> Erecting and repairing buildings of all types. Construction and repairing roads and bridges, erecting steel and reinforced concrete structures, other civil engineering work such as laying sewers, gas or water mains and electricity cables, erecting overhead lines and live supports and aerial masts, extracting cool from open cast workings etc. The building and civil engineering establishments of government departments, local authorities and new town corporations and commissions are included. Establishments specializing in demolition work or in sections of construction work such as asphalting, electrical wiring, flooring, glazing, installing heating and ventilation apparatus, painting, plastering, plumbing, roofing. The hiring of contractor, plant and scaffolding is included.

The main differences which have the greatest bearing *on the style of construction site managers are*

1. The construction firm's operations are spread over several sites, which gives site managers more opportunity to practice their own styles in managing work forces than they would have in industrial organizations where rules, regulations, and procedures are more established.

2. Construction organizations at the field (project) level are short-lived, in the range of 1 or 2 years, compared with stable manufacturing organizations. This results in fluctuation of work forces; workers need to be hired and fired in some situations. This temporary relationship between site managers and their subordinates could affect the style of site managers. In contrast, in stable or permanent manufacturing organizations, more time and opportunity is normally available for establishing and improving the relationship between managers and their subordinates, thus improving in turn the style of management.

3. The process of construction in the Middle East is affected by the summer season, as most of the work is carried out in exposed outdoor situations rather than indoors. Workers cannot work effectively during the midday period because of the high temperature. This situation will affect the time schedule of a construction project; frequent delays can be expected, and such circumstances can have an impact on the style of a site manager who is under pressure to complete projects on time. In other industries, the work is mostly indoors, where, in most cases, air conditioning is provided, thereby reducing the influence of such environmental problems.

4. The products of the construction industry are tangible, such as houses, schools, hospitals, roads, and bridges. This tends to motivate most work forces and increase job satisfaction, as they can directly see the result of their work. This in turn might shift the managerial style of construction site managers. Often, in manufacturing and service industries, the work output is either intangible or unidentifiable.

DIMENSIONS OF LEADERSHIP STYLE

Early Approaches of Leadership Studies

The Ohio State Leadership Studies.   The aspiration of Ohio State researchers was to identify leadership behavior dimensions. Two main categories of leadership behavior have been observed: initiating structure and consideration.[5-8] The definitions of consideration and initiating structure have been given by Fleishman and Peters[9] as:

> Consideration reflects the extent to which an individual is likely to have job relationships characterized by mutual trust, respect for subordinates' ideas, and consideration for their feelings.
> Initiating structure reflects the extent to which an individual is likely to define and structure his role and those of his subordinates toward goal attainment.

The Ohio State leadership studies had been criticized for lack of a conceptual basis and overlooking the effect of situational variables.[10,11]

The University of Michigan Studies.   Scholars of the Michigan Study Center elicited two major leadership behavior dimensions, namely employee-centered, which emphasized the human relations within the work group, and production-centered, which concentrated on the technical aspect of the job and productivity.[12] The Michigan Study approach draws heavily on the Ohio dimensions of consideration and initiating structure.[11]

There are some methodological differences between Ohio State and the Michigan Center. For example, in the Michigan School, the leadership dimensions were derived from supervisors by using structured interviews, whereas the Ohio researchers drew their leadership classification largely by using subordinates' views of leadership behavior through a self-administered questionnaire.

Bales' Socioemotional and Task Orientation.   Bales derived two types of leadership dimensions, namely task orientation and socioemotional orientation. His distinction was based on observational rating techniques of real leader behavior. He observed that some leaders can have and practice both characteristics to meet the need of various groups.[11,13,14]

Normative Approaches of Leadership

Blake and Mouton's Managerial Grid.   One of the very best known behavioral leadership theories is the managerial grid, developed in the United States by Blake and Mouton.[15] On the basis of the Ohio State research findings, they developed a grid which encompasses these dimensions of management styles: concern for production and concern for people. Blake and Mouton believed that both concerns are crucial components to management effectiveness. They postulated that a manager could exhibit any of the following five styles:

(*a*) Low task and relations behavior

(b)   Low task and high relations behavior
(c)   High task and low relations behavior
(d)   Medium task and relations behavior
(e)   High task and relations behavior

Blake and Mouton described the last category as the most effective style, where managers build a cohesive and cooperative work team.

Reddin's Tridimensional Model of Leadership Effectiveness.   Reddin[16] elaborated and developed the managerial grid concept by adding another dimension, that is, effectiveness. The model provides an evaluational framework by which an assessment of the suitable style (task or relations orientation) can be made. This model is considered a predictive model for whether a leadership style will be effective or ineffective. This model is known as the most comprehensive within the normative leadership theories, as it provides a forecast of the relationships between leadership behavior and situation.[14]

Likert's System 4.   Another theory of leadership developed by Likert[17,18] is called *Likert's (System 4) management approach.* This postulate was based on the Michigan leadership research. Likert defined the management system as a function of managerial leadership and organizational climate. Managerial leadership, which is assumed to be effective, includes: team building, goal emphasis, help with work, and involvement. Organization climate comprises: decision-making procedures, communications flow, motivation, and technical training.

Likert hypothesized four types of managerial behavior, organized on a continuum ranging from:

1. Exploitative authoritative
2. Benevolent authoritative
3. Consultative
4. Participative group

He concluded that "System 4"—participative group—is the most effective style, as it helps to enhance productivity by developing a cohesive and a dynamic work group and promoting team loyalty.

The Concept of Hi-Hi Leader Behavior.   A high-high leader is the one who combines both high employee and high task orientation. This phenomenon of leadership behavior is called a *hi-hi paradigm.*[10,19] The hi-hi paradigm asserts that leaders should be high in both employee and task orientation to achieve high subordinate performance.

There are many researchers who supported the hi-hi paradigm. They believed that there is a positive association between leaders who are high in both initiating structure and consideration and subordinate's satisfaction and productivity.[20–24]

The hi-hi paradigm can be classified into two categories: hi-hi interactive and hi-hi additive models. The former means that there is an interactive relationship between employee orientation and task orientation. The latter is a relatively more complex hi-hi paradigm; it means that there is an additive association between employee orientation and task orientation.[19,25] In both models, a leader has to be high on both categories. If a leader is not high on both, the direction of the criterion relationship might be shifted. This seems to be a very complex situation from the practical sense, as it is very difficult to measure such phenomena.

The hi-hi leader paradigm has been highlighted by Blake and Mouton.[15] They stated that optimality can be achieved by combining concern for people and concern for task, regardless of situation, e.g., job level, composition of work group, location of project, or size of organization. Stogdill[26] reported that several leadership studies advocated the hi-hi paradigm. In their empirical study, Fleishman and Simmons[27] concluded that leaders who combine both high consideration and initiating structure are likely to optimize the effectiveness of supervisors, hence the subordinate's performance. The findings of Lansley et al.[28] support the hi-hi leader paradigm, and they concluded that the successful building companies are characterized by a high task/high people orientation in management style.

On the other hand, several researchers found that the agreement concerning the hi-hi paradigm is not universal, and a number of exceptions to the general rule that a hi-hi leadership style is the most effective have been identified.[29] Consideration has been found, occasionally, to have a negative correlation with high management level,[30,32] and initiating structure associates negatively with subordinate satisfaction and turnover[31] and may adversely affect performance.[34,35] Fleishman[33] stated that recent research has concluded that leadership is, to a great extent, situational, and that what is effective leadership in one situation may be ineffective in another. Hunt and Liebscher[29] pointed out that the preference for and attitudes toward consideration and initiation were found to differ widely as a function of the individual and the situation.

Nystrom[25] replicated and extended the Larson et al.[19] study of the hi-hi leader paradigm. He examined both the managers' own leadership point of view and their perception of their bosses' leader orientation. Although Nystrom used a different scale of measurement, his research did yield the same finding as Larson et al., namely he did not give support to the hi-hi leader paradigm. He suggested, also, that the supposed utility of the hi-hi paradigm myth should be abandoned.

Contingency Approaches

The premise of these approaches is that the effectiveness of leadership styles is situationally contingent. In other words, a particular style which was successful in one circumstance might not be so in other situations. This means that there are no specific leadership styles which can be universally considered. Several contingency approaches have been developed to give a theoretical justification to the thought that a particular situational factor moderates the relationship between a leader's style and the performance of a group.

Fiedler's Contingency Model. Fiedler's approach[36–38] was a self-administered questionnaire, where responses are gleaned from people in leadership positions. In Fiedler's model, which is known as *least preferred coworker* (LPC), leaders were asked to think of a coworker with whom they least preferred working.

Fiedler[36] believes that the characteristics of a situation (e.g., location of project, nature of tasks, and composition of group) will moderate the relationship between leader orientation and performance; therefore leaders are required to adapt themselves according to a situation's features in order to enhance the group performance. Fiedler added there is a strong association between leadership style and the favorableness of the work team being led. According to Fiedler's definition, situational favorableness encompasses three variables:

1. *Leader-member relations.* The group atmosphere has a feasible impact on a leader's effectiveness; if subordinates accept and like their leader, then the leader will have no difficulties in leading the group and getting high performance.

2. *Task structure.* Clear, structured, and well-planned tasks facilitate a leader's function in directing and monitoring work performance. In addition, work teams will have clear task definitions, and consequently they will accomplish their tasks with fewer problems.

3. *Position and power.* This refers to a leader's ability and capacity to give rewards and impose punishment to team members. If a leader has a strong position, more respect will be paid to him, as he will, probably, be able to reward good performance and punish bad performance.

Unfavorable situations are the opposite of the above-mentioned.

In recent research, Fiedler[38] concluded that task-oriented leaders are likely to be very effective in both favorable and unfavorable situations, whereas relationship-oriented leaders are predicted to be more effective in situations of moderate favorability. Fiedler's contingency model of leadership effectiveness received support as well as criticism. Barrow (Ref. 14, p. 234) reported that:

> The contingency model methodology does not lend itself to analysis of the leadership process nor does it appear flexible enough to allow for incorporation of new variables. Consequently, Fiedler's model is not often predictive of leadership effectiveness....

House's Path—Goal Theory of Leadership.   As the Fiedler approach has been criticized by several scholars[14,39,40] for the lack of a theoretical framework, House[41] built his approach on the *expectancy theory* for work motivation.

Bryman (Ref. 11, p. 137) defined the expectancy theory as follows:

> ...Expectancy theorists propose that people choose levels of effort at which they are prepared to work. The choice of a high level of effort is contingent upon their assessment of whether it leads to good performance and value (called "valence" in the language of expectancy theory) of good performance to them.

House believed that the leader is a crucial source of motivation, as he can improve the performance of a work group through his behavior toward his subordinates. Four classifications of leadership behavior have been established[42]:

1. *Instrumental leadership:* Leaders in this case tell the employees what to do and how work should be executed, and put emphasis on productivity.

2. *Supportive leadership:* Leaders tend to be friendly and approachable and they develop good relations with the group members.

3. *Participative leadership:* all parties are involved in decisions.

4. *Achievement-oriented leadership:* leaders place confidence in the group members to achieve a high performance.

There are two main situational factors, upon which the appropriate leadership is dependent:

• Personal characteristics of the group member

- Environmental factors which include the subordinates' tasks and the organization of formal responsibility

House and Mitchell[42] believes that the performance of subordinates can be enhanced if the leader clarifies goal paths. He concluded that leaders who practice a considerate approach in situations characterized by low ambiguity of subordinates' roles can be expected to be more effective. On the other hand, in situations characterized by high job complexity and high role ambiguity, the initiating structural leadership style, which puts more emphasis on tasks' accomplishment than on human aspects of the employees, will be more effective.

Hersey and Blanchard's Life Cycle Theory of Leadership. Hersey and Blanchard[43] developed a concept, which they named *life cycle theory of leadership,* based on the integration of managerial grid theory, Reddin's 3-D leadership effectiveness, and Argyris's maturity-immaturity theory. In 1977, Hersey and Blanchard renamed their theory to *situational leadership model.*[44] They employed the traditional dimensions of leadership style, namely task orientation, which is equivalent to initiating structure, and employee orientation, which is equivalent to consideration; and established their theory in which the task-relevant maturity of subordinates was the main situational factor.

Subordinate task-relevant maturity is suggested to include two components: job maturity and psychological maturity. Job maturity refers to the groups members' knowledge, experience, willingness, and ability to take responsibility. Psychological maturity refers to the level of members' motivation toward task completion.

The criterion of this theory is that, when the maturity of subordinates increases, leadership style should be characterized by a reduced emphasis on the task structuring dimension and an extended emphasis on the consideration dimension. If the maturity continues to rise, there should be an eventual decrease in consideration style used.

Hersey and Blanchard's situational leadership theory has been criticized by Graeff[45] because it considered subordinates as the primary situational factors and neglected other situational factors affecting leader-subordinate relationships. It provides also a definition of maturity which is hard to measure and has not yet been empirically tested.

LEADERSHIP STYLE AND CULTURE

Differences and Similarities in Leadership Style across Cultures

Hundal[46] concluded from his research that, although leadership principles are universal, the methods and procedures by which they are adapted to each culture and work location decide their success and failure. Gonzales and McMillan[47] stated that management practice in multinational organizations is culturally bound and must be closely associated to the influx of technology moderated by cultural norms of behavior.[48]

In their review of literature, Barrett and Bass[49] found that several cultures were still in favor of an authoritarian managerial style despite the conclusions of several studies which support a tendency toward participatory managerial attitudes.

Child[50] stated that there are several factors associated with the divergence in management: organizational, political, and cultural factors. Schein[51] lent more support to the impact of cultural factors on the leadership process when he reported that culture and leadership practice cannot be separated. He added that the majority of leaders are influenced by their cultural background.

Whyte and Williams[52] concluded from their survey that the following supervisory characteristics of leadership have been found to hold true in United States industry:

1. Workers prefer a supervisor who practices general supervision to one who practices close supervision.

2. Workers do not prefer a supervisor who places very high emphasis on the level of productivity achieved.

3. Workers prefer a supervisor who allows them to participate in any decision regarding changes and who takes their opinion into consideration.

4. Workers prefer a supervisor who lets superiors know about the work group problems and requirements.

Although Whyte and Williams did not label the above preferred characteristics of leadership practice, it can be noticed that such leadership style can be categorized as an employee-orientation style. This corresponds to the general point of U.S. companies who are known for their progressive personnel policies, and their good relations with their workers.[52]

Furthermore, Whyte and Williams had tested the validity of the above-mentioned four criteria in an electric light and power company in Peru. They found that there is a common satisfaction and agreement between the U.S. and Peruvian workers regarding a supervisor who understands the needs and problems of the workers, and who is interested in improving their skills by training.

On the other hand, they found that, for the Peruvian workers, there was a positive correlation between closeness of supervision and general satisfaction with the supervisor. The Peruvian workers expressed their preference and satisfaction for a supervisor who tells them what to do and observes their progress rather than one who offers general supervision and leaves the workers on their own most of the time. This is exactly the opposite of the U.S. workers' preference.

While the Peruvian workers tend to have a high respect for a supervisor who puts emphasis on the level of productivity achieved, rather than one who gives lower emphasis, the U.S. workers give no regard at all to the supervisor who puts emphasis on productivity. Finally, there was a positive correlation between participation and U.S. workers' satisfaction; for the Peruvian workers, although there was a positive correlation, it was comparatively low.

It can be clearly seen from the above discussion that culture does affect workers' expectations and perceptions. Consequently, any supervisory leadership style that is successful in one culture will not necessarily be so in another culture.

However, Whyte and Williams did not elaborate on the nature of cultural differences which affect workers' expectations.

One of the largest and most thorough cross-cultural studies of subordinates' perceptions of leadership style was carried out by Hofstede.[53,54] As a result of his survey, which covers 40 countries around the globe and includes 116,000 respondents, Hofstede found distinct differences across countries in the perception of and preference for leadership style. He reported that participative management approaches which were suggested by American theories and are applicable in the United States were found not appropriate to all countries.

Another study has been conducted by Maier and Hoffman[55] in which they compared British and American managers; they found that the British managers seemed more accustomed to an authoritarian style than their American counterparts. This suggests that, although countries can be categorized into distinct groups, such as Western

European, Mideastern, Asian, etc.,[53] differences do exist within each group. On the other hand, Heller and Porter[56] concluded from their study that the operational practices of English and American managers are very similar; this does not support Maier and Hoffman's result.

Negandhi[57] concluded from his research of 56 American subsidiaries in Latin American and Far Eastern countries, and 55 companies in the same region, that culture has an influence on employee morale and interpersonal relationships. Negandhi reported that there were variations in the management practices and methods between the United States subsidiaries in developing countries and the local corporations there; the leadership style in the U.S. subsidiaries was found to be democratic, whereas in the local firms it was an autocratic style. Having said that, Negandhi pointed out that several management practice, e.g., planning and decision making, are not influenced by cultural diversity but are contingent upon technological and market conditions. Negandhi[57] lends support to the belief that management styles vary among cultures, tending to refer such variation to technological and economic discrepancies rather than cultural variables.

A comprehensive cross-cultural study on managerial attitudes, motivation, and satisfaction among 3000 managers in 14 countries was conducted by Haire, Ghiselli, and Porter.[58] They concluded that there were many similarities in managerial attitudes and motivations among managers in all countries. Thiagarajan and Deep[59] concluded from their survey, which included 700 managers in several countries (e.g., France, the Netherlands, United Kingdom, and America), that generally, irrespective of their cultural backgrounds, all managers preferred democratic supervision. This result seems to be consistent with Haire, Ghiselli, and Porter's result.

Massie and Luytjes[60] reviewed the literature with respect to managerial practices in Western and Eastern Europe, South America, Asia, and Africa, and found that there is a tendency toward a similar management practice. They referred to several factors which contributed to such a result, among them the increase of management education across cultures, management functions (e.g., planning and organization), and participation in decision making.

Some cross-cultural scholars tested and advocated the applicability of American leadership theory. Other researchers supported such applicability but in a different interpretation. Most American managers, for instance, support McGregor's theory, which relates to the motivation of employees by satisfying their higher-order needs, as the basic needs for safety and security in the American environment have been met.[61]

The Characteristics of Management Style in the Middle East

A study exploring selected managerial issues (e.g., decision-making style, interpersonal relations, attitudes toward time and change) was conducted by Muna.[62] He interviewed 52 top Arab executives from six countries: Egypt, Jordan, Kuwait, Lebanon, Saudi Arabia, and United Arab Emirates. Some of those executives were the owners of their firms, while others were top managers of business organizations involved in trade or services, or conglomerates. Muna concluded from his research that Arab executives' style is influenced by traditional consultative style and not participative leadership style. Muna gave several reasons for such a choice, for instance: employees expect, normally, to be consulted in decision-making and not to be included in the final decision; their taking part in the decision might reflect a weakness of the executive.

Another empirical study in which the styles of Mideastern managers have been investigated was by Badawy.[63] His sample covered 251 managers who were attending executive development programs for midlevel managers; they came from Saudi Arabia, Kuwait, Abu Dhabi, Bahrain, Oman, and United Arab Emirates. They were working for chemical, petroleum, and transportation industries. Badawy's study led to several conclusions:

1. The Mideastern managers were found to be in favor of a traditional approach influenced by their culture. This gives support to Muna's findings.[62]

2. Badawy found that no significant differences existed within the six subnationalities represented in his sample. All managers from those six countries seem to share the same view toward a general management style.

3. The data suggested that, despite the variation of Mideastern managers' length of experience, they have fairly democratic attitudes toward control of rewards and punishments.

4. A democratic management style was found in the small-sized organizations rather than in the large ones.

Finally, Badawy (Ref. 63, p. 58) stated that: "...fundamental differences in management style exist between Mideastern managers and their western counterparts. Arab people differ from one country to another, but there are remarkable similarities among them."

More recently, Al-Jafary and Hollingsworth[64] carried out a study into managerial practices in the Arabian Gulf region. Their sample covered 381 managers in the top levels of 10 multinational organizations involved in production, petrochemicals, and marketing and was collected from Saudi Arabia, Bahrain, Qatar, and United Arab Emirates. Al-Jafary and Hollingsworth's study was based on Likert's[17] management system classification and found that managers in the Middle East operate within the management System 3 mode (consultative). Moreover, they reported that, although managers in the Middle East are practicing a consultative style, they have a tendency toward the participative style (System 4).

Previous Leadership Style Studies in the Construction Industry

Two research projects have been carried out in the U.K. construction industry, studying management style and its association to the company performance.[3,65] Lansley (Ref. 65, p. 3) defined management style as follows:

People orientation:

> The extent to which managers give priority to the welfare, and involvement of staff as compared to enhancing their own personal standing in the organization (i.e. employee centered v. self centered).

Production orientation:

> The level of concern of managers for efficiency, productivity, and system of management control.

Lansley et al.[28] have conducted research in small- and medium-sized printing and building firms, and examined the relationship between organization structure, management style, and their influence on company performance. They measured the differences between companies' management style by adapting Blake and Mouton's model. They concluded that poor performance was related to low task orientation combined with low people orientation, and high performance was related to high task orientation, disregarding the differences in task between the printing and building industries. Although they generalized their findings, they added: "The only important difference between the two industries was a tendency, in the building sample only, for high performance, particularly commercial performance, to be associated also with a high people orientation."

This, to some extent, supports the argument that management style in construction is different from other manufacturing and service industries. Moreover, Lansley et al.[28] supported the hi-hi paradigm (high task-oriented management, and high people-oriented management), as they found in the building firms that poor performance is associated with low-task/low-people orientation, and high performance is associated with high-task/high-people orientation.

Recently, Brensen et al.[3] have investigated the issue of leader orientation of British construction site managers. Fiedler's contingency model was employed to examine the relationship between site manager's orientation (task orientation or relationship orientation) and the performance of the project. Brensen et al. concluded that site managers' orientations have an impact on project effectiveness, and that this relationship is dependent on project length, contract values, and labor-force composition. In particular, they found that the implementation of a people-oriented management style on a construction site is more likely to improve the project performance than the practice of a task-oriented management style, irrespective of the degree of favorability in this situation.

Favorability here is taken to mean situations where the task undertaken is routine, the positional power of the leader is high, and leader-member relations are good; unfavorability is the opposite of these conditions.[3] Having said that, Brensen et al. found that site managers appeared to be strongly task-oriented. This is an interesting point which needs to be investigated in relation to multicultural construction firms in the Middle East.

On longer-duration, larger-scale projects, the relationship between site managers' orientation and performance increased significantly; whereas in shorter, smaller projects, there is no recognizable relationship.[3] Borcherding and Garner[66] raised this observation previously, stating that on large construction sites, the leadership style is a very important issue because of the site complexity and the problems of morale and motivation.

With respect to work force composition, it was found that there is an association between leadership orientation and increase in the performance level when the majority of work force were employed directly by the main contractor. There is no real association between performance level and the orientation of the site manager on sites employing a great deal of subcontract labor.[3]

Bryman et al.[67] reported that construction site managers (as leaders in temporary organizations) have stronger task orientation than their counterparts in more permanent organizations. They found also that the time span of the construction project moderates the relationship between leadership orientation and performance. Bryman et al. (Ref. 67, p. 18) stated that:

> There is an evident irony that, while site managers as a whole are markedly task-oriented, the greater the emphasis on relationships (especially on longer projects) the better the performance.

PRODUCTIVITY OF MULTICULTURAL WORK GROUPS

The level of a group's productivity depends on various factors, for instance, climate conditions, availability of materials, availability of plant and its modernity, financial liquidity, work practices, planning, style of management, and the composition of the work group. In this section the level of a group's productivity is related only to the composition of work forces, and other factors are considered to be constant. Each ethnic group has different abilities and skills, which can be expressed in terms of productivity ranging from totally unproductive to highly productive. From the site manager's view, the level of productivity in the case of multicultural work groups fluctuates more than in the case of homogeneous groups.

The cultural diversity of a group has both positive and negative impacts on productivity. According to Triandis et al.'s experimental investigation,[68] the best way of maximizing the efficiency of a group is to ensure that group members are homogeneous in ability level to ease accurate communication, and heterogeneous in attitudes to assure a variety of solutions and alternatives to problems. Woodcock (quoted in Ref. 61) reported that the primary factors for the success of Japanese automobile industries was the cultural homogeneity of the Japanese work force. He argued that the homogeneity of a work team fosters a greater solidarity and improves morale, which then results in an increase in productivity and the quality of the product. Cartwright[69] and Anderson[70] supported such logical argument, stating that the likeness between work team members reinforces the cohesion of the team, which will lead to an increase in productivity level.

It has been found that a more diverse work group enhances productivity despite the resulting greater complexity of the work process because such diversity provides more job specialization.[71] Moreover, culturally diverse teams tended to be most or least effective, whereas homogeneous teams tended to be average. Adler explained the reason behind the tendency of culturally diverse teams toward productivity. She stated that the performance level of work teams was related to how they manage their diversity, not, as commonly believed, to the presence or absence of cultural diversity. When well-managed, cultural diversity of individuals becomes a productive resource to the team. When ignored, cultural diversity causes problems that diminish the team's productivity. Having briefly discussed the various perceptions regarding the level of heterogeneous work groups' productivity, it should be mentioned that these views were not based on empirical data; the following section, based on empirical research, therefore will provide quantitative data on this subject.

EMPIRICAL WORK

Methodology

The sample consisted of 79 construction site managers (site mangers are considered as individuals who are in charge of running and directing construction projects toward the achievement of a set of goals through their subordinates) of 62 construction firms which acted as main contractors. The subjects for this study were drawn from six Middle Eastern countries, namely Saudi Arabia, Kuwait, Oman, Bahrain, Libya, and United Arab Emirates (Fig. 24.1). Forty-seven projects were being undertaken by international construction firms, and 32 projects, by local construction organizations (Fig. 24.2). The focus was on medium- and large-sized contracts (Fig. 24.3).

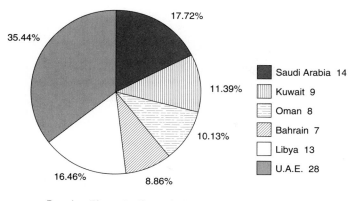

Based on 79 construction projects

FIGURE 24.1   The locations of construction projects.

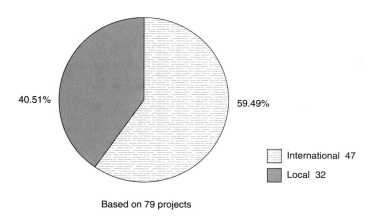

Based on 79 projects

FIGURE 24.2   Categories of construction projects.

The selection of site managers and contractors for the sample was basically contingent on several criteria; namely, the construction site must contain a work force composed of several cultures; the managers had to be accessible; and they had to be willing to cooperate. The data to be reported were obtained by means of questionnaires distributed to construction site managers working in Mideastern projects. The sample included not only Western, but also local site managers and third-country nationals (managers who are not from the country of the contractors or the host country) (Fig. 24.4). Table 24.1 illustrates the site managers' nationalities. Of the 180 questionnaires distributed, 79 were returned completed.

Site managers' style was measured by using a questionnaire (Table 24.2) consisting of 14 statements which were rated on a five-step scale, labeled from 1 (strongly agree) to 5 (strongly disagree). These 14 items consisted of 7 statements ($a–n$) that

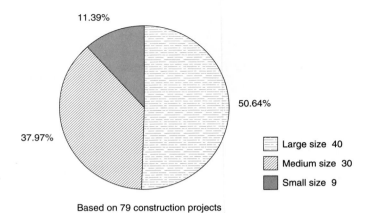

11.39%

50.64%

37.97%

☐ Large size 40

▨ Medium size 30

■ Small size 9

Based on 79 construction projects

FIGURE 24.3   Size of projects.

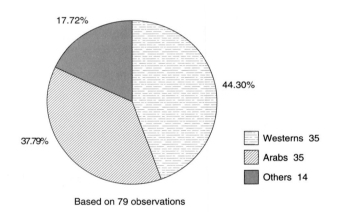

17.72%

44.30%

37.79%

☐ Westerns 35

▨ Arabs 35

■ Others 14

Based on 79 observations

FIGURE 24.4   Cultures of site managers.

referred to task-oriented managers' style and 7 (*c–m*) that referred to employee-oriented managers' style. The statements were mixed, but would be combined into groups at the analysis stage. It was hoped that such an approach would avoid any possible bias and reflect the real orientation of construction site managers (see question 1, Table 24.2). Some of the statements were developed from Whyte and Williams[52] and Anderson,[70] and the other statements were derived from a pilot study.

Productivity measurement was based on a subjective criterion, namely the assessment of site managers in charge of the construction projects. Therefore, no claim can be made regarding objective measurement, which would have necessitated an extended research project in its own right. A direct measurement was made by asking the respondents whether they thought that the productivity of work forces changes when they are composed of several cultures. The aim of this question was to compare the

TABLE 24.1   Nationalities of Construction Managers Who Participated in the Research Survey

Nationality	Number in sample	Percentage of sample
British	20	25.3
Jordanian	12	15.2
Indian	11	13.9
Palestinian	6	7.6
Lebanese	3	3.8
Egyptian	3	3.8
Sudanese	3	3.8
Turkish	3	3.8
Japanese	2	2.5
American	2	2.5
German	2	2.5
Dutch	2	2.5
U.A.E.	2	2.5
Greece	2	2.5
French	2	1.3
Italian	2	1.3
Pakistani	2	1.3
Irish	2	1.3
Bangladeshi	2	1.3
Sri Lankan	2	1.3

Sample size = 79

rate of productivity for a specific job or activity executed by a one-culture work force with that of a multicultural work force carrying out a similar job. Such a judgment can be based on the estimated time allowance for each activity in the construction process; this estimated time allowance is normally calculated according to a single cultural work force's productivity and not to a multicultural work force's. The answers to this question were rated on a five-point Likert scale (see Table 24.2). Another indirect measurement of the level of productivity was that the respondents were asked whether the size of a work force should be increased or decreased when drawn from several cultures if the same level of productivity is to be achieved, so that the project is completed on schedule.

The data were analyzed by cluster and discriminant analysis, cross-tabular analysis, and Kruskal-Wallis one-way analysis of variance tests available as a package. Cluster analysis was used to classify the observation scores which represent the construction site managers' style into groups of similar individuals. The aim of cluster analysis is to place individuals with similar attitude patterns into groups suggested only by the available data, independently of any prior definitions. Each group contains respondents who tend to be similar to each other in some feature, and dissimilar to respondents in different groups.

In interpreting and labeling each cluster, the mean score of each question in each

TABLE 24.2 Questionnaire for Site Managers

1. To what extent would you agree or disagree with each of the following statements:

	Strongly agree	Agree	Unde-cided	Dis-agree	Strongly disagree
a. I try to practice close supervision to reduce unexpected errors.	1	2	3	4	5
b. I try to put emphasis in getting out a lot of work.	1	2	3	4	5
c. I try to be as friendly and approachable to my subordinates as possible.	1	2	3	4	5
d. I try to hold meetings for discussing work-force problems.	1	2	3	4	5
e. I try to use threats and punishments to encourage good work.	1	2	3	4	5
f. I try to insist that the work force come to work exactly on time.	1	2	3	4	5
g. I try to make my subordinates as satisfied as possible with their work.	1	2	3	4	5
h. I try to encourage good work through my friendship with my employees.	1	2	3	4	5
i. I try to be as fair and equal as I can in my dealings with subordinates.	1	2	3	4	5
j. I encourage subordinates to feel that they can come to me with their personal problems.	1	2	3	4	5
k. I expect subordinates to follow instructions without a debate.	1	2	3	4	5
l. I try to keep a close eye on my subordinates' work, to make sure that they understand the instruction.	1	2	3	4	5
m. I allow subordinates to adopt work methods as they see fit, to get the job done.	1	2	3	4	5
n. I have little tolerance of subordinates who question instructions or deviate from instructions.	1	2	3	4	5

2. Does productivity change when a multinational work force is on one site?

(Please tick for "yes")

1. Decreases significantly
2. Decreases slightly
3. No change
4. Increases slightly
5. Increases significantly

TABLE 24.2   Questionnaire for Site Managers (*Continued*)

3. If the productivity decreases, do you think the reasons are: (Would you please rank each item in order of priority with top priority 1.)

	Strongly agree	Agree	Unde- cided	Dis- agree	Strongly disagree
a. Lack of skill	1	2	3	4	
b. Problems arising from cultural variation	1	2	3	4	
c. Misunderstandings and communication problems	1	2	3	4	
d. Others, please specify					
....	1	2	3	4	
....	1	2	3	4	
....	1	2	3	4	
....	1	2	3	4	

4. If the productivity increases, do you think the reasons include: (Please rank each item in order of priority; top priority 1.)

	Strongly agree	Agree	Unde- cided	Dis- agree	Strongly disagree
a. Mutual respect between supervisor and work force	1	2	3	4	
b. Good site engineer	1	2	3	4	
c. National pride	1	2	3	4	
d. Effective communication	1	2	3	4	
e. Separation of the work groups on the construction site according to their nationality	1	2	3	4	
f. Others, please specify					
....					
....					
....					

cluster was studied and compared, then these mean scores were referred to the actual questions which were used to measure the management style to give more information for each cluster. Also, discriminant analysis was used to check whether the classifications of managers' style which were suggested by cluster analysis are accurate and meaningful.

Results

Site managers' style has been categorized into four groups, as has been suggested by the cluster analysis (Table 24.3): high task/low employee, low task/low employee, low task/high employee, and high task/high employee. Figure 24.5 provides a visual indication of the location of each manager's style and the relationship between both dimensions, namely task and employee orientation. It exhibits the two main manageri-

TABLE 24.3   Classification of Site Managers' Style

Cluster	Managers' style	Number in sample	Percentage of sample
1	High task/low employee	14	17.72
2	Low task/low employee	24	30.38
3	Low task/high employee	19	24.05
4	High task/high employee	22	27.85

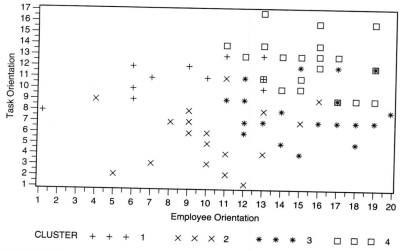

FIGURE 24.5   Task/employee orientation of site managers.

al styles and the various interactions between them, the horizontal axis referring to employee orientation and the vertical axis to task orientation. The results of the discriminant analysis, in other words the four groups identified by the cluster analysis, are meaningful. Looking at Table 24.4, it can be seen that there is hardly any overlap between the clusters. There are only two observations which are misaligned, namely one observation between clusters 2 and 3 and another observation between clusters 3 and 4. This would appear to be an extremely good result which supports the validity and usefulness of these four management style classifications.

The results show that 45.5 percent of site managers believed that productivity in construction projects decreases with mixed cultural work forces (Table 24.5). In addition, 29.1 percent of site managers thought that there is no change in productivity whether the work forces are composed of one or more than one culture. In contrast, 25.3 percent of respondents believed that heterocultural work forces' productivity increases.

Among the several reasons given by respondents for the decrease of productivity were

TABLE 24.4   Classification Summary for Calibration Membership in Each Cluster Using Discriminant Analysis

Cluster	Number of observations (and percentages) classified into cluster				
	1	2	3	4	Total
1	14	0	0	0	14
	(100)	(0.00)	(0.00)	(0.00)	(100)
2	0	23	1	0	24
	(0.00)	(95.83)	(4.17)	(0.00)	(100)
3	0	0	18	1	19
	(0.00)	(0.00)	(94.74)	(5.26)	(100)
4	0	0	0	22	22
	(0.00)	(0.00)	(0.00)	(100)	(100)

TABLE 24.5   Site Managers' Perceptions toward Work Force Productivity

Productivity	Frequency of respondents	Percentage in sample
Decreases significantly	5	6.3
Decreases slightly	31	39.2
No change	23	29.1
Increases slightly	17	21.5
Increases significantly	3	3.8

Sample size = 79

1. Problems which arise from the cultural diversity of the work team. Examples of problems given by site managers are the differences in attitudes toward ways of handling matters on the site and differences in training and educational background which led in several cases to arguments regarding the correct and the quickest method of performing activities.
2. Language and communication problems which led to a loss of time.
3. The lack of subordinates' skills, as most of those in work forces that come from developing countries have low levels of training and education compared with those from developed countries. With regard to productivity increase, 25.3 percent of respondents stated that a multicultural work force's productivity increases compared with that of one culture. It is worthwhile mentioning the reasons which were given. Respondents put emphasis on the overall quality of site managers as a key factor in building a harmonious work team. The mutual respect between supervisors and their subordinates was placed as a second important factor, followed by the benefits which could be derived from recognizing the national pride of each culture and from an effective communication system. With respect to the size of work forces, 67 percent of respondents believed that the number of workers must increase when the work force is composed of several cultures, if a similar standard of productivity to that of a one-culture work force is to be achieved.

TABLE 24.6   Association between Managers' Style and Work Force Productivity

(Wilcoxon scores—rank sums)

Style	Sum of scores	Expected under HO	Mean score	Standard deviation under HO
1	453.00	560.00	32.26	74.09
2	776.00	960.00	32.33	89.23
3	741.00	760.00	39.00	82.92
4	1190.0	880.00	54.09	86.97

Sample size = 79
Degree of freedom = 3
Chi square = 13.88
Significance of Kruskal-Wallis test = 0.003

The results of the Kruskal-Wallis one-way analysis of variance test show that the site managers' style is highly related to the work forces' productivity as perceived by respondents; the level of significance p was found to be 0.003 (Table 24.6). That is, a multicultural work force's productivity increases when managers are highly task- and employee-oriented. One can conclude that it is unlikely that both variables (managers' style and productivity) are independent. As can be noted in the contingency Table 24.7, there is a tendency for productivity of work groups to increase when managers are highly task- and employee-oriented. The majority of managers who believe that the productivity of multicultural work forces decreases are grouped in style 2, and most managers who think that the multicultural work forces' productivity increases are categorized under style 4.

TABLE 24.7   Cross Tabulation of Managers' Style by Work Force Productivity

Style	Productivity		
	Decreases	No change	Increases
1	9	2	3
2	14	9	1
3	9	6	4
4	4	6	12

Sample size = 79

DISCUSSION

It appears from the results that there is no clear emphasis on any particular leadership style category. This might be traced to the complexity of enterprises which comprise several cultures. In a study of site managers' orientation which was conducted in the British construction industry, it was found that site managers tend to be strongly task-oriented. However, the results of the empirical work suggest that in the Middle East, site managers do not appear to be strongly categorized in a single style. A possible interpretation is that construction site managers who come from different cultures to manage construction projects in the Middle East are inclined to practice the same style as in their original culture. In addition, the majority of scholars strictly classified the leadership as initiating structure/consideration or task-/employee-oriented.

Drawing on the presented empirical data and linking them with existing leadership or management style research, it can be argued that it seems an oversimplification to label leadership style in the clear-cut dimensions of task/employee orientation. Moreover, based on the available data, it can be suggested that it is inappropriate to apply completely the manufacturing and service industries, which have been tested in one culture or across cultures, to the construction industries, especially when more than one culture is involved in a project. This does not mean that previous findings of behavioral science research should be ignored. Rather, they must not be applied to the construction industry in the Middle East without first evaluating the effects of the differences in both types of industry and its work situation.

The results also suggest that the level of the multicultural work forces' productivity decreases in this sample in construction projects in the Middle East. This can be traced to cultural differences which exacerbate communication problems between work force groups on the one hand, and between site managers and the work forces on the other.

This result lends support to the proposals of Cartwright and Anderson that the homogeneity of a work team fosters a great solidarity and improves morale, which then results in an increase in productivity and the quality of the product. The results are congruent with Triandis et al.'s suggestion, which indicated that cultural diversity produces poor communication and misunderstanding, hence reducing productivity.

Moreover, the present result seems to support Adler's interpretation, which implies that the cultural diversity of group members may raise the degree of ambiguity, miscommunication, and misinterpretation which will probably reduce the level of productivity. A statement made by Koehn and Brown is found to be in line with the finding. They stated that, in the Middle East, the use of third-country nationals affects strongly the productivity rate.

On the other hand, the results do not lend high support to Steiner's suggestion that the diversity of a work group enhances productivity despite the high complexity of the work process which it may create, because such diversity provides job specialization. Having discussed the present findings in relation to previous work, one notes that there are important differences. The present findings are based on empirical data, whereas previous researchers substantiate their criteria on logical arguments or speculations, rather than factual evidence.

While the present research suggests that the level of multicultural work forces' productivity decreases, it would be erroneous to conclude that the high level of specialty which can be found in a mixed-culture work team can be ignored. If sound management principles are applied and training programs are provided, conflict between the work forces and their management, or between individuals within gangs, is expected to disappear.

The results indicate that there is a very strong correlation ($p = 0.003$) between the style of management and the level of productivity in a multicultural setting. It can be inferred from this association that the style of site managers can create conditions conducive to the enhancement of a mixed-culture work group's productivity. However, this conclusion has to be treated with caution, as it is hazardous to attribute cause-and-effect relationships to such type of data. It would be incongruous to ignore the fact that causal direction could be the opposite. In particular, the results indicate that a high level of productivity can be achieved when site managers are highly task- and employee-oriented. By contrast, a low task/employee managerial style can lead to a decrease in the level of productivity.

In the Middle East, where employees may be drawn from different countries, speaking several languages, managers need to be highly employee-oriented in order to understand the cultural diversity of individuals and groups, so that any destructive and unnecessary conflict can be avoided or at least minimized. Moreover, managers in such an environment need to be highly task-oriented, through the planning, organization, and supervision of their subordinates' work, giving particular attention to these features which would help to avoid misunderstandings and reworking, so that a high level of productivity of such a complex team can be maintained.

It has been observed that most effective managers tend to combine employee orientation for establishing friendliness and harmony among the group members and task orientation for improving productivity and job requirements. Figure 24.6 shows diagrammatically the relationship between managers' style and the managers' effectiveness and the productivity of work forces.

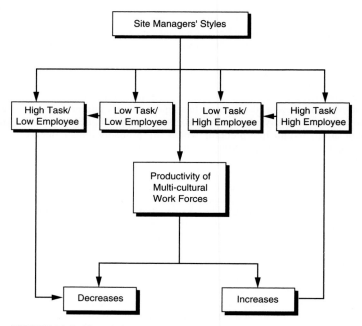

FIGURE 24.6 The relationship between managers' style and work force productivity.

REFERENCES

1. J. D. Borcherding, *An Exploratory Study of Attitudes that Affect Human Relations in Building and Industrial Construction,* Ph.D. dissertation, Stanford University, 1977.

2. P. M. Hillebrandt, *Analysis of the British Construction Industry,* Macmillan, London, 1984.

3. Brensen et al., "Leader Orientation of Construction Site Managers," *Journal of Construction Engineering and Management,* ASCE, vol. 112, no. 3, pp. 370–386, 1986.

4. M. Casson, *The Firm and the Market,* Basil Blackwell Ltd., Oxford, 1987.

5. J. K. Hemphill, *"Leader Behavior Description,"* Bureau of Business Research, Ohio State University, Columbus, 1950.

6. A. W. Halpin and B. J. Winer, "A Factorial Study of the Leader Behavior Descriptions," in R. M. Stogdill and A. E. Coons (eds.), *Leader Behaviour: Its Description and Management,* Ohio State University, Columbus, 1957.

7. R. M. Stogdill and A. E. Coons (eds.), *Leader Behavior: Its Description and Measurement,* Ohio State University, Columbus, 1957.

8. E. A. Fleishman, "Twenty Years of Consideration and Structure," in E. A. Fleishman and J. H. Hunt (eds.), *Current Development in the Study of Leadership,* Southern Illinois University Press, Carbondale, 1973.

9. E. A. Fleishman and Peters, "Interpersonal Values, Leadership Attitudes and Managerial Success," *Personnel Psychology,* vol. 15, pp. 127–143, 1962.

10. S. Kerr et al., "Toward a Contingency Theory of Leadership Based upon the Consideration and the Initiating Structure Literature," *Organizational Behavior and Human Performance,* vol. 12, pp. 62–82, 1974.

11. A. Bryman, *Leadership and Organisations,* Routledge & Kegan Paul plc, London, 1986.

12. D. Katz et al., *Productivity, Supervision, and Morale in an Office Situation,* Institute for Social Research, Ann Arbor, Mich., 1950.

13. R. F. Bales and P. E. Slater, "Role Differentiation in Small Decision-Making Groups," T. Parson and R. F. Bales (eds.), *Family, Socialization and Interaction Process,* Free Press, New York, 1955.

14. J. C. Barrow, "The Variables of Leadership: A Review and Conceptual Framework," *Academy of Management Review,* pp. 231–245, April 1977.

15. R. R. Blake and J. S. Mouton, *The Managerial Grid,* Gulf Publishers, Houston, 1964.

16. W. J. Reddin, "3-D Organizational Effectiveness Program," *Training and Development Journal,* pp. 22–28, March 1968.

17. R. Likert, *New Patterns of Management,* McGraw-Hill, New York, 1961.

18. R. Likert, *The Human Organization,* McGraw-Hill, New York, 1967.

19. L. L. Larson et al., "The Great Hi-Hi Leader Behavior Myth: A Lesson from Occam's Razor," *Academy of Management Journal,* vol. 19, pp. 628–641, 1976.

20. A. W. Halpin, "The Observed Leader Behavior and Ideal Leader Behavior of Aircraft Commanders and School Superintendents," in R. M. Stogdill and A. E. Coons (eds.), *Leader Behavior: Its Description and Measurement,* Ohio State University, Bureau of Business Research, Columbus, 1957.

21. B. T. Keeler and J. A. M. Andrews, "The Leader Behavior of Principals, Staff Morale and Productivity," *Alberta Journal of Educational Research,* vol. 9, pp. 179–191, 1963.

22. C. J. Cunningham, *Measures of Leader Behavior and Their Relation to Performance Levels of Country Extension Agents,* unpublished doctoral thesis, Ohio State University, 1964.

23. G. Misumi and J. A. Toshiaki, "A Study of the Effectiveness of Supervisory Patterns in a Japanese Hierarchical Organization," *Japanese Psychological Research,* vol. 7, pp. 151–162, 1965.

24. D. B. Hooper, "Differential Utility of Leadership Opinions in Classical and Moderator Models for the Prediction of Leadership Effectiveness," *Administrative Science Quarterly,* vol. 16, pp. 321–338, 1968.

25. R. C. Nystrom, "Managers and the Hi-Hi Leader Myth," *Academy of Management Journal,* vol. 21, no. 2, pp. 325–331, 1978.

26. R. M. Stogdill, *Handbook of Leadership,* Free Press, New York, 1974.

27. E. A. Fleishman and J. Simmons, "Relationship between Leadership Patterns and Effectiveness Ratings among Israeli Foremen," *Personnel Psychology,* vol. 23, pp. 169–172, 1970.

28. P. Lansley et al., "Organisation Structure, Management Style and Company Performance," *Omega, the International Journal of Management Science,* vol. 2, no. 4, pp. 467–485, 1974.

29. J. G. Hunt and V. K. C. Liebscher, "Leader Performance, Leader Behavior and Employee Satisfaction," *Organisational Behavior and Human Performance,* vol. 9, pp. 59–77, 1973.

30. E. A. Fleishman et al., *Leadership and Supervision in Industry,* Ohio State University, Bureau of Education Research, Columbus, 1955.

31. E. A. Fleishman and E. F. Harris, "Patterns of Leadership Behavior Related to Employee Grievances and Turnover," *Personnel Psychology,* vol. 15, pp. 43–56, 1962.

32. G. Graen et al., "An Empirical Test of the Man-in-the-Middle Hypothesis among Executives in a Hierarchical Organization Employing a Unit-Set Analysis," *Organizational Behavior and Human Performance,* vol. 8, pp. 262–285, 1972.

33. E. A. Fleishman, "The Description of Supervisory Behavior," *Journal of Applied Psychology,* vol. 37, pp. 1–16, 1953.

34. R. C. Cummins, "Relationship of Initiating Structure and Job Performance as Moderated by Consideration," *Journal of Applied Psychology,* vol. 55, no. 5, pp. 484–490, 1971.

35. R. C. Cummins, "Leader-Member Relations as a Moderator of the Effects of Leader Behavior and Attitudes," *Personnel Psychology,* vol. 25, pp. 655–660, 1972.

36. F. E. Fiedler, *A Theory of Leadership Effectiveness,* McGraw-Hill, New York, 1967.

37. F. E. Fiedler, "Personality Motivational Systems, and Behavior of High and Low LPC Persons," *Human Relations,* vol. 25, pp. 341–412, 1972.

38. F. E. Fiedler, "A Contingency Model of Leadership Effectiveness," in L. Berkowitz (ed.), *Group Process,* Academic Press, New York, 1978.

39. C. A. Schreisheim and S. Kerr, "Theories and Measures of Leadership: A Critical Appraisal of Current and Future Directions," in J. G. Hunt and Larson (eds.), *Leadership: The Cutting Edge,* Southern Illinois University Press, 1977.

40. A. S. Ashour, "The Contingency Model of Leadership Effectiveness: An Evaluation," *Organizational Behavior and Human Performance,* vol. 9, pp. 334–355, 1973.

41. R. J. House, "A Path Goal Theory of Leader Effectiveness," *Administrative Science Quarterly,* vol. 16, pp. 321–338, 1971.

42. R. J. House and T. R. Mitchell, "Path-Goal Theory of Leadership," *Journal of Contemporary Business,* vol. 3, pp. 81–97, 1974.

43. P. Hersey and K. H. Blanchard, *Management of Organizational Behavior,* Prentice-Hall, Englewood Cliffs, N.J., 1969.

44. P. Hersey and K. H. Blanchard, *Management of Organizational Behavior: Utilizing Human Resources,* Prentice-Hall, Englewood Cliffs, N.J., 1977.

45. C. L. Graeff, "The Situational Leadership Theory: A Critical View," *Academy of Management Review,* vol. 8, pp. 285–291, 1983.

46. P. S. Hundal, "A Study of Entrepreneurial Motivation: Comparison of Fast- and Slow-Progressing Small Scale Industrial Entrepreneurs in Punjab, India," *Journal of Applied Psychology,* vol. 55, no. 4, pp. 317–323, 1971.

47. R. F. Gonzales and C. McMillan, "The Universality of American Philosophy," *Journal of Academy of Management,* pp. 33–41, April 1961.

48. W. A. Starbuck, "Organisations and Their Environment," in M. D. Dunnette (ed.), *Handbook of Industrial and Organizational Psychology,* Rand McNally, Chicago, 1976.

49. B. W. Barrett and B. M. Bass, "Cross Cultural Issues in Industrial and Organizational Psychology," in M. D. Dunnette (ed.), *Handbook of Industrial and Organizational Psychology,* Rand McNally, Chicago, 1976.

50. J. Child, "Culture, Contingency and Capitalism in the Cross-National Study of Organizations," in B. M. Staw and L. L. Cummings (eds.), *Research in Organisational Behaviour,* jai, vol. 3, 1981.

51. E. H. Schein, *Organizational Culture and Leadership,* Jossey Bass Publications, San Francisco, California, 1985.

52. W. F. Whyte and L. K. Williams, "Supervisory Leadership: An International Comparison," *Proceedings of the Thirteenth International Management Congress,* pp. 481–488, 1963.

53. G. Hofstede, "Motivation, Leadership, and Organization: Do American Theories Apply Abroad?," *Organisational Dynamics,* pp. 42–63, Summer 1980.

54. G. Hofstede, *Culture's Consequences: International Differences in Work Related Values,* Sage Publications, Newbury Park, California, 1984.

55. N. R. Maier and L. R. Hoffman, "Group Decision Making in England and the United States," *Personnel Psychology,* vol. 15, pp. 75–87, 1962.

56. F. A. Heller and L. W. Porter, "Perceptions of Managerial Needs and Skills in Two National Samples," *Occupational Psychology,* 1966.

57. A. R. Negandhi, "Cross-Cultural Management Research: Trends and Future Directions," *Journal of International Business Studies,* pp. 17–28, Fall 1983.

58. M. Haire, E. E. Ghiselli, and L. W. Porter, *Managerial Thinking: An International Study,* Wiley, New York, 1966.

59. K. M. Thiagarajan and S. D. Deep, "A Cross Cultural Study of Performance for Participative Decision-Making by Supervisors and Subordinates," technical report no. 33, University of Rochester, 1969.

60. J. L. Massie and J. Luytjes (eds.), *Management in International Context,* Harper & Row, New York, 1972.

61. N. J. Adler, *Organizational Behavior,* Kent, Boston, 1986.

62. F. A. Muna, *The Arab Executive,* Macmillan, New York, 1980.

63. M. K. Badawy, "Styles of Mideastern Managers," *California Management Review,* pp. 51–58, Spring 1980.

64. A. Al-Jafary and A. T. Hollingsworth, "An Exploratory Study of Managerial Practices in the Arabian Gulf Region," *Journal of International Business,* pp. 143–152, 1983.

65. P. Lansley, "The Flexibility of Construction Management," *Proceedings of CIB Symposium on Organization and Management of Construction,* 1981.

66. J. D. Boreherding and D. F. Gaerner, "Work Forces Motivation and Productivity on Large Jobs," *Journal of the Construction Division, ASCE,* vol. 107, no. C03, pp. 443–453, 1981.

67. A. Bryman et al., "Leader Orientation and Organizational Transience: An Investigation Using Fiedler's LPC Scale," *Journal of Occupational Psychology,* vol. 60, pp. 13–19, 1987.

68. H. C. Triandis et al., "Some Cognitive Factors Affecting Group Creativity," *Human Relations,* 1964.

69. D. Cartwright, "The Nature of Group Cohesiveness," in D. Cartwright and Zander (eds.), *Group Dynamics,* Harper & Row, New York, 1968.

70. L. R. Anderson, "Small Group Behavior," in S. E. Searchore and McNeill (eds.), *Management of the Urban Crisis,* Free Press, New York, pp. 69–112, 1971.

71. I. D. Steiner, *Group Process and Productivity,* Academic Press, New York, 1972.

CHAPTER 25

THE WORLD HEALTH ORGANIZATION HEALTHY CITIES PROJECT*

Ilona Kickbusch and Agis D. Tsouros

WHO Regional Office for Europe,
Copenhagen, Denmark

Ilona Kickbusch is a Director of the World Health Organization's Regional Office for Europe in Copenhagen, Denmark. She was the instigator of initiatives such as the Ottawa Charter for Health Promotion and the Healthy Cities project, which created the framework for new public health. She supervises Europe-wide projects in schools, cities, and regions and health policy development for women, nutrition, drugs, and AIDS. She has lived in Asia, Africa, and Europe and holds a Doctorate in Social Sciences from the University of Konstanz, Germany.

Agis D. Tsouros is coordinator of the Healthy Cities project at the WHO Regional Office for Europe and has been responsible for the strategic development and implementation of the project since joining WHO in 1989. In 1984 he initiated the first local-level application in Europe of the WHO strategy Health For All in Bloomsbury, a health district in central London. He is a Greek national with a medical degree from the University of Athens and a Ph.D. degree in public health from the University of Nottingham, U.K. He is an accredited specialist in public health medicine.

INTRODUCTION

The WHO Healthy Cities project is an international long-term development project that seeks to put health on the agenda of decision makers in the cities of Europe and to build a strong lobby for public health at the local level. Ultimately the project seeks to enhance the physical, mental, social, and environmental well-being of the people who live and work in the cities of Europe.

The Healthy Cities project contributes to changes in how individuals, communities,

---

*Since this chapter was written, a new phase of the project has started with a network of project cities covering all parts of Europe and with 20 national networks of cities in Europe with about 500 participating municipalities in all. The basic approaches and operational methods of the project remain as described here.

private and voluntary organizations, and local governments throughout Europe think about, understand, and make decisions about health.

The Project applies the WHO strategy for "health for all by the year 2000," the principles of health promotion outlined in the Ottawa Charter for Health Promotion, and the principles of the European Charter on Environment and Health at the local (city) level.

The Healthy Cities project attempts to strike a balance between placing health high on the strategic political agenda of cities and carrying out applied technical and operational measures. This includes developing and implementing specific plans to improve health in the city and taking the structural, organizational, and financial steps to make this possible.

Cities are challenged to reduce inequalities of health status and inequities in access to the prerequisites for health, to develop healthy public policies at the local level, to create physical and social environments that support health, to strengthen community action for health, to help people develop new skills for health compatible with these new approaches, and to reorient health services in accordance with the strategy for health for all and the principles of health promotion.

BACKGROUND

Concern about the health of people in cities has changed over the centuries. The rapid urbanization that accompanied industrialization and the health problems that resulted from so many people being crammed into appalling living and working conditions sparked a powerful response—the nineteenth-century public health movement. Based on a concern for health, new techniques were created and laws developed to counter the health hazards of urbanization and industrialization. Sewers were built, water supplies improved, factory laws and housing codes adopted, antipollution measures taken, food inspection and control instituted, and parks created, while urban planning (an offshoot of public health) grew in importance.

Nevertheless, the advent of vaccines and antitoxins in the late nineteenth century and the later development of effective antibiotics in the 1930s helped to shift the emphasis from the environmental and social conditions of the community to the personal behavior of families and individuals and the power of the therapeutic approach.

In recent years, however, the limits of relying on the medical model have become self-evident. In response to these limitations, the emphasis has shifted again to a broader model of health that reflects a modern understanding of the determinants of health.

WHO has pioneered in broadening the idea of health. From its inception it has recognized that health is more than the absence of disease, and this was confirmed when the World Health Assembly, in 1977, adopted the challenge of achieving health for all by the year 2000 (Table 25.1). Based on an extensive review of health and health problems in Europe, the Member States of the European Region of WHO adopted 38 specific regional targets in 1984 in support of the European strategy for health for all, which had been adopted by the Member States in 1980. These targets provide a framework for action in such areas as equity, preventing disease, life-styles conducive to health, supportive environments, and health-care services. Health promotion, the backbone of the new public health, is a key concept in the strategy of health for all. In

TABLE 25.1   Principles for Health for All

Health for all implies equity. This means that the present inequalities in health between countries and within countries should be reduced as far as possible.

The aim is to give people a positive sense of health so that they can make full use of their physical, mental, and emotional capacities. The main emphasis should therefore be on health promotion and the prevention of disease.

Health for all will be achieved by people themselves. A well-informed, well-motivated, and actively participating community is a key element for the attainment of the common goal.

Health for all requires the coordinated action of all sectors concerned. The health authorities can deal only with part of the problems to be solved, and multisectoral cooperation is the only way of effectively ensuring the prerequisites for health, promoting healthy policies, and reducing risks in the physical, economic, and social environments.

The focus of the health-care system should be on primary health care—meeting the basic needs of each community through services provided as close as possible to where people live and work, readily accessible and acceptable to all, and based on full community participation.

Health problems transcend national frontiers. Pollution and trade in health-damaging products are obvious examples of problems whose solution requires international cooperation.

1986 the Ottawa Charter for Health Promotion further defined health promotion in terms of policy and strategy.

The Ottawa Charter recognized that policy decisions in areas other than health make a key contribution to health, that supportive physical and social environments are important in establishing the conditions for health and the parameters for health behavior, that the community can and must play a crucial role in undertaking actions for health, that a broad range of personal skills for health needs to be developed, and that existing health services need to be reoriented.

The Healthy Cities project was thus launched in the framework of health for all and health promotion, as a testing ground for developing and implementing these new public health approaches at the local level, where the somewhat abstract and global concepts of health promotion and health for all can most easily be concretized. As the lowest level of government, and thus closest to the people, cities can and should play a central role in achieving health, a role they had fulfilled historically but that has been generally neglected in recent years.

NEW STRATEGIES, NEW STYLES

The Healthy Cities project challenges cities to take seriously the process of developing health—enhancing public policies that create physical and social environments that support health and strengthen community action for health. These new strategies for promoting health are explicitly political, environmental, and social in nature, complementing the primarily behavioral and medical strategies employed by public health in the past few decades.

The Healthy Cities project advocates not only new strategies but also the new style of action described in the Ottawa Charter. Health promotion cannot work if people simply take charge, assume responsibility, and direct others. Instead, Healthy Cities projects are challenged by the principles of health promotion to develop new styles of enabling, facilitating, mediating, advocating, and building new partnerships and coalitions for health. Projects should:

- Enable individual people and communities to increase control over and assume more responsibility for health—without victim blaming, without dumping problems on them, and without abdicating societal responsibility

- Facilitate the political, social, and community processes involved in negotiating new ways of doing things

- Mediate between the various and often conflicting interests of the various public, private, voluntary, and community sectors involved in creating the conditions for better health

- Advocate with other social forces and on behalf of people who are powerless the changes necessary to promote health (This advocacy must be directed to the public and private sectors at the local, regional, national, and even international levels that have authority over and the responsibility for actions that protect and promote health.)

- Create new partnerships and coalitions for health, sometimes with less powerful people and groups, sometimes with more powerful people and groups, ideally with both (Such partnerships need to bring together the public, private, voluntary, and community sectors, united in the common purpose of promoting the health of the cities' people.)

These new approaches mean that cities must be managed differently. The old system of organization by professional department and by sector has to be complemented by new approaches to such health issues as equity, sustainability, safety, and mobility. These issues cut across the old departmental lines and indeed across the different sectors—public, private, voluntary, and community. None can be addressed by one department of government alone, nor indeed by city government alone. The whole community has to be mobilized and the efforts of all sectors and departments have to be combined and focused.

New structures and processes should enhance collaboration rather than competition, analyze issues holistically rather than sectorally, and use "both/and" rather than "either/or" approaches. These are profound changes from the way things have been done in the past.

Thus from the start it was recognized that this would be no ordinary project for the cities, given the tremendous changes in strategies, styles, and values implicit in health, health promotion, and the urban environment.

It also became clear that this was no ordinary venture for the European Office of WHO either. The prevailing culture in the organization has been to develop and implement activities on a unit basis with very little interunit or interdepartmental collaboration.

Healthy Cities established a new approach based on concepts of project development and project management. *Partnership* became the key feature of the Healthy Cities project, both in-house and outside.

An international health project (such as Healthy Cities) mainly implies a direct link (partnership) with a target clientele—in this case in the cities of Europe. This clientele is committed to jointly develop and implement a plan of action. Other important aspects of the Healthy Cities project approach include defining its life span, goals, and mission, developing a strategic framework that covers both process and time schedule (including starting-up activities), establishing teamwork methods, introducing explicit consultation processes with partners, networking and building political and technical alliances, and defining clear outcomes and product.

Flexibility and responsiveness to emerging needs and new developments are essential. This means that within an agreed strategic framework some activities can be pre-defined (such as annual business meetings, developing health profiles) and some will be developed according to need in the course of its development.

WHAT IS A HEALTHY CITY?

A city is a living, breathing, growing, and changing complex organism that too often has been considered as only an economic entity. Cities are players in promoting and maintaining health and have a unique capacity to implement ecological health plans.

In the first Healthy Cities Paper, Hancock and Duhl define a healthy city as one that is continually creating and improving those physical and social environments and expanding those community resources which enable people to mutually support each other in performing all the functions of life and in developing to their maximum potential.

A healthy city is defined by a process and not just an outcome. A healthy city is not one that has achieved a particular health status level; it is conscious of health and striving to improve it. Thus any city can be a healthy city, regardless of its current health status. What is required is a commitment to health and a structure and process to achieve it. The Healthy Cities project defines a set of qualities a city should strive to provide (Table 25.2).

TABLE 25.2   Qualities of a Healthy City

A city should strive to provide

A clean, safe, and physical environment of high quality (including housing)

An ecosystem that is stable now and sustainable in the long term

A strong, mutually supportive, and nonexploitative community

A high degree of participation and control by the public over the decisions affecting their lives, health, and well-being

The meeting of basic needs (for food, water, shelter, income, safety, and work) for all the city's people

Access to a wide variety of experiences and resources, with the chance for a wide variety of contact, interaction, and communication

A diverse, vital, and innovative city economy

The encouragement of connectedness with the past, with the cultural and biological heritage of city dwellers, and with other groups and individuals

A form that is compatible with and enhances the preceding characteristics

An optimum level of appropriate public health and sick care services accessible to all

High health status (high levels of positive health and low levels of disease)

MISSION AND OBJECTIVES

The ultimate goal of the Healthy Cities project is to improve the health of the people in the cities of Europe. Within this, the mission is to build a new public health movement in the cities of Europe and to make health the business of everyone at the city level.

The project strives to realize the vision of a healthy city through a process of political commitment, visibility for health, institutional change, and innovative action for health and the environment:

- Political commitment and leadership can provide the necessary legitimization, direction, and resources for the project.

- Visibility for health is necessary to promote wide appreciation and recognition of the major health issues in the city and the economic, social, and physical factors that influence health.

- Institutional change is a fundamental prerequisite for change in cities. The process must be intersectoral in nature, given that the prerequisites for health depend on many different sectors. Community involvement should be promoted to support and direct the political leadership and to ensure community ownership of this process.

- Innovative action for health initiates and supports activities that aim to promote equity, sustainability, supportive environments, community action, and healthy municipal policies.

There are no simple solutions or recipes for such a process. Strategies at the local level must be compatible with the cultural, social, and organizational traditions of a city.

The WHO project office pursues the project objectives and the broader objectives of health promotion and health for all in five ways:

- *Innovation.* Using an international network of cities to generate and test innovation and to build political support for the new public health in cities

- *Dissemination.* Using national networks to promote and disseminate applied innovation and to build national and international coalitions for the new public health

- *Developing leadership.* Facilitating the development of skilled and knowledgeable decision makers who are committed to the ideas of Healthy Cities and the new public health

- *Influencing international organizations.* Introducing health considerations into the urban programs of other international organizations

- *Developing and coordinating resources.* Establishing effective managerial and support systems for the project

Four broad challenges have emerged for project cities:

- To generate visibility at the local level for health issues and the strategy for health for all

- To move health high on the social and political agendas of the city and contribute to the development of healthy municipal public policies

- To facilitate organizational and institutional changes that encourage cooperation between departments and key city sectors and promote community participation

- To create innovative action for health that emphasizes the interaction between people, environments, life-styles, and health

HISTORICAL OVERVIEW

In 1985 the idea of a Healthy Cities project was first put forward within WHO/EURO by the then officer for Health Promotion. A developmental phase took place from 1986 to 1987, and the first implementation phase was from 1988 to 1992. The second phase of the project will last from 1993 to 1997.

In the developmental phase the initial idea was operationalized by a WHO advisory group and further discussed and developed in meetings and symposia with delegations from European cities. By the end of this phase the project could develop a realistic action plan and WHO had designated 11 cities (first round of selections, September 1987) that were committed to work in the Healthy Cities network.

Phase I started with the designation of another 14 project cities (second round of selection, February 1988). The official endorsement by all the parties involved a five-year planning framework (1988–1992), the establishment of a project office at the WHO Regional Office for Europe to coordinate the project, and a method for joint work with the designated cities. From 1988 to 1990 the project also developed and endorsed an information exchange and consultation strategy and a multicity action plan, which will be described later. The project was expanded (third round of selection, November 1989 and March 1990) with the designation of five project cities. The WHO Healthy Cities network now consists of 35 project cities (see Fig. 25.1).

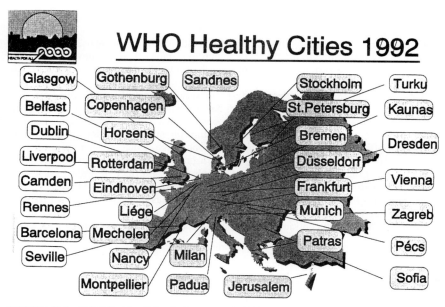

FIGURE 25.1 WHO Healthy Cities, Project network, 1992.

TABLE 25.3   Commitment to Healthy Cities

The project cities were designated based on a commitment to:

Formulate and implement intersectoral health promotion plans with a strong environmental health component, based on WHO policies and strategies and with active community involvement

Secure the necessary resources to pursue and implement the plan

Report back regularly on progress achieved and share information and experience from practice with other project partners

Support the development of national networks of Healthy Cities

Establish an intersectoral political committee to act as a focus for and to steer the project

Establish a project office with full-time staff and resources and a technical committee that brings together professionals from different disciplines and departments to develop and implement health plans

Establish mechanisms for public participation and strengthen health advocacy at city level by stimulating visibility for and debate on public health issues and by working with the media

Carry out population health surveys and, in particular, assess and address the needs of the most vulnerable and underserved social groups

Involve and encourage local research institutions to support the activities of the project

Develop active working links with other project cities, fostering technical and cultural exchange, and hosting Healthy Cities meetings and events

OPERATION OF THE PROJECT

The project involves a network of European cities who have endorsed the principles and policy directives of the WHO strategy for health for all by the year 2000, and who are politically committed to the five-year planning framework and to developing jointly and implementing action strategies for health. Table 25.3 shows the criteria for the selection of project cities.

Cities "franchising" the project need to establish a project office and full-time staff to run the project. A senior local politician has the political responsibility for the project. The project coordinator is generally responsible for the development, management, and implementation of the project. The leadership role and the functions performed by the project office team are the single most important factors determining the success of each Healthy Cities project. Colin Hastings and Wendy Briner of the New Organisation, London, ran a workshop in March 1990 on effective project leadership for project coordinators. The purpose of the workshop was to clarify the leadership role of the project office and to specify leadership action areas through which each Healthy Cities project will increase its chances of success. The activities the project office needs to mix and match continuously to establish leadership were defined as follows:

- Preparing the soil—creating the conditions for success
- Raising awareness of the project
- Planting and fertilizing small projects
- Influencing the formation of healthy policy
- Working on intersectoral levels
- Sustaining interest and energy
- Transplanting successful practices

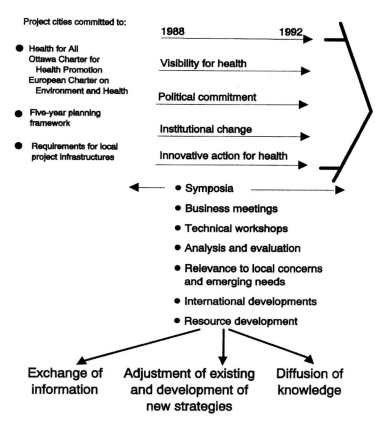

FIGURE 25.2   Operation of WHO Healthy Cities Project.

- Helicoptering—seeing the whole picture

The project is based on a spirit of true partnership and has well-established mechanisms that promote information exchange, sharing of experience, mutual support, adjustment of existing plans, development of new strategies, and dissemination of ideas and products (Fig. 25.2).

In two annual business meetings (spring and fall), project city coordinators, city politicians responsible for the project, and the staff of the WHO Healthy Cities project office present progress reports, discuss ongoing efforts, and determine new strategies and tactics based on experience and emerging needs. The project is relevant to local concerns and responsive to international developments. The experience of implementing the project provides a solid basis for the agendas of these business meetings.

The annual Healthy Cities symposia (which after 1988 have been open to project cities only) are integral to project strategy and development. They have three main functions:

- Providing the basis for sharing experience, learning, and mutual support.
- Providing a focus for making progress visible, maintaining commitment, and planning for the future.

- Providing an ideal forum to present and discuss models of good practice and case studies in the action areas of the project. The emphasis is on translating the ideas of the Ottawa Charter into practical policies and concrete activities.

The Healthy Cities symposia bring together project city politicians, administrators, town planners, architects, public health and care professionals, environmental officers, community activists, researchers, and many others.

The project also organizes specialized technical workshops on such issues as indicators, the needs of different social groups, environmental issues, and training and briefing courses. Open Healthy Cities symposia and conferences are organized with different international partners to enable wider audiences and other cities to learn about the project and to engage and stimulate interest and debate in specific areas, such as research on urban health. Project cities also initiate and organize similar events for local audiences or on local themes of interest and concern.

The project involves visibility for health, political commitment, institutional change, and innovative action for health. Figure 25.3 shows how this process develops from policy to operation. For example, getting political commitment for the project is a strategy to ensure that city politicians are actively involved. This decision (choice of strategy) can then be further elaborated into a number of strategies for action, such as preparing and endorsing a declaration stating that inequalities in health will be reduced in the city. The process of arriving at this desired outcome needs to be further defined and needs specific action to be put into operation.

The action areas of the Ottawa Charter are linked to the process of the project. Cities are expected to initiate activities for each of these areas. For example, action on equity requires that the inequalities in health in the city be documented, exposed, and understood (visibility); that decision makers recognize the inequalities and allocate resources to reduce them (political commitment); that strategies and structures be developed that promote collaboration between different disciplines and agencies and promote strong community involvement (institutional change); and that projects be initiated that address the specific needs of different vulnerable groups such as homeless or elderly people (innovative action for health).

The Healthy Cities project office at the WHO Regional Office for Europe in Copenhagen has been established to:

- Coordinate the project
- Provide international leadership in innovative action for health
- Link and diffuse the ideas to national Healthy Cities networks
- Build coalitions of international bodies for Healthy Cities

The activities that define the joint work of the project cities and the WHO project office are described in documents that were developed and adopted by the partners involved: a five-year planning framework (adopted at the 1988 spring business meeting in Bremen), an information exchange and consultation strategy (adopted at the 1989 fall business meeting in Pécs), and a Multicity Action Plan (adopted at the 1990 spring business meeting in Belfast).

The Five-Year Planning Framework

The five-year planning framework describes the goals and aims of the project, the roles and tasks of the main actors, the role, functions, and responsibilities of the proj-

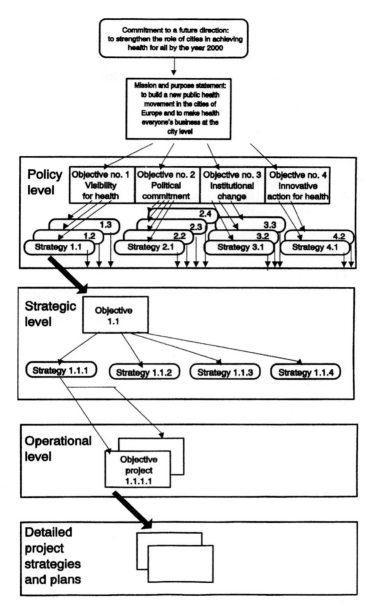

FIGURE 25.3 WHO Healthy Cities Project. (*Adapted from R. D. Archibald, "Managing High-Technology Programs and Projects," 2d ed., New York, Wiley, 1992, p. 9.*)

ect office, a framework for action from 1988 to 1992, and the required support systems and resources.

The framework for action integrates the five key elements of the Ottawa Charter with the regional targets for health for all to provide annual themes for the project.

The annual theme provides a focus for the project cities in their activities, data collection, and reporting; for the project office in organizing workshops and conferences and in its communication activities; and for other cities, national networks, and partner organizations in their activities.

The themes are:

1988   Inequities in health

1989   Strengthening community action and developing personal skills

1990   Supportive environments for health

1991   Reorienting health services and public health

1992   Healthy policies for healthy cities

The Information Exchange and Consultation Strategy

The information exchange and consultation strategy can support the project partners in implementing the project and assessing their efforts. The strategy is a joint effort between the project cities and WHO, based on shared responsibility and shared benefits. The strategy consists of seven elements. Each element is described in terms of tools, activities, resources, output, and schedules.

Assessing the Process: Project Progress and Evaluation.   A general framework to assess the project was first described in "A Guide to Assessing Healthy Cities." The two main evaluation tools so far are annual progress report forms and in-depth interviews of city project staff and partners. Both tools were developed to formats that are structured and adaptable to the project's stage of development and the focus of evaluation. The results are analyzed by independent researchers and other specialists. The evaluation of project progress from 1988 to 1990 provided the basis for "Healthy Cities Project: Analysis of Project Progress," individualized city consultation reports, and the review in this report. The project encourages research institutions in the project cities to participate in in-depth research on the policy and process of the project.

Assessing Components of the Project.   Tools are being developed and used to assess such components of the project as information and activities on health inequalities and strengthening community action. Nine such components are described. The tools and activities include developing project theme indicators (a group of project cities is currently working on this) and completing and analyzing a questionnaire on project city action on the annual theme.

Providing Models of Good Practice.   Tools and activities are being developed to create practical means of documenting, storing, and accessing information about ongoing and new projects, to contact people in various fields in project cities, and to produce bibliographies in different languages. A computerized Healthy Cities database is being considered.

Consultation.   There are two different types of city visits: formal visits focusing on policy and technical visits focusing on program and development. Project cities are usually consulted during city interviews.

Information Exchange between Cities.   Cities exchange information, consult, and make bilateral agreements. An electronic wide-area network linking the cities, training seminars, and other ways to share experience, resources, and learning are being developed. A first draft of a communication strategy was presented at the 1990 autumn business meeting.

Developing Indicators of a Healthy City.   The development and use of indicators in the project has not yet been resolved satisfactorily in terms of process of semantic content. In 1987 a Healthy Cities workshop on indicators agreed on the necessity of and the requirements for indicators, but not on what the specific indicators should be. The indicators should be relatively few and should identify changes in both process and outcome that are useful to each city and to the project as a whole. One of the major questions is whether a single set of indicators can meet the unique needs of diverse project cities.

In addition to the work being done by project cities, the WHO project office has negotiated the active involvement of specialist research institutions. In May 1990 a French-speaking working group on indicators for a healthy city met in Nancy and agreed on a set of health, environmental, and social indicators for healthy cities. These indicators were endorsed by all WHO project cities at the 1990 healthy cities symposium in Stockholm.

Population Health Surveys.   This refers to developing and using health surveys in such areas as subjective health, perceptions of and knowledge about health, access to and use of services and social support systems, health inequalities, environmental assessment, social deprivation and vulnerability, as well as more traditional aspects such as morbidity and life-styles.

Evaluating the Project (1990–1992).   The information and analysis resulting from the evaluation for 1990–1992 will generate the material needed to prepare eight distinct products or resources. Outside institutions will have to support or perform the necessary analytical and editorial work. The material that could be produced includes:

- A consultation report for each city, describing the key information obtained in each interview, analyzing its significance, and recommending future action
- Training materials to be used by project cities, national networks, and academic and professional institutions on such areas as Healthy Cities management and developing the new public health at the local level
- A Healthy Cities handbook based on examples of action at the city level that would identify the main issues that arise in creating new structures and processes, gaining visibility for health, and introducing new policies and programs
- A computerized file giving examples of new policies and programs at the city level and approaches taken to solving the various problems that arise in creating Healthy Cities projects
- A computerized file identifying personnel in the Healthy Cities network who can be used to advise and consult on project development

- An assessment guide for national networks in carrying out interviews and consultations similar to that being prepared for project cities
- A monograph identifying the research and data that are needed to develop Healthy Cities based on the experience of network cities
- An assessment of the significant developments in the WHO network from 1987 to 1992, including political issues, changes in structures and processes, and new policies and programs for health for all initiated through the project.

THE MULTICITY ACTION PLAN

The Multicity Action Plan involves groups of cities working together to address common concerns. The initial projects include equity, traffic, housing, young people, elderly people, tobacco, AIDS care, nutrition, mental health, and the health-promoting hospital. The plan will enable smaller groups of cities to work together on issues of high priority to them and will expand the number of partners and the resources made available both in the cities and from WHO. The goal of this plan is to develop jointly, implement, and disseminate innovative models of good practice at the local level. The plan involves a business partnership of project cities committed to work together on one of the action areas for at least 2 years, and open-market events which are open to other cities and provide the forum to present models of good practice, exchange information, and monitor the progress of the plan.

The Multicity Action programs bring in new social forces who contribute their time and energy to the project. For example, programs on housing and health should involve the city's housing, town planning, and environmental health departments; the social services departments responsible for caring for groups with special needs such as people with disabilities, elderly people, and the homeless; local housing cooperatives and neighborhood associations; and public and private construction firms. At the international level, WHO can provide experience and expertise and, together with the cities, can also bring in resources and expertise from national and international organizations (such as the European Foundation for the Improvement of Living and Working Conditions, the European Communities, the Council of Europe, and the Organisation for Economic Cooperation and Development) that are concerned with housing and urban environmental health.

Involving Other WHO Programs in Achieving Health for All at the Local Level

The project is one of WHO's major vehicles for achieving the strategy for health for all. It has provided the testing ground for applying new strategies and methods of operation in cities, which is completely new to the work of WHO. The project has established political, professional, and technical alliances for health and fertile ground for change and innovative action. New channels of communication and potential entry points to other active programs of the Regional Office for Europe and other international agencies have been opened up.

The project is a joint enterprise of the health promotion program and environmental health program of the Regional Office. The possibilities for joint work with other programs include technical support for preparing project events such as symposia and workshops, participation in the Multicity Action Plan, technical support and consultation with cities requesting help in various areas, promotion of action packages prepared by different programs—from nutrition guidelines for local caterers to housing standards and aspects of primary and hospital care, and networking with different local professional groups and institutions and establishing contacts with local people with relevant experience and expertise. To accomplish this, the Regional Office needs to address the resources needed to enable and facilitate this collaboration.

PROJECT DISSEMINATION

Although the Healthy Cities project was always considered to be broader than just the project cities, the scale and speed of national and international dissemination have been breathtaking. As of mid-1991 there were 18 national Healthy Cities networks (Tables 25.4 and 25.5) and an international French-language network. Subnational

TABLE 25.4   National Networks in Europe on July 1, 1993

Austria	Israel
Belgium	Italy
Croatia	Netherlands
Czech Republic	Poland
Denmark	Portugal*
Finland	Slovak Republic*
France	Slovenia
Germany	Spain
Greece	Sweden
Hungary	United Kingdom

*National networks in the process of being established.

TABLE 25.5   Strategic Objectives of National Healthy Cities Networks

To advocate widely and mobilize resources for the Healthy Cities idea within a country at all levels

To mediate between partners relevant to Healthy Cities development (such as associations of local authorities, schools of public health, universities, professional groups, community action groups)

To enable cities to start and maintain a Healthy Cities process at the local level

To provide information and support on national and international Healthy Cities developments

To assist with experience and information exchange between cities

To maintain links with other national Healthy Cities networks

To maintain close links with WHO and WHO project cities

networks (regional or local) have also been developed in six countries, involving numerous small to medium-size towns.

In 1989 a European network of national Healthy Cities networks (EURONET) was established. EURONET is an intermediary structure that serves the following purposes:

- Mutual exchange between national Healthy Cities networks to strengthen the Healthy Cities movement and to develop a joint European perspective
- Linkage to information on developments in health promotion and health policy
- Linkage to WHO and the WHO Healthy Cities Project Office

This is done through the coordinators of the national Healthy Cities networks. The WHO Regional Office for Europe assists EURONET through the Healthy Cities Project Office by:

- Giving political, technical, and strategic support
- Facilitating information and data exchange
- Organizing regular meetings of national Healthy Cities network coordinators
- Negotiating with Member States of WHO for support to national Healthy Cities networks
- Ensuring global links accessible to WHO

National and regional networks have been established in Australia, North Africa (Maghreb countries), the United States, and Canada, involving over 700 cities. There is also a strong interest in taking up similar initiatives in African, Middle Eastern, and Latin American cities.

Healthy Cities was one of the main themes discussed in the May 1991 World Health Assembly Technical Discussions on strategies for health for all in the face of rapid urbanization. The project was widely endorsed and recognized as a successful approach to improve health and the environment in cities in both the developed and the developing world. (Technical discussions on strategies for Health For All in the face of rapid urbanization. World Health Organization, Geneva, May 1991. A 44/Technical discussions background documents 1–7.)

PROJECT ACHIEVEMENTS

The project is now becoming global as interest grows and the project expands in developing countries. The major strength of the project is its attractiveness to many different groups and professions and the political and community leadership in many cities.

The project has been very successful in accumulating practical knowledge about the strategies and structures that can help to promote the Healthy Cities idea. It is now possible to prepare a composite picture of the organizational structures and managerial processes that predict success in developing new approaches to public health at the local level.

Health has been put high on the political agenda of cities. About half of the project cities have successfully developed a new organizational model and strategy for addressing health.

TABLE 25.6   The Healthy Cities Project: A Taste of Success

It has put health high on the political agenda of cities.

It has inspired thousands of people in hundreds of cities and towns all over the world.

It is a proven example of how to implement the strategy for health for all by the year 2000 at the local level.

It has laid the basis for a new European public health movement and helped restructure and recreate public health at the city level.

It has created a network of cities working together in a spirit of international cooperation.

It has built a body of experiential knowledge that can be used to develop new practical ideas.

It has created structures for communication between sectors in the cities.

It has raised the issue of supportive environments as a central problem for health.

It has emphasized the importance of equity for health at the local level.

It is visible proof that improving health requires the active participation of many different groups, disciplines, and professions and that such collaboration is possible.

It has, for the first time, forged links between WHO and local governments and has made WHO more visible, credible, and relevant at the community level.

It has provided good experience for a new generation of potential city administrators.

It has brought international agencies together under city auspices.

The mayors and senior political representatives of the WHO project cities have issued a strong declaration of political support (the Milan Declaration on Healthy Cities), and several cities have already devoted significant resources to the project. The project is based on a spirit of true partnership and has well-established mechanisms that promote information exchange, sharing of experience, mutual support, adjustment of existing plans, development of new strategies, and dissemination of ideas and products.

The project is one of WHO's major vehicles for achieving the strategy for health for all. It has provided the testing ground for applying new strategies and methods of operation in cities, which is completely new to the work of WHO. The project has established political, professional, and technical alliances for health and fertile ground for change and innovative action. New channels of communication and potential entry points to other active programs of the Regional Office and other international agencies have been opened up.

The project has laid the basis for a new European public health movement, forged links between WHO and local governments for the first time, and has made WHO more visible, credible, and relevant to local needs and concerns.

There are few other comparable examples in the history of public health in which a relatively small investment has paid off so well (Table 25.6).

FUTURE CHALLENGES

The success of the project to date in providing an effective vehicle for developing and applying the strategy for health for all at the local level and in stimulating local political support for this strategy has prompted the Regional Director for Europe to extend WHO's support to the project at least until 1997.

The practical experience being accumulated from the cities involved in the project is becoming a source of inspiration, guidance, and legitimacy for all the cities in the movement. Assessment and documentation of project progress, handbooks and manuals of practical strategies and methods, and training programs are urgently needed to support and fulfill the expectation of hundreds of cities.

The project review for 1987–1992 will emphasize local policies and community action for health and will analyze the effects of previous changes in structure and management in project cities.

The Healthy Cities project must now find new organizational forms and begin, for example, to establish Europe-wide Healthy Cities resource centers that will provide information, analysis, and guidance on strategy in the efforts to establish a new, locally based, international public health movement. National networks and links to the international organizations that will significantly affect Europe's future development will be of great importance in establishing this movement.

The project must address the needs and opportunities for public health action in central and eastern Europe. The globalization of the project, expanding into cities in the developing countries, requires new approaches for healthy cities that are compatible with the health status and the social, cultural, economic, and environmental circumstances of these countries.

BIBLIOGRAPHY

"A Guide to Assessing Healthy Cities," FADL, Copenhagen, 1988 (WHO Healthy Cities Papers, No. 3).

Draper, R., "Making Equity Policy," *Health promotion,* vol. 4, no. 2, pp. 91–95, 1989.

European Charter on Environment and Health, Nov. 1989.

"Five-Year Planning Framework," FADL, Copenhagen, 1988 (WHO Healthy Cities Papers, No. 2).

Giroult, E., "Equity and the Urban Environment," *Health promotion,* vol. 4, no. 2, pp. 83–85, 1989.

Hancock, T., and L. Duhl, "Promoting Health in the Urban Context," FADL, Copenhagen, 1988 (WHO Healthy Cities Papers, No. 1).

Kickbusch, I., "Good Planets Are Hard to Find," FADL, Copenhagen, 1989 (WHO Healthy Cities Papers, No. 5).

Kickbusch, I., "Healthy Cities: A Working Project and a Growing Movement," *Health promotion,* vol. 4, no. 2, pp. 77–82, 1989.

Milan Declaration on Healthy Cities, Apr. 1990.

National Healthy Cities Networks in Europe. Healthy Cities Project, World Health Organization Regional Office for Europe, 1992.

Ottawa Charter for Health Promotion, *Health promotion,* vol. 1, no. 4, pp. iii–v, 1986.

"Targets for Health for All," WHO Regional Office for Europe, Copenhagen, 1985.

The Process of an Established Multi City Action Plan: The Example of the AIDS Care and Services MCAP. Healthy Cities Project, World Health Organization Regional Office for Europe, 1992.

Tsouros, A. D., "Equity and the Healthy Cities Project," *Health promotion,* vol. 4, no. 2, pp. 73–75, 1989.

Tsouros, A. (ed.), "WHO Healthy Cities Project: A Project Becomes a Movement (review of progress, 1987 to 1990)," WHO/SOGESS, Milan, 1990.

Twenty Steps for Developing a Healthy Cities Project, World Health Organization Regional Office for Europe, 1992.

WHO Healthy Cities Project: Review of the First Five Years (1987–1992). Draper, R., Curtice, L., Hooper, J., and Goumans, M. Healthy Cities Project, World Health Organization Regional Office for Europe, 1993.

CHAPTER 26

CONSTRUCTION PROJECT MANAGEMENT IN JAPAN AND THE UNITED KINGDOM

N. J. Smith

*Department of Civil and Structural Engineering,
UMIST, Manchester, United Kingdom*

Dr. Nigel Smith, BSc, MSc, PhD, MICE, MAPM, CEng, is a Senior Lecturer in the Project Management Group at the University of Manchester Institute of Science and Technology, Manchester, UK, and a Project Director of the European Construction Institute. He has supervised project management research from a number of Japanese postgraduates from the Shimizu Corporation. His current funded research interests include work into the project management of infrastructure maintenance and a major study of BOOT projects.

INTRODUCTION

Project management has had many differing interpretations and definitions over the years. Some were very narrow in their application, while others attempted to encompass every possibility. Attention has been diverted away from semantics and is now concentrated on the practice of project management as a philosophy of managing by projects to realize the internal and external objectives of an organization.

One of the earliest users of project-management methods was the construction industry and related process-plant projects, which commenced with a major design and construction phase prior to the operation phase, and this industry is still in the forefront of project-management development and implementation. The construction industry is not a single industry, but an agglomeration of the building industry; housing, commercial, and light industrial work; the civil engineering industry; major infrastructure projects; heavy engineering; the process-plant industry; and manufacturing and operational facilities.

Developed countries like Japan and the United Kingdom have construction industries that are split into about 25: 65: 10 percent civil engineering: building: process, and they are becoming increasingly involved in repair and maintenance work. These industries are key contributors to their respective gross domestic products and are major employers, who make significant contributions to overseas earnings.

The building sector, while large in total, tends to consist of a large number of small contracts, using existing technology and incorporating little risk, and is almost entirely based in the domestic market. Consequently the main thrust of this chapter is concerned with the civil engineering sector of the construction industry which encompasses major contracts, inherent risk, project management, and domestic and international markets.

Within the civil engineering sector of the industry there are three major roles—client, consultant, and contractor—which are often undertaken by separate organizations. Typically the client, sometimes referred to as the owner, employer, or principal, is the organization requiring the project. Some large clients have in-house design teams, but most engage consultants. Consulting engineers concentrate on design, but may also offer contract administration skills, site supervision, and project management skills, as needed. Contractors are employed by clients to be primarily responsible for construction, but increasingly are able to offer a combination of some or all of the above functions and the additional ability to organize project finance if required.

CONSTRUCTION PRACTICES

A comparison of the construction practices in different countries has to take into account the different cultural perceptions of project management and to assess this in terms of the current industrial procedures. However, as both Japan and the United Kingdom have been relatively successful in their own domestic construction markets and, often in competition with each other, in international construction markets also, the comparison of project-management methods should be particularly interesting.

In order to develop this comparison, the traditional practices in each country are first investigated, and then the current market position is described. This is followed by a comparison of the relative performance of the two countries in the international construction market. From this base an assessment of the likely future international markets for these industries and the associated cultural implications for project management is undertaken, and then predictions are made for the subsequent impact on project management training and development in Japan and the United Kingdom.

CONSTRUCTION IN JAPAN

The geographic position of Japan and its social development are thought by many to be major factors in the formation of current industrial and managerial practices. Being an island nation, self-contained, and with no significant immigration for about 2000 years, a distinct national and cultural homogeneity has evolved. Within this national structure the prime social unit is the family group. This has engendered a sense of family loyalty, which transfers into company loyalty and a greater concern for extended family welfare than exists in most other developed countries.

As Tokunaga[1] has shown, the effects of the traditional collective decision-making process can be seen in all Japanese management methods. This concept of "Nehmi-Washi," whereby all discussion with all parties continues until agreement is reached and then there is acceptance of responsibility for the consequences of that decision, ensures good communication and joint responsibility for project decisions.

The construction industry in Japan accounts for a large sector of a growing economy, and construction output per annum is of the order of £220 billion*, which is about

---

*All billions are 1,000,000,000.

seven times greater than in the United Kingdom. Work is obtained through the "Dango" system, whereby a contractor builds up good relations with a specific client over a long period of time and receives special appointments without the need for competitive bidding.

In these circumstances the contractors are motivated to perform to protect and enhance their reputation and goodwill. Many companies may obtain up to three quarters of their domestic revenues under this system, and an organization which has to tender for the majority of its work is viewed with some suspicion. This philosophy has meant that the use of "Conditions of Contract" for resolving construction disputes has become largely irrelevant and has reduced the demand for innovative forms of contract strategy. It has also motivated companies to produce work of a high quality without recourse to claims.

In Japan the concept of managing major construction projects has been developed over hundreds of years. In the 1850s clients directly appointed gangs of carpentry or masonry workers to undertake construction. These gangs became relatively powerful and a few developed into a type of general contractor. Since the Meiji restoration in 1868 there has been government control of major civil engineering projects. However, the consequences of the wars with China in 1894 and with Russia in 1905 were that a demand for overseas construction was created. One of these contracts was undertaken by Kajima in 1899, constructing a railway in Korea.

These traditional cultural and industry practices underwent a transformation at the end of World War II, when constraints and restrictions were placed on the development of Japanese industry, and direct exposure to the techniques in use in developed countries outside Japan took place. The current industry has therefore emerged as a blend of Japanese tradition and culture with practices that were originally imposed from outside.

Domestic Construction Market

Large contracting groups now dominate the industry, with 0.02 percent of the over 500,000 contracting companies accounting for 28.9 percent of the value of construction works, and the top 10 companies having 13 percent of the market. In the domestic market long-term relationships have been established between the contracting organizations and their clients, often in the public sector. The construction organizations can be broken down into a small number of major companies, offering a wide range of design, construction, management, and financial expertise and a large number of very small full-time or part-time organizations. The five largest organizations are major multinational, multidisciplinary companies. The use of specialist subcontractors is not widespread, as the major companies prefer to train staff in new skills and then to include the new service as part of an enhanced construction capability offered to clients. The turnover of staff is also very low, with people joining the company and expecting a life-time career within the same organization.

A small number of government contracts are awarded using a public tendering system. However, the prequalification procedure is very restrictive. It is based upon companies having the necessary license and has rigorous conditions regarding previous performance, financial status, and other factors, with the effect that the domestic market has attracted very little competition from other countries.

Design

Design organizations in Japan developed from a long tradition of public sector monopoly, as from the time of the Restoration in 1868 the government had control of major projects. After 1945 there were a small number of independent design consultancies, working mainly as subcontractors to the construction majors. By 1954 the Japanese economy had regained its prewar levels, and the Japanese Institution of Engineers, which had been established in 1951, encouraged a growth in consulting organizations, with most being hybrid private-sector and government firms. The first qualified engineers appeared in 1958 as a result of the Engineer Act of 1957, and about this time overseas work as part of the war reparations helped to develop the industry.

In 1966 the Japanese Official Development Assistance (ODA) was launched and made yen credits available to assist Japanese consultants in obtaining work in developing countries. By 1990 the consultancy sector had grown to support over 2000 consulting organizations responsible for about ¥400 billion per year. The largest and best known firm is Nikai Koei with some 1600 employees, but many of the major construction groups undertake design work internally. Nevertheless, from the 1989 figures Japanese consultants had only a 3.5 percent share of the international consultancy market.

CONSTRUCTION IN THE UNITED KINGDOM

Background

The industry in the United Kingdom was established on the concept of the separation of design and construction, free and fair competitive tendering, and the use of Model Conditions of Contract based upon case law, usually administered by an independent engineer. The conventional basis of procurement is sometimes known as the "contract of conflict," with a client engaging a consultant on a fee basis to design the works, administer the contract, and supervise construction and then using competitive tendering to award a construction contract to a contractor, usually on the basis of the lowest acceptable price.

Most of these contracts are admeasurement contracts, whereby the client, with the assistance of a consultant, prepares a Bill of Quantities, which contains the principal quantities of materials required in the permanent works as measured in accordance with a particular method of measurement. The bill is then priced by the contractor at tender stage, providing the core of the tender bid submission and evaluation, and ultimately is used as the basis for payment as the work progresses. Long-term relationships between client and contractor are not encouraged, as this could be viewed as collusion and a barrier to free and fair competition. The annual revenues in construction are about £30 billion.

Domestic Construction Market

Civil engineering works are undertaken predominantly for public sector clients, although since 1980 it has been U.K. government policy to move several major client organizations into the private sector or into hybrid public and private sector organizations. This change has increased the emphasis on the commercial aspects of tendering, but has not significantly affected the size or operation of the construction market. Like most developed countries in the world, the U.K. civil engineering sector suffered a major recession from the mid 1970s until the late 1980s, and although there have been improvements, the 1990 figures have not yet regained 1974 levels in real terms.

Contracting companies range in size from multinational contractors to small specialist firms. The major national and international companies dominate the market, with the top 100 contracting firms being responsible for 25 percent of the domestic construction market and the top 0.06 percent of companies undertaking 19.6 percent of the work. Nevertheless, the companies in the United Kingdom are not large on the world scale, and although construction is a high-risk business, the groups, unless they continue to grow, many become targets for acquisition, particularly in times of recession.

In the private sector and in certain exceptional cases in the public sector, some tenders are awarded by negotiation. New trends have been welcomed in recent years, and new strategies and new documentation continue to evolve to service the different needs of the clients, consultants, contractors, and subcontractors in a changing domestic market. An increasing proportion of work is awarded using a range of contract strategies to meet particular client requirements, including Package Deal and Management Contracts, both of which require additional services from the contractor.

The use of subcontractors for specialized sections of the works is well established, and most major contracts involve several subcontractors, some of whom may have been nominated by the client. Although subcontractors are invited to tender for work, in practice informal relationships between some main contractors and particular subcontractors have been established based upon successful performance.

Design

In the United Kingdom firms of consulting engineers offering design skills, contract administration, and site supervision have existed independently for over 100 years. Their origin can be traced back to the 1890s with the demand for additional expertise to help the fledgling electricity and water companies deliver their services to the general public. Originally these consulting organizations operated as small independent partnerships, with the profits and the liabilities being the responsibility of the partners. This continued into the 1960s, but the increasing use of litigation by clients and the greater demands for a more multidisciplinary approach has led to many partnerships merging to form larger organizations, most of which are limited-liability companies.

There are now a few large organizations with about 4000 employees, but the average number of between 50 and 100 staff reflects the bulk of the 600 consulting practices. U.K. consultants have a well-established record of working overseas. In 1974 the revenues were £4 billion, by 1980 they had increased to £35 billion, and currently they are on the order of £55 billion.

INTERNATIONAL CONSTRUCTION MARKETS

International construction markets have existed since the days of colonial expansion, but they have become increasingly competitive since 1945. As the world population increases, there is a demand for improved infrastructure and higher expectations of the speed of development of industry, commerce, and of general living standards, which many countries are unable to satisfy internally. Developing countries in particular may have difficulties with the financial consequences of this demand, or with the supply of the necessary technical construction capabilities. The developed countries have responded with a combination of grant aid and competitive international tendering to satisfy these needs. Ultimately there will be a limit to this market as populations stabilize and domestic engineering skills become common, but the market is currently expanding and is likely to do so for some years to come.

The six leading nations in the international market are the United States, Japan, France, Italy, Germany, and the United Kingdom. In 1988 the world market for building, civil engineering, and process engineering was \$94,100 billion.[2] It showed a trend away from working in developing countries and toward working in developed countries.

Japanese International Construction

After 1945 some Japanese companies were involved in international construction works as part of the war reparations and Kajima, among others, was involved in the construction of hydroelectric power schemes in Burma in the early 1950s. These projects were not commercial, but did provide some experience until they were finished in 1964. With these exceptions, Japanese companies worked almost exclusively in their strong domestic market, which from 1961 to 1972 grew at more than 10 percent annually. However, in 1973 increasing oil prices caused a recession of 5 percent per year in the domestic construction market, which caused the industry to expand into new markets rather than suffering severe cutbacks.

Figure 26.1[3] traces the development of Japanese international construction markets since 1973. Initially the market was divided between the nearby Asian countries and North America, with a major involvement in the oil-rich countries of the Middle East. By 1980/1981 the Asian market had expanded dramatically and was accounting for

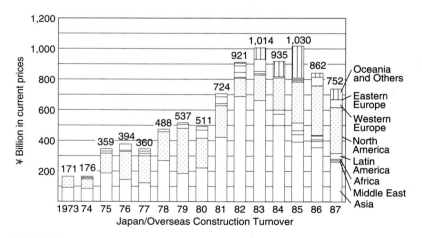

FIGURE 26.1 Japanese international construction revenues, 1973 to 1987.

more than half of a total Japanese overseas market, which was about seven times greater than that of 1973. In common with other international contractors, the Japanese companies had largely withdrawn from the Middle East market, which now accounted for less than 2 percent per year. From 1987 the Asian market sector has remained relatively constant, partly as a result of competition from the newly industrialized countries of Korea, Taiwan, Singapore, and Hong Kong, as defined by the Organisation for Economic Co-operation and Development, and partly due to the effects of recession in the host markets. Consequently Japanese companies turned their attention to new markets, and by 1990 North America accounted for about 30 percent and Europe for 10 percent of their international construction market.

International contracts executed by the large Japanese contractors have been partially financed by the Japanese Overseas Development Administration (ODA) and the Foreign Direct Investment (FDI). The FDI had originally been founded to procure the raw materials needed by Japanese domestic industries, but evolved into a much more complex organization using capital movements to improve the overseas manufacturing base of Japanese companies, who in turn tended to employ Japanese contractors. Indeed by 1989 the massive support from these two organizations accounted for over one-third of Japan's international construction market.[4] However, the structure of Japanese international contracting is changing, and while Japanese companies have taken part in the merger and acquisition market in host countries, on a relatively small scale, they have also been very successful in establishing new internationally based subsidiary companies.

U.K. International Construction

A long tradition of international construction, from the days of colonial expansion, has established a knowledge of the English language and of U.K. methods and standards of construction in many parts of the world. This basis, combined with the fluctuations in demand in a restricted domestic market, has meant that the United Kingdom has had a significant stake in the world construction market for many years. The geographic location of the work has changed with the variations in national finances and demand for services. In the 1940s the prime international market was in the commonwealth countries, particularly of Africa and Oceania.

Figure 26.2[5] shows the international market for U.K. companies from 1973 until 1988, illustrating clearly that markets change to match economic strength. In the 1970s the bulk of the work was located in the Middle Eastern countries based on the oil economies, with surplus capital due to the quadrupling of the price of oil. By 1980 the Middle East market was down to about one-third and large increases in work in Africa and the Americas had occurred. As we approach the 1990s, the major market for U.K. companies working overseas is in the United States and Canada, which accounts for about two-thirds of the international market. It also shows a small but nevertheless significant growth in international earnings from EC and other European countries.

The major U.K. construction companies have tended to adopt a divisional organizational structure, with a board of directors of the parent or holding company under the auspices of which are grouped a number of operating companies, at least one of which would be responsible for international construction. A number of these international divisions have formed temporary or permanent mergers with companies in host markets or have acquired local firms to establish an indigenous base. The other EC countries have been specific targets for the U.K. companies, but in addition some companies have acquired companies based in the United States. An alternative

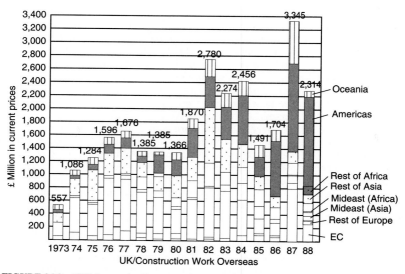

FIGURE 26.2   U.K. international construction revenues, 1973 to 1988.

approach, which has been used by many organizations, is to form temporary consortia or joint venture organizations with both U.K. and non-U.K. companies to share risk and expertise for particular projects. As international practice differs significantly from the conventional U.K. system, organizations have had to adapt to be able to offer the packages that the client requires.

Summary

International markets have become more competitive, and both countries have seen a fall in revenues compared with the mid 1980s. The relative performance of the yen against the pound over this period shows that Japan has been more successful in international markets than the United Kingdom.

SURVEYS OF JAPANESE AND U.K INDUSTRIES

In 1983 the Overseas Construction Association of Japan, Inc., instigated a survey, undertaken by the Nomura Research Institute, among the major Japanese civil engineering contractors to highlight their perceptions of their own competitive strengths and weaknesses in international contracting, compared with their counterparts in other industrialized countries such as France, Germany, Italy, and the United States.[6]

In 1989, as part of the research program of the Project Management Group at UMIST, it was decided to repeat the approach of the Japanese researchers, contacting leading U.K. companies engaged in international contracting to ascertain their views on project management and other perceptions of their strengths and weaknesses. The results of the new U.K. survey were then compared with the original Japanese survey results published in 1984. The respective project-management perceptions and abilities are related to the traditional background of the industry in each country, and the results are reviewed in the following.

The comparison suffers from two major weaknesses. First, there is a time difference between the two surveys so that the Japanese survey, if repeated at the same time, might have reflected some improvement in perceived problem areas, although it would appear that there have been no major changes in the basic characteristics of major Japanese contractors over the period. Second, it is difficult to be certain that the terminology and question interpretation are exactly consistent. However, it was felt that the results would be useful in highlighting any key or strategic differences between Japanese and U.K. industries.

A total of 163 companies in the United Kingdom were approached and 48 responded, which is slightly lower than the earlier Japanese survey of 69 companies. The survey was carried out by postal questionnaire, with respondents ranking their perceptions of their own performance on a graduated scale between "weak" and "strong" when compared with other industrialized countries. A summary of 23 items from the survey relating to international construction is shown in Fig. 26.3.[5]

It is difficult to draw detailed conclusions from Fig. 26.3, especially when the differences in culture, industry, commercial environment, and market could all be significant factors. Nevertheless, general comparisons can be made, which reveal important factors. Interestingly the survey revealed that the U.K. companies perceived themselves to be strong or equal in all categories, while the Japanese perceived themselves to be particularly strong in category 2 but equal or weak in all the others. This result is more surprising when compared with the international revenues discussed earlier, which showed that the Japanese had performed better than the United Kingdom in the last 10 years. However, the survey is a self-perception survey rather than an objective assessment of performance, and the Japanese perceived weakness is likely to be due to a limited experience of competitive market conditions where long-term relationships between client and contractor are viewed as possible collusion or unfair practice rather than as representative of good performance.

As Fig. 26.3 shows, the items have been grouped into five categories. The first deals with project-management skills, including contracts and negotiation. The second category considers construction technology and site management, and the third deals with estimating, procurement, and the successful administration of the works. The fourth concerns the interface between the international contractor and the host country. Finally two important features about the domestic resources of the contractor are addressed, the quality of staff and the support of the home government.

Examining Fig. 26.3 in more detail reveals the key areas of relative strength and weakness. The Japanese would appear to be strong in the technology, quality, time, and cost control elements, but weak in negotiation, risk management, and contracts. This can be matched against the experience and skills gained from working in the domestic market in Japan with its high proportion of noncompetitive contract awards and the emphasis on early completion of a high-quality job without claims. The U.K. responses indicated strengths in project management, construction, risk management, and negotiation and weaknesses, in particular, in collaboration with other industries. This result also corresponds to the situation in the U.K. domestic market, where years of using "contracts of conflict" have stressed the importance of negotiation and risk management and where the construction industry tends to be isolated from the process-plant industry and other engineering disciplines. However, there are some areas of agreement; both Japanese and U.K. companies regard their own employees highly and are dissatisfied with the support provided by government.

The general summary can be broken down into specific sectors, and Fig. 26.4[5] shows the findings relating to project-management activities. The elements contribut-

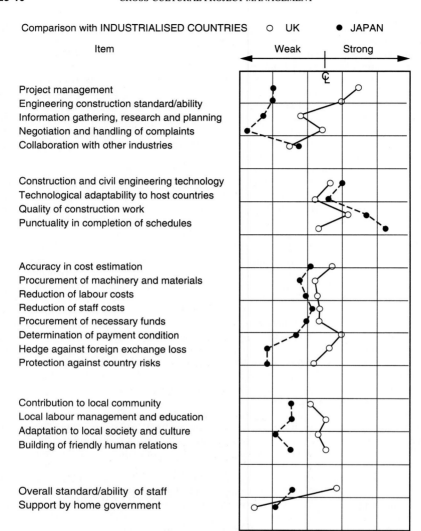

FIGURE 26.3 Comparison of strengths and weaknesses of Japanese and U.K. construction companies with that of companies from other industrialized countries.

ing to the general perceptions of strong by the U.K. respondents and weak by the Japanese are identified separately. Although the U.K. companies perceived themselves to be strong in project management and the Japanese weak, a study of the other elements in Fig. 26.4 shows that both groups perceive themselves to be strong in four elements, although not the same four, and weak in three and four elements, respectively, with the rest being perceived as equal. Consequently the apparent differences are not as great as it may appear.

Considering the findings in more detail, the strengths perceived by the Japanese are

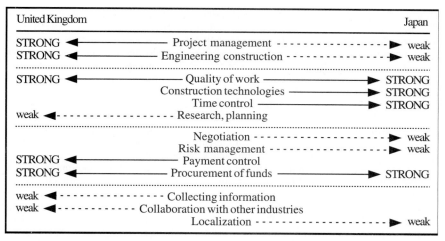

United Kingdom		Japan
STRONG ◀─────── Project management -------------▶ weak		
STRONG ◀─────── Engineering construction ------------▶ weak		
STRONG ◀─────── Quality of work ───────▶ STRONG		
Construction technologies ───────▶ STRONG		
Time control ───────▶ STRONG		
weak ◀----------------- Research, planning		
Negotiation ------------------▶ weak		
Risk management -----------------▶ weak		
STRONG ◀─────── Payment control		
STRONG ◀─────── Procurement of funds ───────▶ STRONG		
weak ◀---------------- Collecting information		
weak ◀----------- Collaboration with other industries		
Localization ------------------▶ weak		

FIGURE 26.4 Comparison of self-perceptions of strengths and weaknesses of Japanese and U.K. construction companies in project management.

in construction technology, quality, procurement of funds, and time control, with weaknesses in risk management, engineering construction, negotiation, and localization. Localization is the process of operating efficiently under the procedures in use in a particular country, employing local people, and operating the company in line with accepted practice. Many Japanese companies are making great efforts in this area, and one of the important steps being taken by some companies is to decentralize decision making from Tokyo to the head office in the host country.

In contrast, the U.K. companies had perceived strengths in engineering construction, quality, procurement of funds, and payment control, with weaknesses in research, collaborating with other industries, and collecting information. The relative lack of investment in research and development has been highlighted in a number of reviews of the funding and structure of research in the United Kingdom, which shows research investment to be about one-tenth of the levels typically obtained in Japan.

The Japanese and U.K. perceptions of strength and weakness shown in Fig. 26.4 can be traced back to the respective domestic markets. The perceived strengths generally reflect the current industrial practice in the respective countries and the weaknesses tend to indicate areas which are important to international clients, but in which companies lack professional experience. Both countries contain major companies which are advanced and experienced in international construction, but the procedures, skills, and areas of specialization required to continue to compete successfully will necessitate further training and development.

FUTURE MARKETS

The domestic markets of both countries are in the process of major change. Agreements signed between the United States and Japan indicate the beginning of the relaxation of the Japanese domestic market. The domestic market in Japan is opening up to international competition in an increasing way, although by European standards the market is still restricted. The entry of the United Kingdom into the Single European Market (SEM) in 1992 was conditional upon accepting common practices for the procurement of public sector construction contracts and the common treatment of competing bidders from all EC member states. This will considerably increase the level of competition in the U.K. domestic market, but it will allow the United Kingdom greater access to the domestic markets in other EC countries.

The United States–Japan Construction Accord, signed in 1988, concerns the so-called "third-sector" joint public and private projects such as Kansai International Airport. Initially 14 third-sector projects were identified and a further three were added later. In 1990 the agreements with the United States permitting access to certain categories of public and private sector work were widened to include the United Kingdom and other EC countries. However, the accord is aimed at removing obstacles and not at ensuring free and fair competition, and consequently any significant market penetration is likely to take many years.

The Japanese domestic market still enforces many conditions and restrictions on companies wishing to compete. The most successful penetration of the market to date by non-Japanese contractors has been by major Korean companies. The United Kingdom has a small involvement in the Japanese market, principally from U.K. consultants in collaboration with corresponding Japanese companies undertaking design work for major projects.

However, international penetration of the Japanese market is still at a low level, as demonstrated by the fact that although the United States currently has construction work valued at $195 million in Japan, the Japanese have contracts worth $1.77 billion in the United States. All tenderers have to have Japanese partners and to satisfy other procurement criteria, but as international pressure continues to be exerted on the Japanese to make their domestic market more accessible, competition is likely to increase in the long term.

For the United Kingdom since 1992 the SEM, which permits the free movement of goods, people, services, and capital, has produced a combined internal construction market seven times larger than the current U.K. domestic market. At present there is only a small amount of international competition between the organizations within the EC member states. About 2 percent of construction work is undertaken by nondomestic contractors. However, the SEM will permit companies from other EC countries to compete inside the U.K. domestic market and may even encourage an increased level of competition from outside the market. This competition will be regulated by strict procedures imposed on the procurement of major construction contracts to ensure free and fair competition within the market. Ultimately the goal is to harmonize the procurement of construction throughout the SEM. However, this is likely to take some time, given the different legal systems among the 12 countries together with the likely addition of several countries from eastern Europe in the near future.

The U.K. domestic market has not experienced major penetration from international competition. Several companies which have formed semipermanent or temporary joint venture companies with EC neighbors or with firms from outside the EC have undertaken more work in recent years. The U.K. market is dominated by the principle of accepting low tender prices and using the process of claims during the progress of the works to reimburse the contractor for risks which are beyond the contractor's control. The small size and the unique nature of the market do not make it very attractive to international competitors. However, a number of countries have become established in the United Kingdom, and hence will be within the SEM after 1992, including several Japanese organizations. These Japanese companies have a relative small revenue at present and are working predominantly for Japanese clients. Interestingly, they are operating with different client packages, sometimes including no claims tenders, with lower overheads, and with less waste of resources, which would provide a strong basis for attracting U.K. and EC clients. The SEM should increase the levels of construction, and despite the fierce competition from the domestic construction industries, this economically and politically stable market is attractive to both the United Kingdom and Japan.

A new market started to show its potential in 1990, that of eastern Europe, in particular the more developed countries of Russia, the Czech Republic, Hungary, and Poland. The policy of restructuring a state economy into a market economy is difficult to implement, but as markets become established, there appears to be tremendous potential if funding can be arranged. Some financial assistance is being made available from the European Bank for Reconstruction and Development, and the Japanese ODA has awarded ¥1000 billion over 10 years to help Japanese firms become established in this market. The possibilities of adopting build-own-operate-transfer (BOOT) contracts for international tender, hence avoiding hard-currency problems, are also being investigated by the authorities of Russia and the Commonwealth of Independent States and many U.K. organizations.

There is also an area of high economic growth closer to Japan, consisting of the Newly Industralised Countries (NIES) as classified by the Organization for Economic Co-operation and Development, and some Asian countries, including South Korea, Taiwan, Singapore, Hong Kong, Thailand, Indonesia, the Philippines, Malaysia, and Brunei. This region is forecast to produce growth over the next 10 years about two and a half times greater than the rate forecast for the United States. Due to their proximity to this market, Japanese companies have revenues about three times greater than the combined revenues of the western industrialized countries. However, the United Kingdom, with its long established connections with Hong Kong, Singapore, Malaysia, and Brunei, has established a presence in this market.

In addition there are also a number of countries in this region moving from communist, planned economies to market economies, with differing rates of progress, which would offer new markets, particularly to Japan. The countries include China, North Korea, Laos, Cambodia, Vietnam, Mongolia, and Burma. It will be interesting to note the development of these countries in the near future.

Both Japan and the United Kingdom will need to retain a competitive advantage in their domestic markets, compete in each other's markets and in their main international markets in North America, and become competitive in new markets. It is always difficult to predict the future and, as has been demonstrated, international markets change with economic performance, but in the short term the North American market, although the largest market, is in recession and there has been overconstruction of certain types of facilities with subsequent losses for developer-contractors. However, the United States do have the capacity to emerge from this recession, and the free-trade area with Canada and Mexico may also provide a boost to the construction market. It seems likely that both Japan and the United Kingdom will want to retain a sizable stake in this major market.

Another option to increase the construction market is to make use of the BOOT contract, which allows the contractor to manage the market rather than responding to fluctuations in economic performance.

BOOT CONTRACTS

The BOOT strategy is not new and has been used in France, where it has been known as a Concession Contract, for over 100 years. The concept of the strategy is that the client or principal grants a concession to an organization, or more usually a joint venture company or consortium, usually known as the promoter or concessionaire, for the design, construction, financing, operation, and maintenance of a facility for a specified period of time before returning the facility to the principal.

There are two types of BOOT contract, those proposed by the client and for which tenders are invited, and those proposed as speculative bids by the contractor. These projects require a high degree of pretender work, and consequently speculative tendering, where the contractor is not competing with other bidders, is attractive. Recent research shows that about 60 percent of BOOT contracts are speculative bids proposed by a single contractor and accepted by the client, while the remaining 40 percent are the result of competition between two or three bidders invited to tender by the client. One of the key features of the BOOT contract is the ability of the contractor to "create" work where no proposal currently exists, and consequently this strategy is likely to become more attractive to both Japanese and U.K. companies in the future.

The comparative survey of the Japanese and U.K. companies showed that companies in both countries perceived themselves to be strong in the area of "procurement of funds," which is an important component of the BOOT contract. Traditionally the contractor's role was to act as an intermediary for clients and bankers and to negotiate financing loans. Currently more emphasis is being placed on making equity investments in projects and in conducting investment appraisals for funding authorities, both of which are useful abilities to companies considering BOOT contracts.

The main organizational relationships in a BOOT contract are shown in Fig. 26.5, which indicates the fundamental role of the concession agreement between the principal and the promoter. The agreement is the basis of the contract. It addresses the legal, commercial, and environmental risks associated with the project and defines the scope of the project in commercial terms. In countries which do not have enabling legislation and the client is a part of the government, concession agreements are statutory in nature, but where legislation does exist, the agreement is contractual. The agreement

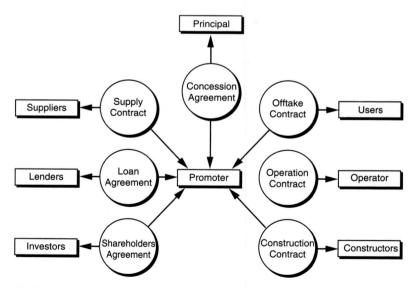

FIGURE 26.5   Typical corporate structure of BOOT contract.

would also include the terms of the concession period, including details of exclusivity and of any concessions on existing facilities to be incorporated.

Not all of the relationships shown in Fig. 26.5 would be required in every BOOT contract, although there would be cases when this would be necessary. One of the determining factors would be the composition of the promoter, which would influence the need for external shareholders, lenders, or construction contractors. Briefly, the relationships in the figure relate to the design and construction, the financing, and the operation of the concession.

The construction contract would only be needed if the concessionaire did not include companies capable of undertaking the work and of sharing the construction risk. The contract could be awarded to a single company if the project was for infrastructure development, or it could be a series of package contracts in the case of a more complex process facility. Often the contract would be negotiated and awarded on the basis of a lump sum, fixed-price turnkey contract.

The loan agreement would be used if the concessionaire did not already include organizations with sufficient financial expertise and resources. Most loan agreements use a form of nonrecourse financing, whereby money is loaned against the revenue from the project cash flow rather than the security of the concessionaire. The limits on the debt-to-equity ratio would have been contained within the concession agreement. Shareholders are not part of every BOOT contract, but are often used to increase the equity capital on major projects.

An operational contract is adopted when the promoter does not wish to operate and maintain the facility after construction. The two contract options shown in Fig. 26.5 for the supply and offtake only apply for certain types of projects. For example, in a water-treatment contract there could be an agreement about the quantity and quality of the inflow and an agreed tariff and quality standard for the output. The important feature of these contracts is that they would determine the financial viability of the project.

Under the BOOT system and its variants, the build-operate-transfer (BOT) and the build-own-operate BOO, the contractor is responsible for the design, construction, finance, and operation of a facility. For long-term assets the facility is transferred back to the client after the end of the concession period, but for short-term assets the BOO option may be adopted, whereby the concessionaire retains the facility, which will probably have little residual value. The World Bank, through the International Banks for Reconstruction and Development, has tended to favor package deals and is now considering providing equity loans for BOOT contracts.[2] This could help to mitigate against the adverse effects that developing countries with a weak economy and restricted access to hard currency experience in funding the development of major projects.

Many U.K. companies are involved in BOOT contracts in the United Kingdom and around the world, primarily for infrastructure projects. Japanese companies have constructed the Canadian embassy in Tokyo using a BOOT strategy and are currently engaged upon BOOT schemes in Turkey, Hong Kong, and Thailand. There is little experience with these types of projects, but if the national governments, who are the financing houses, continue to favor the BOOT contract, both Japanese and U.K. companies will compete in this sector.

IMPLICATIONS FOR PROJECT MANAGEMENT

Despite the large volume of international construction being undertaken by Japan and the United Kingdom, both countries are modifying their traditional practices without fundamentally changing their approach to project management. However, an appreciation of the problems of the management of culture and an awareness of the need for new skills are becoming apparent.

The management of culture has not yet become the major component of project-management development that may ultimately be needed. This need has been partially avoided by companies from both countries, the Japanese tending to undertake work for Japanese-owned client companies and the U.K. companies working largely in countries with a background of colonial, commonwealth, or trading relations. Nevertheless both countries are aware of the need to tackle the management of culture in the pursuit of new markets.

A feature of the management of culture is the process, already started by many Japanese companies, of localization. This introduces a blend of personnel from the host country at all levels in the organization and matches the management style more closely to the domestic practice. A variety of skills can be provided to assist with this problem, and the Japanese have adopted training programs in languages and overseas postgraduate study in project management and business subject areas. Education coupled with experience in international projects is viewed as the basis for developing a new approach. The United Kingdom has well-established project-management education at the first degree level, and the main efforts are being made in client education to U.K. methods.

Project management training is needed to emphasize both project skills and human skills, together with linguistic and cultural awareness and the ability to manage change. This policy would naturally lead toward a higher proportion of qualified personnel in an international company. The Japanese have been seen to be ahead in this area, with most of the major organizations having qualified professionals as almost one-quarter of the total work force.

The recognition of the need for new skills in the future is an important factor in project management development. The approach to clients may also have to be revised as competition increases. It will no longer be sufficient to offer clients simple packages, and deals effectively establishing longer-term relationships may be more productive in terms of future work and reduced overhead. The new product-management skills might require a blend of technical abilities and management science to encourage the balance of competition and cooperation. Competition will be likely to increase, but there seems to be more evidence to indicate an increase in cooperation also. The reciprocity in opening domestic markets is evident in the signing of the Joint Anglo-Japan Aid Policy in Tokyo in 1989 to work jointly on major projects in Africa and longer-term joint ventures or consortia. Japan has a long tradition in this area, but lacks experience of competition, whereas the United Kingdom, which has always operated under the influence of market forces with a belief in competition, is becoming more interested in long-term agreements, joint ventures, and partnership contracts.

SUMMARY

The main findings from this review are based on the value of the self-perception survey and the significance of the delay in the timing of the surveys on the relevance of the outcome. Nevertheless, a general indication of the key strengths and weaknesses of Japanese and U.K. practice in construction project management has emerged, and it would appear that the findings are related to traditional domestic market procedures.

There are major changes occurring in the domestic and international construction markets, and the ability to compete, to manage change, and to manage culture is likely to be increasingly important. Competition and collaboration will be key components in future project-management strategy. Under conditions of increasing competition in domestic and international markets there may be an approach toward self-funded contracts, such as BOOT, to enable organizations to gain some control over market fluctuations and hence over the stability of the corporate finances.

New organizational structures are being considered to reduce the wasteful use of scarce and expensive engineering resources involved in unsuccessful competitive tendering. Joint ventures are well regarded by both Japan and the United Kingdom, and are also attractive to many host countries in potential market growth areas. The emphasis on cooperation as well as competition would also lead to a closer integration and flexibility within construction and manufacturing. This may ultimately lead to modified corporate hierarchical structures more in the form of flat, flexible structures to manage both internal and external projects.

Both Japan and the United Kingdom will need to invest in project-management education, on-the-job training, and research and development to meet these challenges. The current company structures in Japan are well suited to this task, but the U.K. companies will have to make a greater investment in personnel. Both countries are in pursuit of the same goals, but different backgrounds necessitate different approaches to project-management development.

ACKNOWLEDGMENT

The author would like to acknowledge the assistance provided by Shinichi Miyamura, Kengo Hirose, and Hideaki Umayahara of the Shimizu Corporation, who during their postgraduate work at UMIST translated papers and explained the traditions and methods of Japanese project management, which form the basis for many of the figures included in this chapter.

REFERENCES

1. T. Tokunaga, "Japanese Traditions and Project Management Implications in General," presented at the 7th INTERNET Congress, Copenhagen, Denmark.

2. ECGI, "Review of the International Construction Market in 1988," Export Group for the Construction Industry, U.K., 1990.

3. *Journal of OCAJI,* vol. 12, no. 6, June 1988.

4. S. J. Maswood, "Japan and Protection," Routledge, London, 1989.

5. S. Miyamura, unpublished M.Sc. dissertation, UMIST, 1989.

6. OCAJI, "Basic Directions and Corresponding Measures for Overseas Construction," Nomura Research Institute, Tokyo, Japan, Oct. 1984.

BIBLIOGRAPHY

Hasegawa, F., et al., *Built by Japan,* Wiley, New York, 1988.

Hirose, K., unpublished M.Sc. dissertation, UMIST, 1991.

Japanese Economic Newspaper, Nihon Keizai Shibea Inc., Mar. 14, 1989.

Smith, N. J., and A. Merna, *Guide to the Preparation and Evaluation of Build-Own-Operate-Transfer (BOOT) Project Tenders,* UMIST, Manchester, United Kingdom, 1993.

Smith, N. J., and S. Miyamura, "Project Management in the Construction Industry in Japan and the UK," presented at the 10th INTERNET Congress, Vienna, Austria, 1990.

Spon's International Construction Cost Handbook, E & FN Spon, London, 1988.

CHAPTER 27
THE PROJECT IMPLEMENTATION PROFILE: AN INTERNATIONAL PERSPECTIVE

Jeffrey K. Pinto

College of Business Administration,
University of Maine

Dennis P. Slevin

Katz Graduate School of Business,
University of Pittsburgh

Dr. Pinto is Associate Professor of Management in the University of Maine's College of Business Administration and Research Associate with the Graduate Center for the Management of Advanced Technology and Innovation at the University of Cincinnati. He received both his Ph.D. and M.B.A. from the University of Pittsburgh and holds a B.A. in History and a B.S. in Business Administration from the University of Maryland. His research interests include the study of project management and the processes by which organizations implement innovations and advanced technologies. He is a member of the Project Management Institute, the Academy of Management, and the Engineering Management Society.

Dr. Slevin is Professor of Business Administration at the Joseph M. Katz Graduate School of Business, University of Pittsburgh. Dr. Slevin holds the B.A. degree from St. Vincent College, the B.S. degree from M.I.T., the M.S. degree from Carnegie-Mellon University, and the Ph.D. degree from Stanford University. He is the author of numerous refereed journal articles, book chapters, and books. His research and consulting interests focus on project management, the implementation of organizational innovation, and entrepreneurship.

Many examples of project management tools used for tracking the "harder" technical aspects of projects throughout their development and implementation exist today. While there are great advantages to the use of these techniques, they have some potential drawbacks for successful project implementation. Often, some of the longer run and more subjective factors are overlooked. The Project Implementation Profile (PIP) was developed with two purposes in mind. First, it allows the project manager to assess the "softer" behavioral side of the project management process to determine the status of the project in relation to its human elements. Second, the PIP gives project managers the opportunity to focus some of their attention on the strategic issues of project development. This chapter discusses the project implementation process, focusing on 10 critical project success factors identified in a recent study. It further outlines the ways in which the PIP can be used by project managers. An illustrative example of a project that used the PIP to assess the success of implementation is given. Finally, we report on the results of a recent study that used the PIP to assess the different approaches for innovation employed by Crown versus private corporations in Canada.

INTRODUCTION

Project monitoring and control is a difficult and often inexact process. A variety of different cues and a large amount of information are constantly confronting project managers as they attempt to track their projects throughout the various implementation stages. Further, the more complex the project, the more likely that project managers are faced with making sense of the wide variety of technical, human, and budgetary issues (or project critical factors) with which they must contend. As a result of the complexity involved in project management and the demands on the project manager's time, the project management process has seen the rise and increasing acceptance and use of a wide variety of tracking systems. These systems are both computer-driven and manual, and are intended to aid the project manager in keeping track of the myriad of variables that must be accounted for to help ensure project success.

Well-established project monitoring aids have been in existence for some time. Systems such as the project evaluation and review technique (PERT), Gantt charts, and critical path methodologies can be extremely useful in helping project managers untangle the various project activities with which they must contend, including the careful tracking of costs, schedules, performance of project subassemblies and subcontractors and so forth. However, project managers, in focusing on this minute level of detail may be drawn away from some important "larger picture" aspects of project management necessary for success. In other words, overattention to the specific, tactical "fire fighting" and detail management activities related to project management often prevents the project manager from developing clear, periodic assessments of the overall project strategy. The project manager must constantly ask questions such as:

- Is this project solving the right problem?
- Will the project be used by the clients?
- Is top management truly supportive of this project?
- Are the client's needs adequately understood?
- Is the basic project mission still on target?
- Does the company have the necessary project team personnel to succeed?

Questions of this sort emphasize another aspect of project management. Although the project manager may possess numerous detailed reports that provide careful tracking of the "hard" project numbers, there is another major component of project management that should not be overlooked. The so-called soft side of project management involves key behavioral variables that are crucial to project success. Issues for project success such as quality of project team personnel, top management support, and client acceptance are as equally important as the harder technical detail management and must be attended to by both project managers and upper management.

What has been needed is a project management tool to allow project managers to sort through the information they receive, to more accurately and systematically monitor and assess the current state of those factors that have been shown to be critical for project implementation success. Further, a tool that would allow project managers to gain an overall strategic sense of the project implementation process would be of great benefit in efforts to exert successful comprehensive project control. It was in order to address these points that the Project Implementation Profile was originally created.

DEVELOPMENT OF THE 10-FACTOR MODEL

In developing the PIP, projective information was obtained from 54 managers who had experience with a variety of projects. Participants of this study were asked to consider a successful project with which they had been recently involved. Assuming the role of project manager, they were asked to indicate activities in which they could engage or important issues they could address that would significantly increase the likelihood of project success. This process was repeated until a set of critical activities or criteria was uncovered.[11] These identified activities resulted in the creation of a set of 10 critical success factors for project implementation and resulted in a 50-item instrument which can be used to measure project implementation performance in relation to the critical factors.

These 10 critical success factors were subsequently validated and found to be generalizable to a wide variety of project types in a recent study of over 400 projects.[10] Further, the PIP allows project managers to systematically monitor these 10 critical success factors in relation to their specific projects. On both an individual and collective basis, the following factors are strongly correlated with project success.

Project Mission. *Mission* refers to the initial clarity of goals and general directions for the project. In a sense, the first step in the project development process is to know what it is one wishes to develop, what the project's capabilities are, why the project is needed, and how it will benefit those who use it. The decision to develop and implement a new project often signals the commitment by the organization of a large amount of time, monetary, human, and material resources. Before such a commitment occurs, it is vital to have a clear, well-acknowledged vision of the goals or mission underlying the project.

Top Management Support. An important question to be asked once the mission of the project is determined is whether there exists a willingness on the part of top management to truly support the project. It may be easy for top management to pay lip service to the "importance" of the new project, but, often, this support can be reduced or not provided over time. Is top management committed to providing the necessary resources throughout the development and implementation process? Will top managers use their authority to help the project? Will top management support the project team in the event of a crisis? Answers to these and similar questions frequently indicate the true degree of support the project manager can expect to receive from top management.

Project Schedule/Plans.   In order for a project to proceed successfully, it requires a well-laid-out and detailed specification of the individual action steps required. All necessary activities must be scheduled. Further, there must be plans in place to determine when vital resources (human, budgetary, and material) will be required. Finally, it is important that a measurement tool be in place to assess the actual progress of the implementation against schedule projections.

Client Consultation.   The *client* refers to anyone who is the ultimate, intended user of the project. Clients can be the firm's customers but may also be internal to the organization. Because a project is intended for the client's benefit, it is vital that communication and consultation with the clients occur not only at the beginning of the development process, but throughout the project's implementation. Projects are often subject to a variety of changes throughout their development; as a result, clients must be kept apprised of the status of the project and its capabilities, rather than surprised at the end when the project is transferred to them. The project team must remain aware of the client's needs.

Personnel.   The parent organization's people represent a very important ingredient for successful project implementation. Simply put, does the organization have the necessary personnel to staff the project team? Is it necessary for the company to recruit or provide additional training to personnel in order for them to function effectively on the project team? Too often organizations ignore the importance of this factor, sometimes assigning individuals to project teams on the basis of convenience or their nonusefulness to other, current organizational activities. As a result, the project team may be staffed with the castoffs of other departments, a formula sure to result in potential future project difficulties and possible failure.

Technical Tasks.   This factor refers to the assessment of the availability of the required technology or technological resources to aid in the project's development. Does the organization possess the technological resources to develop the project? Further, it requires a determination of whether or not those individuals who are developing the project understand it from the technological standpoint. For successful project development, skilled people and adequate technology are equally important.

Client Acceptance.   The penultimate question that must be asked as the result of any project development is, "Is the client satisfied with the project and making use of it?" One finding that has come through time and again is that it is not enough to simply create a project, transfer it to a client, and assume that it will be accepted and used. In reality, client acceptance is a critical factor that must be handled just like any other criterion for project success. In addition to performing the technical and administrative activities necessary to develop the project, the project team must also function in a marketing/selling role in working to gain client acceptance.

Monitoring and Feedback.   It is important that at each step in the implementation process, key project team members receive feedback on how the project is progressing. These control mechanisms allow the project manager to be on top of any real or potential problems, to oversee corrective measures, and prevent deficiencies from being overlooked. The better the control processes, the more likely that the final project will retain high quality.

Communication.   Communication is a key component for project success throughout the development process. Project teams routinely engage in a three-way pattern of communications with clients and the parent organization. It is vital that these commu-

TABLE 27.1   Ten Project-Critical Success Factors

1. *Mission*—initial clarity of goals and general directions

2. *Top management support*—willingness of top management to provide the necessary resources and authority/power for project success

3. *Project schedule/plans*—a detailed specification of the individual action steps required for project implementation

4. *Client consultation*—communication, consultation, and active listening to all affected parties

5. *Personnel*—selection, recruitment, and training of the necessary personnel for the project team

6. *Technical tasks*—availability of the required technology and expertise to accomplish the specific technical actions

7. *Client acceptance*—the effect of "selling" the final project to its ultimate intended users

8. *Monitoring and feedback*—timely provision of comprehensive control information at each stage in the implementation process

9. *Communication*—the provision of an appropriate network and necessary data to all key actors in the project implementation.

10. *Troubleshooting*—the ability to handle unexpected crises and deviations from plan.

nication channels remain open to ensure the transfer and exchange of relevant information among these three major players in the project implementation process. Consequently, the project manager needs to ensure that there is an appropriate network to transmit all necessary data concerning the project to each project stakeholder.

Troubleshooting.   It is safe to say that few projects are developed without problems occurring along the way. Projects require constant fine tuning and readjustment throughout their creation in order to address these trouble spots. As a result, the final critical success factor refers to the availability of contingency plans, systems or procedures that are in place in order to handle unexpected crises and deviations from plan (see Table 27.1).

HOW THE PROJECT IMPLEMENTATION PROFILE WORKS

Since its development, the PIP has been completed by over 2500 project managers responsible for a wide variety of projects, from construction to high-technology R&D. Use of the PIP suggests that project managers should engage in *regular,* periodic reviews of their projects, throughout the project's entire life. Both the project manager and members of the project team perform an audit of the current status of the project, based on an assessment of the project's health as measured by the 10 critical success factors. Each critical factor is composed of a set of five subitems, making a total of 50 questions to which the project manager must respond. Based on a national sample of over 400 projects, percentile scores have been developed so that the project manager can monitor and track project performance in comparison to a database of other successful projects (see Appendix to this chapter).

The reason for periodic reviews of project status should be readily apparent. Project development and implementation is a dynamic, ongoing process, requiring constant review and reassessment in order to gain an accurate picture of the project at any point in time. In terms of deciding when best to reassess project status, the author's experience has been to employ a combination of two methods: elapsed time and critical incidents. For example, in a project expected to have a 1-year duration, the authors have found that project reviews should take place at regular intervals of not longer than one month. Not only does this method provide regular project assessments, but it puts the project manager and team members in the habit of performing regular strategic implementation monitoring activities.

The second assessment method concerns using the PIP following specific critical incidents. Examples of critical incidents may include the achievement of important target dates, the entrance into a new state of the project life cycle, or the development and/or resolution of a crisis situation. These incidents do not neatly follow a 1-month project monitoring program, but signal important changes are taking place in the project which require an updated project status audit.

FOUR-STEP PROCESS

In addition to a discussion of the theory underlying the development and inclusion of each of the ten critical success factors, the Appendix to this chapter contains an abbreviated copy of the PIP. As a result, project managers can make practical use of this instrument by following a simple four-step process that is outlined below.

Periodically Monitor the 10 Factors over Time

At each project monitoring point, the project manager and significant members of the project team each fill out the PIP. Collecting data from as many people on the project team as possible provides a wide range of perspectives of the status of the project. It also eliminates the possibility that one or two key members of the team may be overly optimistic about the project's progress.

Use Consensus to Develop a Collective Picture of the Project

After everyone has filled out the profile, indicating their perception of the current status of the project, review the results and discuss the likely causes of disagreements among project team members in the scores on the critical success factors of the PIP. Perhaps administrative and technical personnel disagree in their assessments of project status because of their more specifically focused backgrounds. Use consensus to establish as accurate an assessment of project status as possible.

Pay Close Attention to Low Factors

"Low" scores are those that have a ranking below the 50th percentile. These low scores indicate likely future problem areas which may have an adverse effect on successful project implementation. Project managers should start developing action plans for improving these factor scores. For example, if the factor score of personnel is rated as low, it is sensible to critically examine the project team to see if present team members have the necessary skills to perform their tasks. This low score may signal the need to locate and enlist additional project team personnel.

Visually Emphasize the Critical Success Factors

Putting the current profile on the bulletin board or in memos can be a powerful tool for indicating to members of the project team and upper management both the current status of the project and where problem areas exist. These problem areas would the suggest obvious "pressure points," which require extra consideration. In one instance, a project manager would attach a copy of the project profile scores to the office doors of members of his project team overnight so that this status report was the first thing they saw when arriving in the morning. These profile scores represent an excellent visual reinforcement and an alternative feedback mechanism to the reams of budget and schedule data that are generated on a daily basis.

ILLUSTRATIVE EXAMPLE

A major engineering corporation was recently involved in the development, testing, and commercialization of a nuclear reactor monitoring and diagnostic system. The goals of this project were to develop and install a diagnostic system in reactor coolant pumps in order to identify problems before potentially dangerous leakages occur.

Figure 27.1 shows a sample copy of the Project Implementation Profile for this particular project, as filled out by the project manager, indicating the 10 critical factor scores. The raw scores obtained from answering the 50 questions covering the 10 critical factors of the PIP have been converted to percentile scores through use of the database of over 400 projects.

Interpreting this profile would suggest that six of the ten critical factors for the project can be rated as strong or well-handled. One of the factors could be considered marginal and the other three critical success factors can be interpreted as being weak, representing potentially serious problem areas.

The six strong factors include project mission, project scheduling/plans, client consultation, client acceptance, monitoring and feedback, and communication. Each of these critical factors has a percentile ranking of over 60 percent. The practical interpretation of these scores for the project manager would be that these are critical areas that are currently being handled well; that is, there are no problems with these aspects of the implementation process that are being exhibited at the present time. As a result, project managers can turn their attention toward other, more immediate, problem areas.

The marginal factor is top management support, with a percentile score hovering near the 50th percentile. This score would suggest the potential for future difficulties. In the case, a signal should be sent to the project team that they may not have the backing of top management to the degree that is sufficient to successfully implement the project. It would be advisable to look for possible causes or reasons for this potential lack of support, as it could be the likely source of future problems.

Finally, the three weak factors are those of personnel, technical tasks, and troubleshooting. These three factors scored below the 50th percentile. While the marginal score discussed above suggests future potential for trouble, weak factors such as these indicate present difficulties. Left unattended or unresolved, these factors can have a seriously debilitating effect on the implementation of the project. In this case, it seems apparent that this project has suffered from a lack of trained personnel, supported by the adequate technology, to successfully develop and install the monitoring and diagnostic system. Further, perhaps as a result of the lack of adequate technology, the troubleshooting mechanisms needed to deal with deviations from plan and unexpected difficulties with developing and installing the system were inadequate. As a result of thorough analysis of the profile shown in Fig. 27.1, action plans were developed by the project team which stressed: (1) the need to maintain better linkages with top management, (2) the recruitment of additional project team personnel to ensure greater technical expertise, and (3) the acquisition of state-of-the-art technical equipment to aid project team members in the system development.

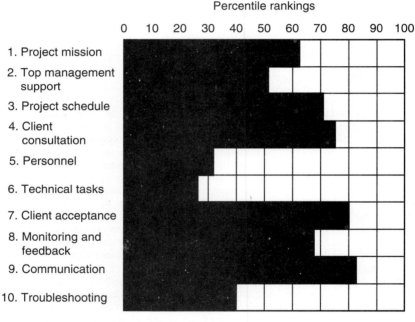

FIGURE 27.1   Percentile scores for sample project.

A CANADIAN EXAMPLE USING THE PROJECT IMPLEMENTATION PROFILE

The previous example was used to illustrate how the PIP can be used by an individual project manager to assess the current status of one particular project. The PIP has also been used in larger-scale research studies to investigate aspects of corporate innovative practices. As an example of one such use of the PIP in industrial research, we have been involved in an exploratory study of the differences in innovative practices within Canadian firms. In particular, we sought to determine how these innovative practices differed between Crown (public) and private corporations in Canada. In studying this research question, we employed the PIP as a research tool to help shed light on the different approaches to innovation used by these two sections of Canadian industry.

Background

The Canadian government's policy to divest Crown corporations was initiated over 10 years ago and it continues to gather considerable momentum. Traditionally, Canadian Crown corporations operated under governmental financial protection. Due to privatization policies, many of these corporations are now being forced to operate in an increasingly competitive marketplace.[1,4] In addition, these newly privatized ventures are confronted with many uncertainties resulting from the Canadian-American Free Trade Agreement that removed government protections and tariffs from large new sectors of the economy. While many questions remain unanswered about the impact of privatization and free trade, there is no doubt that operating strategies of many Canadian companies, both Crown and private, have seen irrevocable changes.[7]

One of the most important outgrowths of privatization has been forcing companies that had been under Crown protection to become more innovative and competitive in the marketplace. When coupled with the demands of free trade, these Canadian firms are being required to develop and maintain ever higher levels of technological and product innovation.[5] Two fundamental questions remain to be answered: (1) will the majority of Canadian companies, both Crown and private, develop the competitive approaches necessary to remain innovative in this expanded marketplace, and (2) how do the innovative practices between the two segments of Canadian industry (Crown and private) differ in their approaches to innovation?

The Canadian government instituted the policy of divestment to inject a new spirit of competition, innovation, and fiscal responsibility into the previously protected firms.[2,8,12] As a governmental policy, privatization was expected to lead to a change in old corporate philosophies and to instill a new sense of innovativeness in many firms that had operated within a financial safety net. As several years have now elapsed since the initiation of privatization, it seems appropriate to assess the desired versus actual changes that have been introduced into Canadian business.

The purpose of this example is to report on the results of a recent exploratory study, using the Project Implementation Profile, that investigated the effects of divestment/privatization on the relative importance of the 10 project-critical success factors for many Canadian firms. Specifically, we attempted to assess some of the differences in perceived importance of these critical success factors between Crown and private corporations as they implemented 55 innovative projects.

THEORETICAL BACKGROUND

The concept of organizational innovation has been defined in a number of ways.[3,6,9] Generally, innovation is viewed as "the development and adoption of an idea or behavior that is new to the organization's industry, market, or general environment."[3] The importance of innovation has been underscored in recent years with the tremendous changes in the international business arena. Drastically reduced project life cycles, increased product obsolescence, and shorter research and development time have forced many companies to adopt a more proactive, innovative outlook to remain competitive in the marketplace. In essence, innovation is no longer a luxury, but rather a necessity for corporate survival.

Innovation and the PIP

Previous research using the Project Implementation Profile has suggested that it has considerable power in predicting the degree of success of a project's implementation process.[11,15] The 10 critical success factors have been found in other research studies to account for well over 70 percent of the total causes of project success.[10,14]

In this study, we sought to attempt to determine whether some subset of the success factors were more relevant to private sector innovations than those that were most critical for Crown corporation innovation.

In defining an innovative project's success, we made use of two separate measures. Past research and anecdotal evidence from other project managers has suggested that two distinct aspects of project success are relevant to a company implementing a new project. The first element refers to the classic *efficiency* measures of success. These efficiency measures correspond to the internal organizational concerns with a new project development: adherence to schedule and budget, and basic performance expectations. In other words, getting the project out "on time, on budget, and on spec."

The second determinant of project success concerns the external factors that impact a new project. These external factors constitute a project's *effectiveness* and refer to user satisfaction with and use of the project. A project that is put out on time, under budget, and meets performance specifications is still useless if it is not well-received by the client for whom it is intended. The most efficiently run project development cannot guarantee that a project will be successful unless regard is given to the clients. On the other hand, a project that is well received may have only marginal usefulness to a company if its development went way over budget and took too long to introduce.

Clearly then, when considering exactly what defines a "successful" project, it is important to look at both aspects of project outcomes: internal efficiency (doing it right) and external effectiveness (doing the right thing). In this study, we included measures in the PIP that look at both elements of project success.

METHODS

Sample

Questionnaires were mailed to 130 Canadian members of the Project Management Institute—an international organization of project managers. Usable questionnaires were returned by 55 individuals, for a response rate of 42.3 percent. Of the 55 innovation projects described by the respondents, 28 (51 percent) were from individuals employed in Crown corporations and 27 (49 percent) were from private organizations.

Respondents reported on an *innovative project,* or a project designed to implement an innovation within their organizations. From the private sector, some typical examples of innovative projects described in this study were (1) development of an automated train dispatching and control system, (2) expansion and rehabilitation of a pollution control center, and (3) creation of a programmable industrial control computer simulation. Examples of innovative projects within Crown corporations included (1) creation of an offshore oil and gas development project, (2) development and installation of a computerized personnel/payroll system, and (3) development and/or expansion of product lines in the fishing industry.

Table 27.2 describes some additional respondent and project information. The majority of respondents had the title or duties of project manager for the particular innovation on which they were reporting. Further, more than half of the respondents (58 percent) had been employed by their companies for 10 years or more. Finally, these innovation projects had required a substantial commitment of resources and time, with a large majority (79 percent) taking between 1 and 2 or more years to complete. These individual responses compose the unit of analysis for our study.

TABLE 27.2   Demographic Characteristics of Sample

Characteristic	Frequency	Cumulative percentage
Respondent's title		
Project manager	30	55
Manager on project team	3	60
Project team member (technical)	3	65
Project team member (administrative)	4	72
Member of business unit affected by project	6	82
Other	9	100
Length of employment		
1 to 3 years	11	21
4 to 6 years	7	35
7 to 10 years	4	40
More than 10 years	33	100
Project duration		
1 to 3 months	1	2
4 to 6 months	6	13
6 to 12 months	5	23
1 to 2 years	16	51
2 years or more	27	100

Total sample N for all characteristics = 55.

A wide assortment of both the Crown and private corporations were sampled for this research. Firms in the Crown sector ranged from energy and transportation industries to telecommunications and fishing. Private corporations sampled were also highly diverse, including research and development laboratories, oil and energy exploration, and small high-technology ventures. Sampling a broad variety of organizations should enhance the generalizability of the findings across the population of Canadian private and Crown corporations.

In addition to the 10 critical success factors, the PIP also included a set of dependent measures of project success, designed to measure the respondent's perception of project efficiency and effectiveness.

DATA ANALYSIS AND RESULTS

Multiple regression analysis was used to test the relative effects of the 10 project-critical success factors for innovative projects within sets of both private and Crown corporations. The results reported include those factors that were found significant as well as the total variance explained by these factor sets.

Results

Table 27.3 reports the correlation matrix of all research variables included in the study. Reliability (coefficient alpha) estimates are included in parentheses on the diagonal. Since the correlations between the constructs are generally low to moderate, it is unlikely that results will be confounded by the potential for multicolinearity among the scales. Further, note that the subscales for each of the critical success factors show strong internal consistency, indicating that the PIP is highly reliable.

Tables 27.4 and 27.5 give the results of the stepwise multiple regression analysis for the sets of private and Crown corporations. Three analyses were run. In the first set of analyses, the dependent measure was an overall, aggregate measure of project success, in which both internal efficiency and user satisfaction (external effectiveness)

TABLE 27.3   Correlation Matrix of Research Variables

Variables	1	2	3	4	5	6	7	8	9	10	11	12	13
1. Mission	(.82)												
2. Top management support	.41	(.85)											
3. Project schedule	.16	.41	(.77)										
4. Client consultation	.35	.17	.16	(.82)									
5. Personnel	.47	.58	.54	.23	(.83)								
6. Technical tasks	.54	.24	.45	.29	.68	(.64)							
7. Client acceptance	.42	.52	.59	.31	.42	.41	(.78)						
8. Troubleshooting	.53	.32	.46	.17	.53	.57	.39	(.77)					
9. Monitoring	.36	.10	.48	.22	.43	.50	.52	.46	(.65)				
10. Communications	.35	.22	.19	.06	.44	.44	.27	.67	.30	(.76)			
11. Aggregate success	.45	.25	.43	.36	.36	.43	.50	.56	.43	.36	(.91)		
12. User satisfaction	.52	.23	.32	.45	.39	.45	.41	.59	.41	.40	.73	(.76)	
13. Internal efficiency	.19	.25	.45	.11	.25	.29	.47	.35	.33	.16	.82	.55	(.89)

$p < .05$ for coefficients greater than .25.
$p < .01$ for coefficients greater than .33.
$N = 55$; tests are one-tailed.
Scale reliabilities (coefficient alpha) are listed in parentheses.

TABLE 27.4   Stepwise Regressions: Predictors of Project Success in Private Canadian Corporations

Step	Variable	Cum. r^2	T value
	Dependent variable: aggregate measure of success		
1	Client acceptance	0.24	3.01
2	Client acceptance		2.71
	Troubleshooting	0.31	1.85
	Dependent variable: internal efficiency		
1	Client acceptance	0.26	3.15
	Dependent variable: user satisfaction		
1	Client acceptance	0.17	2.54

Note: P < .01 for *t* values shown greater than 2.70.
    P < .05 for all others shown.

TABLE 27.5   Stepwise Regressions: Predictors of Project Success in Canadian Crown Corporations

Step	Variable	Cum. r^2	T value
	Dependent variable: aggregate measure of success		
1	Mission	.35	2.21
2	Mission		2.01
	Client consultation	.41	2.85
3	Mission		2.35
	Client consultation		2.54
	Troubleshooting	.46	1.96
	Dependent variable: internal efficiency		
1	Schedules/plans	.23	2.75
	Dependent variable: user satisfaction		
1	Mission	.44	4.38
2	Mission		3.22
	Client consultation	.60	3.14
3	Mission		1.96
	Client consultation		3.82
	Communication	.65	2.02

Note: P < .01 for *t* values shown greater than 2.70.
    P < .05 for all others shown.

measures were included. The other two analyses employed each aspect of success individually.

For private corporations reporting in this study, the most striking finding was that the preeminent critical success factor was that of client acceptance. Client acceptance was critical across all measures of project success, including the internal efficiency measures. It is interesting to note this emphasis, which strongly suggests that for private corporations in Canada, the sole determinant of a project's efficacy has to do with whether or not its intended client will, in fact, make use of it.

The results were a bit more varied for the Crown corporations. Note that when an aggregate measure of success is used, project mission, client consultation, and troubleshooting are the three most important predictors of project success, accounting for almost half of the total causes of success ($r^2 = 0.46$). On the other hand, when an internal measure of success is used, only project scheduling and plans are considered most critical. This finding is not surprising in that efficiency concerns are often best addressed by devoting considerable attention to developing comprehensive plans and project schedules. Finally, when user satisfaction is the determinant of project success, mission, client consultation, and communication are found to account for 65 percent of the total variance. Note that the factor "communication" refers to communication between the project team and both top management of the company as well as communication with external clients as needed.

DISCUSSION

As a wide variety of economic pressures and events act on Canadian industries, it becomes increasingly important to focus attention on some of the results of past policy decisions, particularly in the divestment of Crown corporations. One of the stated purposes of divestment was to inject a greater sense of competitiveness in the marketplace.[2,13] In other words, a goal was to allow the free market to regulate business practices and encourage these companies to become more proactive and innovative in their competitive approaches. Several years have passed, divestment has hit high gear, and continues to be a major force for change within Canada.[7] While some research has attempted to assess the economic impact of these divestment efforts,[1,8] little has been done from a purely managerial-oriented perspective to determine divestment's impact on Canadian business—i.e., on Canadian firms' ability to develop and maintain managerial practices that are supportive of an innovative corporate outlook. The purpose of this study was to gain an inside look at how the sector of the economy can affect the relative importance that a company places on various issues in attempting to successfully develop new, innovative projects. In effect, we sought to use the PIP to shed some light on the approaches to and emphasis for innovation that are being taken by Crown and private Canadian corporations.

The results of this study offer some interesting insights into the differences in project development between Crown and private companies, as well as the relative importance of critical success factors needed for successful innovation. While both sets of organizations are able to offer examples of recent innovations in terms of technology, product development, and/or administrative procedures, the findings suggest that some fundamental differences do exist in the way in which corporations develop these new projects and in the emphasis they place on some critical success factors over others. Within private companies in Canada, it is clear that the emphasis is centered on the user. Private firms appear to have adopted the outlook that effectiveness in the marketplace is the true deciding factor in whether a project is successful or not.

Crown corporations have taken a somewhat different tack. While it appears that the client is also important for their measures of success, the emphasis seems to be that clients are contacted early in the process and then the project is developed in house with little additional input from intended users. Crown corporations do not go to the same extremes to ensure client acceptance as do private companies. Rather, they may devote more of their attention and emphasis to handling the project development as efficiently as possible, focusing on the overall project mission and adhering to schedules.

It is arguable that if any prescriptions are to be taken from this study, they might have to do with the internal/external focus of the firm. We might speculate that as

Crown corporations become more privatized they must try to move from an internal focus on project schedule and budget to an external focus on clients. With revenues coming from sales in a free market rather than government funds, the key to survival of the private firm is the customer response rather than internal efficiencies. Perhaps project managers of the future should not only focus on successful project management, but also on managing organizational transformations—metamorphoses moving from an internal to an external focus.

LIMITATIONS AND CONCLUSIONS

Limitations

While this study offers some interesting and potentially useful findings regarding the use of the Project Implementation Profile as a method for better understanding the differences in approaches to innovation by the two primary sectors of the Canadian economy, some potential limitations of the study need to be addressed. First, because of the exploratory nature of this study, the sample was limited. While we sought to develop matched sets of responses (28 from Crown corporations and 27 from private), the findings need to be considered in light of the total sample size of 55 individuals from 46 different companies. Clearly, this study is not definitive, and future research should attempt to a more in-depth and comprehensive look at the total population of Canadian businesses.

As a second potential limitation, we should note that the approaches to innovation and relative importance of the project-critical success factors of most of the firms sampled in this research were represented by one respondent per firm (and never more than two per firm). Obviously, this sampling procedure increases the potential for bias because only one individual from each organization is asked to fill out the PIP for one specific innovative project within the firm. While this research approach would lead to highly questionable findings if we were to conduct an industry-by-industry analysis with very small sample sizes, the concern is somewhat lessened by "collapsing" the results into two broad categories, representing Crown and private companies. Still, it is important to note that the results are preliminary and need to be interpreted in light of the small sample sizes.

Conclusions

This chapter has made use of two examples to illustrate how the PIP can be used by project managers and researchers. In the first example, the purpose was to demonstrate how project managers can use the PIP as a single project diagnostic instrument to help them do a better job attending to the critical factors that can make or break their project. The second example was offered as a large-scale demonstration of how the PIP can be used by researchers to gain greater understanding of larger, industry-wide emphases and trends for new project development.

Use of the PIP has been shown to provide an excellent additional monitoring and tracking system, stressing both the human side of the project implementation process and those factors that have been found, from an overall strategic perspective to be crucial to project success. Developing a regular program for use of the PIP will not only indicate current status of a project, but will also suggest specific action steps and alternatives in the cases where critical success factors have been found to be substandard or lacking. Successful project managers will find this tool a useful complement to their more traditional project tracking systems.

REFERENCES

1. S. F. Borins and B. C. Boothman, "Crown Corporations and Economic Efficiency," Canadian Industrial Policy in Action, Royal Commission, vol. 4, Toronto University Press, Toronto, 1985.

2. D. W. Caves and L. R. Christensen, "The Relative Efficiency of Public vs. Private Firms in a Competitive Environment: The Case of Canada's Railroads," *Journal of Political Economy,* vol. 5, 1980, pp. 958–976.

3. R. L. Daft, "Bureaucratic versus Nonbureaucratic Structure in the Process of Innovation and Change," in S. B. Bacharach (ed.), *Perspectives in Organizational Sociology: Theory and Research,* JAI Press, Greenwich, Conn., 1982, pp. 129–166.

4. A. Ginsberg and A. Buchholtz, "Converting to For-Profit Status: Corporate Responsiveness to Radical Change," *Academy of Management Journal,* vol. 33, no. 3, 1990, pp. 445–477.

5. C. Green, Canadian Industrial Organization and Policy, 3d ed., McGraw-Hill Ryerson, Toronto, 1990.

6. W. Gruber and J. Niles, "How to Innovate in Management," Organizational Dynamics, vol. 3, no. 2, 1974, pp. 31–47.

7. H. Hardin, *The Privatization Putsch,* Institute for Research on Public Policy, Halifax, 1989.

8. D. McFetridge, "Commercial and Political Efficiency: A Comparison of Government, Mixed, and Private Enterprises," *Canadian Industrial Policy in Action,* Royal Commission, vol. 4, Toronto University Press, Toronto, 1985.

9. J. L. Pierce and A. L. Delbeqc, "Organizational Structure, Individual Attitudes and Innovation," *Academy of Management Review,* vol. 2, 1977, pp. 27–37.

10. J. K. Pinto, "Project Implementation: A Determination of Its Critical Success Factors, Moderators, and Their Relative Importance Across the Project Life Cycle," unpublished doctoral dissertation, University of Pittsburgh, 1986.

11. J. K. Pinto and D. P. Slevin, "Critical Factors in Successful Project Implementation," *IEEE Transactions on Engineering Management,* vol. EM-34, no. 1, 1987, pp. 22–27.

12. Royal Commission on Financial Management and Accountability, Canadian Government Publishing Centre, Hull, Quebec, 1979.

13. R. Schultz, "Selling the Jewel in the Crowns: The Privatization of Teleglobe," in G. B. Doern and A. Tupper (eds.), *Public Corporations and Privatization in Canada,* Institute for Research on Public Policy, Montreal, 1988.

14. D. P. Slevin and J. K. Pinto, "The Project Implementation Profile: New Tool for Project Managers," *Project Management Journal,* vol. XVIII, no. 4, 1986, pp. 57–71.

15. D. P. Slevin and J. K. Pinto, "Balancing Strategy and Tactics in Project Implementation," *Sloan Management Review,* vol. 29, no. 1, 1988, pp. 33–41.

THE PROJECT IMPLEMENTATION PROFILE

Jeffrey K. Pinto
Dennis P. Slevin

How is your project progressing? How can you possibly attend to all the demands made upon your time in getting the project implemented? Do you get so caught up in putting out fires that you have little time to consider the larger picture? What are the specific aspects of managing a project which can determine whether or not it will succeed?

It was in an effort to answer these questions that the Project Implementation Profile (PIP) was originally developed. The PIP allows you to make periodic assessments of the current status of key factors concerning your project throughout the implementation process. The philosophy of the PIP implies that project management is a complex task, requiring that the project manager and team members attend to a wide variety of factors in attempting to successfully implement their projects. In addition to the "hard" factors related to the technical, operational aspects of the project, managers must also consider the "softer" human side of managing the project and the project team. The so-called soft side of project management involves key behavioral variables that are equally crucial to project success.

An additional benefit of the PIP is in allowing the project team to take an occasional "step back" to develop an overall picture of the current status of the project. Often project team members become so immersed in the daily "fire fighting" that they can lose sight of important elements crucial to successful project development. The PIP identifies for you and helps you to measure ten critical success factors for project success. These factors were uncovered as the result of a series of in-depth studies and interviews with practicing project managers.

For those who may be interested in obtaining a copy of the full Project Implementation Profile, please write to Xicom Inc., Woods Road, Tuxedo, NY 10987. Phone: (914)351-4725.

TEN CRITICAL SUCCESS FACTORS

1. *Project mission*—initial clarity of goals and general direction.
2. *Top management support*—willingness of top management to provide the necessary resources and authority/power for project success.
3. *Project schedule/plan*—a detailed specification of the individual action steps required for project implementation.
4. *Client consultation*—communication, consultation, and active listening to all impacted parties.
5. *Personnel*—recruitment, selection, and training of the necessary personnel for the project team.

6. *Technical tasks*—availability of the required technology and expertise to accomplish the specific technical action steps.

7. *Client acceptance*—the effect of "selling" the final project to its ultimate intended users.

8. *Monitoring and feedback*—timely provision of comprehensive control information at each stage in the implementation process.

9. *Communication*—the provision of an appropriate network and necessary data to all key actors in the project implementation.

10. *Troubleshooting*—ability to handle unexpected crises and deviations from plan.

SURVEY INSTRUCTIONS

Please think of a project in which you are currently or have been recently involved. Use the PIP to evaluate this project.

PROJECT IMPLEMENTATION PROFILE

Project name: _____

Project manager: _____

Profile completed by: _____

Date: _____

Briefly describe your project, giving its title and specific goals:

_____

_____

_____

_____

Think of the project implementation you have just named. Consider the statements on the following pages. Using the scale provided, please circle the number that indicated the *extent* to which you agree or disagree with the following statements as they relate to activities occurring in the project about which you are reporting.

Factor 1— Project Mission

	Strongly Disagree		Neutral			Strongly Agree	
1. The goals of the project are in line with the general goals of the organization............................	1	2	3	4	5	6	7
2. The basic goals of the project were made clear to the project team...	1	2	3	4	5	6	7
3. The results of the project will benefit the parent organization..	1	2	3	4	5	6	7
4. I am enthusiastic about the chances for success of this project...	1	2	3	4	5	6	7
5. I am aware of and can identify the beneficial consequences to the organization of the success of this project..	1	2	3	4	5	6	7

Factor 1—Project Mission Total	

Factor 2—Top Management Support

	Strongly Disagree		Neutral		Strongly Agree		
1. Upper management will be responsive to our requests for additional resources, if the need arises................................	1	2	3	4	5	6	7
2. Upper management shares responsibility with the project team for ensuring the project's success..............................	1	2	3	4	5	6	7
3. I agree with upper management on the degree of my authority and responsibility for the project	4	5	6	7			
4. Upper management will support me in a crisis 1	2	3	4	5	6	7	
5. Upper management has granted us the necessary authority and will support our decisions concerning the project........................	1	2	3	4	5	6	7

Factor 2—Top Management Support Total	

Factor 3—Project Schedule/Plan

	Strongly Disagree		Neutral		Strongly Agree		
1. We know which activities contain slack time or slack resources that can be utilized in other areas during emergencies.....................	1	2	3	4	5	6	7
2. There is a detailed plan (including time schedules, milestones, manpower requirements, etc.) for the completion of the project..................	1	2	3	4	5	6	7
3. There is a detailed budget for the project............	1	2	3	4	5	6	7
4. Key personnel needs (who, when) are specified in the project plan	1	2	3	4	5	6	7
5. There are contingency plans in case the project is off-schedule or off-budget........................	1	2	3	4	5	6	7

Factor 3—Project Schedule/Plan Total	

Factor 4—Client Consultation

	Strongly Disagree			Neutral		Strongly Agree	
1. The clients were given the opportunity to provide input early in the project development stage	1	2	3	4	5	6	7
2. The clients (intended users) are kept informed of the project's progress	1	2	3	4	5	6	7
3. The value of the project has been discussed with the eventual clients	1	2	3	4	5	6	7
4. The limitations of the project have been discussed with the clients (what the project is *not* designed to do)	1	2	3	4	5	6	7
5. The clients were told whether or not their input was assimilated into the project plan	1	2	3	4	5	6	7

Factor 4—Client Consultation Total

Factor 5—Personnel

	Strongly Disagree			Neutral		Strongly Agree	
1. Project team personnel understand their role on the project team	1	2	3	4	5	6	7
2. There is sufficient manpower to complete the project	1	2	3	4	5	6	7
3. The personnel on the project team understand how their performance will be evaluated	1	2	3	4	5	6	7
4. Job descriptions for team members have been written and distributed and are understood	1	2	3	4	5	6	7
5. Adequate technical and/or managerial training (and time for training) are available for members of the project team	1	2	3	4	5	6	7

Factor 5—Personnel Total

Factor 6—Technical Tasks

	Strongly Disagree			Neutral		Strongly Agree	
1. Specific project tasks are well managed	1	2	3	4	5	6	7
2. The project engineers and other technical people are competent	1	2	3	4	5	6	7
3. The technology that is being used to support the project works well	1	2	3	4	5	6	7
4. The appropriate technology (equipment, training programs, etc.) has been selected for project success	1	2	3	4	5	6	7
5. The people implementing this project understand it	1	2	3	4	5	6	7

Factor 6—Technical Tasks Total

Factor 7—Client Acceptance

	Strongly Disagree		Neutral			Strongly Agree	
1. There is adequate documentation of the project to permit easy use by the clients (instructions, etc.)	1	2	3	4	5	6	7
2. Potential clients have been contacted about the usefulness of the project	1	2	3	4	5	6	7
3. An adequate presentation of the project has been developed for clients	1	2	3	4	5	6	7
4. Clients know who to contact when problems or questions arise	1	2	3	4	5	6	7
5. Adequate advance preparation has been done to determine how best to "sell" the project to clients	1	2	3	4	5	6	7

Factor 7—Client Acceptance Total

Factor 8—Monitoring and Feedback

	Strongly Disagree		Neutral			Strongly Agree	
1. All important aspects of the project are monitored, including measures that will provide a complete picture of the project's progress (adherence to budget and schedule, manpower and equipment utilization, team morale, etc.)	1	2	3	4	5	6	7
2. Regular meetings to monitor project progress and improve the feedback to the project team are conducted	1	2	3	4	5	6	7
3. Actual progress is regularly compared with the project schedule	1	2	3	4	5	6	7
4. The results of project reviews are regularly shared with all project personnel who have impact upon budget and schedule	1	2	3	4	5	6	7
5. When the budget or schedule requires revision, input is solicited from the project team	1	2	3	4	5	6	7

Factor 8—Monitoring and Feedback Total

Factor 9—Communication

	Strongly Disagree		Neutral		Strongly Agree		
1. The results (decisions made, information received and needed, etc.) of planning meetings are published and distributed to applicable personnel	1	2	3	4	5	6	7
2. Individuals/groups supplying input have received feedback on the acceptance or rejection of their input	1	2	3	4	5	6	7
3. When the budget or schedule is revised, the changes *and* the reasons for the changes are communicated to all members of the project team	1	2	3	4	5	6	7
4. The reasons for the changes to existing policies/procedures are explained to members of the project team, other groups affected by the changes, and upper management	1	2	3	4	5	6	7
5. All groups affected by the project know how to make problems known to the project team............	1	2	3	4	5	6	7

Factor 9—Communication Total	

Factor 10—Troubleshooting

	Strongly Disagree		Neutral		Strongly Agree		
1. The project leader is not hesitant to enlist the aid of personnel not involved in the project in the event of problems..	1	2	3	4	5	6	7
2. Brainstorming sessions are held to determine where problems are most likely to occur..................	1	2	3	4	5	6	7
3. In case of project difficulties, project team members know exactly where to go for assistance..	1	2	3	4	5	6	7
4. I am confident that problems that arise can be solved completely	1	2	3	4	5	6	7
5. Immediate action is taken when problems come to the project team's attention	1	2	3	4	5	6	7

Factor 10—Troubleshooting Total	

Project Performance

	Strongly Disagree		Neutral		Strongly Agree		
1. This project has/will come in on schedule	1	2	3	4	5	6	7
2. This project has/will come in on budget	1	2	3	4	5	6	7
3. The project that has been developed works (or, if still being developed, looks as if it will work)	1	2	3	4	5	6	7
4. The project will be/is used by its intended clients	1	2	3	4	5	6	7
5. This project has directed benefited/will directly benefit the intended users through either increasing efficiency or employee effectiveness	1	2	3	4	5	6	7
6. Given the problem for which it was developed, this project seems to do the best job of solving that problem—i.e., it was the best choice among the set of alternatives	1	2	3	4	5	6	7
7. Important clients, directly affected by this project, will make use of it	1	2	3	4	5	6	7
8. I am/was satisfied with the process by which this project is being/was completed	1	2	3	4	5	6	7
9. We are confident that nontechnical start-up problems will be minimal, because the project will be readily accepted by its intended users	1	2	3	4	5	6	7
10. Use of this project has led/will lead directly to improved or more effective decision making or performance for the clients	1	2	3	4	5	6	7
11. This project will have a positive impact on those who make use of it	1	2	3	4	5	6	7
12. The results of this project represent a definite improvement in performance over the way clients used to perform these activities	1	2	3	4	5	6	7

Project Performance Total	

Percentile Scores

Now see how your project scored in comparison to a data base of 409 projects. If you are below the 50th percentile on any factor, you may wish to devote extra attention to that factor.

Percentile Score	Raw Score				
% of Individuals Scoring Lower	Factor 1 Project Mission	Factor 2 Top Management Support	Factor 3 Project Schedule/ Plan	Factor 4 Client Consultation	Factor 5 Personnel Recruitment, Selection, Training
100	35	35	35	35	35
90	34	34	33	34	32
80	33	32	31	33	30
70	32	30	30	32	28
60	31	28	28	31	27
50	30	27	27	30	24
40	29	25	26	29	22
30	28	23	24	27	20
20	26	20	21	25	18
10	25	17	16	22	14
0	7	6	5	7	5

Percentile Score	Raw Score					
% of Individuals Scoring Lower	Factor 6 Technical Risks	Factor 7 Client Acceptance	Factor 8 Monitoring and Feedback	Factor 9 Communi- cation	Factor 10 Trouble- shooting	Project Performance
100	35	35	35	35	35	84
90	34	34	34	34	33	79
80	32	33	33	32	31	76
70	30	32	31	30	29	73
60	29	31	30	29	28	71
50	28	30	29	28	26	69
40	27	29	27	26	24	66
30	26	27	24	24	23	63
20	24	24	21	21	21	59
10	21	20	17	16	17	53
0	8	8	5	5	5	21

INTERPRETATION OF SURVEY RESULTS

It is crucial that the successful project manager attend to all 10 critical success factors. Use the flow chart in Fig. 27.2 to identify and sort the factors according to those which are *critical* (below the 50th percentile); *fair* (50–80th percentile) and *good* (greater than the 80th percentile).

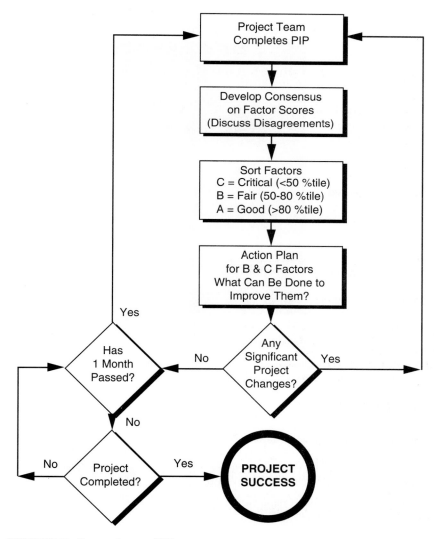

FIGURE 27.2   Program for use of PIP.

PROJECT TRACKING

In the management of ongoing projects, you may wish to complete the profile once a month or to track the results. For that purpose an extra copy of the profile graph Fig. 27.3 is included. Track your projects on a regular basis.

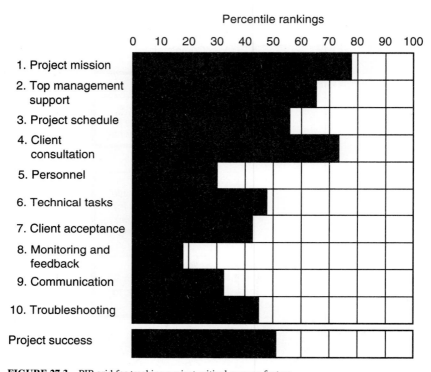

FIGURE 27.3   PIP grid for tracking project critical success factors.

BIBLIOGRAPHY

Pinto, J. K., and D. P. Slevin, "Critical Factors in Successful Project Implementation," *IEEE Transactions on Engineering Management,* vol. EM-34, no. 1, February 1987, pp. 22–27.

Pinto, J. K., and D. P. Slevin, "Critical Success Factors in Effective Project Implementation," in D. I. Cleland and W. R. King (eds.), *Project Management Handbook,* Van Nostrand Reinhold, New York, 1988.

Pinto, J. K., and D. P. Slevin, "Critical Success Factors in R&D Projects," *Research*Technology Management,* January–February 1989.

Pinto, J. K., and D. P. Slevin, "Critical Success Factors Across the Project Life Cycle," *Project Management Journal,* June 1988.

Pinto, J. K., and D. P. Slevin, "Project Success: Definitions and Measurement Techniques," *Project Management Journal,* February 1988.

Schultz, R., D. P. Slevin, and J. K. Pinto, "Strategy and Tactics in a Process Model of Project Implementation," *Interfaces,* June 1987.

Slevin, D. P., and J. K. Pinto, "Balancing Strategy and Tactics in Project Implementation," *Sloan Management Review,* Fall 1987.

Slevin, D. P., and J. K. Pinto, "Leadership Motivation, and the Project Manager," in D. I. Cleland and W. R. King (eds.), *Project Management Handbook,* Van Nostrand Reinhold, New York, 1988.

Slevin, D. P., and J. K. Pinto, "The Project Implementation Profile," *Project Management Institute Proceedings,* 1989, pp. 174–177.

Slevin, D. P., and J. K. Pinto, "The Project Implementation Profile: New Tool for Project Managers," *Project Management Journal,* September 1986, pp. 57–70.

P · A · R · T · 7

NEW INTERNATIONAL CHALLENGES FOR CONSTRUCTION PROJECT IMPLEMENTATION

More complex perceptions of projects and new methods are required to successfully implement international construction projects.

In Chap. 28 Michael Jelinek describes how the strategy of management by projects is used for the design process of the new airport at Munich. The author makes the important point that large-scale construction projects can be treated as a network of simultaneously performed subprojects.

Henry F. Padgham describes, in Chap. 29, a model organization and management control strategy for large public works construction projects.

Per Willy Hetland presents, in Chap. 30, specific project controlling techniques applied by the project owner organization for the implementation of North Sea proects.

Hans Knoepfel describes, in Chap. 31, the (physical) systems of construction projects. He analyzes the relationship between the systems' characteristics and the project management functions to be performed.

CHAPTER 28

MANAGEMENT BY PROJECTS AS A MANAGEMENT STRATEGY FOR LARGE-SCALE CONSTRUCTION AT THE NEW MUNICH AIRPORT

Michael Jelinek

Institute for Project Management
Vienna, Austria

Dipl. Ing. Dr. techn. Michael R. Jelinek was educated in civil engineering at the University of Munich. From 1981 to 1985 he was project manager at construction sites in Germany, Africa, and Saudi Arabia. Since 1985, he has been leading engineer on the General Project Coordinator Team for the New Munich Airport, and since July 1991 Vice-Director of the DREES + SOMMER Ingenieurgesellschaft für Projektmanagement, Wien.

INTRODUCTION

The planning process of an entirely new airport plant—such as that under construction at the new airport of Munich, with about 100 separate facilities and the same number of different planning teams, nearly all of which are related to each other—can no longer be controlled and managed by the traditional boss-subordinate management structure.

From the very beginning of the project, it was obvious that, for successful management, it would be necessary to have an adequate and powerful project-management structure that can coordinate the various project members, control the project's progress, and rapidly react to problems in order to reach the project's main objectives. This structure should realize the possibility of flexible and effective access to the project members' skills and resources. Project responsibility should be delegated in an appropriate manner to minimize bureaucratic procedures.

GENERAL MANAGEMENT STRUCTURE

From Fig. 28.1, which shows the main layout of the New Munich Airport, the schematic diagram of the organization structures (Fig. 28.2) and the detail (Fig. 28.3)

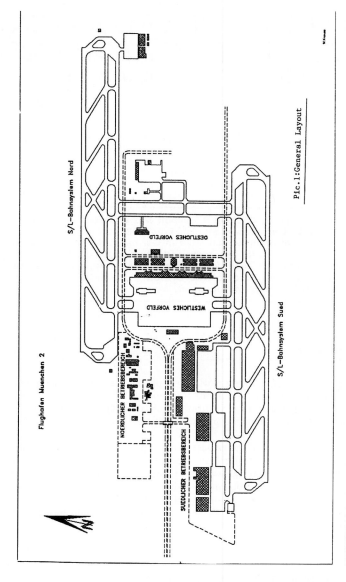

FIGURE 28.1 General layout.

of the main work-breakdown schedule, it is obvious that large-scale construction projects can be treated as a network of simultaneously performed subprojects. According to the principles of *Management by Projects,* the management of a network of projects is organized by a steering committee for projects (see also Ref. 1).

In the case of the New Munich Airport, there was therefore established a *Project Coordinator* position between the top management and the planning project team members, composed of architects and expert planners for various technical systems.

Also, according to the above-mentioned principles and to allow Management by Projects to become second nature to every team member, it was decided that project-management tasks should be delegated to the team members as necessary, to increase responsibility and flexibility of the team members. The problem in implementing the decision was not the smaller project-team members, but rather larger-department team members. It was not easy to make them aware of the need to play their new role in a project with as little bureaucratic behavior as possible.

In detail, the project teams had to fulfill the following, additional project-management functions:

- Scheduling and controlling their project contribution within the boundaries set by the project coordinator
- Analyze and prepare necessary decisions for the project coordinator
- Coordinate their project contribution with all related and affected project members

All the team contributors are supported by the project coordinator. Besides such standard project-management functions as scheduling and controlling, the project coordinator has to

- Analyze the quality of project members' output
- Maintain and control project communication between the project members
- Analyze and prepare all necessary top-management decisions
- Integrate and maintain communication flow with team members in the postproject phase
- Analyze and control external and internal project influences caused by dynamic changes of the project environment[2]

It is essential that the project coordinator does not plan and design. Nevertheless the coordinator can establish task force teams to do intensive, task-oriented planning when project problems make it necessary.

The main progress information is gathered by the project coordinator in monthly progress reports from all team members and condensed to quarterly progress reports to the top management.

CONTRACT MANAGEMENT

Undoubtedly the flat organization structure is possible only when all project members are able and willing to replace the conventional management structures by teamwork, responsive project contributions, and flexibility. This also means that a lot of the usual staff support functions are cut back and responsibilities have more and more to be delegated.

On the other hand, all project team members are bound together by their contracts to the client's authority. These contracts therefore have to regulate the boundaries in which their new responsibilities and their duties are fixed. It is obvious that the degree of flexibility is limited by the contractual boundaries. This is especially true for all activities that are necessary for the project member's work and have to be done by others. It is therefore necessary to enlarge the contractual agreements with additional coordination tasks for each team member to ensure adequate project progress. This means, of course, additional costs for the client at about 1 percent of the project amount.

MANAGEMENT OF THE INTERACTION OF PROJECT MEMBERS

Successful interaction of the team members is possible only when result-dependent project members are working simultaneously in adequate design steps and scales to ensure providing the required quality of results. For technical project planning, this is especially important because many of the required results are largely dependent on the finally chosen system (that is, a particular company's system). Providing these results is therefore always a part of the company's contract. That means these contracts have to be signed in an early project phase. Unfortunately, there is often very low project security in early phases.

Concerning the schematic diagram of the organization structure for the project's design process (Fig. 28.2), it is obvious that the complete design project can be structured in several subprojects, depending on the functional programs of each object.

Besides the architect's so-called general planning team, including several technical planning subteams, which are contractually awarded by the architect, there are several, mostly technically oriented, planning contributors, who have to involve projectwide system planning in nearly each of the objects, for example, projectwide information systems.

While the interaction of the diverse project members has to be arranged and pre-

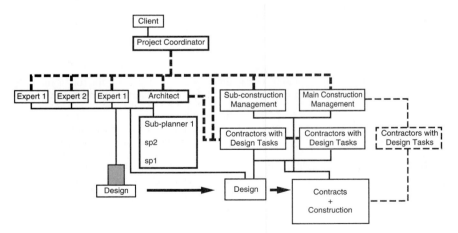

FIGURE 28.2   Project relations of all design contributors and construction managements.

pared by the project coordinator by means of the management tools described in the following chapter, the responsibility for successful interaction between the project members involved in the complete object-design task is delegated to the project members, and the general object planner is the object coordinator for all objects concerning coordination and management tasks.

MANAGEMENT TOOLS

Common Tools

Scheduling.   The project coordinator's main management tool is the general work-breakdown schedule in the form of a bar-chart diagram. With this general work schedule, the project coordinator has to set up the overall time schedule for each object by a detailed time schedule, as shown in Fig. 28.3.

At this scheduling level, the project coordinator arranges all design phases, the preparation phase of all contracts, and all milestone dates for top-management decisions defining the functional program of the object. As a matter of fact, it has to be guaranteed that all project members really can interact with each other. What this means is that:

- The functional and technical program is agreed to by the client
- All necessary planning works are under contract
- The investigation sum is agreed to by the client

Using the above-mentioned schedules, the project members have to set up more detailed schedules in order to make visible important milestones for the internal planning project such as the client's decisions and key events dependent on the interaction of project members. The responsibility of working out all necessary milestones for the ongoing planning process is by the project members.

If possible, there should be the support of a projectwide, personal computer–based datalink system.

Progress Control.   It is not necessary to explain the general use of continuous and intensive progress control. Normally this is done by monthly reports. In case of long-term, complex projects, it is useful to prepare a structured list which is used by all project members in the same way. From this, the project coordinator gets the desired information about progress and is able to prepare the next steps as necessary.

As the usable information thereby depends on the contents of the progress report list, the project coordinator may have to make great efforts to get complete reports. Complete reporting should therefore be required in the project members' contracts.

Because of the multiproject character of the New Munich Airport, with many different project members, it is nevertheless usually necessary to interpret the contents of the reports in an individual way. We therefore did not integrate progress control into a PC-supported system.

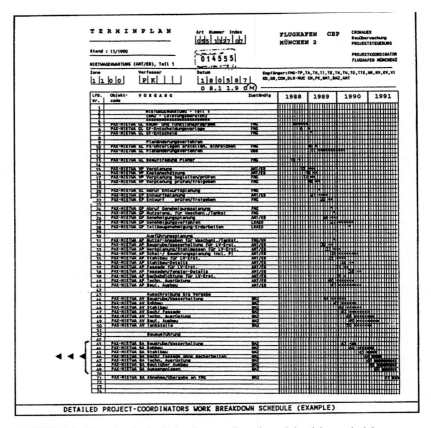

DETAILED PROJECT-COORDINATORS WORK BREAKDOWN SCHEDULE (EXAMPLE)

FIGURE 28.3　Example of a detailed project coordinator's work-breakdown schedule.

Expert Tools

Management of Decisions.　All necessary client decisions for keeping the project running, such as

- Contract awards for designing works
- Definition of functional programs for each object
- Definition of investigation programs
- Contract awards with later users to define the commercial structure of the building

are combined by the project coordinator within monthly decision lists so that top management is able to make the necessary decisions. The use of PC support is possible and offers the ability to issue monthly decision lists.

Communication in Top Management. By means of weekly or monthly project meetings between top management and the concerned project members, communication can successfully be organized to get a steady flow of information among the project members. Most of the necessary project decisions are prepared and released here.

PROFESSIONAL DATA MANAGEMENT

Data management with PC-supported systems is an important tool in project management of large-scaled, complex design projects. For nearly all of the mentioned tasks in scheduling and preparing management decisions, it is useful to work with data-processing systems.

But due to the fact that there is no efficient, overall management software system that is flexible and powerful enough to enable reliable data management for all cases, we decided to use only standard software such as database systems and spreadsheet programs, and adapted these to the desired function with self-written routines.

The use of network scheduling became too complex and ineffective as the project developed. We therefore created a flexible schedule-data pool from which all team members could gain the necessary scheduling dates. Even when this pool soon expanded to several thousands of activities, it never became a problem to handle and use this instrument.

CONCLUSION

Experienced project managers know that it is impossible to manage large-scale, highly complex design projects, such as the planning of an entirely new airport-plant, without any failure. Sometimes near-fatal problems come from external and internal sources, caused by:

- Dynamic external project environments such as laws, rules, and general project acceptance by affected groups
- Business-oriented interests, financing, and marketing aspects
- Resource availability of project teams
- New technical developments

While these can partly be handled by the evolutionary planning approach (planning—acting—reflecting), mistakes still happen.

Nevertheless, by establishing the described flat organization structure for the management of the network of projects, it is possible to

- Move responsibility more and more to the directly affected project members to improve flexibility and quality
- Improve project culture by more intensive communication in the network-oriented structure
- Reflect more effectively about the process of planning in the "third level" by establishing a project-steering committee.

But is was also not possible to integrate all project teams completely into the organization structure to the desired degree.

Due to the fact that the new model works only when all project teams are accepting their new level of project responsibility and project-management culture, it still will take great efforts to generally define a suitable environment for establishing the Management by Projects culture. These efforts have to be done also in defining new planning contracts, so that the higher degree of project coordination is also remunerated, for each project team, in an adequate way.

As one of the main objectives of management by projects, the traditional top-down relationships have to be replaced. And, as this organization structure is a very practical attempt to establish more project culture by delegating responsibility to all project members, it will become eventually the main organization model, accepted by the new type of flexible and quality-conscious project managers.

The experiences of the management organization described here suggest that the structure still offers a lot of opportunities for improving project-management success.

REFERENCES

1. R. Gareis, in *Proceedings of the 10th Internet Conference,* Vienna, 1990.
2. J. Brandenberger and E. Ruosch, in *Proceedings of the 10th Internet Conference,* Vienna, 1990.

CHAPTER 29

ORGANIZATION AND CONTROL FOR LARGE PUBLIC WORKS CONSTRUCTION PROJECTS

Henry F. Padgham

Vice President, CH2M Hill,
Oakland, California

Mr. Padgham has 30 years experience in the field of management of large multiple-contract public works design and construction projects which includes service as chief cost engineer on the San Francisco and Atlanta Rail Transit Programs, and overall project director on the Milwaukee Water Pollution Abatement Program and the Interstate-880 Freeway Widening Project. He is Vice President of CH2M Hill, Oakland, Calif.

The large size and multiple-element characteristics of public works projects challenge the owner to effectively organize and control these undertakings. This chapter begins by providing a description of the kinds of publicly owned project work that often is of significant size and the public agencies that undertake this work. It then goes on to demonstrate how the management organization structure for the project can be aligned with the project phases, and it concludes with a discussion of the management controls that are needed by the public owner as the project progresses through its four phases.

PROJECT DESCRIPTION

The organization and control of public works projects described herein applies to projects that are usually in excess of U.S. $1 billion construction costs, have multiple subprojects or major elements within them, and require management resources far beyond the owner's present staff capabilities.

These projects include the rebuilding or development of new public infrastructure facilities such as:

- Pollution control
- Airports

- Hazardous waste cleanup
- Wastewater conveyance and treatment systems
- Potable water treatment and distribution systems
- Highway systems
- Rail transit
- Irrigation supply systems
- Port development

Projects of this magnitude are undertaken by public owners, a group that can include public agencies such as the following:

- Operating sewerage agencies
- Transportation authorities specially constituted for specific infrastructure improvements
- Operating port and airport authorities
- State transportation departments
- Irrigation districts

Some of these public agency owners have staffs capable of performing planning, design, and construction. Some are legally constrained from having any of these staff capabilities. But all owners face the following common concerns when they undertake a large infrastructure design and construction project:

1. Both their internal management capability and internal design or construction resources range from significantly inadequate to totally nonexistent for the project.

2. They are legally faced with accomplishing the projects in three fairly distinct steps: planning, design, and then construction. The planning process is often prescribed by law or regulations, and may embody complex environmental impact assessment requirements. In almost all instances in the United States, the planning phase must culminate in a government agency-approved Environmental Impact Report. Without this report, the project cannot proceed into design, applicable project permits cannot be obtained, and the always time-critical function of right-of-way acquisition (if required) cannot begin.

3. These public agency owners must obtain their construction contractors through the legally prescribed public bidding process. This work must be awarded in accordance with the legally prescribed definitions to the lowest *responsible* and *responsive* bidder. So the scope of work must be clearly defined, and the public agency owner is thus faced with producing complete sets of design plans and specifications on which the bids are to be based.

4. Because of the legally mandated process described above and often the mandates of special social programs, the public agency owner cannot use the project delivery systems often available to private sector owners (these include negotiated contracts for construction and the associated use of guaranteed maximum or cost-plus-fee arrangements for construction). Even the use of design/construct arrangements by a single construction contractor with design capability is generally precluded. Although this latter project delivery process is now being reviewed for more general use in the United States by some public agencies, it is generally precluded because the agency must relinquish considerable control over how its project may ultimately perform.

ORGANIZATION

Public owners when undertaking major programs use various techniques to obtain the resources to accomplish the projects. These include staff augmentation, engagement of multiple design firms, bidding the construction to many construction contractors and/or equipment suppliers, use of specialty firms to do planning work and construction inspection services, etc. The important point, however, is that certain basic management functions must be performed in order for the project to be successful. And the project should be organized to ensure that these management functions are performed in an orderly and satisfactory manner.

The owner's organizational approach for the project must be established first because it then dictates the management functions of the project organization that must be fulfilled. The mode that the owner selects to provide the management resources for the management functions can then take the form of increasing staff (with the proper experience and project management qualifications), hiring a single management firm, hiring several specialty management firms, blending owner personnel and consultant personnel, etc.

The organization foundation for the public works undertaking recommended herein is based on the phased project approach shown in Fig. 29.1. These projects have the distinct phases of:

1. Planning
2. Design and bidding
3. Construction and/or equipment procurement and installation
4. Start-up and commissioning

Each phase requires unique technical skills related to the management of that phase. Failure to recognize this can cause serious project management inadequacies. For example, as previously mentioned, the planning phase involves working in accordance with legal and regulatory agencies' required formats and processes. Often, highly visible and structured public involvement programs are required. Also, the conceptual scope of work is developed during the planning phase, and this scope establishes the basic facilities' configuration and operational features of the project. Finally, the overall project Work Breakdown Structure (WBS), master program schedule, and overall cost estimate for the entire project should be developed as an end product of the planning phase. Inadequate and incomplete planning-phase scope, cost, and schedule descriptions will cause serious control problems during the follow-on design and construction phases. The planning-phase work should be done by a single team that, in essence, can define the program's overall goals and make sure its diverse elements are united. The project scope this team develops must function as a unified whole and meet the overall legal, environmental, and functional objectives of the entire project.

During the design and bidding phase, final design may be assigned to several design teams who prepare plans and specifications for distinct elements of the project. Follow-on construction is usually done by many construction firms.

The overall project Work Breakdown Structure, one of the most important elements of control for a project, must be developed during the planning stage. The WBS establishes the owner's organizational and managerial project structure. The WBS is the framework against which the project scope is separated into manageable elements, and it establishes the basic cost and schedule accounts for the project.

Figure 29.2 is the hypothetical WBS that will be used to demonstrate the organization and control aspects of the four project phases. It assumes that the project can be

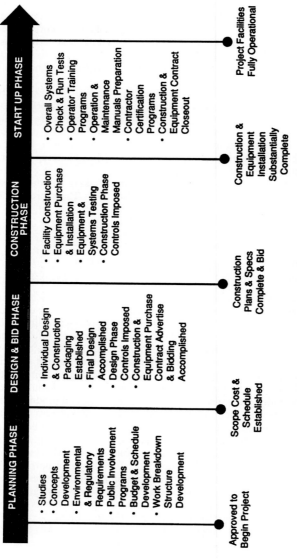

FIGURE 29.1 Major public works construction project phases.

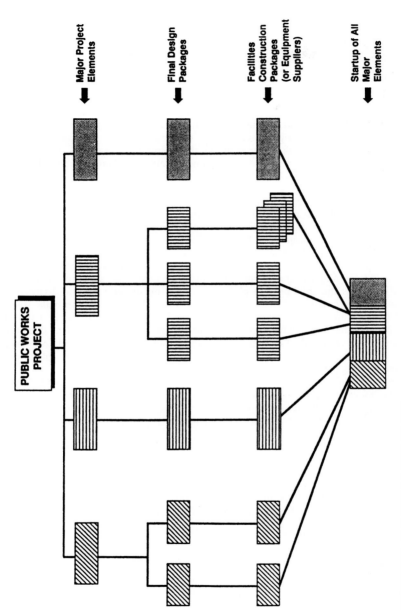

FIGURE 29.2 Basic Work Breakdown Structure.

logically subdivided into major program elements, which is the case for most public works infrastructure programs. For example, in a large wastewater pollution control project, the principal project elements could be

1. Treatment plants
2. Solids disposal facilities
3. Area-wide wastewater conveyance systems
4. Local collection systems

The scope of each of these major elements is defined during the planning stage of the project, but each element retains its separate identity as the design and construction stages proceed. These separate elements are ultimately knit together as a composite operating system during final program start-up. Or these elements may go into operation independently of each other (for example, the operation of segments or lines of a transit system). Each major program element, once established, is then subdivided into separate design packages which, after bidding, become one or more separate construction packages.

As depicted in Fig. 29.2, the four major program elements evolve into seven final design packages, and then nine facilities construction contracts. Of course, the actual large public works project often has many more design and construction packages than this example. For example, the US $2 billion Wastewater Program in Milwaukee had nearly 100 design packages and 300 construction packages.

Figure 29.3 shows how the physical project itself, as defined by the WBS, will logically establish the major management functions of the project organization. The organization indicates the four project phases. In addition, it indicates the extremely important controls function. This function is *not* phase-related; it represents the overall need to maintain project control, from beginning to end, on all phases. Control techniques will differ from phase to phase, but the owner's organizational structure must contain this important function. As shown on Fig. 29.3, the Controls Manager operates on an equal level with the managers of the other four phases (planning, design and bidding, construction, and start-up).

An organizational structure that aligns with the phases of large public works projects offers the owner a good opportunity for achieving project success, and it provides organizational flexibility. It is also easier to staff, regardless of whether the owner provides the Project Director and first-tier management positions, or if contract consultants staff these positions.

The organizational structure has been expanded in Fig. 29.4 to show the specialty manager positions that each Phase Manager would generally staff to accomplish that phase of the project.

This phase management approach, deriving from the WBS structure of Fig. 29.2 and elaborated in Fig. 29.4, means that the program, during its life cycle of all four phases, will have a Project Director, to whom a Phase Manager reports for each of the following functions:

- Control
- Planning
- Design and bidding
- Construction management
- Start-up

This approach is effective and flexible for the following reasons:

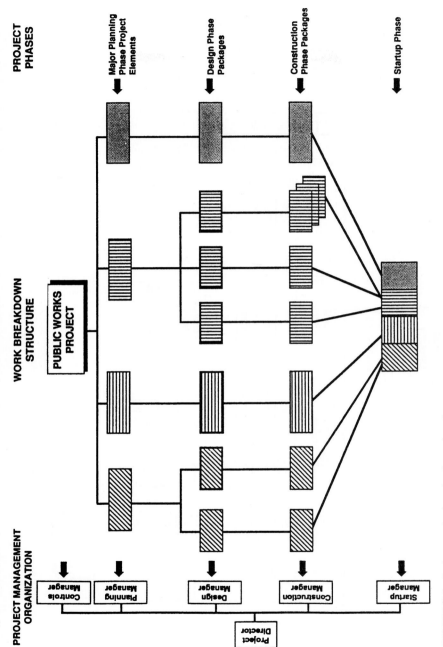

FIGURE 29.3 Project organization related to project phases.

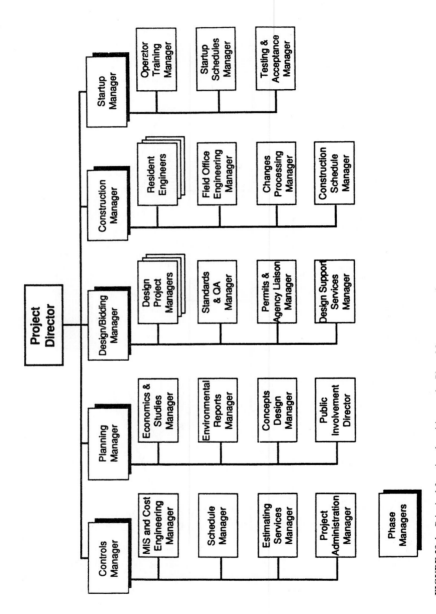

FIGURE 29.4 Principal functional positions under Phase Manager control.

1. Each of the five functions requires a manger who has experience unique to that phase; individuals and/or specialty firms are generally most readily available for that unique work. If the owner elects to assume the role of Project Director, then Phase Managers are obtained who know their end of the business. If the owner elects to retain a multidisciplined firm to provide all management services, then that firm should have personnel with the above five, separately defined functional skills on its staff. The owner can furnish the Project Director and engage specialty firms to perform the five separate functions, but managers of these functions *must* be assigned as part of the owner's staff and take their direction and authority from the Project Director.

2. As the project progresses through its phases, the emphasis of the phase functions shifts, and the organization should shift in emphasis accordingly. Figure 29.5 shows the organizational change in management emphasis as the phases of the project advance. Personnel can be shifted in or out of the project accordingly, and reorganization, when required, will not be disruptive to the project team because the need for reorganization (shifting of emphasis) will be planned for and, thus, expected. The planning phase personnel, for example, can understand from the beginning that they play an important role in the early phase of the project but that their skills are not applicable in the latter stages. It is important to note that the only function common to all phases is the control management function.

3. This organizational structure places a Phase Manager in a direct reporting relationship to the Project Director; each Phase Manager has equal organizational status and brings to the project unique and specific experience related to his or her specific phase. This direct reporting relationship provides the proper technical and managerial emphasis for each phase.

4. This organizational structure avoids the handicap of having Phase Managers performing different functions across more than one phase. Projects often get in trouble when planners are assumed to make good designers, designers to make good construction managers, etc.

As mentioned previously, the owner may or may not staff the Project Director position. However, the owner *must* retain responsibility for:

- Policy
- Providing a general overview of the project
- Making the basic contracting decisions

The owner needs sufficient staff to accomplish these minimum responsibilities.

It should be emphasized that the more an owner elects to assign responsibility for a project phase to a specialty firm, as often occurs, the more responsibility the owner assumes for overall control of the project as a whole.

Before concluding this section on organization of large public works infrastructure programs, a word about such techniques as fast tracking and concurrent engineering and construction.

Fast tracking is not a project delivery process but rather a technique used to accelerate schedules. Generally, it means that construction begins on a facility before design is complete (i.e., a structure's foundations are constructed before the final design decisions on the superstructure are formulated). Thus, fast tracking as a technique is generally not applicable for use with public works projects because such construction work is not sufficiently scoped to obtain comparable quotations using a public bidding process.

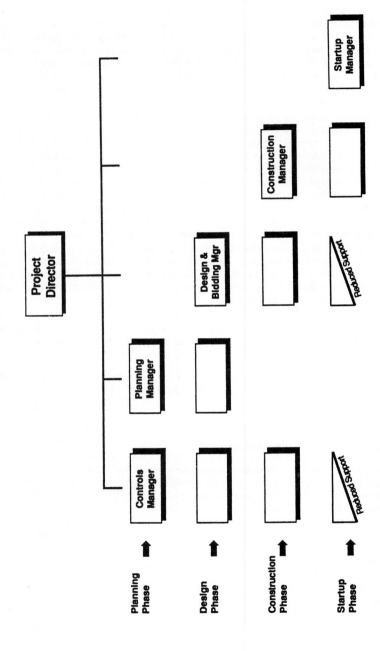

FIGURE 29.5 Organization setup and changes for each phase.

For a private sector owner, fast tracking is usually used with negotiated contract delivery systems, a process not generally available to the public agency owner. On the other hand, the large public works project is readily amenable to having engineering and construction on several project subelements occur concurrently. Usually, the large project must have most of its planning phase finished before other phases can begin. Having accomplished that milestone, however, many project design packages emerge where completed sets of contract plans and specifications can be immediately readied for bidding, and construction started. The scope of the design packages need only meet the test of being a unit of work that can be described by a set of plans and specifications sufficient to satisfy the public bidding requirements, with no material ambiguity of the scope being evident in the bidding documents. Thus, large public works projects are often accomplished by using concurrent design and construction.

Examples of project packages for rail transit and water pollution projects where concurrent design and then construction can begin include:

1. Rail transit system subelements:

   - Roadbed preparation (civil work) for at-grade line sections
   - Tunnels and subterranean station caverns
   - Basic aerial structure
   - Above-grade station heavy construction
   - Station finish construction
   - Track work installation
   - Train control systems
   - Communication systems

2. Urban area water pollution abatement program:

   - Deep storage and transmission tunnels
   - Near surface collection systems
   - Pumping stations
   - Instrumentation and control systems
   - Treatment plants (or subelements thereof)

PROJECT CONTROL INTRODUCTION

The control of the public works project, like any other project, will involve mechanical techniques that are mostly well-established, and there is certainly a plethora of computer project management software systems to meet practically every data processing need. But these techniques and computer data systems are only tools to help the owner and the project team more efficiently manage their job. Control of a project is not a matter of selecting a software system. Efficiently controlling a large public works project means knowing what kind of systems, procedures, and resources are applicable to each of the project's four phases.

Efficiently controlling the project means obtaining and maintaining control of:

- Project scope and quality
- Project cost
- Project schedules

Processes related to establishing and maintaining control in the above three areas are unique to each phase. This reinforces the need for a phased organizational structure.

If project control is to be maintained uniformly and consistently, then the project must be properly staffed with personnel experienced in doing the work of each phase. And the control systems applied during each project phase must be specifically applicable to that phase. Figure 29.6 lists the basic control features required during each phase. It is important to note that the planning phase is the phase of the project related to "creating" and "establishing." The basic scope and master schedule plan are created in this phase, and design standards, budgets, and Management Information Systems are established.

The design and bidding phase is almost all relates to control activities. Designer costs must be controlled, as must construction scope and cost, and design quality control standards must be attained.

Control during the construction phase relates to monitoring the construction quality process, maintaining schedules, and promptly resolving changes. The construction phase is rigidly defined as to process, certainly more so than the planning and design/bid phases. Control during the construction phase is a matter of making sure that construction is carried out in strict accordance with the project's plans and specifications as bid by the contractor. Change orders to the construction work will inevitably emerge. Control here consists of making sure that change orders are for work that is essential to the project, etc.

Control during the start-up phase involves the need for excellent communication systems between many diverse parties (operations personnel, the designer, construction people and equipment suppliers, trainers, and often, human relations specialists).

The control systems that are discussed below relate to the owner's management control of the project and not the working control used, for example, by specialty planning teams, the design room operator, or the construction contractor. In other words, the owner's management control systems are not intended to interfere with or substitute for the means and methods used by, say, a construction contractor to control work. A specialty team or consultant, design firm, or construction contractor must, however, provide regular information so that the owner's management control system can operate smoothly.

The following discussion follows the outline in Fig. 29.6 and provides a detailed review of each of the listed control functions or processes within each project phase. The control functions within each phase are handled jointly by the Phase Manager and the Controls Manager. The Controls Manager sets formats and standards for the MIS, assists in creating budgets and schedules, performs estimates, etc., while the Phase Managers oversee the technical work and control scope and concept development, design quality assurance reviews, construction inspection services, etc.

PLANNING PHASE CONTROLS

Create the Project Scope

The essential end product of the planning phase is the development of the project's work scope. This work scope provides the basis for final design and follow-on construction to proceed. Developing this work scope is often an arduous process costing substantial money and taking considerable time. For example, the planning phase for

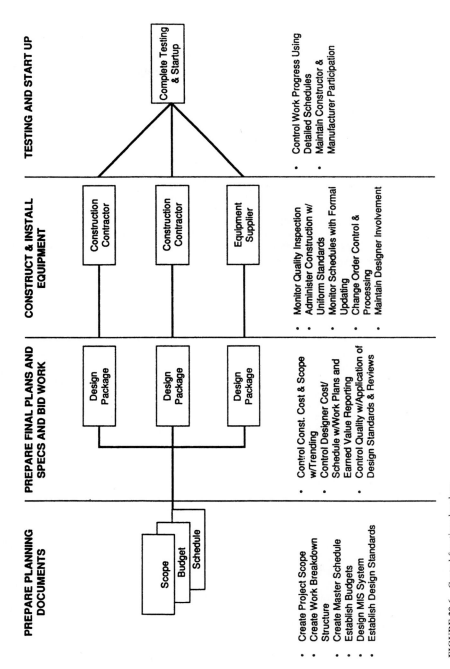

FIGURE 29.6 Central functions by phase.

the Milwaukee Water Pollution Abatement Program cost $30 million and took $2\frac{1}{2}$ years to accomplish. Developing the scope of work entailed reviewing many alternative solutions and selecting a final solution that was cost-effective and met the regulatory agency and legal requirements for the pollution cleanup program in Milwaukee. The scope of work end product from the planning phase was basically a conceptual-level design for the program. This scope of work defined the following by drawings and text:

1. General layout in plan and profile of all the major facilities' elements.

2. System capacities, general operating features, and hydraulic grade lines.

3. Requirements for staging construction to maintain existing wastewater conveyance and treatment systems in operation.

The project scope, created as part of the planning stage of public works programs, must be extensively detailed and thoroughly documented. The project scope must be sufficiently detailed that preparation of biddable plans and specifications can proceed into the final design process with a clear understanding by the designer of what is required. In general, the test of satisfactory scope development from the planning phase is that the scope is sufficiently specific that the final design assignments can be made to design teams on a firm fixed-price basis for their design services.

Create Work Breakdown Structure

A general layout of the WBS is shown in Fig. 29.2. The WBS for the large public works project is, of course, considerably more complex but it should follow this basic format for relationship to the physical work elements of the project.

The major construction packages of the project should emerge from the planning stage as clearly defined and scoped parts of the basic WBS. Figure 29.7 is an example of the basic program element (construction packages) for a treatment plant of the Milwaukee Water Pollution Abatement Program.

The development of the basic WBS during the planning stage should also include the development of the project number identification system for each WBS element that will be used for the duration of the project. The system must be flexible, provide for expansion (more detailed WBS project definition will emerge as the project matures), and still maintain the overall identification parameters originally intended for the program.

The WBS identification of project construction elements that emerges from the planning stage will almost certainly expand during the final design phase.

Create Master Schedule

The master schedule created at the end of the planning phase must include all the basic construction contract elements defined in the WBS. This schedule will expand as the WBS expands (as mentioned above). The planning phase master schedule should contain, as a minimum, the following basic activities for each construction element:

1. Final design

CONSTRUCTION PACKAGES

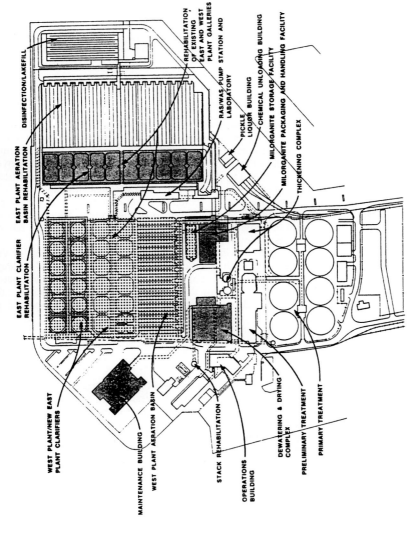

FIGURE 29.7 Basic construction packages for a wastewater treatment plant.

29-15

2. Bid (tender) period
3. Construction period
4. Testing and start-up
5. Equipment procurement
6. Right-of-way acquisition
7. Special studies or planning work that will continue after the general planning phase is completed

The schedule should be in critical path network format and should be set up for computer processing. The software systems should be able to handle very large networks.

Establish Budgets

The completion of the planning phase should include the establishment of a project cost estimate that encompasses cost allowances for every basic project element matched to the WBS account structure. Further, the costs should be subdivided into each of the categories defined in the planning-level master schedule described above. The cost account structure as defined by the WBS forms the project number identification system that is reflected in both the schedule and cost estimate categories.

Costs should be carefully estimated by personnel experienced in developing conceptual-level estimates within the technical disciplines of the project. Significant contingency allowances should be provided, *and* the estimates should be stated in ranges of expected variations at the planning stage of the project.

The estimate completed at the planning stage becomes the initial budget for the public works project. It may be stated in either current costs or inflated to end-of-project costs, depending on the owner's policy requirements. The project budget estimates should be cost-loaded into the activities of the planning-phase master schedule and cash-flowed across each activity. These project cost cash flows should be updated as needed for financial planning.

Design MIS System

The project basic Management Information System should be designed as part of the planning-phase activities. Figure 29.8 provides a general listing of data entry items and report/status output information that the public works project might generally require. The design of the MIS system must include the preparation of a comprehensive project cost/schedule control manual. This manual should contain the basic subject information shown in Table 29.1. Production of the cost/schedule manual should define the type of computer software needed to effectively manage the project.

Establish Design Standards

The project's design standards must be established during the planning phase. These standards include both design manuals and specification standards. In some public work projects, these standards may be well-established. Such is the case with public works projects related to renovation or new construction of state highways or freeways: the design must conform rigidly to the state highway department's established

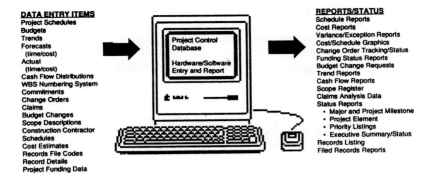

FIGURE 29.8   Management interaction system data and reports.

TABLE 29.1   Table of Contents for Project Cost and Schedule Control Manual

Description of the WBS and Its Numbering System

Scope Control Systems:
  Planning Level Scope Documentation
  Design Period Scope Control (Trend Monitoring System)
  Change Order Processing System during the Construction Phase

Cost Control Systems:
  Report Contents and Formats
  Cost System Database
  Cost Database Input Forms
  Data Entry and Reporting Frequencies
  Computer Systems—Cost

Scheduling Systems:
  Master Schedule—Activity Levels and Format
  Project Schedules—Activity Levels and Format
  Updating Policy and Process
  Contractor Standard Reporting Formats
  Designer Work Plans
  Computer Systems—Schedule

Management Reports:
  Earned Value Reports
  Exception Reports
  Cash Flow Reports
  Budget/Commitments/Forecast Summary Reports

roadway design standards. In other cases, design manuals and standard specifications may have to be developed before final design of the project is allowed to proceed.

Standard specifications, whether existent or undergoing development for the project, should be carefully reviewed by both experienced design personnel and construction phase management personnel.

DESIGN PHASE CONTROL FUNCTIONS

Control Construction Cost and Scope with Trending

The trending program is a system designed to require approvals at the proper management levels on a timely basis for all significant changes made to the project construction work scope and estimate *during* the design phase. The goal of the trending program is to maintain the integrity of the *construction* estimate and scope during final design. Trending, as discussed here, is done during the design phase. It is intended to provide continuous review, discussion, revision, and reporting of estimated *construction costs* as the design evolves. The success of the trending program depends on:

1. Establishing an initial baseline construction cost estimate and initiating the trending program to monitor any changes in the estimate. The conceptual-level design is usually sufficiently detailed for a baseline estimate.
2. Holding regular meetings to probe, review, and report potential design changes that cause changes in the cost estimate.
3. Resolving potential design changes quickly.
4. Documenting and updating the construction cost estimate as a result of adopted scope changes.

Management action should consist of the owner's thoroughly reviewing the necessity of each change. The owner must accept or reject the change.

Control Designer Costs and Schedule with Work Plans and Earned Value

Control by the owner of designer costs and schedules during the design phase is a matter of having information early enough to know if problems in these two areas are emerging. Of these two control parameters, the failure to meet the schedule for completion of design should be viewed by the owner as having the most detrimental impact on the overall project. Delays in bidding the construction work mean increased construction costs by the annual inflation amount in the project's locale. Such construction cost increases are often greater than the marginal increases in the initial design costs that would have prevented the design schedule delay in the first place.

Usually public agencies assign a significant amount of a large project's final design work to several design firms. These firms may do dozens of individual construction contract plans and specification designs for the owner for one project. The owner's Design Phase Manager must require that these designers furnish their work plans and a regular assessment of earned value, and project status reports in S-curve graphic format. Both these requirements go together to provide the Design Phase Manager with an accurate assessment of the status of the design work from a cost and schedule perspective. Work plans alone are inadequate because the designer's failure to update them cannot be detected without the application of earned value reporting.

It should be reiterated that the Design Phase Manager is not managing the methods and means by which the actual design is being accomplished. He or she is simply insisting that the designer provide regularly reported information in prescribed formats so that the designer's cost and schedule performance can be tracked. Designer work plans and earned value S-curve graphics track schedule and cost performance simultaneously. The use of network scheduling for design progress tracking is not necessary, and is often ineffective.

Work plan information should be developed as part of the initial negotiation discussion and budget setting with the designer. The work plans should contain the basic information shown in Fig. 29.9. These work plans should be updated regularly throughout the designer assignment and include:

1. Forecast of projected value of work to complete the assignment in the line item detail of the work plan.

2. Assessment of progress (physical percent complete), for each line item. This should be done using a rational basis of accessing design progress by drawing completion or milestone attainment, etc.

The data for items 1 and 2 above can be run on any number of available hardware/software project management systems, and earned value S-curve graphics like Fig. 29.10 should be produced. The information on these graphics tells a manager where the project is projected to be complete in terms of both cost and schedule. Or, alternatively, it can quickly warn if the work plans and progress assessments are not being done currently or completely by the designer.

Control Quality with Design Standards

The owner should insist that the design standards established in the planning phase be adhered to by all parties who produce final design plans and specifications for bidding. These standards are intended to establish a level of design quality that ensures a consistently uniform quality of design product. They are not intended to substitute for the designer's means and methods needed to guard against design defects; that remains the designer's responsibility. Owners should review the design process, however, to determine that calculations, etc., are being reviewed regularly. However, the Design Phase Manager is responsible for ensuring that quality review teams are established which regularly review the design product for its conformance to the design standards before the work is released to bidding. These quality review teams should consist of personnel with the technical experience to provide expert review. The same team should review all design work of a common technical nature, to ensure product uniformity across the entire project. The Design Phase Manager should require the designer to respond in writing to *every* review comment that is made by the Quality Review Team. The response must indicate if a change was made and how, or if a change was not made and why.

Establishing effective design standards, and having all those concerned adhere to them, will have a significant positive effect on the contractors' bidding process and the consistency of the product that they produce. These additional reviews reduce ambiguity in bidding documents and produce consistency that may result in construction cost savings that many times exceed the cost of the design review itself.

CONSTRUCTION PHASE CONTROL FUNCTIONS

Monitor Quality with Inspection

The objective of the owner's inspection program during the construction phase is to obtain a constructed product that complies with the designer's specifications for the project. Construction cannot change the basic decisions made during the design phase

FREEWAY IMPROVEMENT - ALVARADO-FREMONT BOULEVARD INTERCHANGE

ESTIMATE TO COMPLETE AS OF FEBRUARY 29, 1992

Monthly columns: **1991** — AUG SEP OCT NOV DEC; **1992** — JAN FEB MAR APR MAY JUN JUL AUG SEP OCT NOV

TASK No./DESCRIPTION

1. PROJECT MGMT. & ADMIN.

	Monthly values
Plan	17,000 17,000 17,000 17,000 17,000 17,000 17,000 17,000 17,000 17,000 17,000 17,000 17,000 14,000 11,000 11,000
ACTUAL•FCS	10,000 20,000 18,900 13,300 21,650 20,000 18,000 9,000 9,000 9,000 18,000 21,000 21,000 21,000 21,000 11,000

2. DATA COLLECTION AND UTILITY COORDINATION

	Monthly values
Plan	3,000 3,000 3,000 17,000 17,000 17,000 2,000 2,000 2,000 2,000 2,000 2,000 2,000 2,000 2,000
ACTUAL•FCS	1,160 1,290 3,570 5,290 5,180 3,000 4,000 5,000 5,000 3,500 3,500 1,000 1,000 21,000

3. PRELIMINARY DESIGN AND PS&E PREPARATION

	Monthly values
Plan	15,000 20,000 33,000 33,000 38,000 38,000 40,000 32,000 10,000
ACTUAL•FCS	18,650 28,600 38,000 38,000 32,000 32,000 19,000 10,000

4. GEOTECH INVESTIGATION

	Monthly values
Plan	7,000 7,000 7,000 8,000 8,000 2,000 2,000 2,000 2,000 2,000 7,000 7,000 4,000
ACTUAL•FCS	7,560 1,290 6,290 5,290 5,180 3,000 5,630 5,000 5,000 3,500 1,000 21,000 1,000 21,000

5. HYDROLOGY/HYDRAULICS ANALYSIS

	Monthly values
Plan	4,000 4,000 4,000 4,500 4,500 3,000 4,000 4,000
ACTUAL•FCS	1,590 3,870 4,800 4,900 2,500 2,500 1,500

6. FINAL PS & E

	Monthly values
Plan	100,000 100,000 100,000 150,000 150,000 150,000 150,000 130,000 80,000 72,000
ACTUAL•FCS	100,000 100,000 150,000 150,000 150,000 130,000 80,000 72,000

TOTAL TASK COST

	Monthly values
Plan	27,000 27,000 42,000 51,000 65,000 69,000 71,500 163,500 137,000 181,000 176,000 153,000 181,000 176,000 153,000 192,000 180,500 158,000 105,000 97,000 85,000
ACTUAL•FCS	18,660 28,210 30,700 44,070 64,550 69,970 69,000 64,800 158,900 155,500 192,000 180,500 180,500 158,000 105,000 93,000 72,000

Summary columns

TASK	Actual Cost	FCST is Comp.	Total Est. Cost	Budget	Act. % Comp.
1. PROJECT MGMT. & ADMIN.	121,850	138,000	259,850	257,000	44%
2. DATA COLLECTION AND UTILITY COORDINATION	23,490	21,500	44,990	38,000	36%
3. PRELIMINARY DESIGN AND PS&E PREPARATION	123,250	131,000	254,250	221,000	11%
4. GEOTECH INVESTIGATION	51,610	58,000	109,610	110,000	41%
5. HYDROLOGY/HYDRAULICS ANALYSIS	5,460	19,200	24,660	33,000	15%
6. FINAL PS & E	0	932,000	932,000	932,000	0%
TOTAL TASK COST	325,660	1,299,700	1,625,360	1,591,000	12.64%

EARNED VALUE AS OF FEBRUARY 29, 1992 = $1,591,000 × 12.64%
= $201,120

FIGURE 29.9 Designer work plan.

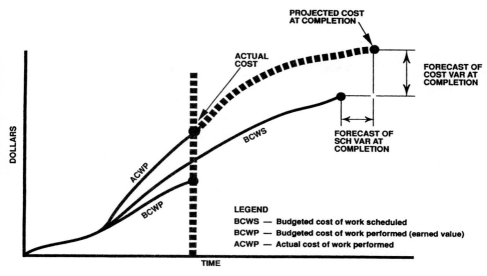

FIGURE 29.10   Earned value S curves.

that relate to the finished project overall end use and cost-effectiveness. The owner's inspection program is basically quality assurance oversight of the construction contractor's (or equipment manufacturer's), quality control program. Inspection involves observing contractor operations for compliance with the specifications and, as necessary, performing appropriate testing, to measure this compliance.

The owner's inspection program should be under the overall control of the Construction Phase Manager. The inspection must be done by qualified personnel. Further, it must be continuous enough so that all critical contractor operations are observed. The type of specification tends to dictate the manner of inspection: prescriptive specifications may require less inspection time, whereas performance specifications tend to require more.

Inspectors may communicate inspection or test results to the contractor's employees, but inspectors are not responsible for quality control. This is solely a contractor's prerogative.

The inspection program as discussed means that a complete program for construction contract administration by the owner exists. Such contract administration should be assigned to qualified Resident Engineers, and their responsibilities and overall contract administration requirements should be clearly defined in a Resident Engineer's operations manual. This manual can be prepared by the Construction Phase Manager for use across all the construction activities of the project. The Resident Engineer administers the contract on behalf of the owning agency. He or she is charged with enforcing specifications so that the work is substantially in compliance.

Maintain Designer Involvement

The designer of record must be maintained in the loop during construction of his or her designs. Most important, he or she should review and provide designer input on all key element contractor submittals or changes to the work. Systems must be established that ensure the designer's prompt response in processing submittals and changes and providing design interpretation information.

Administer Construction with Uniform Standards

Large public works projects often have many construction contractors working on different project areas. If different Resident Engineering staffs administer contract items inconsistently, their action may result in significant criticism from the construction contracting community, give rise to claims, and produce unnecessary additional administrative workloads. Contract administration differences can occur in such areas as:

1. Payment for stored materials
2. Variation in inspection standards or frequency of inspection
3. Administration and processing of change orders
4. Schedule updating standards

Monitor Schedules with Formal Updating

The owner must specify the use by the contractor of modern scheduling techniques and see to it that the contractor's schedules are updated regularly. As a construction contract package is readied for bidding, a scheduling requirement tailored to that package is inserted into the specifications. Basically that specification will stipulate to bidders the following:

1. The contractual elapsed time length of the project.
2. Key intermediate milestone dates that the owner/engineer require adherence to. These are generally only required where contractors must interact with other contractors.
3. The contractor must furnish a detailed schedule of work for the project. The specification will generally require this schedule to be in critical path method (CPM) format for complex work, or in bar graph format with production rates for sewer and tunnel construction, for example.
4. The contractor must participate with the Resident Engineer in regular schedule updating sessions and must maintain accurate updating of this schedule.

The following three basic requirements are essential to the owner's program for schedule monitoring of contractor operations:

1. Sufficiently experienced scheduling staff on the Construction Phase Manager's

team are essential. They are the ones who can make update sessions meaningful and help the Resident Engineer detect and deal with contractor's rigged schedules.

2. In all cases the owner must insist that the construction contractor provide a schedule and regularly update it. If the contractor does not, then the owner has no documentation with which to defend potential time delay claims.

3. The owner *must* be involved in the contractor's scheduling process, to the extent of reviewing and accepting that schedule. The owner must insist that contractors provide accurate and workable schedule plans, and the owner must review those schedules and reject them if inadequate.

Control and Process Change Orders Efficiently

Since practically all public works contracts are fixed-price or unit-price, cost control during the construction stage from the owner's perspective is principally a matter of controlling the change order process and handling claims promptly and equitably. Change orders are inevitable, and claims can never be ruled out, but change orders will increase for a project where an incomplete or unreviewed design was rushed into the bidding stage.

Changes arise mainly as a result of:

1. Differing site conditions
2. Errors and omissions in plans and specifications
3. Changes instituted by regulatory agencies
4. Design changes
5. Overruns/underruns in quantities
6. Factors affecting time of completion

The large public works projects with many construction contractors will have many change orders in various degrees of processing. For example, at the peak of construction on the US $2 billion project in Milwaukee, more than 1000 outstanding change orders were being processed at any one time. These included claims where settlement involved either compensation or schedule adjustments for the contractor because such settlements must be documented through the change order process. With this number of changes, and this is a typical amount for large public works projects, the change order handling process and system must be orderly, and status reports must be comprehensive and timely.

Change order handling should embody the following key features:

1. A system to screen the need for the change orders, with designer and owner involvement.
2. A sound process for negotiating the changes with the contractor equitably and promptly that includes thorough documentation of the settlement.
3. A system to obtain prompt designer input to the changes so that the contractor's operations are not impeded.

START-UP PHASE CONTROL FUNCTION

Control Work Program Using Detailed Schedules

The start-up phase requires extremely detailed scheduling by the owner, more so, in fact, than any other phase of the project. Start-up and commissioning of large and technically complex projects often require network schedules of several hundred to a few thousand activities. The reason for this is that the activities of many parties contributing to the start-up must all come together in a timely and organized fashion.

The start-up phase uses the detail schedule as the basic communication tool to bring operator, trainer, designer, contractor, and equipment supplier together. This program requires an employee (the Start-up Phase Manager) who is experienced in operating the unique aspects of this program. The owner must not assume that the Design Phase and Construction Phase Managers will be able to make the start-up phase happen without outside coordination, since at the end of the project their attention is on other design work and on closing construction contracts, etc.

Maintain Construction Contractor and Manufacturer Participation

The construction contractor's and manufacturers' responsiveness and timeliness must be controlled during the start-up phase. Often, the construction contractor has reached substantial completion perhaps months and maybe even years before the constructed facility is brought on line for start-up. Callbacks to assist in operator training and completion of punch list items may require incentives such as retaining payments and extended warranty periods. Equipment furnished by the owner and installed by others may likewise pose a special problem. This equipment must be exercised regularly, if not in service, for extended periods of time before start-up. Extended warranties again may be necessary to ensure responsive callbacks for equipment service, both during start-up and after the facility is commissioned into operation.

Owners can also improve control of the start-up of complex systems by retaining a specialty contractor on an on-call basis. This way owners can make adjustments and minor corrections to work completed by others during the start-up phase. Also, this provides the owner with more scheduling control than waiting for the original contractor to respond to fixing minor items.

SUMMARY

The purpose of this chapter is to suggest that when public owners undertake the implementation of large infrastructure design and construction projects that there is a natural linkage between the Work Breakdown Structure and phases of the project and the companion organization structure and control systems for management of each phase.

The four project phases (planning, design/bid, construction, and start-up) each have unique technical and managerial characteristics that require management personnel with the unique skills, experience, and training for each respective phase. The projects then lend themselves to having a Phase Manager reporting to the Project Director. In addition, there must be a Controls Manager who has equal organizational standing (and project authority) with the other Phase Managers.

This organizational approach provides the owner with the flexibility to make orderly and easily understood organizational shifts by management discipline as the project matures from planning through the start-up phase.

It is also simpler from an organizational perspective to obtain management personnel with the appropriate skills unique to accomplishing each phase successfully than to obtain personnel with wide enough general experience to be responsible for all phases that a major element of the project must pass through (including having the requisite project controls experience).

Each of the four project phases has a project controls aspect (cost, schedule, scope, quality) that is also unique to that phase. Control during the planning phase focuses on the creation of the overall project's scope, budget, schedule, and basic Work Breakdown Structure.

Control during the design/bidding phase focuses on keeping the construction scope from drifting off course, maintaining owner design standards and review processes, and maintaining the design process and schedule.

Control during the construction phase focuses on maintaining the standards established for the contractor (builder) in the plans and specifications upon which he or she bid, and on handling the inevitable changes to the construction work in an orderly and timely matter.

Finally, control during the start-up phase focuses on the need for clear communication between many diverse parties having a role in successfully concluding the project as part of the start-up process. The owner must often create and maintain very detailed schedules, and effectively communicate them to all parties for adequate control during this final and very important phase of the project.

CHAPTER 30

TOWARD ULTIMATE CONTROL OF MEGAPROJECTS IN THE NORTH SEA

Per Willy Hetland

*Norwegian Association of Project Management,
Oslo, Norway*

Per Willy Hetland is a civil engineer and economist. He has extensive experience of project management in the civil, mechanical, offshore, and petroleum industries, the last 15 years from megaprojects in the North Sea. Mr. Hetland is a cofounder of the Norwegian Association of Cost and Planning Engineers, and a cofounder and past president of the Norwegian Association of Project Management. He is a board member of INTERNET and he has chaired several workshops and seminars on project management. Mr. Hetland has written two books and numerous articles on various aspects of project management.

INTRODUCTION

The last 20 years have added a lot to our general understanding of project controllability.[1] This knowledge is, however, not yet widely spread. The objective of this article is to illustrate the growth of project control science in this period. I will use the background of North Sea projects and the exploitation of the Continental Shelf of Norway as references.

CHARACTERISTICS OF NORTH SEA PROJECTS

Norway proclaimed sovereignty over the Norwegian Continental Shelf in May 1963. Two years later, licenses to explore the Continental Shelf were awarded, and drilling activities commenced the following year. After some disappointing initial results the first commercial reservoir of hydrocarbons was found—the Ekofisk field. In the years to come, several encouraging wells were drilled on other locations, resulting in other major fields such as Frigg and Statfjord.

The total capital cost of a typical North Sea platform project is about 1 billion Great Britain pounds (GBP) and construction takes 3 to 5 years. Up to now 30 billion

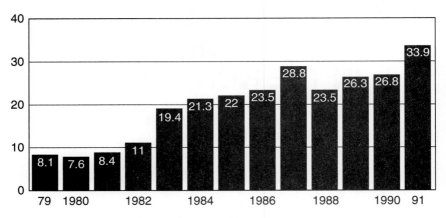

FIGURE 30.1　Investment profile for Norwegian North Sea projects (billion NOK).

GBP has been invested in Norwegian North Sea projects, and current spending is 3 billion GBP yearly. The investment profile over the last 12 years is shown in Fig. 30.1.

Currently, the largest producer of crude oil in the North Sea is Den norske stats oljeselskap a.s, Statoil. Statoil, founded in 1972, is solely owned by the Norwegian state and manages the Norwegian government's commercial interests in the petroleum sector.

Statoil is a fully integrated oil and gas group mainly rooted in the Norwegian Continental Shelf. The Statoil group is involved in all areas of the petroleum business, from upstream exploration and production via transportation to downstream refining and marketing. Statoil is divided into four business units:

- Exploration and production
- Refining and marketing
- Natural gas
- Petrochemicals and plastics

Apart from Norway, Statoil is active in Sweden, Denmark, Finland, the United Kingdom, West Germany, the Netherlands, France, Belgium, the United States, China, and Thailand. The Statoil Group employs 11,000 people and had an operating revenue of 72 billion Norwegian kroner (NOK) in 1991.

The water depth, the hostile environment, and the distance from shore created certain difficulties in the design of facilities bringing the hydrocarbons from the reservoir to the market. A process plant had to be built close to the reservoir, which meant it had to be supported by a huge substructure. As the first production platforms needed a storage volume for stabilized crude oil, substructures made of concrete became dominant, of which a Norwegian design, Condeep, came to play a major role (Fig. 30.2).

At Ekofisk the water depth was about 50 meters; at Statfjord, about 150 meters. The technology of concrete platforms has been stretched, and today gravity-base platforms of concrete can be applied even beyond a water depth of 300 meters. For platforms where storage capacity is not required, steel platforms have often been pre-

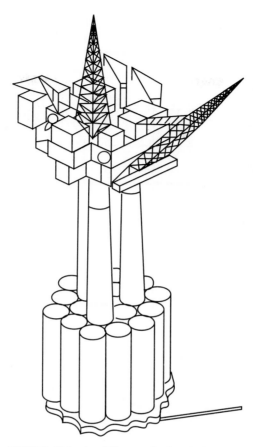

FIGURE 30.2   An offshore production platform with
a Condeep substructure.

ferred. To get an impression of a steel platform located at a water depth of 300 meters,
think of the Eiffel Tower planted in the middle of the North Sea supporting 1 to 2
acres of process plant with a total weight of 30,000 to 50,000 tonnes.

As the North Sea now has a widespread infrastructure, field developments are
today often based on other concepts such as floating facilities and wellheads placed
directly on the sea bed.

Exploitation of the Continental Shelf is well organized and controlled by the
authorities. For each license, a consortium of oil companies is established, one of
which, elected by the authorities, acts as an operating company on behalf of the consortium. The operating company appoints a venture management group accountable
for that particular license. The venture management acts as if it has the sole business
responsibility for the venture, reporting to a board consisting of the oil companies in
the consortium.

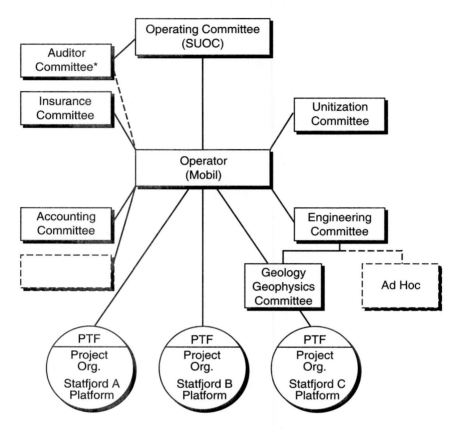

*) The operator is not represented.

FIGURE 30.3   Overall organization of the Statfjord field project.

As project activities commence, the venture management establishes a special project task force (PTF), accountable for project planning and execution. Typically a PTF is formed when a planning permit is given and the project funds are authorized by the consortium. The physical work (engineering and construction) is normally contracted, while the PTF retains the overall project management role, including interface management between different contracts.

The organization structure for a typical project, the Statfjord field, is shown in Fig. 30.3.

The exploration license for Statfjord was given to the group of licensees in Table 30.1.

Originally Mobil was appointed to the operating role, a role that was later transferred to Statoil.

As we can see from the license group, exploration activities in the North Sea are conducted at an international level. The same is true if we look at who is actually

TABLE 30.1   Licensees for Stratfjord Exploration

Licensee	Share before October 1979, %
Den norske stats oljeselskap a.s.	44.44
Mobil Exploration Norway Inc.	13.33
Norske Conoco A/S	8.88
Esso Exploration and Production Norway, Inc.	8.88
A/S Norske Shell	8.88
Saga Petroleum a.s.	1.67
Amoco Norway Oil Company	0.93
Amerada Petroleum Corporation of Norway	0.93
Texas Eastern Norwegian Inc.	0.93
Conoco North Sea Inc.	3.71
Gulf Oil Corporation	1.85
Gulf (U.K.) Offshore Investments Ltd.	1.85
British National Oil Corporation (BNOC)	3.71

doing the job. Figure 30.4 shows a breakdown of the Statfjord A platform project. Here we find internationally well-known contractors.

THE HISTORY OF PROJECT CONTROL

Our Ancestors' Approach to Project Control

Our knowledge of how ancient projects were performed is limited. Historians have, however, given us some clues upon which we can build to create a vision of how things may have happened.

The oldest projects that seem to have attracted current project managers are the pyramids of Egypt.[2] Built about 5000 years ago, they still exist today. The interesting thing is that they were designed to last. The life cycle of a pyramid was meant to be eternity. Construction work commenced as soon as a new Pharaoh came to power so that it should be ready for use on his death and function forever. Little, however, is known as to how the project team estimated the lifetime of a Pharaoh to learn the time available for construction work.

We know the Egyptians gave some consideration to the use of resources. The outer blocks were cut at some remote site where stone materials of an excellent quality could be found. The inner blocks were, however, cut as close as possible to the erection site of the pyramid. The supply of labor seems to have been just about unlimited, but restricted to a particular season when waters of the Nile were not otherwise in use for irrigation purposes. Beyond this, we know very little about the ancient Egyptians' concern for cost, time, and resources.

A very different story is recorded in China about 1000 years ago.[3] The Chinese Wall may have attracted most interest, but the reconstruction of the Ding Imperial Palace could be more significant from a project control point of view. The old palace had been destroyed by fire, and Ding Wei Engineering was asked to accomplish the Emergency Rebuilding Palace Project with the "shortest time, highest quality and limited money." To do this, Ding Wei first excavated a canal from the Brian River to the work site by the palace. The excavated soil was then used to bake huge bricks on the

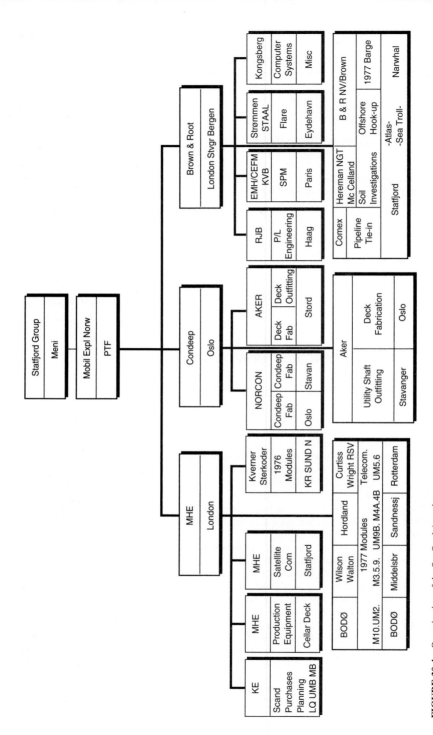

FIGURE 30.4 Organization of the Statfjord A project.

site. Other materials were shipped in via the canal from North and South China. This cut the time and costs dramatically. Finally the canal was filled with fired waste and construction waste and a road was made on top, cutting the costs even further as transport of debris became unnecessary.

We do not know precisely what "shortest time" or "limited money" really implied, but certainly people of that time must have had some understanding of what it meant. If the objectives were not met, the project manager was likely to be decapitated by the Emperor. We do not know how many project managers actually were beheaded, but we do have some evidence that this was more than just a threat. Such an event is recorded after the building of Conway Castle in North Wales in the 13th Century.[4] The project manager was made accountable for a cost overrun of 14 GBP and consequently had to pay with his own life.

As we have seen, emperors and kings of the past did worry about time and cost of construction projects. Some projects were even postponed, waiting for better times to come. This was the case of the Royal Opera at Versailles.[5] As music played an essential role in Court life, Louis XIV wanted a large theater. The event that finally triggered the project was the wedding of the Dauphin (Louis XIV) to Archduchess Marie-Antoinette. To meet the set wedding day, the schedule was tight and the money scarce after many years of war. To save cost and time, the traditional building material of the time, marble, was substituted by wood, painted in a marble-like color. Gabriel, the project manager, completed the theater in 21 months, in due time for the wedding to take place in Spring 1770.

Project Disasters

It is always interesting to read about project disasters, often far more interesting than reading about the successful ones. Kharbanda and Stallworthy[6] have gathered an interesting selection of such disasters. Ironically, it may be easier to learn from the unsuccessful rather than the successful ones, as it is often easier to see what went wrong than what went right.

Kharbanda and Stallworthy reserve the right to "Project Disaster" citation for projects that have overrun the original budget more than 10 times. Leading the disasters, according to the quoted book, are the Sydney Opera House and the Vienna General Hospital. The Opera House overran the budget 16 times, while the hospital had overrun the budget 12 times by 1986, but was still not reported finished.

Construction projects are overrun all over the world, today as they were in earlier days. The North Sea had its share of unsuccessful projects in terms of cost overruns as well as schedule delays. Extensive investments, however, made overruns in the North Sea vulnerable indeed. In particular the first couple of projects gave us unexpected surprises. The cost overruns were not in any way disastrous, but still created a storm of protests against the operating companies. Official investigations took place in Norway as well as in the United Kingdom. The results for Norway are summarized in Ref. 7. The following excerpts from the main conclusions should illustrate what went wrong:

- All completed projects have been underestimated with regard to size and complexity. Of a total cost increase of 178% for completed projects on the Norwegian shelf, the inflation represents about 48%, while increase in resource consumption represents about 52%. For some of the development projects presently in progress, costs seem to be under better control.

- Some of the projects have been weakly managed by the operator. This is partly due to underestimation of the projects and partly to inadequate organizational arrangements. In some of the projects a major share of the assignments have been left to consultants which in some cases have used an alarming amount of resources in engineering and management.

- Incomplete planning before fabrication was started led to a shortage of drawings and materials, changes of the plans along the way, delays, acceleration, and cost increases. The delays have added to the considerable inflation share mentioned above.

- It is the Committee's opinion that cost control has not been satisfactory in the projects completed so far. This aspect seems to be better taken care of in some of the projects presently in progress.

- Cost consciousness has partly been unsatisfactory, particularly in cases of engineering and management having been awarded as reimbursable work to consultants which were not controlled by the operator in a satisfactory manner.

- Costly technical solutions. This applies to Frigg and Statfjord B in particular.

- Offshore hook-up work became considerably more expensive than planned, partly because the work became more comprehensive as a result of deficient completion by contractors, and partly as a result of the rapidly rising costs of offshore work execution. The cost per hour of work has risen 4–5 times in the 1970s at the same time as it seems as if productivity has decreased.

- Fabrication for offshore activities involves high quality demands the yards sometimes had difficulties in meeting.

- Weather conditions were more difficult than expected, and caused breaks in the work, delays, and partly also wreckages.

- In the middle of the 1970s the North Sea market was very tight with regard to deliveries of important investment goods for the development projects on the Norwegian and British shelves. This situation was due to a number of major development projects occurring at the same time, while the traditional industrial activity also was high.

- The market was particularly tight for crane barges and other special vessels, a fact which caused high costs and tight contract terms for transportation and installation work on jackets, decks, and modules.

- New safety specifications introduced by the authorities caused sizable additional costs in the case of some projects, but not of such magnitude that they can be said to have been of significant importance to the cost picture.

- Laws and regulations concerning work environment, worker protection, marine environment protection, etc., have also caused considerable costs without having been of decisive importance for the total cost picture. An exception is hook-up work which, because of new provisions in laws and regulations as well as a stricter enforcement of existing laws, has become about 15% more expensive. Questions linked to the locationing and standard of quarters have also had cost effects.

- The Norwegian goods and service requirements has caused cost increases, but not to any particular extent as compared to total costs. As estimated by the Committee, indirect effects have been of greater significance....

Introduction of Quantitative Methods

Little is known as to how the master builders of the past presented their plans or documented their controls. Most likely the use of quantitative techniques commenced around the turn of this century. The "Scientific Management" school is the first recorded attempt to apply "scientific methods" to physical work. Taylor's work to set and improve productivity standards ought to be well known. Unfortunately, the Scientific Management School fell into disrepute because some extreme applications considered human beings to function as machines. However, one direction originated by Taylor and explored by Gantt is still widely applied—the Gantt diagram. The Gantt diagram was originally developed and applied during World War I. However, the real breakthrough in the development of quantitative techniques first commenced during World War II. The introduction of *Operations Research* (OR) dramatically advanced planning and control of military operations. After the war, OR techniques were successfully applied to manufacturing processes. Operations Research has also given rise to application of quantitative methods for managers, a school to be known as *Management Science* (MS).

The development of the Network Schedule originated from OR and MS. Originally two different concepts were developed independent of each other, PERT and CPM. PERT, Program Evaluation and Review Technique, was developed by the U.S. Navy, faced with the challenge of producing the Polaris Missile system in 1958. CPM, Critical Path Method, was initiated a year earlier by du Pont and Remington Rand Univac. While PERT was designed to estimate the expected duration of an uncertain development project, CPM focused on a cost-time tradeoff in connection with plant overhaul, maintenance, and construction work.

In the '60s PERT and CPM grew into a common network scheduling technique, leaving the three estimates and tradeoffs behind. In the years to come, several network variants were developed for specific purposes like PERT-COST, GERT, and VERT.

UNCERTAINTY AND ITS COMPLICATIONS

The Contingency Schools

Maybe the most fuzzy word ever used in project control is "contingency." Contingency covered up bad estimates, hid financial reserves, and fed project funds when appropriate. Whatever the use, contingency funds largely contributed to major misunderstandings of what the true budget was and what it was for.

Though several companies had their own systems of estimate classification, linking budgets to some performance standard, it was not until the early '80s it became *the* hot issue in professional forums. A significant breakthrough in the understanding of contingencies was made by Richard Westney through the introduction of contingency schools.[8] The "schools" represented different ways to view the meaning of contingency.

The Backyard Bone-Burying Approach. In this view, estimators tend to "bury" contingency where only they can find it (Fig. 30.5). When requested to cut a budget proposal or when faced with a potential budget overrun, they simply head for the "backyard" and pick up a sufficient number of bones to reestablish the balance. The approach may be convenient for the estimator but hardly for management who must make decisions based on estimates with "an unknown number of buried bones."

The Blank Check. The blank check view of contingency simply states that contingencies are there to be spent. Items not specifically expressed in the estimate are lumped together into one line and estimated as a lump sum, either based on a percent-

FIGURE 30.5   The backyard bone-burying approach.

age markup (often 10 to 15 percent) on the base estimate or simply based on professional judgment without backup calculations.

The Piggy Bank.   The piggy bank view says that contingency funds are not to be used under ordinary circumstances. This is a rather complicated view, as we have to distinguish between ordinary and extraordinary circumstances. Though this is a frequently used principle in technically related issues, it still is troublesome to make a reasonable distinction between the two circumstances. In technical standards environmental loads, e.g., wave height, are based on a certain frequency. In the design of North Sea platforms a 100-year wave is considered ordinary, while a 200-year wave is not. But the basis for selecting 100 instead of 80 or 130 is a rather tricky question to answer, in particular since observations of actual wave heights in situ are limited to the last 20 to 30 years.

Though the actual basis for the standard is somewhat fuzzy, the meaning is that the structure should be robust enough to withstand environmental forces that are likely to occur during the lifetime of the structure. Transferred to budgets, contingency funds are not meant to be used under ordinary circumstances, i.e., the budget should be robust enough to withstand events that are likely to occur during the duration of the project. However, from time to time, events that are not likely to happen still happen. Contingency funds, according to the piggy bank view, are meant to be used under such unlikely circumstances only.

The Traaholdt Corrective to the Cost Study

I have discussed elsewhere the results of a cost study in which the first-generation projects were found to have overrun the initial budgets by 178 percent, and the second- and higher-generation projects, by gradually smaller amounts. The common interpretation of these figures is that the control efforts improved considerably as experience data became known to the estimators. But were the previous generations' projects better controlled? They certainly represented fewer surprises in terms of budgets overruns. But were they beter controlled in the sense that they cost less "per unit or project"?

This core question was not raised until one day Hans Traaholdt started to explore the as-built unit costs. He was not particularly interested in comparing actual to budgeted cost, but actual to actual cost per ton of physical work. By comparing as-built costs for different generations of projects, he came to an astonishing conclusion. The as-built unit costs in constant money remained approximately the same; independent of how tight the initial budget was, the projects cost what they cost.[9]

The Monte Carlo Approach

Major discrepancies between actual and planned performance on the early North Sea projects triggered a search for better ways to assess uncertainties in estimates and schedules. Originally estimate uncertainty was expressed as accuracy bounds. Different types of estimates were assumed to have different accuracy as shown in Fig. 30.6.

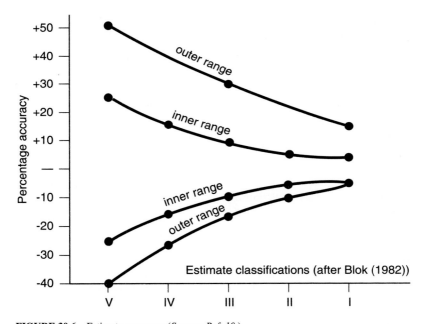

FIGURE 30.6   Estimate accuracy. (*Source: Ref. 10.*)

These accuracy boundaries were never accepted for North Sea projects. If true, no project could be overrun by more than 15 percent, but as we have already seen, the first-generation projects were overrun by 178 percent—on average! Some interests were vested in applications of the Beta distribution as originally used in classic PERT-scheduling techniques to calculate expected values for total project costs.

For different reasons, interests moved swiftly to the Monte Carlo simulation approach. Several computer programs soon entered the market, some directed toward estimating, some toward scheduling. The most advanced analyzed time and cost impact simultaneously.[11] When assessed for time uncertainty, the Monte Carlo simulation technique was superior to the PERT technique as shifting critical paths were considered, a phenomenon giving rise to a special discrepancy referred to as *nodal bias.* A

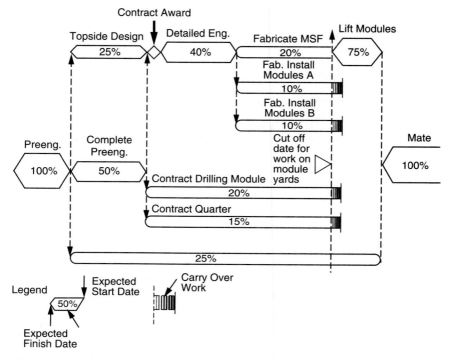

FIGURE 30.7   A stochastic network schedule.

stochastic network schedule analyzed by a Monte Carlo simulation is shown in Fig. 30.7. Each activity has a criticality index which expresses how frequent the activity appears on the critical path.

This example is an excerpt from an analysis of an early Sleipner platform alternative. The network technique applied is advanced as it handles conditional branching in addition to uncertainty distributions. If a module is not fully outfitted by the target move date, the module is transported regardless of the state of completion as long as it technically is movable. Remaining outfitting work, if any, is carried over to the succeeding phase—in the Sleipner example, to the deck outfitting site. *Carry over* is a frequently used strategy to recover delays. Carry over of work to the latest phase—offshore completion—should, though, be limited, as to the overall unit costs increase 6 to 7 times compared with work taking place in a workshop.

Setting the Performance Standard

The result of a Monte Carlo simulation process is shown in Fig. 30.8. Comparisons between traditional deterministic methods and stochastic methods of cost estimating show that the deterministic estimates have more than 50 percent chance of being overrun; frequently the chance is as high as 60 to 80 percent. The major reason for the dis-

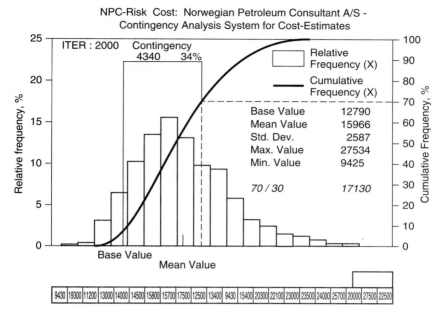

FIGURE 30.8   The result of a Monte Carlo simulation process.

crepancy is that deterministic methods do not handle dependencies between stochastic variables in a proper way. Consequently we do not know how tight a deterministic estimate is unless we carry out a stochastical analysis.

Over the years a sort of understanding between major companies has developed, that the cost estimate to be used for control purposes—the Control Estimate—should have an equal chance of being overrun as underrun. The Control Estimate therefore has come to represent the median—the 50 percent mark on the cumulative probability curve.

INTEGRATING TIME AND COST

The CTRs

The major overruns in the '70s forced the oil companies to improve their project control skills. Several models were tried out, of which Shell Expro's cost, time, and resources (CTR) concept come to play a major role in the years to come. The three parameters, Cost, Time, and Resources, are tied in to the same piece of work. The basic philosophy is to integrate time, cost, and resource information, as they are closely related. The bottom-line control is the CTR sheet, a piece of paper which contains a description of a specific piece of work with a resource estimate and a cost estimate. Possible logical dependencies with other CTR sheets are indicated, as well as time for early start and latest finish. The CTR sheets are grouped hierarchically to encompass a gradually increasing level of aggregation. The CTR sheets are similar to the Work

Package Description Sheets (WPDS) previously introduced by the U.S. Department of Defense (DOD). The aggregation of CTRs to CTR Catalogs is similar to the aggregation of Work Packages up the Work Breakdown Structure (WBS) to Contract Packages, Major Tasks, and so forth. Though the U.S. DOD's requirements for Work Breakdown Structuring had been known for some time in defense circles, the CTRs represented something new in control of North Sea projects. The CTR concept was originally linked directly to payment for contracted work.[14,15]

CTR sheets were typically grouped into a *Payment Milestone*. As soon as all the CTRs encompassed in a Payment Milestone were completed, the agreed lump sums for the CTRs were paid. If, however, work on a CTR or two were not complete, payment for all CTRs in the same Payment Milestone was withheld.

Even though WBS/CTR techniques have been used for quite some time, we continue to see profound misunderstanding of how to apply the techniques properly. A WBS is the selected way of executing the project, nothing less, nothing more. One way to get started is to look at a physical breakdown of the project into pieces suitable for issuing contracts. Figure 30.9 is a schedule originally prepared for planned contracts on the Gullfaks A project.

The contract approach can be further refined by assigning pricing forms to typical contracts (see Fig. 30.10).

Dealing with Change

To be applied successfully, the CTR concept requires work to be defined at a high level of precision. The work logic and time schedule will also have to be fixed at a fairly detailed level. The up-front planning is considerable and the work requires a high degree of definition before being contracted. The concept is certainly not designed to cope with major changes. Rather, it is used under the assumption that an important control function is to prevent project scope from being changed.

Project changes have to be dealt with at two different levels: Project Level and Contract Level. At the contract level, changes to the formal contract between the Project Owner and the Contractor have to be handled. At the project level, changes to the agreement between the Owner's Venture Management and Project Task Force have to be addressed.

Changes at the contract level encompass any agreed change to the formal contract regardless of the origin of the change. Contracted work is controlled by establishing a specific reference Contract Baseline (Fig. 30.11).

Contract work scope is broken down through a Work Breakdown Structure, which, for the client, is an appropriate level of detail. The elements of the lowest level are normally referred to as *Work Packages*. The contractor, on its side, continues to break down the work scope to much smaller pieces of work, to suit its own control of contract work. The lowest level is referred to as CTRs to keep them separate from the client's work packages.

At the time of contract award, the work scope definition is rarely 100 percent complete. Different contract forms are applied, depending on the type of work and work scope definition. For control purposes it is important to state the *volume of work* to be carried out by the contractor. In addition, it is important to define the *available time frame*. Based on this information, contractors compete on *lowest costs*. A key issue, often forgotten, is the relationship between the volume and time information the client gives. Often this is based on the Net-Scope/Gross-Time concept.[12] Assuming that contract work is not 100 percent defined at contract award, the contractor should plan for

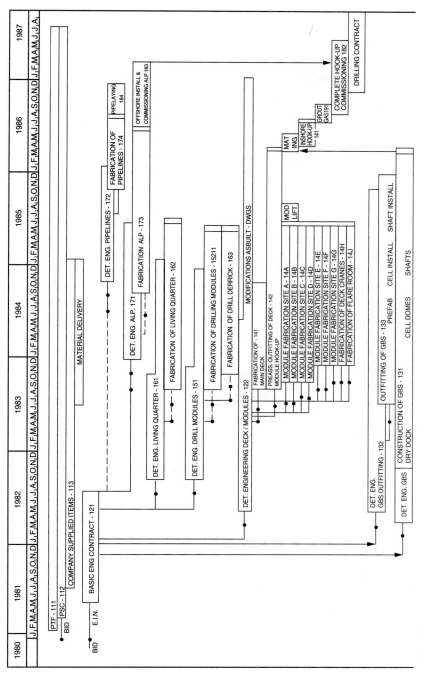

FIGURE 30.9 Contract schedule.

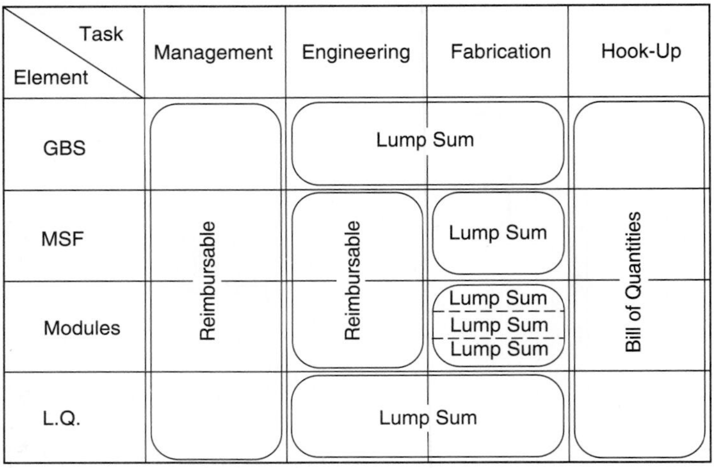

Task Element	Management	Engineering	Fabrication	Hook-Up
GBS		Lump Sum		
MSF	Reimbursable	Reimbursable	Lump Sum	Bill of Quantities
Modules			Lump Sum Lump Sum Lump Sum	
L.Q.		Lump Sum		

FIGURE 30.10   Project breakdown in contracts. GBS = gravity base structure; MSF = module support frame; LQ = lining quarters.

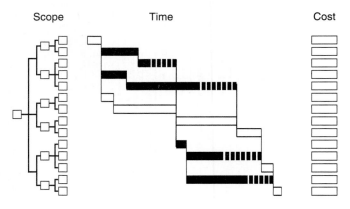

Scope   Time   Cost

FIGURE 30.11   Contract baseline.

the Net Contract Scope—the scope stated in the original contract—to take place within the (for the client) gross total time available for work on the particular Contract Work Package in the client's WBS structure. Additional work scope, which will have to be added later, must be carried out without extending the original contract delivery date, as no slack is allowed for in the schedule. Increases in scope can only be executed by increasing the resource intensity over and above the planned level, which frequently means increased unit costs. A way to avoid this problem is to use the net/net or gross/gross concept (Fig. 30.12).

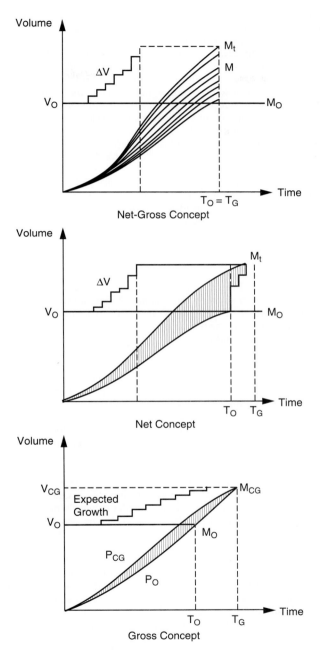

FIGURE 30.12 Volume of work, available time-frame concepts. (*Source: Ref. 12.*)

In the net/net or net concept the contractor plans for the net scope—the original work scope—to take place in the net time frame. The client informs the contractor that a scope increase is anticipated and that a certain time buffer is sized sufficiently to allow the contractor to work efficiently all the time. As work scope is being added, the delivery time will be extended to secure efficient working conditions.

In the gross/gross or gross concept, the contractor is informed about the client's expectations of the total work scope and the total available time frame. The contractor is then asked to plan for net scope and set a net time, considering it has to carry out the expected total work scope in an efficient manner. Both concepts have been tried out to some extent on the Gullfaks B project with some success. In the net concept case, the summer vacation was used as a time buffer. The gross concept was indirectly used by asking the contractor to quote *band rates* for different levels of work intensity.

At the project level, agreed changes between venture and project management are documented and the project budget and/or time frame amended accordingly. As for control of contracts, a baseline is used as a reference for the control of work performed, time spent, and costs consumed.

Performance Measurements

Traditionally cost, time, and resource performance were measured separately. This is probably a major reason why so many projects have failed to meet their objectives—in the North Sea and elsewhere. This is why Papucciyan[13] formulated his famous recipe for a project disaster, "separate the responsibility for time and cost."

The earned value concept and the need for integrating time and cost is assumed to be well-known and will not be discussed in this paper. A convenient way of presenting earned value or, more generally, cost efficiency is illustrated in Fig. 30.13. Note that the budget itself is not part of the calculation of cost efficiency.

Forecasts of total costs are partly based on recorded cost efficiency factors and partly on other sources of information. A complete picture of the overall cost performance for the project is illustrated in Fig. 30.14.

The current control estimates (CCEs) are forecasts of expected total costs made at different points in time. To be consistent, an analysis of uncertainty is carried out and the CCE value is defined at the 50 percent level of significance on the cumulative distribution graph. As the project progresses, changes are made to the agreed project work scope. The changes are represented in terms of volume and the corresponding revised budgets are referred to as master control estimates—MCE(1), MCE(2), etc. To be consistent, the budget revisions are based on the project's original performance norm. If CCE(i) is less than MCE(i), this means that you are forecasting a budget underrun. This difference can be split in two, BCWP(i) − ACWP(i) and PRF(i), of which the first is *already achieved*.

The example in Fig. 30.14 is from the Gullfaks B project. The changes in the MCEs are to some extent a result of moving work scope to and from other projects. The reduction from MCE(2)—10.4 billion NOK—to MCE(3)—9.8 billion NOK—was a result of an option on the Gullfaks A project to fabricate an additional loading buoy at a very favorable price. Consequently, the work packages concerned were transferred and so were the approved work package cost estimates.

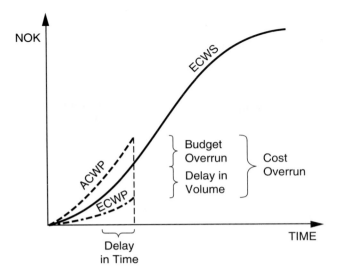

$$\text{Budget Overrun } ACWP_t - ECWS_t$$
$$\text{Cost Overrun } ACWP_t - ECWP_t$$
$$\text{Cost Efficiency } (ACWP_t) / (ECWP_t)$$

FIGURE 30.13 Cost efficiency. (*Source: Ref. 14.*)

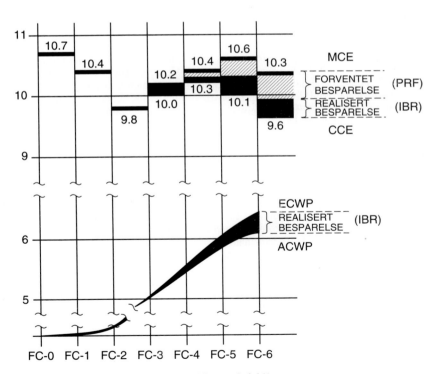

FIGURE 30.14 Total cost performance. (*Source: Ref. 14.*)

THE ULTIMATE CHALLENGE—CONTROLLABILITY

Future Scenarios

I have several times already described the future as being uncertain. Uncertainty has been expressed by an uncertainty spread in the form of a frequency distribution or a cumulative distribution graph. When we talk about total project costs, the distribution is assumed to be symmetrical. As a consequence, the mode (the most likely value), the median (the 50/50 percent value), and the expected value coincide. Normally we act as if we believe the calculated expected value will occur in the future. From a theoretical point of view, expected values will never happen in real life because the probability of occurrence is practically equal to zero. The future will either be better or worse than expected, never as expected. It is only in statistical calculations and Monte Carlo simulations that expected values occur.

During the lifetime of a project, some activities turn out below expected values of time and cost and some above. The belows and the aboves will normally *not* equal out for any single project. Project control efforts have traditionally been directed toward the bad news—time delays and cost overruns. This is a familiar concept of classic cybernetic control. A deviation from a norm is considered bad and, when detected, a corrective measure is triggered to bring the project process back to the norm—the agreed project plan. The principle could be extended to a *control loop* which also feeds information back to the people planning the next generation's projects.

At this point we have already left the basic assumption of the Monte Carlo simulation process. The Monte Carlo process progresses to make random draws assuming there is no one there looking after the project performance. This assumption contradicts the normal way of running projects, as we deliberately appoint project managers and project teams to monitor the project performance and to deliberately take corrective measures as required.[15]

At this point it is appropriate to make a note that interference by the project manager and crew is not necessarily adequate corrective steps. The project manager may panic or misjudge the situation and create a deviation which was not there or jeopardize an inadequate trend. Maybe the most tricky point with human behavior is that it is difficult to predict, as it does not always follow predictable rational patterns.

The Good News

So far we have focused on the bad news. However, not all deviations from the norm are bad. What if an activity finishes ahead of time? Can we take advantage of such a deviation?[16,17] Of course we can start the succeeding activity earlier than scheduled, which in turn brings forward the complete chain of succeeding activities. But, if we do so, do we improve the project performance or not?

If construction work finishes earlier than scheduled, production may start earlier—that is, if we have prepared for such an event to take place. If production starts earlier, the revenue flow starts earlier, and the project's net present value increases, resulting in improved project profitability. Generally a shorter construction time increases the profitability of the project, that is, if the shortening can be done without increased construction costs.

REFERENCES

1. P. W. Hetland, "What did 20 years of North Sea activities add to our general understanding of Project Management," *Proceedings INTERNET/NORDNET,* Reykjavik, 1987.

2. A. F. El-Marashly, "Project Management as perceived from ancient Egyptian projects," in H. Reschke and H. Schelle (eds.), *Dimensions of Project Management,* pp. 275–289, Munich, Springer-Verlag, 1990.

3. Z. Nairu, "Project Management and Project Manager in China," *Proceedings of the 14th International Expert Seminar,* Zürich, 1990.

4. M. Barnes, "Conclusions of the Seminar," *Proceedings of the 14th International Expert Seminar,* Züurich, 1990.

5. G. Kemp and D. Meyer, *Versailles,* Fonteney-sous-Bois, Editions d'Art Lys, 1982.

6. O. P. Kharbanda and E. A. Stallworthy, *How to Learn from Project Disasters,* Aldershot, Gower Publishing Company Ltd., 1983.

7. Ministry of Petroleum and Energy, "Cost Study—Norwegian Continental Shelf," Oslo, 1980.

8. R. Westney, "Current concepts of contingency," The Energy Technology Conference, New Orleans, 1984.

9. J. Huslid, "Effektiv Prosjektstyring," NPF Conference, Bergen, 1984.

10. F. G. Blok, "Contingency: Classification, Definition and Probability," Trans. 7th International Cost Engineering Congress, London, 1982.

11. S. Skogen, Aa. Helgeland, and A. Jacobsen, "Integrated Risk Analysis of Estimates and Schedules," Ninth International Cost Engineering Congress, Oslo, 1986.

12. O. Granli, P. W. Hetland, and A. Rolstadaas, *Applied Project Management: Experience from Exploitation on the Norwegian Continental Shelf,* Trondheim, Tapir forlag, 1986.

13. T. L. Papucciyan, "The Functional Roles in Project Management," *Proceedings Sixth International Cost Engineering Congress,* Mexico City, 1980.

14. P. W. Hetland, "Control Perspectives on Project Management," in H. Reschke and H. Schelle (eds), *Dimensions of Project Management,* pp. 159–164, Munich, Spinger-Verlag, 1990.

15. I. Jordanger and P. W. Hetland, "A Humanized Approach to Intelligent Project Risk Management," *Proceedings INTERNET 90,* Vienna, 1990.

16. P. W. Hetland, "Målsøkende strategier for investeringsprosjekter," *Proceedings NORDNET 90,* Copenhagen, 1990.

17. P. W. Hetland, "Avdekking og utnyttelse av forbedringspotensialer," *Proceedings NORDNET 90,* Copenhagen, 1990.

CHAPTER 31
PROJECT MANAGEMENT AND PHYSICAL SYSTEMS

Hans Knoepfel

Rosenthaler Partner AG,
Zurich, Switzerland

Hans Knoepfel received his civil engineering diploma in 1965 and graduated in structural analysis and design. From 1973, he built up the project management research and teaching team at the Engineering and Construction Management Institute of the ETH Zurich. Dr. Knoepfel was a consultant for several large construction projects and taught as a visiting associate professor in the U.S.A. He is acting as a Vice President for INTERNET. Since 1991, he is a partner of a consulting company for management and information technology in Zurich.

INTRODUCTION

Approaches to Project Management

The substance of management and project management is composed of many aspects and many types of work tasks and responsibilities. Therefore, a considerable number of approaches have been found, for example:

a. The core, and common denominator, of all projects is the activities. These activities are interrelated. Thus, the network of activities is the basis for project management.

b. All human decisions are based on having and comparing information or, in other words, data. These data are interrelated. Thus, relational databases are suitable as the primary background for project management and other management.

c. Values are the real issue for motivation and activities. The values of goods and services are best generated or identified on the market. Thus, the cost or value of the goods and services is the key to project management and other management.

d. Nothing is certain, the world is stochastic. All facts and figures must be seen in terms of their probability of occurrence. The best manager manages *risk* best.

e. All activities, information, values, and judgments are based on human attitudes, motivation and judgment. Without people working together, nothing can be moved or done. Engaging and organizing people is the most fundamental work of project managers and other managers.

f. Projects should transform an unsatisfying existing or future state to a better state within a certain time, using a limited effort. The crucial points are the final state to be achieved and the way (intermediate states) to this target. Thus, the project objectives are what project management basically has to be concerned with.

g. Changes are taking place less and less in independent fields. They are realized within a competitive world market, social groups asking questions, and many existing organizations, facilities, and parts of nature. Therefore, project management must not primarily build on the objectives of its own people and organization, but on the conditions of a complex environment.

Basic Definitions

The general concept that is used and defined in this paper is called *Systemic Project Management*. A *project* is an undertaking that is limited in scope and time. *Management* is the direction of an organization, including the functions of planning, motivating, and controlling the systems and operations of the organization. Project management means directing an organization that is responsible for an undertaking that is limited in scope and time. Systemic management is management based on thinking in terms of systems. Systemic project management is directing a project organization based on thinking in terms of systems.

Reasons for Systemic Project Management

The following reasons led to developing systemic project management:

- The conclusion of the author's situation analysis, at the start of his work in project management in the '70s, was that the feasibility study is the key to successful project management. If a project is not sound itself, management control cannot help. The most important criterion of the feasibility test was whether the system established by the project cycle continued to be successful during the subsequent operations.

- Many professional fields were created in the past and dozens or hundreds of specialists may be working on a project. There is a need for managing specialists to coordinate the work of all these specialists. But also, the clients want the project manager to cover the overall responsibility.

- Time and again, the method of controlling only some aspects (e.g., the time schedule or the costs) has failed. Time schedules were not acceptable due to an uneconomic use of the resources. The real reason for delays and cost overruns were technical and human problems. Thus, the project manager should have controlled the economic, technical, and human fields, foremost.

- No method was available for the overall planning and control of complex projects. Often, details and major parameters are mixed in the planning and control of project managers. Not even the specific project objective is clear and well understood.

- For advanced research and development work, the structure of the respective professional field should be known. Such work is undertaken to create in-depth knowledge and experience concerning a specific subject for a long time, i.e., for many applications. Theorems showing the rules of the professional field and allowing tests in practice should be developed.

- Every project is different, and every project is the same. Project management should be able to use good theory and good practice in those fields that are known, and to detect the aspects that are new. New aspects may be mastered by learning or by engaging new staff or consultants, for example.

The concise conclusion from the whole analysis was that a basic structure and logic was necessary for both managing projects in a responsible way and expanding project management knowledge.

BASIC STRUCTURE AND LOGIC

Characteristics of Systems

The end product of a project cycle is a new state of one or several systems that is more satisfying than the initial, preproject situation. Projects create new potential for the clients and the society, or solve one of their major problems.

A system is composed of a number of subsystems and components that have defined relations to each other and to the environment of the system (Fig. 31.1). Systems are produced and changed by other systems. For example, a product is generated by a company. Or a company is established by one or several founders. For project management, we can differentiate:

- The systems to be created or changed or liquidated
- The systems producing the new state
- The goal systems of the parties involved in the project.

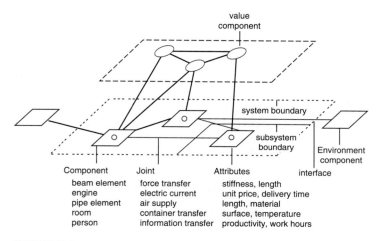

Component	Joint	Attributes	
beam element	force transfer	stiffness, length	
engine	electric current	unit price, delivery time	
pipe element	air supply	length, material	
room	container transfer	surface, temperature	
person	information transfer	productivity, work hours	

FIGURE 31.1   Basic structure of a system.

Most systems encountered are dynamic (i.e., changing in time) and open (i.e., related to their environment). The systems keep changing both during their design and construction and (in a different sense) during operation and maintenance. The environment may change or may be changed by the project as well. However, it is not assumed to be controlled by project management.

Organizations and technical and natural systems are the most frequent types of systems. The types of management can be derived from the system and work types as follows:

- Managing organizations (corporate management, company management)
- Managing the creation or change or liquidation of systems (project management)
- Managing goods and services (product management).

If the management is systemic, it is based on thinking in terms of systems. This is the common denominator of all three types of management. Organizations, projects, and products are interrelated. Thus, the possibility of interacting presents very substantial demands.

Basic Model of a Facility

A facility can be broken down into physical entities:

- Operational units (buildings, bridges, road or railway sections, tunnels, sectors of buildings, sections of tunnels, etc.)
- Subsystems (bearing structure, sewerage system, maintenance system, electrical supply, signaling, etc.)
- Components (elements) (reinforced concrete slabs, engines, pipes, repair workers, etc.).

These three levels are typical for constructed facilities. However, the respective units can be integrated to higher-level systems (e.g., the road section is a component of a road system) or they can be differentiated to more detailed systems (the engine is a system itself with subsystems and components).

This structure serves for defining the basic numbering of subsystems and components in the planning, design, construction, and start of operation phases of the project. At a certain time, the actual state as well as past and intended future states of each part of the system can be defined (Fig. 31.2). This structure can be used as well for operation and maintenance management. One of its big advantages is that it can be seen in reality and shown on drawings. Therefore it is an excellent basis for reliable and efficient communication and data administration.

The components, subsystems, and systems have attributes or properties that can be used for engineering and management purposes. For example, a structural component has a certain stiffness, an anchor needs a certain time to be set, a human-machine group needs a certain amount of space, a road section costs a certain amount of money.

The joints between the components, subsystems, and operational units are again locational physical relationships and therefore generally observable. Examples of joints or interfaces are a change of responsibility at the interface between a road section and a tunnel section, a force transfer between two structural components, and a cost information exchange between two positions in the project organization.

The subsystems and the entire constructed facility have boundaries which cut relationships, i.e., are at joints.

The behavior of the entire system, the subsystems, and the components should allow the fulfillment of the objectives of the constructed facility. The attributes should contribute to achieving these objectives. For example, the quality of a component

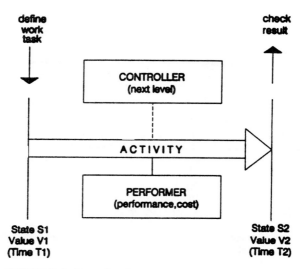

FIGURE 31.2   Dynamics of a system.

should be according to standards and its price should be within budget. Project management is concerned with the goals to be set and the functions to be defined.

For data administration, the structure of the physical system can be used as a primary key, the attribute structures can correspond to secondary keys, and views can represent the structure of objectives.

Management Model for Complex Systems

A system is defined by its components and their topology and attributes. Relational databases can be used for defining systems and for providing management data.

Complex systems are easier to design and control if the multisystem, multilevel, multidiscipline, and multiphase structure shown in Fig. 31.3 is applied.

Several systems allow one to distinguish different purposes and geographical ranges

Several levels display larger and smaller coordination ranges in the organization

Several disciplines refer to a number of professionals involved who are responsible for a type of subsystem or component

Several phases indicate the development stages of the system parts from initiation to completion

The system structures make integration as well as differentiation much easier. They are defined in the project structure (planned, actual, past) lists, graphs, and drawings.

The more complex a system is, the more important a well-designed data model and

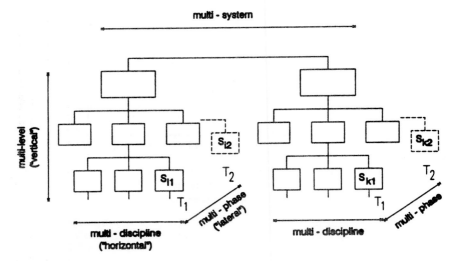

FIGURE 31.3   Management approach for complex systems.

a well-organized data administration become. Both poorly structured heaps of data and sophisticated data accumulation normally do not stand the test.

Theorems of Systemic Project Management

In this section, an attempt is made to define the concept of systemic project management by using theorems. The theorems should allow one to communicate, discuss, test and improve all parts of this concept. For further explanation, examples can be added. Roman numerals are used to identify the theorems. Additions to the theorems are described in brackets.

    I. The primary basis of systemic management is the scope and structure of the systems managed. (Often, physical systems are a suitable primary basis.)

    II. Physical systems serve for producing goods and services (products) by performing work. (These products themselves are systems which can be subjects for project management, i.e., when they are being created or changed.)

    III. The driving force behind the actions of systems is their objectives. (Usually, these objectives are set by persons and organizations or by nature.)

    IV. Systems are created and operate in a certain environment interacting with them. (All systems should be seen in their context—the physical, economic, and sociopsychological framework conditions which depend on time.)

    V. All artificial systems are produced, operated, and maintained by organizations. (Organizations may be taking care of natural systems as well.)

    VI. The organizations are systems as well. (The human beings who are parts of organizations can be considered as systems too.)

VII. Systems keep changing in time due to their properties and interactions with their environment. (The changes can be shown by the altered states of a system during time.)

VIII. The purpose of any activity is to change the state of one or several systems or to maintain a state, by intention. (All activities are related to the systems managed.)

IX. The most important function of the management is to direct all activities toward common goals. (The common goals are derived from the objectives of the systems managed.)

X. The decision of how to change a system depends on the attributes and values of the system before and after the action. (The values are defined by the people and organizations making the decision.)

XI. Systems can be represented by data. (These data are descriptions, figures, photographs, etc. which, however, always are simplifications of the system represented.)

XII. The primary data structure of a model is defined by its subsystems and components (which possess attributes) and time. (Many other data structures can be defined using views on attributes of the system and its parts and environment.)

CREATION, MANAGEMENT, AND LIQUIDATION OF SYSTEMS

Creation and Liquidation of Systems

The design and fabrication of new and adapted systems as well as the liquidation of existing systems is done by other productive systems (organizations). Such new or adapted systems may be constructed facilities, products, companies and other organizations, proposals, laws, and so on.The earth is more and more populated by people and their manmade systems. Therefore, the issue of system liquidation and materials recycling is becoming more and more relevant.

Systems that are operated should produce an assortment of goods within a certain time period. The creation and liquidation of systems are value-adding actions in a multilevel and multidimensional environment of organizations, products, infrastructure, and parts of nature. The systems are interrelated. An example of such a new system is shown in Fig. 31.4.

The systems, subsystems, and components are usually not designed or fabricated or liquidated by the project management. Its central mission is to assure the planning, motivation, and control functions of the design and fabrication or liquidation project. However, the project management should understand the important aspects of these physical systems and may possibly intervene when controlling the project objectives.

The design engineers develop, display, and communicate the information about the physical systems schematically and verbally, by drawing components and subsystems and by writing reports. They create an abstract system which defines the real system that shall be fabricated.

The structure of the whole system allows a project to be realized through intermediate targets. The project management concentrates its attention on a certain subsystem for a limited time in order to drive that subsystem to a certain level of study, investigation, design, realization, or operation. Indistinct or complicated technical sys-

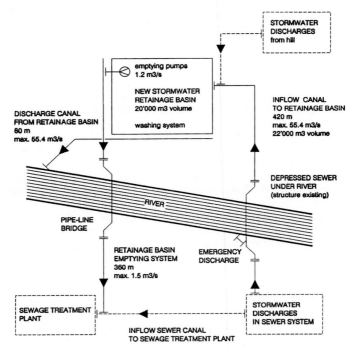

FIGURE 31.4   New stormwater retainage facility.

tems are often an indication of an inappropriate design, which results in problems of meeting quantity, quality, time, cost, and satisfaction targets in the stages of design, coordination, production, installation, and operation.

The usual phases for creating a new or adapted system are

- The conceptual studies and project definition
- The design
- The fabrication or implementation
- The start of operation and project closedown

The design and design coordination of a constructed facility are governed by the following aspects:

- The desired production output of the facility is defined.
- The geometrical structure and shape of the facility are designed.
- The type and properties of the component (processing units etc.) with their space and media usage are determined.
- The flows of materials, energy, and information are calculated.
- The feasibility and optimality of the design are checked with reference to the objectives.

The three fundamental elements for the creation, adaptation, and liquidation of systems are

- The systems that are going to be created
- The organizations (productive systems) that create or adapt or liquidate them
- The work that is required

Figure 31.5 shows the main systems involved in the creation of a new or adapted physical system.

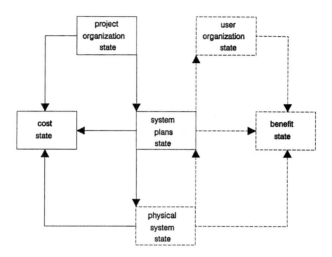

FIGURE 31.5   Creation of new or adapted physical system.

Operation and Maintenance

The desired output (product assortment) and the selected methods and conditions for producing it are the basis for the operation and maintenance concept of a system. The operation and maintenance concept itself includes, for example, in the case of a constructed facility, the technology, processes, and organization for production and maintenance. This basis is designed and selected by the industrial engineers. An example of an operation concept and layout is shown in Fig. 31.6. The corresponding activities do not take place in one phase or one stage of the project. The general production concept is determined in the conceptual phase; the details of the operation and maintenance of a component may be defined in the design phase and may even be adapted in the start-up-operation phase.

The project management should carefully build up this basis for the creation of systems by the project organization and keep it under control. Otherwise, the operation concept cannot be a reliable starting point for the design and construction of the facility, but is a starting point for change orders propagating throughout the project.

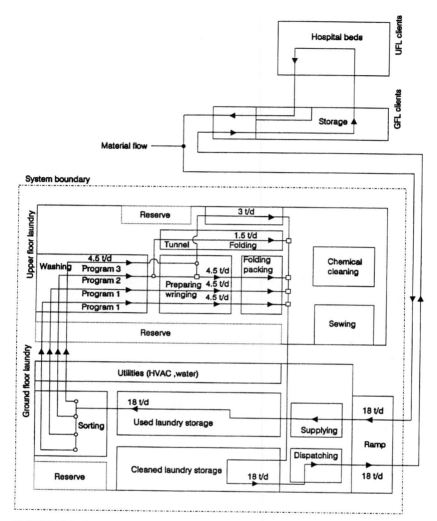

FIGURE 31.6   Example of an operations concept and layout.

Environment of Systems

Most of the systems are, in practice, open systems. They are affected by or have an effect on the environment of the constructed facility. This is in contrast to physical systems tested in a laboratory where external factors are kept constant. Because the scope of a construction project is often a subject for physical, economic, and organizational design, the system's boundary can be an important issue in engineering.

The three main types of framework conditions for systems are:

- Physical conditions (technical standards, surrounding facilities and nature, connections to public infrastructure, etc.)
- Economic conditions (market prices, financial limits and conditions, taxes, etc.)
- Sociopsychological conditions (rules of society, attitudes of stakeholders and opponents, objectives and experience of persons in the project organization, etc.).

The first class of framework conditions can be enforced by law (legal conditions), a second class binds the members of a company or an organization, and a third class represents informal conditions.

The systems created or adapted or liquidated are often interacting with other systems. The project management must identify the interfaces to these systems and allocate the responsibilities for taking them into account in the project organization.

It is essential to be aware of the organizations and persons that may have an impact on the project's success. Tasks are fulfilled by persons belonging to an environment external to the projects. Each organization and person in this environment, as well as each person working on the project, has an identity and a personal interest. The project management should observe potential influences at all times and, if necessary, take steps to maintain the motivation for achieving the project objectives.

Organizations

The project organization is the group engaged in planning and realizing the project. This organization should be tailored to the requirements of the specific project and phase. Organizing means defining the work tasks with the responsibilities, allocating them to positions, designing the procedures in the organization, and selecting adequate performers for the positions.

The relationship between project and permanent organizations has been a substantial concern for project managers since the beginning. Good organizational solutions allow a smooth integration of the project work into the company organizations, as well as a high responsibility and independence of the project management and project organization (see Fig. 31.7).

Specialized labor generally leads to increased quality and productivity of a complex production. The division of work is normal for the creation, adaptation, and liquidation of systems and for their operation and maintenance. The definition, coordination, and allocation of work packages and functions to appropriate organization groups and positions are therefore necessary.

Information flows assist the transfer of intermediate products and support the coordination of the work, the allotment of work tasks, and the control and updating of the project objectives.

In Fig. 31.5, two types of organizations were distinguished:

- The project organization (creates a new or adapted system or liquidates an existing system, with its own and purchased resources)
- The user organization (operates and maintains an existing system and is usually considered a part of it)

The physical system that is created or adapted or liquidated can be used as a basis for planning, developing, controlling, and finally reducing the project organization. An example of a project organization for the detailed design phase is shown in Fig. 31.8.

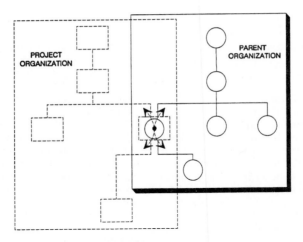

FIGURE 31.7 Project and parent organizations.

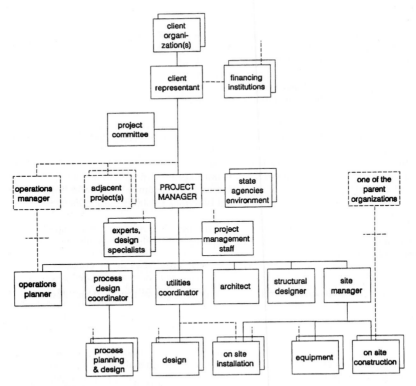

FIGURE 31.8 Example of project organization for the detailed design and construction phase.

A responsible person or group can be assigned to each component, to each subsystem, and to the entire constructed facility.

The organization chart, together with the corresponding job descriptions, the responsibility matrix, and the definition of important information flows, is the typical abstract model for developing, displaying, and communicating the information about organizational systems. Responsible persons (components), groups (subsystems), and organizations (systems) are working on certain subsystems and components of the system that is the subject of a project. This fact relates the project organization to the physical system. The project management tries to allocate each work package to the company or group or person offering or assuring the most favorable conditions.

Applying again the idea of realizing a project through intermediate objectives, the project management tends to maintain a constant organization pattern and responsibility through a stage leading to the next set of results. This means that essential reorganization is normally undertaken just before or just after the beginning of a new development stage of a project (for example, the beginning of the main design phase). An evaluation of the preceding phase of the project, a plan for the next phase, a badge of organizational work, and an improvement or initiation of contracts can be combined at that time by preparing and realizing an initiation meeting.

The response of organizations to work tasks that are given to or found by these organizations can be shown by using the feedback model. In this model, processing activities and information, material, and energy flows are assumed to start according to defined criteria.

The management of the resources (crews, equipments, materials, and third-party goods or services) is not logical for one project alone. The resources should be managed for a whole organization unit or company (multiproject management).

Value of Systems

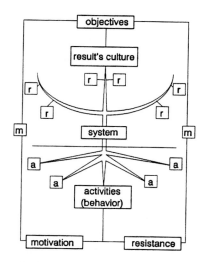

FIGURE 31.9   The culture tree.

Physical systems are seldom an end in themselves. They are designed and built for producing a set of goods and values. This set motivates the owner to allocate the financial resources to the project. However, creating, adapting, and liquidating a system may have effects on other objective systems. New jobs and opportunities for making money, new financial commitments, and new qualities of life may be made available and new environmental problems may be created.

Culture is normally related to creating, producing, and building precious works. It is developed and promoted by intention; it is keeping up a set of customary knowledge, beliefs, and behaviors (see Fig. 31.9). Culture is

- Intentional, therefore it is most important to define the objectives for the project
- To be realized, therefore the culture of the results shall be evaluated

- Built, thereby activities and patterns of behavior lead to results
- Evaluated; nothing happens without motivation which is directly related to the objectives of those who are involved in the project

Objective functions demonstrate the effect of design and implementation variations on the preference parameters of the client. In the field of cost planning, the client wants answers to three questions: "What is it?," "What is it worth?," and "What does it cost?" (see Fig. 31.10). Special attention is directed to the question "What is it worth?" Answering it means determining the subjective preferences of the client. Without knowing the answers, a general optimum cannot be reached.

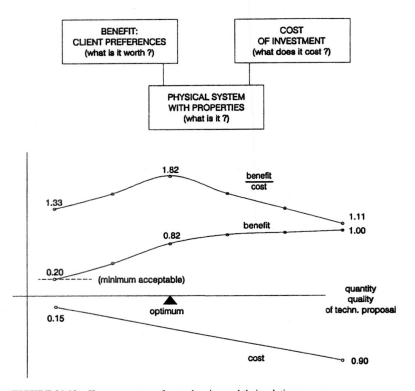

FIGURE 31.10   Key parameters of cost planning and their relations.

The physical system is again used as a starting point. Price tags and performance data are attached to the components and subsystems. The structure of the cost estimate and the cost control elements should coincide with the structure of the physical system. Subsystems can be used as cost centers of the cost accounting for the operated systems.

The general and top-level objective is the performance of the system that is created or adapted during its operation and maintenance time. For an economic evaluation,

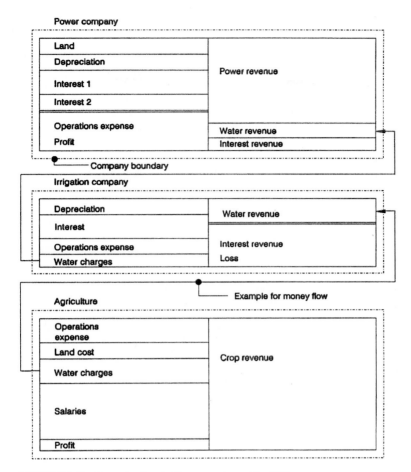

FIGURE 31.11 Example of project income statement for 1995.

planned income statements of the organizations involved and the operated system can be used (see Fig. 31.11).

The investment costs show up in the expense accounts as interest and depreciation. An alteration in the quality required may change the expenses for operation and maintenance, the depreciation rate, and the revenue (through the quality of the goods produced by the system). Planned balance sheets and financial plans demonstrate the usage and origin of the financial resources as well as the liquidity as a function of time.

In a capital-intensive system such as a hydroelectric power station, the investment costs incurred during the project cycle have a very significant impact on the facility's economy. On the other hand, investment costs have a relatively small impact on the total income statement of a shopping center. The operation manager should therefore turn his or her attention more to sales effects and material handling solutions.

Technical uncertainties, changes in market and procurement conditions, inadequate working methods, unrealistic time and cost estimates, not observing the wishes of stakeholders, and the influence of low-cost, poorly qualified consultants can all reduce the probability of achieving intermediate and final targets of a project. In order to achieve the project objectives, the project management must, during the project cycle, successively reduce the probability of outcomes situated outside an acceptable range of variance. This can be achieved by making a risk analysis and by finding risk-reducing solutions like design reviews, repetition of work items, prefabricated components, and insurance.

Displaying and Communicating Information

An ideal information system would provide, during the whole project cycle, well-presented, understandable, correct, and technically consistent actual data in several degrees of detail for the systems, subsystems, and components and their attributes, at a reasonable cost. New information technology allows one to store, process, display, and transfer data more and more easily and economically. However, (temporary) project information systems are not easy to establish.

The first condition is that each member of the project organization be equipped with a computer processing, display, and memory installation and data transfer unit. In addition, a hard-copy printer and the traditional noncomputer communication equipment (mail, telephone, telefax) should be available.

The second condition is that the data that the user is referring to be available at his or her working place. For example, it should be possible to use standard specifications, contract patterns, cost figures from other projects, standards, and regulations, if possible, in a multitask processing and display mode.

The third condition is related to the decentralized way of working. If the information system should function for a complex, multidisciplinary project structure (Fig. 31.12), the users need to have a common understanding of definitions (semantics) regarding the data transferred. Clear data structures are necessary for an efficient and reliable data administration as well.

Standard data structures and common terms down to the most detailed level are necessary for

- Developing, implementing, and applying software efficiently
- Being able to interpret correctly incoming information from another company working on the project
- Avoiding the danger of errors propagating over the whole information system

The hierarchical physical system structure can be used as the primary key for work and cost data, for example (Fig. 31.13). Both specifications and cost data for a project usually are related to project-independent databases as well. Often, this information (e.g., standard specifications or unit cost values) must be adapted and completed according to the size, location, and other project-related attributes and framework conditions before becoming actual project information.

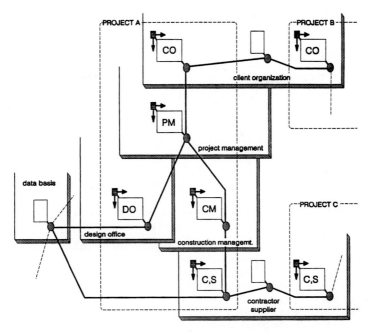

FIGURE 31.12   Project information system and its environment.

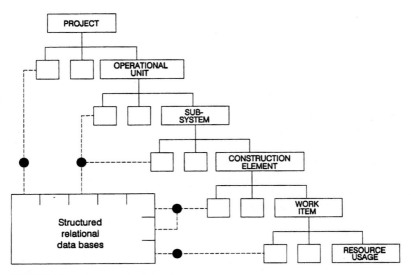

FIGURE 31.13   Structured, relational databases.

SUMMARY

Project management aims to allocate the resources during the project cycle in such a way that the planned operations of the finished system can be run optimally. The following aspects are predominant:

- The situation which is to be changed and its environment is known as a composition of systems, subsystems, and components.
- The scope and planned operational functions of a system are defined.
- The technical concepts are developed, coordinated, and improved or maintained.
- The work tasks, procedures, and durations are designed over the whole project cycle.
- The resources and the tasks and responsibilities are allocated within the project organization.
- The system of objectives and the framework conditions to which the optimization of requirements and costs shall be referred are determined.

The methodology of systems engineering emphasizes the procedure of designing and realizing a system. In a general methodology of systemic project management, the structures of this system will be given at least an equal importance. The relationship between the physical system and its development and change process is given by the transition from one state of the system to another state. This change is normally achieved by activities.

By comprehending the physical system, project management provides the technical basis for answers to economic, scheduling, organizational and other questions. Therefore, competent project managers or other managers in a line function should understand the important technical aspects of their projects.

BIBLIOGRAPHY

D. Scheifele, "Bauprojektablauf," Schriftenreihe der GPM, Verlag TÜV Rheinland, Köln, 1991.

M. Barnes, "Summary of the day and some new thoughts," *Proc. Int. Symposium on Promoting and Managing Projects without Failures,* INTERNET, Zürich, 1991.

H, Knöpfel, "Tendenzen für das Projektmanagement der neunziger Jahre," *Forum 90,* GPM, Aachen, 1990.

H. Knöpfel, H. Notter, A. Reist, and U. Wiederkehr, "Kostengliederung im Bauwesen," Vereinigung Schweiz. Strassenfachleute, Zürich, 1990.

H. Knoepfel, "A Concept for Project Management in the Nineties," PMI/INTERNET Seminar/Symposium, Atlanta, 1989.

R. Berger, "Bauprojektkosten," Schweiz. Zentralstelle für Baurationalisierung (CRB), Zürich, 1988.

R. Burger, "Bauprojektorganisation," Institut für Bauplanung und Baubetrieb, ETH, Zürich, 1985.

H. Knoepfel, "Systematic project management," *Int. Journal of Project Management,* vol. 1, no. 4, Butterworth, 1983.

W. Daenzer, *Systems engineering,* Verlag Industrielle Organisation," Zürich, 1982/83.

S. Lichtenberg, "Real world uncertainties in project budgets and schedules," Joint PMI/INTERNET Symposium, Boston, 1981.

D. W. Halpin, and R. Woodhead, *Construction Management,* Wiley, New York, 1980.

A GLOBAL HISTORICAL PROJECT

In Chap. 32, Fred N. Bennett provides a fascinating review of one of the great explorers of all times. Captain James Cook's first voyage of exploration was a project in the truest sense of the word—a project that was at the cutting edge of science as well as imperial and commercial expansion.

CHAPTER 32

THE GREAT SOUTH LAND PROJECT

Frederick Nils Bennett

Australian Institute of Project Management

Frederick Nils Bennett, AM, B.Ec, F.A.I.M., M.A.I.P.M., was born in Sydney, Australia, in 1930 and educated at Sydney High School and Sydney University, Mr. Bennett lives in the national capital, Canberra. He is Honorary National Development Officer of the Australian Institute of Project Management and consults, writes, and lectures on project management. He is a Member of the Order of Australia for services to Australian industry and defense, including posts of Deputy Secretary in the Departments of Defense and Industry and Commerce.

INTRODUCTION

I sometimes use a voyage of exploration as a metaphor when teaching project management. This chapter takes the opposite approach by interpreting an actual voyage through concepts of project management. In the beginning it was to be only part of a chapter about project management in Australia, but the spirit of James Cook would not be so contained. His first voyage, which was only the first phase of a project, took up the whole chapter. That project led to the British settlement in New South Wales in 1788 and ultimately to the creation of the nation of Australia. We shall call it the Great South Land Project because the discovery of an unknown continent was one of its principal objectives.

Is this a book about project management or history, you may ask. The answer is that the two go together. Although the existence of the project management profession has only been recognized in recent times, it is an ancient one. The case histories of great projects contain many lessons for project managers and the Great South Land Project is no exception.

THE CONTEXT

Project management is the art of managing change. The 18th century was an era of sweeping change that molded much of the world as it is today. It was the age of enlightenment marked by ascendancy of the idea that "... human beings could comprehend the operation of physical nature and mold it to the ends of material and moral improvement"[1]; an age of progressive application of the fruits of the scientific revolution of preceding centuries; the age of the industrial revolution; the age in which the British Empire and the United States of America were born. Despite these sweeping changes "... the eighteenth century remained the age of the aristocracy."[2] The Great South Land Project was a product of all these influences.

COOK'S FIRST VOYAGE AS A PROJECT

James Cook's first voyage of exploration was a daring and ambitious project at the cutting edge of science and imperial and commercial expansion. It had three main goals: to enable measurement of the distance of the earth from the sun, to discover a continent the size of Asia, and to claim it for the British Crown.

Stuckenbruck has defined a project as "... a one-shot, time-limited, goal-directed, major undertaking, requiring the commitment of various skills and resources."[3] How does Cook's first voyage measure against those criteria?

One-Shot

There is no doubt that the voyage meets this criterion. The stimulus for the expedition came from the pinnacle of British science, the Royal Society, which wished to make a more accurate estimate of the distance of the earth from the sun. That distance was the fundamental baseline for astronomical observations and was very imperfectly known. Some 50 years earlier Edmund Halley had explained that the distance could be calculated by the parallax method by timimg the beginning and end of the transit of Venus over the disk of the sun as observed from several widely separated places at known distances. A site in the South Pacific was needed and the recently discovered King George III Island (now known as Tahiti) was chosen. The transit was predicted to occur on June 3, 1769 and would not occur again for 105 years. Since the observation had to be made at a precise moment, Cook literally had only one shot at performing that task.

Time-Limited

For the same reason, the initial phase of the project was strictly time-limited. When appointed as commander of the Endeavour in May 1768, Cook had just over one year to prepare the expedition, sail to Tahiti, and set up an observatory in readiness for the observation. The overall duration of the project was limited by the stock of provisions, by the deterioration that would occur in the condition of the hull and rigging of the vessel, and by the high crew attrition rate then common to long voyages.

---

[1]D. Kagan, S. Ozment, and F. M. Turner, *The Western Heritage,* 2d ed., Macmillan, New York, 1983, p. 619.
[2]Op. cit., p. 578.
[3]Linn C. Stuckenbruck, *The Implementation of Project Management: The Professional's Handbook,* Project Management Institute, Addison-Wesley Publishing Company, Reading, Pa., 1981, p. 1.

Goal-Directed

The voyage was clearly goal-directed. It had multiple objectives and a hidden agenda, a combination that has wrecked many a modern project. The hidden agenda, set out in secret, sealed instructions that were not to be opened until the transit had been observed, was "... to discover whether the Pacific contained a great continent to the south of Tahiti and whether Tasman's New Zealand was part of this unknown continent." Cook was also directed to "... take possession for his Majesty..." of promising new lands or establish good relations with their inhabitants.[4] These objectives were for the purpose of establishing strategic bases, opening up trade and establishing new colonies. These "secret" instructions are reproduced in Annex A. They are specific and demanding and leave no doubt that if anything went wrong, the project manager would be to blame.

ANNEX A

Secret   By the Commissioners for executing the office of the Lord High Admiral of Great Britain &[c]

Additional Instructions for L[t] James Cook, Appointed to Command His Majesty's Bark the Endeavour.

Whereas the making Discoverys of Countries hitherto unknown, and the Attaining a Knowledge of distant Parts which though formerly discover'd have yet been but imperfectly explored, will redound greatly to the Honour of this Nation as a Maritime Power, as well as to the Dignity of the Crown of Great Britain, and may trend greatly to the advancement of the Trade and Navigation thereof; and Whereas there is reason to imagine that a Continent or Land of great extent, may be found to the Southward of the Tract lately made by Capt[n] Wallis in His Majesty's ship the Dolphin (of which you will herewith receive a Copy) or of the Tract of any former Navigators in Pursuits of the like kind; You are therefore in pursuance of His Majesty's Pleasure hereby requir'd and directed to put to Sea with the Bark you Command so soon as the Observation of the Transit of the Planet Venus shall be finished and observe the following Instructions.

You are to proceed to the southward in order to make discovery of the Continent above-mentioned until you arrive in the Latitude of 40°, unless you sooner fall in with it. But not having discover'd or any evident signs of it in that Run, you are to proceed in search of it to the Westward between the Latitude before mentioned and the Latitude of 35° until you discover it, or fall in with the Eastern side of the land discover'd by Tasman and now called New Zeland.

If you discover the Continent above-mentioned either in your Run to the Southward or to the Westward as above directed, You are to employ yourself diligently in exploring as great an Extent of the Coast as you can; carefully observing the true situation thereof both in Latitude and Longitude, the variation of the Needle, bearings of Head Lands. Height, direction and Course of the Tides and Currents, Depths and Soundings of the Sea, Shoals, Rocks, &c. and also surveying and making Charts, and taking Views of such Bays, Harbours and Parts of the Coast as may be useful to Navigation.

You are also carefully to observe the Nature of the Soil, and the Products thereof; the Beasts and Fowls that inhabit or frequent it, the fishes that are to be found in the rivers or upon the Coast and in what Plenty; and in case you find any Mines, Minerals or valuable stones you are to bring home specimens of each, as also such Specimens of the Seeds of the

---

[4]A. Grenfell Price (ed.), *The Explorations of Captain James Cook in the Pacific as Told by Selections of His Own Journals,* Angus and Robertson, Sydney, 1969, p. 17.

Trees, Fruits, and Grains as you may be able to collect, and Transmit them to our Secretary that We may cause proper Examination and Experiments to be made of them.

You are likewise to observe the Genius, Temper, disposition and Number of the Natives, if there be any, and endeavour by all proper means to cultivate a friendship and Alliance with them, making them presents of such trifles as they may Value, inviting them to Traffic and Shewing them every kind of Civility and Regard; taking Care however not to suffer yourself to be surprized by them, but to always be on your guard against any Accident.

You are also with the Consent of the Natives to take possession of Convenient Situations in the Country in the Name of the King of Great Britain; or if you find the Country uninhabited take possession for His Majesty by setting up Proper Marks and Inscriptions, as first discoverers and possessors.

But if you should fail of discovering the Continent before-mentioned, you will upon falling in with New Zeland carefully observe the Latitude and Longitude in which that land is situated and explore as much of the Coast as the Condition of the Bark, the health of her Crew, and the State of your Provisions will admit of, having always great Attention to reserve as much of the latter as will enable you to reach some known Port where you may procure a Sufficiency to carry you to England, either round the Cape of Good Hope, or Cape Horn, as from circumstances you may judge the Most Eligible way of returning home.

You will also observe with accuracy the Situation of such Islands as you may discover in the Course of your Voyage that have not hitherto been discover'd by any Europeans, and take possession for His Majesty and make Surveys and Draughts of such of them as may appear to be of Consequence, without Suffering yourself however to be thereby diverted from the Object which you are always to have in view, the Discovery of the Southern Continent so often Mentioned.

But forasmuch as in an undertaking of this nature several Emergencies may Arise not to be foreseen, and therefore not particularly to be provided for by Instruction before hand, you are in all such Cases, to proceed, as upon advice with your Officers you shall judge most advantageous to the service on which you are employed.

You are to send by all proper Conveyances to the Secretary of the Royal Society Copys of the Observations you shall have made of the Transit of Venus; and you are at the same time to send to our Secretary, for our information, accounts of your Proceedings and Copys of the Surveys and drawings you shall have made. And upon your Arrival in England you are immediately to repair to this Office in order to lay before us a full account of your Proceedings in the whole Course of your Voyage, taking care before you leave the Vessel to demand of your Officers and Petty Officers the log books and Journals they may have Kept, and to seal them up for our inspection, and enjoyning them, and the whole Crew, not to divulge where they have been until they shall have Permission to do so.

Given under our hands the 30th of July 1768.

E^d. HAWKE
Piercy BRETT
C. SPENCER

By Command of their Lordships
P. H. STEPHENS[5]

Various Skills and Resources

The Project Team.   Cook was at the head of what we would now call a multidisciplinary team. The disciplines included in that team were seamanship, navigation, cartography, astronomy, botany, zoology, and art, and the expedition also made contributions in the field of anthropology. The ship's company included all the skills

---

[5]Reproduced in Price, pp. 18–20.

necessary to defend and sustain the vessel and the party on board including a surgeon, boatswain, carpenter, sailmaker, gunner, company of marines, quartermaster, clerk, and cook.

The scientists were prominent in their fields. Apart from the astronomer Charles Green, they included the "naturalist" Joseph Banks, who was a Fellow of the Royal Society, and a well-credentialed Swedish botanist, Dr. Solander. The unusual way in which this team came together is described under the heading "The Role of Chance."

PLANNING THE PROJECT

Risk Analysis

The outstanding characteristic of the project was the enormous risk involved. A voyage to the Antarctic in the 18th century was as adventurous, and far more dangerous, than a voyage to the moon in the 20th. The Pacific claimed the lives of many of its explorers from Magellan to La Pérouse, and ultimately of Cook himself. In common with a modern project manager, the planners of the Great South Land Project would have considered the multiple risks and devised ways of managing them. Roman[6] has suggested a procedural model for risk identification and analysis as follows:

Defining the goal

Identifying risks

Assessing consequences

Assessing degree of risk

Manipulating consequences

Manipulating degree of risk

Assessing results

Developing a best plan of attack to achieve a goal

The goal having been defined, the risks would have been identified as:

Failure of the vessel or crew under the effects of the destructive marine environment and hardships of the voyage

Shipwreck when sailing in the world's stormiest ocean and close to land, and ice, in unknown and uncharted waters

Exhaustion of supplies of food, water, wood, and other necessities

Disease

Navigational failure (fixing longitude was a particular problem that exposed ships to danger even in supposedly charted waters)

Inaccurate mapping (at that time, because of the inability to determine longitude accurately, many islands and reefs could not be found at their previously charted locations)

---

[6]Daniel D. Roman, *Managing Projects: A Systems Approach,* Elsevier, New York, 1986.

Degradation of scientific equipment necessary for the project's scientific tasks

Conflict with native populations (the eventual cause of Cook's death)

Disharmony among the crew and the scientific party on board

A risk that Cook failed to identify and manage effectively was attack by those jealous of his appointment and achievements and in defense of their unsound theories.

The approaches adopted by Cook, the Admiralty, and the Royal Society show that, whether or not they conducted an explicit risk analysis in the way it is defined today, they were well aware of the risks and managed them effectively.

Endurance and Avoidance of Shipwreck

Management of this risk began with the choice of leader. Cook was a very robust man with a remarkable capacity to endure hardship. He was also a highly accomplished seaman, particularly in the dangerous art of coasting in uncharted waters, which would be necessary for accurate surveys of new lands, and navigational hazards such as reefs and shoals. He was also an outstanding navigator, having taught himself mathematics, navigation, and astronomy in the winter months when the coal ships on which he spent his early maritime career were inactive. Unknowingly, Cook had spent his whole life in preparation for the Great South Land Project.

The choice of vessel was also crucial. The ship chosen by Cook provides a fine example of the use of appropriate technology. It was a three-masted bark, the *Earl of Pembroke,* a 3-year-old coastal collier built in the port of Whitby, close to Cook's home village in Yorkshire. Cook learned seamanship working on such vessels in the coal trade from Newcastle to London. It was purchased by the Admiralty and renamed *HMS Endeavour.*

The *Endeavour* was small (only 370 tons) and "cat built" in the nordic style: "...bluff bowed, shallow drafted and necessarily slow."[7] hardly the type of ship, it may seem, in which to attempt to penetrate the Southern Ocean and sail round the world. Subsequent events showed the *Endeavour* to be the ideal choice. It was "immensely strong" and well able to withstand the rigors of a long ocean voyage,[8] and the shallow draft enabled it to be beached and repaired by its own crew. Those features were to save the project from disaster when the ship ran aground on the Great Barrier Reef.

The endurance of the crew was also important. Chosen by the Admiralty, they were young, mostly less than 30 years of age, and better able to cope with the rigors of the voyage than the average run of British seamen at that time.

Supplies

The answer to this problem also began with the choice of vessel. The *Endeavour*'s capacious holds provided space for a large volume of stores. In addition, because of the design of its rigging, the crew required to work the ship was relatively small (some 80 odd in all), reducing the volume of food and drink that had to be carried. In the

---

[7]Price, p. 16.
[8]Price, p. 16

modern idiom one might say that *HMS Endeavour* was a very efficient vessel. Adequate financial resources were also required and they needed to be well-spent. The British government provided a grant of 4000 pounds sterling, which in today's values would be worth a fortune. The victualing was directed by Cook. Annex B is a list of the initial supplies for Cook's second voyage which illustrates what the *Endeavour* would have carried. These were augmented at Madeira and again at Rio de Janeiro.

Prevention of Disease

One of the project's greatest achievements stemmed from Cook's rigorous application of available medical knowledge to conquer the scourge of sailors on long ocean voyages, the terrible disease of scurvy. The capacity of scurvy to devastate a ship's crew had been demonstrated on Commodore Anson's circumnavigation of the world on a raiding expedition against Spanish treasure ships in 1741–44. Within the first year, he lost 200 of the 500 on his own ship. Anson returned with "...the biggest booty that ever returned to England in a single vessel."[9] but lost over 600 men out of his combined ships' complements of less than 1000; most of them died from scurvy.

ANNEX B

Provisions on Board HMS Resolution

Biscuit	59531 pounds
Flour	17437 Do.
Salt Beef	7637 four pd. pieces
Do. Pork	14214 two pd. pieces
Beer	19 Ton
Wine	642 gallons
Spirit	1397 Do.
Pease	358 Bushels
Wheat	188 Do.
Oatmeal	300 gallons
Butter	1963}
Cheese	797 }Pounds
Sugar	1959}
Oyle olive	210 Galln^s
Vinegar	259 Do.
Suet	1900} Pounds
Raisins	3102}

[9]G. M. Badger, *Cook the Scientist,* in G. M. Badger (ed.), *Captain Cook Navigator and Scientist,* ANU Press, Canberra, 1970, p. 43.

Salt	101 Bush[l]
*Malt	80 Bushels
*Sour Krout	19337}
*Salted Cabbage	4773}
*Portable Broth	3000} Pounds
*Saloup	70}
*Mustard	400}
*Mermalde of Carrots	30 gallons
Water	45 Tons
Experl Beef	1384 pounds
Inspisated Juce of Beer	19 half Barrels

*The articles marked thus (*) are antiscorbutics and are to be issued occasionally.
Source: A. Grenfell Price (ed.), *The Explorations of Captain James Cook in the Pacific as Told by Selections of His Own Journal,* Angus and Robertson, Sydney, 1969, p. 105.

In making preparations for the voyage, Cook provisioned the ship with a wide variety of antiscorbutics and induced the crew to consume them to such good effect that success in conquering scurvy became the achievement for which he was most acclaimed. He did not know what caused scurvy and may have been mistaken about which preventative was most effective, but by assiduously applying them all he conquered the disease.

The antiscorbutics were all preserved by some means or other and included sauerkraut, condensed orange and lemon juice, and malt. The original provisions contained no fruit or vegetables whatever. Cook overcame this by acquiring green vegetables from every port and island where they were available and by collecting wild vegetables, grasses, and berries in other places where landings were made. Cook described the process this way:

> We came to few places where either the art of man or nature did not afford some kind of refreshment or other, either of the animal or vegetable kind. It was my first care to procure what could be met with of either by every means in my power, and to oblige our people to make use thereof, both by my example and authority; but these benefits arising from such refreshments soon became so obvious, that I had little occasion to employ either the one or the other.[10]

In addition to the dietary regime, close attention was paid to the personal hygiene of the crew and to keeping their clothing and quarters dry. These measures helped to maintain general good health at sea.

Navigation

This was a most important matter. Although it was possible to measure latitude, and hence distance from the equator, there was up to that time no satisfactory way of determining longitude. This was obviously a matter of great importance to a trading

[10]Quoted in Badger, p. 48.

nation and maritime power like Britain, so important that it was one of the first tasks undertaken when the Royal Observatory was established at Greenwich in 1675. By 1768, Maskelyne, the Astronomer Royal, had produced tables of the motions of the moon. This was a kind of ready reckoner which could be used in the estimation of longitude by observing the angular distance of the moon from the sun. The method was still laborious, the calculations requiring about 4 hours for a single observation, but it worked. Cook obtained these tables and achieved an unprecedented degree of accuracy in determining longitude.

It was, perhaps, fortuitous that the scientific party included the professional astronomer Charles Green who, free from the responsibilities of a ship's officer, had the time to make meticulous daily observations. Typically Cook managed to engage and sustain Green in assiduous application to this task, which was not really his responsibility. Accurate determination of longitude ensured the lasting scientific value of the project's discoveries and is reflected in the accuracy of the charts produced, many of which were in use a century or more later. What a contrast to the previous situation in which islands discovered by navigators simply could not be found again because their position was not accurately known.

Accurate determination of the longitude of Tahiti was also critical to the possibility of measuring the distance of the earth from the sun. This interdependence of navigation and astronomy is summed up by Sir Richard Woolley, a latter day Astronomer Royal:

> ... the determination of the longitude of Fort Venus represented a sort of culmination of the entire campaign to determine the longitude of very distant places by astronomical means, which started with the foundation of the Royal Observatory in 1675...[11]

As other aids to navigation, the *Endeavour* carried a substantial library of journals and charts from previous Pacific voyages, including extracts from the journals of Abel Tasman and the log of Captain Wallis, European discoverer of Tahiti. The former was to play a part in Cook's crucial decision to return home "... by way of the east coast of New Holland."[12]

Mapping

Apart from the benefits of more accurate navigation, the foundations of the incomparable standard of the charts produced by the project were Cook's own genius. Apart from his knowledge of mathematics, navigation, and astronomy and his unflagging application to the task, Cook had the inherent aptitude to produce charts of astonishing accuracy. His selection as leader of the project stemmed in no small measure from his demonstrated skill in this regard. During the Seven Years' War with France, Cook volunteered to serve on a man-of-war and was credited with a crucial contribution to General Wolfe's decisive victory over Montcalme at Quebec. In preparation for the attack and "... with the utmost skill and bravery..." he carried out "... the dangerous task of surveying the river St. Lawrence, so that the safety of the heavy ships of war moving up to the attack might be assured."[13]

---

[11]Sir Richard Woolley, *The Significance of the Transit of Venus,* in G. M. Badger (ed.), *Captain Cook Navigator and Scientist,* ANU Press, Canberra, 1970, p. 126.

[12]Captain W. J. Wharton, *Captain Cook's Journal During His First Voyage Round the World,* Australiana Facsimile Editions, Libraries Board of South Australia, Adelaide, 1968, p. 213.

[13]E. J. M. Watts, *Stories from Australian History,* revised and supplemented by K. R. Cramp, William Brooks and Co., Sydney, 1934, p. 56.

Scientific Equipment

The project was well-supplied with scientific equipment by the Royal Society and by Joseph Banks, who spent 10,000 pounds sterling preparing for the expedition. The equipment included a delicate astronomical quadrant for use in observation of the transit. The only available equipment that was lacking was Harrison's new chronometer, which was to provide an even better method of measuring longitude than Maskelyne's ready reckoner. The chronometer had completed its final trials in 1764 and at the time of Cook's departure was being copied.

All the equipment was carefully packed and stored on board and survived the voyage to Tahiti intact. However, one risk that had been anticipated was underestimated. That was the likelihood of theft by Pacific islanders. The astronomical quadrant was among the many items stolen. The problem this caused and the way it was overcome are discussed later, as is Cook's continuing and ultimately fatal inability to prevent theft.

Avoiding Conflict with the Natives

For this purpose, and to pursue the instruction to cultivate a friendship with inhabitants of the lands he visited, Cook laid down a policy and set of rules at the beginning of the voyage and ensured that they were strictly observed. The policy stressed that natives were to be treated with humanity. The rules strictly governed trade and other relations with natives so as to avoid any unnecessary quarrels. Cook also tried to ensure that members of his crew did not transmit venereal disease, which could spread rapidly among Pacific islanders, given what was known of their sexual mores. Before arrival at Tahiti he had the surgeon examine all the crew for signs of disease and pronounced himself satisfied on that score. A number of the crew did contract disease from local women but whether it was the tropical disease of yaws or a venereal disease transmitted by previous European visitors is not known.

Maintaining Harmony on Board

The answer to this problem lay in Cook's own personality and leadership qualities and, as discussed later, in his skills in human resource management. Whereas other scientific expeditions such as Charles Baudin's later reconnaissance of southern Australia were riven by bitter disputes, Cook held the total loyalty of his crew, the respect of the scientists, and the willingness of both to leave their destiny in his hands.

Protection against Critics of the Project

As will become evident, this aspect of the planning and execution of the project was not successful. Cook could hardly have been unaware of this danger but he was either unconcerned or unable to counteract it. Attacks by his principal rival Alexander Dalrymple damaged his reputation and denied for a time the full credit that his discoveries deserved.

The Project Plan

When the nature of the risks have been considered and the decision has been made to proceed with a project, the next step is to develop a project plan on the basis of the risk management strategies selected. The plan for the Great South Land Project was straightforward. It was to sail to Tahiti by way of Cape Horn; establish the observatory; observe the transit; sail south as far as latitude 40° in search of the Great South Land; if that were not found, sail west between Latitudes 35° and 40° to New Zealand; explore its coast; and finally return to England by Cape Horn or the Cape of Good Hope. The difficulties in this case lay not in determining the broad outline of the plan but in developing it in detail and, even more, in executing it. We begin by considering human resource aspects of the project.

Human Resource Management

In the Project Management Body of Knowledge (PMBOK) the subject of human resources is viewed from the perspective of the project manager, not the human resources specialist, and that is also the perspective from which we examine it in the Great South Land Project. As the PMBOK points out: "The project manager is responsible for developing his project team and building it into a cohesive group to complete the project" but often does so "...within the constraints imposed by the parent organisation."[14] Cook was no exception: like many a present-day project manager, he inherited a team selected by the parent organization—in this case the "consortium" of the Admiralty and the Royal Society plus the self-invited Joseph Banks who chose the scientific party. Cook thus exercised a limited role in the activities shown on the administration wing of the PMBOK work breakdown chart.[15]

The task of personnel placement was well-done but not without controversy over the critical choice of project manager. One of the secret purposes of the project was to test the hypothesis of the eminent geographer Alexander Dalrymple that there was a huge continent in the South Pacific, larger than "...the whole civilised part of Asia, from Turkey eastward to the extremity of China."[16] Dalrymple was recommended to be one of the two observers of the transit but was prepared to go only if given entire control of the ship, which the Admiralty refused. The choice of Cook, who was a non-commissioned officer with no social standing, "...reduced Dalrymple to fury..."[17] and he withdrew from the expedition. Cook was then appointed by the Royal Society as one of the observers as well as commander. Dalrymple's resulting animosity was subsequently increased when Cook contradicted his hypothesis by sailing over a large part of the supposed continent. In response, Dalrymple "...made unscrupulous and lying attacks on Cook."[18]

It is a matter for conjecture whether the Admiralty rejected Dalrymple out of principle or because they regarded Cook as a better choice. Although Cook was not a commissioned officer, he had served with distinction as a marine surveyor under

---

[14]PMBOK, p. F1.

[15]PMBOK, p. F3.

[16]J. C. Beaglehole, *The Exploration of the Pacific*, 3d ed., Adam and Charles Black, London, 1966, p. 193.

[17]J.C. Beaglehole, (ed.), *The Endeavour Journal of Joseph Banks*, The Trustees of the Public Library of NWS in association with Angus and Robertson, Sydney, 2d ed., 1963.

[18]Price, p. 16.

Captain Palliser, a member of the Navy Board, and was well-known to Philip Stephens, the Admiralty Secretary: "...they had no doubt of his ability to command."[19] In the event their choice was amply justified by results.

It is a point of good practice that the members of a project team should have complementary skills. The officers to serve under Cook were chosen with this in mind. Cook had not previously been on a round-the-world voyage nor had he been to Tahiti. Lieutenant Gore had circumnavigated the world twice (under Byron and Wallis) and had been to Tahiti with the latter. Five other crew members had been to Tahiti (not counting the ships goat!).[20]

As has been mentioned, a very young crew was appointed and five of them were already known to Cook, having been under his command in the schooner *Grenville* on his survey of Newfoundland. Cook may have had some say in the choice of crew but certainly not a free hand. He was even forced to accept the extraordinary appointment of a one-handed cook.[21]

IMPLEMENTING THE PROJECT

The Voyage to Tahiti

The voyage to Tahiti was successfully accomplished in good time to establish friendly relations with the indigenous people and build a protected base (Fort Venus) for observation of the transit. Departing Plymouth on August 26, 1768, the *Endeavour* sailed by way of Cape Horn, with stops at Rio de Janeiro and Tierra del Fuego.

The Portuguese Viceroy at Rio de Janeiro suspected that the *Endeavour* was a contraband ship because she was such an unlikely type of vessel for a British naval ship. He imposed restrictions on movement ashore which hampered Cook in obtaining fresh water and provisions. This led to a paper war with Cook which persisted throughout the *Endeavour*'s stay at Rio.

The stop at Tierra del Fuego was a brief one to obtain wood and water before attempting to round Cape Horn and undertake the long Pacific leg of the voyage. It also enabled the scientists to go ashore to gather botanical specimens unknown in Europe and observe the inhabitants of that bleak domain. The scientific party pushed so far inland that they were forced to bivouac overnight. Two of Joseph Banks staff perished from exposure in the freezing conditions.

The infamous Cape Horn was rounded without difficulty and Cook sailed northwest toward what is now known as French Polynesia, arriving at King George's Island (Tahiti) on April 13, 1769, after a voyage of $7\frac{1}{2}$ months.

Relations with the Tahitians

Most projects with an international dimension encounter problems arising from the clash of different cultures. The Great South Land Project was no exception. It was dif-

---

[19]Banks' Journal, vol. 1, p. 22.
[20]Price, p. 17.
[21]Price, p. 16.

ficult to anticipate and prepare to manage the inevitable problems because little was known in Europe about the structure and customs of Tahitian society.

Cook's first concern was to establish good relations with the island people who greeted the *Endeavour* in a friendly and submissive fashion, having felt the power of naval guns during an encounter with Wallis. Two of the ship's company had visited Tahiti with Wallis and were known to one of the first important visitors on board. Cook took advantage of this, made a fuss of the visitor and of other chiefs who came later and gave them presents. The policy of treating the natives with humanity was in the main successful. The exception was the Tahitians' proclivity for stealing and the skill with which they carried it out.

> ...they are thieves to a man, and would steal but everything that came their way, and that with such dexterity as would shame the most noted Pickpocket in Europe.[22]

Nothing was safe from them: a musket was snatched from a marine's hands, Banks and Solander had a telescope and snuff box picked from their pockets on their second day ashore, and Cook's stockings were stolen from under his pillow when he was a guest of the island chief and at a time when he claimed to be awake. A pair of pistols and a sword were other items stolen. The astronomical quadrant with its arrangement of mirrors was an item of particular fascination for the Tahitians and it was soon stolen despite being kept under guard in Fort Venus.

Sexual relations between the Tahitians and the crew were another matter of concern. The stock of nails for use in maintenance of the ship and its boats was threatened by the value the islanders placed on them and the tendency for them to become a kind of sexual currency. Desertion was also a danger. Two marines who had taken local "wives" deserted 4 days before departure from Tahiti. These men were returned to the ship after Cook took some island chiefs as hostages but the incident tarnished the generally good relations that he maintained with the local people.

Observing the Transit

The transit of Venus took place as predicted and was observed by two parties on different islands in the group. The weather was clear and there were no terrestrial hindrances to successful observation, but the results were disappointing. Cook noted an "...Atmosphere or Dusky shade round the body of the planet which very much disturbed the times of the Contact, especially the two internal ones."[23] This difficulty impaired the accuracy of measurement of the parallax. Observers elsewhere noted similar phenomena. The reasons for this are not known, although it can be said that in Tahiti the project faced some particular difficulties. The observations were made with an instrument exposed to air temperatures of 119°F and in the full force of the tropical sun. The clarity of its images may have been impaired by damage sustained when in Tahitian hands, but it was still usable.

A century after the observations were made, analysis of the results from all the different observation points round the globe showed the results to have been more useful than was believed at the time.[24] As a latter day Astronomer Royal, Sir Bernard

---

[22]Journal, p. 92.

[23]Journal, p. 76.

[24]Sir Richard Woolley, *The Significance of the Transit of Venus*, in G. M. Badger (ed.), *Captain Cook Navigator and Scientist*, ANU Press, Canberra, 1960, pp. 133, 134.

Woolley, pointed out in this century, the contribution of the project to astronomical measurement had its limitations but was not in vain. On the basis of the observations

> ...astronomers of the early nineteenth century knew the distance to the sun much better than present-day astronomers know the distance to the centre of the galaxy.[25]

The Search for the Great South Land

To complete his task in Tahiti, and with the enthusiastic participation of Banks and party, Cook explored and described the island and the way of life of its inhabitants. He then opened his sealed orders, although it is clear that he had known their general tenor before departure, and the *Endeavour* sailed from Tahiti on July 13, 1769. On board was a Tahitian chief called Tupia, one of several who had sought to accompany the expedition. Tupia had knowledge of nearby islands that Cook planned to visit (the Society Islands) and could make introductions to Island chiefs.

On Tupia's advice, Cook sailed south to the Society Islands, claiming the island of Ulitea for the Crown, and then farther south to the Austral Group. From there he pressed on into the Roaring Forties. Having found no land as far south as latitude 40° 22′ and encountering the very strong westerly gales and huge swell from which that region takes its name, he changed course to the north to avoid damage to the ship. This southerly excursion of 2400 kilometers (or 1500 miles) shattered the illusion of a large continent in the central South Pacific.

Cook then sailed westward toward New Zealand where he made landfall one month later at Poverty Bay. For once the policy of establishing friendly relations failed and to Cook's great regret three Maoris were killed in one encounter. A number of landings were made at different places in New Zealand and met with varied reactions from the Maoris. Sometimes they were friendly, but on other occasions conflict ensued. Cook maintained the policy whenever possible and meticulously treated the locals humanely and fairly, going so far as to punish crewmen for stealing potatoes from a Maori garden.

In accordance with his instructions, Cook circumnavigated North Island. In the course of doing so he proved the existence of Cook Strait and discovered Queen Charlotte Sound which was to figure in his later voyages. He described the sound as "...a collection of some of the finest harbours in the world..."[26] and remarked on the abundance of timber and fish. He took formal possession of Queen Charlotte Sound on January 31, 1770.

The next stage was the circumnavigation of South Island, proving that it could not be part of a southern continent. In the process of these voyages, Cook charted the entire coast of New Zealand in his meticulous way. He gave a glowing description of New Zealand as a place for settlement:

> It was the opinion of every body on board that all sorts of European grain, fruit, Plants etc., would thrive here; in short, was this country settled by an industrious people they would very soon be supplied not only with the necessaries, but many of the Luxuries of Life.[27]

---

[25]Op. cit., p. 135.
[26]Journal, p. 193.
[27]Journal, p. 216.

In particular he marked down the River Thames and the Bay of Islands as the best places for a colonial settlement. Having completed the tasks set in his sealed orders, Cook was now free to sail home by a route of his own choice.

The Crucial Initiative: Discovery of New South Wales

In choosing a route home, Cook's main consideration was to make the best use of the return voyage to further the objectives of the project, "...being now resolv'd to...bend my thought towards returning home by such a route as might Conduce most to the Advantage of the Service I am upon...."[28] He would have preferred to return by Cape Horn so as to further explore the Southern Ocean for sign of a continent but the condition of the ship after more than a year and a half at sea and the onset of winter were strong arguments against that course. A return direct by way of the Cape of Good Hope was rejected "...as no discovery of any Moment could be hoped for in that route."[29] After consultations described below, Cook decided to return by way of the East Indies, steering first "...to the Westward until we fall in with the E. Coast of New Holland..."[30] and following it north as far as it went.

Cook's intention was to make landfall at the point at which Tasman departed from Tasmania, but the *Endeavour* was driven northward by one of the gales that Sydneysiders call "southerly busters." Landfall was made at Cape Everard to the North of Bass Strait, leaving in doubt for the time being whether Tasmania was an island. Cook named the cape after Lieutenant Hicks who first sighted it but, regrettably, the name was later changed.

After attempting to land near Bulli about 50 miles south of the present city of Sydney, Cook put into Botany Bay, so named because of the great variety of botanical specimens collected by Banks and Solander. (The bay is now encompassed by the city of Sydney and its 3½ million people.) Nine years later, Banks was to suggest to a Committee of the House of Commons that Australia was ideally suited as a "...substitute for the lost American colonies as an overseas penal settlement." One might well wonder at that choice. Unlike other parts of New South Wales, Botany Bay is a windswept, sandy place with low agricultural potential, whereas New Zealand had offered several places that were ideal for colonial settlement.

Confronting Dilemmas

Management of project execution might well be described as a process of resolving a stream of dilemmas. Schedule, cost, quality, and risk are interrelated; action to overcome a problem in one has consequences for the others. Hobson's choice is the project manager's regular diet. Our project provides several illustrations of dilemmas of this kind.

Continuing to sail northward, and charting as he went, Cook passed the present boundary between New South Wales and Queensland, stopping briefly near the present-day port and industrial center of Gladstone. The *Endeavour* then unknowingly

---

[28]Journal, p. 213.
[29] Journal, p. 213.
[30]Journal, p. 213.

entered a navigational trap between the coast and the Great Barrier Reef. At its southern extremity the reef is more than 100 miles from the mainland; in the north that gap is almost completely closed. In that season the southeast trade wind blows steadily at speeds up to 30 knots, ruling out any attempt to reverse course.

It was in these waters that the project came closest to disaster. Some coral reefs rise almost vertically from a considerable depth, a phenomenon outside Cook's experience and it led him, in a moment of indecision, to make a serious mistake which was later used against him by his enemy Dalrymple. On June 11, instead of anchoring for the night as prudence suggested, he sailed on. Cook described his thoughts this way:

> In standing off from 6 until near 9 oClock we deepen'd our water from 14 to 21 fathom when all at once we fell into 12, 10 and 8 fathom. At this time I had every body at their stations to put about and come to anchor but in this I was not so fortunate for meeting again with deep water I thought there could be no danger in standing on.[31]

It came as a nasty shock when on that clear moonlit night, and with successive soundings showing a depth of 20 fathoms, and then 17 fathoms, the *Endeavour* struck before another cast of the lead could be made. The ship was holed and making water faster than the pumps could cope. This was the dilemma. If the ship was left on the reef the calm sea would soon become rough and pound the ship on the coral rocks, breaking it to pieces; on the other hand if the ship were dragged into open water it might quickly sink. Should they take to the boats and try to reach the mainland or try to save the ship with all its provisions, specimens and scientific data at greater risk to their lives (at least in the immediate future)? Cook confronted this dilemma head on. At the first opportunity, a high tide 24 hours later, and having thrown overboard some 50 tons of guns, ballast, casks, and stores to lighten ship, Cook "...resolv'd to risque all, and heave her off...."[32] (The guns were recovered in recent times and form part of Australia's national heritage.)

Anchors were put out and the *Endeavour* was dragged off the reef. All hands worked feverishly at the pumps and the area where the hole was thought to be was covered with a fothering sail—a sail covered with oakum, wool, and sheep or horse manure. The *Endeavour* stayed afloat and was sailed carefully to the relative safety of the nearest navigable inlet, now called the Endeavour River. There the ship was beached. Upon examination of the damage to the hull, it was found that "...a large peice of Coral rock was sticking in one Hole...."[33] As noted by Grenfell Price in his published selections from Cook's Journals, "The casual as well as the causal creates history and but for this piece of coral there might have been no British colonisation of Australia."[34]

The escape from disaster cannot be put down merely to good fortune. At the moment of truth, Cook's resolution to risk all, rather than leave the *Endeavour* on the reef and take to the boats, was crucial. The crew acted with cool professionalism throughout the emergency. The immense strength of the *Endeavour* helped her to withstand the impact and the flat bottom made beaching and repair possible.

The *Endeavour* was beached on June 22, and that part of the bottom that was above water was repaired as well as possible. On July 6, the ship was refloated but the constant southeast winds prevented departure until August 4. The time was spent exploring the nearby countryside, looking for a passage through the reefs, and obtaining

---

[31]Journal, p. 274.
[32]Journal, p. 275.
[33]Journal, p. 280.
[34]Journal, p. 73.

fresh meat and vegetables. There was plentiful supply of edible fish and turtle and greens which in Cook's opinion "...eat as well as, or better, than spinnage."[35] There were also wild yams. In a revealing comment Cook wrote that the tops were edible but "...the roots were so bad that few besides myself could eat them."[36]

There were several encounters with Australian aborigines during this time. Although they were not always friendly, the encounters did not lead to serious conflict. The crew were surprised on one occasion when, in a dispute over turtles, a group of aborigines attacked by encircling a shore party with a grass fire which they luckily escaped. The scientific party took the opportunity to gather specimens of plants and animals unknown in Europe and observe the way of life of the aborigines. The specimens collected included that unusual creature the kangaroo.

Another dilemma soon presented itself. On August 4, Cook set out again. Although immediate disaster had been avoided, the *Endeavour* was still trapped on an unknown shore; parts of the bottom could not be repaired and the ship made water at the rate of one inch in the hour; the pumps were in need of repair; the sails were worn out; and the remaining provisions would last only 3 months even under Cook's careful husbandry. Escape would require safe navigation of an uncharted stretch of coral reefs, among the most treacherous in the world.

For the next week the *Endeavour* crept carefully northward, winding through the maze of coral reefs with boats out ahead to look for channels and warn of danger. After another consultation with the officers, Cook resolved "...to quit the Coast altogether until such time as I found I could approach it with less danger."[37] This still required finding a passage through the outer reef, and for this purpose he climbed a hill on Lizard Island, now a luxury fishing and diving resort. One of the few safe passages was found (now known as Cook's passage), and on August 14, the *Endeavour* sailed through it to the open sea. Although he was reluctant to give up charting the coast, the escape from the reefs gave Cook "...no small joy...."[38]

His relief was shared by all. Banks recorded the irony:

> ...that very Ocean which had formerly been looked upon with terror by (maybe) all of us was now the Assylum we had long wished for and at last found. Satisfaction was clearly painted in every mans face.[39]

The satisfaction was to be short-lived. The ocean swell "worked" the ship and soon it was making water at 9 inches an hour. But worse was to come; 2 days after clearing the outer reef, the *Endeavour* was becalmed, driven toward the reef by huge Pacific Ocean swells which crashed on the reef in "...Vast foaming breakers...."[40] The bottom was too deep to anchor and death for all on board seemed inevitable. At 6 A.M. Cook recorded that:

> ...we were not above 80 or 100 yards from the breakers. The same sea that washed the side of the ship rose in a breaker prodidgiously high the very next time it did rise, so that between us and destruction was only a dismal valley, the breadth of one wave....[41]

---

[35]Journal, p. 293.
[36]Journal, p. 282.
[37]Journal, p. 300.
[38]Journal, p. 301.
[39]Banks' Journal, vol. 2, p. 104.
[40]Journal, p. 303.
[41]Journal, p. 303.

Banks had them "...within 40 yards of the breaker...."[42]

Cook and his team again rose to the occasion. By towing with the ship's boats and rowing with large sweeps, the ship was held off the reef. Agonizingly, with the help of an intermittent faint offshore breeze and the effect of the ebb tide racing through a narrow channel in the reef, the ship edged away from destruction to temporary safety. Being "embayed" by the curving reef, it was probable that an attempt to escape to seaward would fail and the next flood tide and the return of the southeast trade wind would drive the *Endeavour* to destruction. The only alternative was to attempt the dangerous passage through a small opening in the outer reef back into the nightmare of the inner reefs. The second course was chosen and safely executed.

Cook's reaction was an uncharacteristic lament that will strike a chord with many a latter-day project manager:

> It is but a few days ago that I rejoiced at having got without the Reef, but that joy was nothing when compared to what I now felt at being safe within it. Such are the Vicissitudes attending this kind of service....Was it not from the pleasure which naturly results to a Man from being the first discoverer, even if it was nothing more than Sands and Shoals, this service would be unsuportable especialy in far distant parts, like this, short of Provisions and almost every other necessary.[43]

The next stage was the voyage to Batavia. A considerable achievement in itself, and notable for the fact that, despite the deteriorating state of the ship, Cook continued the work of charting the dangerous passage of Torres Strait and the south coast of New Guinea.

Managing the Interface with the Environment

An important aspect of a project manager's responsibility is to manage the interface between the project and its natural, political, social, industrial, and technological environment. It was the first of these that now made its presence felt. Cook managed the microenvironment of the ship exceptionally well, and coped professionally with the wider maritime environment. Managing the land environment at Batavia was a different matter.

The *Endeavour* arrived at Batavia October 11, 1770, more than 2 years after departure from England with "...not one man upon the Sick List."[44] The damage to the ship proved to be more severe than had been thought and a stay of over 2 months was necessary to effect repairs and refit and reprovision the ship. During this time dysentery and malaria and other tropical diseases struck. Two months in Batavia took more toll than 2 years at sea and in Tahiti. Cook recorded that:

> We came in here with as healthy a Ship's Company as need go to sea, and after a stay of not quite 3 months left it in the condidtion of a hospital ship.[45]

Seven died in Batavia and 23 more on the voyage home, including the astronomer Charles Green.

---

[42]Banks' Journal, vol. 2, p. 105.
[43]Journal, p. 305.
[44]Journal, p.354.
[45]Journal, p.364.

Human Resource Management

Cook's responsibilities in this area are among the activities shown under the behavioral wing of the PMBOK work breakdown chart for this function. They include leadership, team building, team motivation, and team decision making.

Leadership

Cook's approach to human resource management was rooted in his humanity and concern for the welfare of his "people" as shown in his method of command:

> ...in a savage age, he pioneered comparatively mild punishments and the wearing of clean and warm dress. there was little of the usual brutal flogging, and, instead of keelhauling any seaman who undressed during the long and wet voyages, Cook encouraged his men to change their damp and filthy clothes.[46]

Understandably he ran happy ships and "...members of his crews enlisted for voyage after voyage."[47]

Cook led by example. Eloquent testimony to this comes from Surgeon Samwell and Lieutenant King, who served under Cook on his third voyage. Note that they served with Cook at a time when the hardships and strain of his voyages, the efforts of preparation, the struggles with enemies at home, and the publication of his journals had exacted a physical and mental toll which showed in "...growing obstinacy and occasional bouts of ill temper."[48] To Samwell, Cook was

> ...the animating spirit of the expedition; in every situation he stood unrivalled and alone; on him all eyes were turned: he was our leading star, which at its setting left us involved in darkness and despair.[49]

and to King,

> His carriage was cool and determined, and accompanied with an admirable presence of mind in the moment of danger.[50]

and further:

> The constitution of his body was robust, inured to labour, and capable of undergoing the severest hardships. His stomach bore, without difficulty, the coarsest and most ungrateful food. Indeed, temperance in him was scarcely a virtue; so great was the indifference with which he submitted to every kind of self denial.[51]

---

[46]Price, p. xvi.
[47]Price, p. xvi.
[48]Price, p. xvi.
[49]Quoted in Price, p. 288.
[50]Quoted in Price, p.287
[51]Quoted in Price, p. 287.

We have alredy seen an example of this in the case of the taro roots that few others than Cook could eat.

Team Building

Illustrations of Cook's skill as a team builder come from the case of the "scientific party" and from events in the Coral Sea. First the scientific party. Joseph Banks, who invited himself and his "suite" to join the expedition, was a typical 18th century gentleman. As Beaglehole put it:

> ...on the scene...steps the figure of Joseph Banks, the gifted, the fortunate youth: enthusiastic, curious, the voyager, the disciple of Linnaeus, the botanist and zoologist, the devotee of savages; not yet, as one examines his early career, a Public Figure, but certainly a Gentleman, certainly a figure typical of his age; and certainly as much as anyone, and more than most, the Gentleman Amateur of Science.[52]

Note that Banks was not a geographer or astronomer, the key scientific specialties of the expedition. He was 25 years old, a member of the landed gentry, "A child not himself born into ermine could hardly hope for more excellent connections."[53] and already a Fellow of the Royal Society of which he was to become President.

Banks persuaded the Royal Society to request the Admiralty to allow him and his seven scientists and assistants to join the expedition. The Navy directed Cook to receive this additional party of eight on his small and already crowded ship. What Cook thought of this at the time we can only imagine, but he accepted "the gentlemen," as he always referred to them in his journal, on board with good grace and gradually made them members of his team. This is a good example of the principle of co-optation that Sapolsky commends in his history of the Polaris project.[54]

By the time of arrival at Rio de Janeiro 3 months after departure, Banks had been drawn into Cook's team. It is apparent that he acted as a draftsman for Cook in the paper war with the colonial governor. Throughout the voyage the two shared their journals and Cook accepted corrections by Banks and sometimes drew on Bank's own journal. It is hard to imagine that Dalrymple would have been so congenial.

By Tahiti we see the whole suite of scientists drawn into the team. As an example, Herman Sporing, the assistant naturalist in Bank's party, repaired the damage done to the astronomical quadrant used for observation of the transit. Sporing had worked as a watchmaker in England, carried with him a set of watchmaker's tools, and was "...clever with his fingers in a mechanical way."[55]

It has been said that Banks' capacity for dealing with people was even more important in Tahiti than Sporing's mechanical skill.

> Cook had it, but perhaps he was a little formal, he was feeling his way: with Banks, on the other hand, all was ease and spontaneity. No one was so able to manage the ship's trade; no one so confident and generous in his proffered friendship, or—in the main—so naturally tactful.[56]

---

[52]Banks' Journal, vol. 1, p. 3.
[53]Beaglehole (1963), vol. 1, p. 4.
[54]H. M. Sapolsky, *The Polaris System Development: Bureaucratic and Programmatic Success in Government*, Harvard University Press, Cambridge, Mass., 1972.
[55]Journal, vol. 1, p.27.
[56]Beaglehole (1963), vol. 1, p. 40.

Even allowing for the role played by Bank's personality, any suggestion that team building of this kind is a simple matter would soon be quickly shattered by the sad story of Henri Baudin's voyage of exploration and scientific investigation in southern Australian waters,[57] the more so as we recall the formidable social class barrier between Cook and the "gentlemen."

The magnitude of Cook's achievement in surmounting this barrier shows in this comment by Banks:

> ...having by a series of public services raised himself to a situation eminently above his most sanguine expectations, his genius broke forth, and enabled him to emerge from obscurity, by giving proofs of the most shining capacity in every qualification that could be required by a discoverer of unknown countries.[58]

Others co-opted included astronomer Green, as has been described, and the Tahitian chief Tupia. Tupia had been taken on board as a guide to the islands south of Tahiti, but Cook also used him as an interpreter with the Maoris of New Zealand. Although the Tahitian and Maori languages differed, a degree of understanding was possible between Tupia and the Maoris. There would have been much more bloodshed in New Zealand in the absence of this ability to communicate with the warlike Maoris.

Team Motivation

Through the example of his own outstanding professionalism, Cook cultivated a similar attitude by his team. This is illustrated by the record of the moment of crisis when the *Endeavour* was aground on the Great Barrier Reef:

> The officers behaved with inimitable coolness void of all hurry and confusion, and indeed, opportunity was taken at noon to determine the latitude of the ship.
> All this time the seamen workd with surprizing cheerfullness and alacrity; no grumbling or growling was to be heard throughout the ship, no not even an oath (tho the ship in general was well furnished with them as most in his majesties service).

The critical nature of the situation is reflected in Bank's own reaction:

> Now in my own opinion I intirely gave up the ship and packing what I thought I might save prepared myself for the worst.[59]

Promoting Health

The most famous example of Cook's motivational skill is the matter of sauerkraut, which exemplifies his ability to motivate his team to eat all manner of unaccustomed and often unpalatable food, a skill which no doubt was the secret of his success in combating scurvy. As usual it is best expressed in Cook's own words:

---

[57]F. B. Horner, *The French Reconnaisance*, Melbourne University Press, Melbourne, 1987.
[58]Price, p. 289.
[59]Badger, p. 30.

> The Sour Krout, the Men at first would not eat it, until I put in practice—a method I never once Knew to fail with seamen—and this was to have some of it dressed every day for the Cabin Table, and permitted all the Officers, without exception, to make use of it, and left it to the Option of the men either to take as much as they pleased or none at all; but this practice was not continued above a Week before I found it necessary to put every one on board on an allowance.[60]

Apart from illustrating his motivational psychology, this passage gives us an insight into the strength of Cook's "persuasiveness." One wonders what might have befallen the officer who broke the rule of permission without exception.

Diet was not the only area of general health promotion. Cleanliness was also a priority: "...cold bathing was encouraged and enforced by example."[61] The decks were scrubbed regularly and dried by stoves, even in hot weather.

Team Decision Making

Although firm, Cook's leadership was not autocratic. As instructed, he made it a practice to consult his officers on important decisions. As a typical example, having circumnavigated New Zealand and decided to leave it in order to seek more discoveries, Cook held a conference with his officers on the question of which route to then follow. The outcome has already been described. There are many other examples in the journal including the dangerous situation off the barrier reef. Cook did not need these conferences to strengthen his own resolve; he used them to strengthen the commitment of his officers to the decision taken.

Controlling the Project

This aspect of project management is best illustrated by Cook's ability to control his budget, which in effect comprised the provisions on board. The *Endeavour* originally carried 18 months' victuals. These were made to last for more than 2 years by a process of supplementation with wild and, where possible, cultivated fresh food and by the strictest husbandry. There was a daily allowance of food, personally controlled by Cook, and those who did not eat it were liable to punishment. Stocks were rigorously accounted for and their condition was periodically checked. Attempts were made to prevent food from going bad, for example, by drying the ship's bread (actually biscuits), and no waste was permitted. Even unpalatable food such as weevilly biscuits and decaying meat was consumed.

PROJECT MANAGEMENT PRINCIPLES

The Great South Land Project illustrates many principles of good project management. The first of these comes from David Cleland. The next five are selected from a

---

[60]Journal, p. 59.
[61]The ship's surgeon, Mr. Perry, quoted by Wharton, p. xxix.

list of preconditions of success in major projects drawn by Morris and Hough from a series of case studies of 20th century projects.[62] The others come from a variety of sources, including the author's own experience.

Strategic Fit

For a project to succeed, it is important that the goals pursued and the strategies adopted are in harmony with the wider goals and strategies of the company or organization in which it is embedded. The Great South Land Project met this requirement. It fitted the desire for greater glory for British science; it pursued national goals of improved navigation and imperial expansion for commercial and strategic purposes.

Positive Client and Senior Management Attitudes

In the above circumstances, this precondition was readily met. The clients were the British Royal Society and King George III. For the reasons given above, their attitude to the project was very positive. The project also gained strength from participation by the young, but socially well placed, naturalist Joseph Banks. As described earlier, this positive attitude soon turned sour so far as one member of the Royal Society was concerned in the person of Alexander Dalrymple.

The senior management was the Admiralty. Their duty lay in support of imperial policy, of which they were an important instrument. They also gave high priority to Pacific exploration, for which they had despatched several earlier missions. They might also have been frustrated by the failure of the expeditions led by Byron and Wallis to achieve the objectives set for them. We might reasonably assume therefore that the Admiralty had a positive attitude to the mission. Some doubt might have arisen at first from the decision to send only one ship, and a coastal collier at that, under the command of a newly promoted lieutenant. In fact these choices reflected high priority and good judgment. Cook had well-placed admirers in senior naval circles, already having a reputation as an astronomical observer as well as a navigator and cartographer, and the collier was his personal choice.

Comprehensive and Clearly Communicated Project Definition

The first step in project definition is to "...establish an objective and a schedule."[63] The Admiralty instructions to Cook did so clearly, unambiguously, and comprehensively, but not to the point where they were unduly constrictive. It was left to Cook to choose whether to return home by Cape Horn or the Cape of Good Hope, a happy provision that led to British discovery of the east coast of Australia.

The project had multiple objectives. The first two objectives were precise, to observe the transit of Venus, and to test whether a large southern continent existed and whether Tasman's New Zealand was part of it. The third objective was broad, to take possession of new lands for the British Crown and scout for land for colonization. All were to be achieved in a single voyage.

---

[62]P. W. G. Morris, and G. H. Hough, *Preconditions of Success and Failure in Major Projects,* Major Projects Association, Templeton College, Oxford, 1986.

[63]Linn C. Stuckenbruch, *The Implementation of Project Management: The Professional's Handbook,* Project Management Institute and Addison Wesley, Reading, Mass., 1989, p. 55.

Beyond the observation stage, the instructions were secret, but Cook must have been made aware at least of their substance because he made such appropriate preparations. On return he was to collect all journals kept on board and keep them secure until delivered to the Admiralty.

A good project definition is only the beginning. Cook's predecessors in Pacific exploration in the 1760s, including Byron and Wallis and the French explorer Bougainville, also had clear and detailed instructions but none managed to carry them out as effectively as Cook.

Good Planning

In any project, the planning done and the decisions taken in the concept and development phases are crucial to success. In those phases the goals are established, the scope is defined, the team is appointed, the strategy is chosen, the risks are assessed, and the master plan is drawn up.[64]

The Great South Land Project was a model of effective planning. As discussed above, the goals and scope were clearly defined, a suitable team was assembled and was led by an ideal commander, the strategy was sound, and the planning and risk management were effective. The choice of vessel showed excellent judgment; the chosen route to Tahiti posed no serious obstacles; the provisioning was adequate especially as regards antiscorbutics; the necessary scientific equipment was provided; the need for navigational aids was recognized and they were provided; and information obtained from previous explorations was collected and carried on board.

Clear Schedules

Scheduling a voyage of exploration is an inherently uncertain task. In this case the critical phase of the schedule covered the voyage to Tahiti and preparations to observe the transit. Adequate time was provided for the voyage, including provisioning stops, and for establishment of the observatory at Tahiti. Equally important was the tight scheduling of food consumption.

Firm Leadership

Morris and Hough point to the need for there to be one person with strong overall authority. As commander of the *Endeavour* and observer of the transit, Cook held that strong authority, and maintained his grip on it. The results that he achieved are ample testimony to the quality of his leadership. What manner of man was he? At the time of the Great South Land project, he was 40 years of age, tall, strongly built, and "...good looking in a plain sort of way."[65] Cook enjoyed good health and was capable of great endurance. He was noted for his unremitting perseverance, firmness, patience, and humanity as well as for courage, coolness in a crisis, and vigilance.[66] One could hardly ask for more appropriate qualities in a project manager.

---

[64]Project Management Institute, *Project Management Body of Knowledge,* Drexel Hill, Pa., 1987, pp. 1–4.

[65]J. C. Beaglehole, "Cook the Man," in G. M. Badger (ed.), *Captain Cook Navigator and Scientist,* ANU Press, Canberra, 1970, p. 12.

[66]Op. cit., p. 11.

To be fair we should mention that Cook had critics who said he was "subject to hastiness and passion"[67] and "a cross grained fellow who sometimes showed a mean disposition and was carried away by a hasty temper...."[68] Few experienced project managers would disagree that at times the display of such temperament is just what is needed. In any event, if Cook sometimes had a short fuse it did not prevent him from leading teams to remarkable feats of endurance and scientific achievement.

He possessed a full measure of a leader's proper concern for the health and welfare of members of his team and went to great lengths to promote both objectives. As we have seen, this involved imposing a regime of diet and hygiene that went much against the grain of their natural inclinations and a route and schedule that was sometimes unpopular. This in no way reflected softness. He persisted in pursuit of objectives in the face of bad weather and discomfort for all on board, and unhesitatingly faced danger where pursuit of the project's objectives called for it.

Effective Teamwork

Cook's ability to mold his officers and crew into an effective team has already been discussed, as has his capacity to win the professional respect of the scientists in his multidisciplinary team. Cook went on to achieve recognition as a scientist as well, being elected to membership of the Royal Society and awarded its Copley Medal.

Looking Forward

It may seem trite to say that nothing a project manager can do will change events that have already taken place. Nevertheless, the point is often lost in the welter of post mortems that tend to follow a project crisis. Cook certainly was not in the habit of looking back or bemoaning bad luck or his own mistakes. At times of crisis, such as the theft of the quadrant, the grounding, entrapment in the barrier reef, and the imminent danger of shipwreck on the outer reef, he took the necessary action to resolve the immediate crisis, then looked forward and replanned in the light of the new situation.

Nor did Cook waste time in idle reflection on achievements. Consider his position on departure from New Zealand. He had completed his scientific task, demonstrated that a vast area of the Pacific was free of the supposed continent, and circumnavigated New Zealand, charting and claiming lands for the Crown as he went. It was a year and a half since the departure from England and the ship was in need of refit. Instead of heading directly for the East Indies for supplies and repairs by a known route, Cook chose to head for the unknown coast of New Holland (Australia) and follow it northward. This led to one of his most important discoveries and also the expedition's greatest danger.

One should not imagine that Cook was insensible to anxiety or immune to the stress induced by crisis. Having at long last escaped from the navigational trap of the Great Barrier Reef, he was forced, at great risk, to reenter it in order to escape disaster. His reaction, and something of his motivation, is revealed in entry in the Journal previously quoted.

---

[67]Op. cit., p. 23.
[68]Ibid.

The Project Manager Is to Blame

When a project succeeds, there is no shortage of claimants to the credit. When it fails or falls into difficulty, the project manager stands alone. Cook's Journal records his lament on this tendency, illustrating a classic dilemma faced by managers of projects which stretch the bounds of existing knowledge.

> If you leave off your work because of its danger you are accused of "Timerousness and want of Perseverance"; if you carry on too long, of "Temerity and want of conduct."[69]

Document the Project's Progress and Pass on Lessons of Experience

Cook was required to maintain a ship's log and a journal and to produce charts of areas explored. He went beyond what was formally required. His charts were of exceptional accuracy and proved to be of inestimable value to subsequent generations of navigators. He wrote full accounts of the geography of the countries he visited and the way of life of their inhabitants, accounts that were to be influential in British settlement of Australia, New Zealand, and other Pacific colonies. He recorded his own management experience, the consequences of his decisions or actions, and his reflections on them for the benefit of future navigators.

Far too few project managers follow this path and pass on the hard-won lessons from their experience. There are current examples, such as the report of the "management experience" in construction of the new Australian Parliament House,[70] but the principle is mostly honored in the breach. As we shall see, Cook did make one serious mistake; he did not take steps to ensure that his journal was fairly and objectively edited and published. Lack of confidence in his own literary ability and the Admiralty's concern for security were to blame. The consequences for his reputation were serious and may not have been overcome except for the sheer magnitude of the project's achievements.

The Role of Chance

> Project management is the art of creating the illusion that any outcome is the result of a series of predetermined deliberate acts when, in fact, it was dumb luck.[71]

Harold Kerzner had his tongue well in cheek when he set up this straw man as a foil for his definition of project management, but it acts as a reminder of the importance of luck. The Great South Land Project had several strokes of extremely good luck that outweighed many of the opposite kind. The first came when Joseph Banks invited himself to join the expedition. Cook could have been forgiven for being resentful of this, but as we have seen it turned out to be a benefit to the project and a blessing, if a somewhat mixed one, to Cook himself.

---

[69]See Note 38.

[70]Parliament House Construction Authority, *Project Parliament: The Management Experience*, AGPS, Canberra, 1990.

[71]H. Kerzner, *Project Management: A Systems Approach to Planning, Scheduling, and Controlling*, 2d ed., Van Nostrand Reinhold Co., New York, 1984.

Cook's own writing style improved from his cooperation with Banks. The world also benefited from the record in Banks' Journal which complemented Cook's by providing dramatic narrative and recording emotions which Cook generally eschewed. Cook began the voyage as self-educated navigator and surveyor with some skill in astronomical observations. By voyage end, his intellect had broadened and encompassed much of natural science. Living at close quarters over a long period with a group of scientists must have contributed to this development.

Finally, the scientific world benefited from the matchless collection of botanical and zoological specimens that Banks' party brought home.

The benefits were not merely a matter of chance. Cook's acceptance of the unwonted intrusion, his ability to win the respect, nay admiration, of the scientists, and the way he integrated them with the project are models for any aspiring project manager.

Banks' presence was not all good fortune for Cook. It was Banks who reaped most of the early harvest of recognition from the achievements of the project. The Admiralty gave Cook's Journal to one Dr. Hawkesworth to edit for publication. Banks did the same with his own journal. Hawkesworth combined the two and because of Banks' social standing gave more weight to Banks. He also imposed his own flamboyant style and religious views and added inaccurate comments. Cook was also attacked by Dalrymple, who maintained his belief in the southern continent. No project manager should assume that he will automatically receive the credit due. That is just one of the reasons why professional status is so important.

Another stroke of good luck was the piece of coral rock that wedged itself in the hole torn in the ship's bottom by Endeavour Reef. Without it the *Endeavour* may well have foundered.

CONCLUSION

The Passionate Professional

Cook was to return to the Pacific twice more in the course of two projects. His second voyage from 1772 to 1775 completed the Great South Land Project. In its course he dispelled forever the myth of a great southern continent; deduced the existence of Antarctica, and may have unknowingly seen it; crossed the Antarctic Circle repeatedly, sailing further south than any ship had ever done in the same longitude. His third voyage, from 1776 to his death in 1779, took him into the Arctic Ocean in search of the fabled northwest passage. In the process he completed the charting of the whole Pacific Ocean.

The foundations for the massive achievements of these projects were Cook's determined and unremitting pursuit of the objectives set for him and his total professionalism. He maintained high professional standards in all aspects of his work, including seamanship, navigation, surveying, mapping, anthropology, astronomical observations, and mathematics, and he set new standards of professionalism in project management. Complete dedication to the objectives of the project and professionalism in the pursuit of them are hallmarks of the successful project manager. Could there be a better phrase to sum up these characteristics than Beaglehole's description of Cook as "...the passionate professional."[72]

---

[72]In Badger, p. 18.

INDEX

ABOUT THE EDITORS

David I. Cleland is the Ernest E. Roth Professor of Engineering and Management in the School of Engineering at the University of Pittsburgh. An internationally recognized leader in the field of project management, Dr. Cleland is the author or editor of 25 books, including *Project Management: Strategic Design and Implementation, Military Project Management Handbook, Shared Manufacturing: A Global Perspective, The Automated Factory Handbook: Technology and Management,* and the award-winning *Project Management Handbook,* second edition.

    Roland Gareis is Director of the post-graduate program on "Project Management in the Export Industry" at the Technical University and the University of Business Administration in Vienna, Austria, the head of Roland Gareis Consulting, and President of Projektmanagement Austria, the Austrian INTERNET Association. A recognized expert on project management, he is the author of *The Handbook of Management by Projects,* and other books.